Quantum Computing without Magic

Scientific and Engineering Computation
William Gropp and Ewing Lusk, editors; Janusz Kowalik, founding editor

Unstructured Scientific Computation on Scalable Multiprocessors, edited by Piyush Mehrotra, Joel Saltz, and Robert Voigt, 1992

Parallel Computational Fluid Dynamics: Implementation and Results, edited by Horst D. Simon, 1992

The High Performance Fortran Handbook, Charles H. Koelbel, David B. Loveman, Robert S. Schreiber, Guy L. Steele Jr., and Mary E. Zosel, 1994

PVM: Parallel Virtual Machine–A Users' Guide and Tutorial for Network Parallel Computing, Al Geist, Adam Beguelin, Jack Dongarra, Weicheng Jiang, Bob Manchek, and Vaidy Sunderam, 1994

Practical Parallel Programming, Gregory V. Wilson, 1995

Enabling Technologies for Petaflops Computing, Thomas Sterling, Paul Messina, and Paul H. Smith, 1995

An Introduction to High-Performance Scientific Computing, Lloyd D. Fosdick, Elizabeth R. Jessup, Carolyn J. C. Schauble, and Gitta Domik, 1995

Parallel Programming Using C++, edited by Gregory V. Wilson and Paul Lu, 1996

Using PLAPACK: Parallel Linear Algebra Package, Robert A. van de Geijn, 1997

Fortran 95 Handbook, Jeanne C. Adams, Walter S. Brainerd, Jeanne T. Martin, Brian T. Smith, and Jerrold L. Wagener, 1997

MPI—The Complete Reference: Volume 1, The MPI Core, Marc Snir, Steve Otto, Steven Huss-Lederman, David Walker, and Jack Dongarra, 1998

MPI—The Complete Reference: Volume 2, The MPI-2 Extensions, William Gropp, Steven Huss-Lederman, Andrew Lumsdaine, Ewing Lusk, Bill Nitzberg, William Saphir, and Marc Snir, 1998

A Programmer's Guide to ZPL, Lawrence Snyder, 1999

How to Build a Beowulf, Thomas L. Sterling, John Salmon, Donald J. Becker, and Daniel F. Savarese, 1999

Using MPI: Portable Parallel Programming with the Message-Passing Interface, second edition, William Gropp, Ewing Lusk, and Anthony Skjellum, 1999

Using MPI-2: Advanced Features of the Message-Passing Interface, William Gropp, Ewing Lusk, and Rajeev Thakur, 1999

Beowulf Cluster Computing with Linux, second edition, William Gropp, Ewing Lusk, and Thomas Sterling, 2003

Beowulf Cluster Computing with Windows, Thomas Sterling, 2001

Scalable Input/Output: Achieving System Balance, Daniel A. Reed, 2003

Using OpenMP: Portable Shared Memory Parallel Programming, Barbara Chapman, Gabriele Jost, and Ruud van der Pas, 2008

Quantum Computing without Magic

Devices

Zdzisław Meglicki

The MIT Press
Cambridge, Massachusetts
London, England

MIT Press books may be purchased at special quantity discounts for business or sales promotional use. For information, please email special_sales@mitpress.mit.edu or write to Special Sales Department, The MIT Press, 55 Hayward Street, Cambridge, MA 02142.

This book was set in LaTeX by the author and was printed and bound in the United States of America.

Library of Congress Cataloging-in-Publication Data

Meglicki, Zdzisław, 1953–
Quantum computing without magic : devices / Zdzisław Meglicki.
p. cm. — (Scientific and engineering computation)
Includes bibliographical references and index.
ISBN 978-0-262-13506-1 (pbk. : alk. paper)
1. Quantum computers. I. Title.
QA76.889.M44 2008
004.1—dc22

2008017033

10 9 8 7 6 5 4 3 2 1

Dedicated to my cats, Bambosz and Sofa, who have a bone to pick with Dr Schrödinger.

Contents

Series Foreword

The Scientific and Engineering Series from MIT Press presents accessible accounts of computing research areas normally presented in research papers and specialized conferences. Elements of modern computing that have appeared thus far in the Series include parallelism, language design and implementation, systems software, and numerical libraries. The scope of the Series continues to expand with the spread of ideas from computing into new aspects of science.

One of the most revolutionary developments in computing is the discovery of algorithms for machines based not on operations on bits but on quantum states. These algorithms make possible efficient solutions to problems, such as factoring large numbers, currently thought to be intractable on conventional computers. But how do these machines work, and how might they be built?

This book presents in a down-to-earth way the concepts of quantum computing and describes a long-term plan to enlist the amazing—and almost unbelievable—concepts of quantum physics in the design and construction of a class of computer of unprecedented power. The engineering required to build such computers is in an early stage, but in this book the reader will find an engaging account of the necessary theory and the experiments that confirm the theory. Along the way the reader will be introduced to many of the most interesting results of modern physics.

William Gropp and Ewing Lusk, Editors

Preface

This book arose from my occasional discussions on matters related to quantum computing with my students and, even more so, with some of my quite distinguished colleagues, who, having arrived at this juncture from various directions, would at times reveal an almost disarming lack of understanding of quantum physics fundamentals, while certainly possessing a formidable aptitude for skillfully juggling mathematics of quantum mechanics and therefore also of quantum computing—a dangerous combination. This lack of understanding sometimes led to perhaps unrealistic expectations and, on other occasions, even to research suggestions that were, well, unphysical.

Yet, as I discovered in due course, their occasionally pointed questions were very good questions indeed, and their occasional disbelief was well enough founded, and so, in looking for the right answers to them I was myself forced to revise my often canonical views on these matters.

From the perspective of a natural scientist, the most rewarding aspect of quantum computing is that it has reopened many of the issues that had been swept under the carpet and relegated to the dustbin of history back in the days when quantum mechanics had finally solidified and troublemakers such as Einstein, Schrödinger, and Bohm were told in no uncertain terms to "put up or shut up." And so, the currently celebrated Einstein Podolsky Rosen paradox [38] lingered in the dustbin, not even mentioned in the Feynman's Lectures on Physics [42] (which used to be my personal bible for years and years), until John Stewart Bell showed that it could be examined experimentally [7].

Look what the cat dragged in!

When Aspect measurements [4] eventually confirmed that quantum physics was indeed "nonlocal," and not just on a microscopic scale, but over large macroscopic distances even, some called it the "greatest crisis in the history of modern physics." Why should it be so? Newtonian theory of gravity is nonlocal, too, and we have been living with it happily since its conception in 1687. On the other hand, others spotted an opportunity in the crisis. "This looks like fun," they said. "What can we do with it?" And this is how quantum computing was born. An avalanche of ideas and money that has since tumbled into physics laboratories has paid for many wonderful experiments and much insightful theoretical work.

But some of the money flowed into departments of mathematics, computer science, chemistry, and electronic engineering, and it is for these somewhat bewildered colleagues of mine that I have written this book. Its basic purpose is to explain how quantum differs from "classical," how quantum devices are supposed to work, and even why and how the apparatus of quantum mechanics comes into being. If this text were compared to texts covering classical computing, it would be a text about the most basic classical computing devices: diodes, transistors, gates. Such books are well known to people who are called *electronic device engineers. Quantum Computing without Magic* is a primer for future *quantum device engineers.*

In classical digital computing everything, however complex or sophisticated, can be ultimately reduced to Boolean logic and NAND or NOR gates. So, once we know how to build a NAND or a NOR gate, all else is a matter of just connecting enough of these gates to form such circuitry as is required. This, of course, is a simplification that omits power conditioning and managing issues, and mechanical issues—after all, disk drives rotate, have bearings and motors, as do cooling fans; and then we have keyboards, mice, displays, cameras, and so on. But the very heart of it all is Boolean logic and simple gates. Yet, the gates are no longer this simple when their functioning is scrutinized in more detail. One could write volumes about gates alone.

Quantum computing, on the other hand, has barely progressed beyond a single gate concept in practical terms. Although numerous learned papers exist that contemplate large and nontrivial algorithms, the most advanced quantum computers of 2003 comprised mere two "qubits" and performed a single gate computation. And, as of early 2007, there hasn't been much progress. Quantum computing is extremely hard to do. Why? This is one of the questions this book seeks to answer.

Although *Quantum Computing without Magic* is a simple and basic text about qubits and quantum gates, it is not a "kindergarten" text. The readers are assumed to have mathematical skills befitting electronic engineers, chemists, and, certainly, mathematicians. The readers are also assumed to know enough basic quantum physics to not be surprised by concepts such as energy levels, Josephson junctions, and tunneling. After all, even entry-level students nowadays possess considerable reservoirs of common knowledge—if not always very detailed—about a great many things, including the world of quantum physics and enough mathematics to get by.

On the other hand, the text attempts to explain everything in sufficient detail to avoid unnecessary magic—including detailed derivations of various formulas, that may appear tedious to a professional physicist but that should help a less experienced reader understand how they come about. In the spirit of stripping quantum computing of magic, we do not leave such results to exercises.

For these reasons, an adventurous teacher might even risk complementing an introductory course in quantum mechanics with selected ideas and materials derived from this text. The fashionable subject of quantum computing could serve here as an added incentive for students to become acquainted with many important and interesting concepts of quantum physics that traditionally have been either put on the back burner or restricted to more advanced classes. The familiarity gained with the density operator theory at this early stage, as well as a good understanding of what is actually being measured in quantum physics and how, can serve students well in their future careers.

Quantum mechanics is a probability theory. Although this fact is well known to physicists, it is often swept under the carpet or treated as somehow incidental. I have even heard it asked, "Where do quantum probabilities come from?"—as if this question could be answered by unitary manipulations similar to those invoked to explain decoherence. In the days of my youth a common opinion prevailed that quantum phenomena could not be described in terms of probabilities alone and that quantum mechanics itself could not be formulated in a way that would not require use of complex numbers. Like other lore surrounding quantum mechanics this opinion also proved untrue, although it did not become clear until 2000, when Stefan Weigert showed that every quantum system and its dynamics could be characterized fully in terms of nonredundant probabilities [145]. Even this important theoretical discovery was not paid much attention until Lucien Hardy showed a year later that quantum mechanics of discrete systems could be derived from "five reasonable axioms" all expressed in terms of pure theory of probability [60].

Why should it matter? Isn't it just a question of semantics? I think it matters if one is to understand where the power of quantum mechanics as a theory derives from. It also matters in terms of expectations. Clearly, one cannot reasonably expect that a theory of probability can explain the source of probability, if such exists at all—which is by no means certain in quantum physics, where probabilities may be fundamental.

This book takes probability as a starting point. In Chapter 1 we discuss classical bits and classical registers. We look at how they are implemented in present-day computers. Then we look at randomly fluctuating classical registers and use this example to develop the basic formalism of probability theory. It is here that we introduce concepts of fiducial states, mixed and pure states, linear forms representing measurements, combined systems, dimensionality, and degrees of freedom. Hardy's theorem that combines the last two concepts is discussed as well, as it expresses most succinctly the difference between classical and quantum physics. This chapter also serves as a place where we introduce basic linear algebra, taking care to distinguish between vectors and forms, and introducing the concept of tensor product.

We then use this apparatus in Chapter 2, where we introduce a qubit. We describe it in terms of its fiducial vector and show how the respective probabilities can be measured by using the classical Stern-Gerlach example. We show a difference between fully polarized and mixed states and demonstrate how an act of measurement breaks an initial pure state of the beam, converting it to a mixture. Eventually we arrive at the Bloch ball representation of qubit states. Then we introduce new concepts of Pauli vectors and forms. These will eventually map onto Pauli matrices two chapters later. But at this stage they will help us formulate laws of qubit dynamics in terms of pure probabilities—following Hardy, we call this simple calculus the *fiducial formalism.* It is valuable because it expresses qubit dynamics entirely in terms of directly measurable quantities. Here we discuss in detail Larmor precession, Rabi oscillations, and Ramsey fringes—these being fundamental to the manipulation of qubits and quantum computing in general. We close this chapter with a detailed discussion of quantronium, a superconducting circuit presented in 2002 by Vion, Aassime, Cottet, Joyez, Pothier, Urbina, Esteve, and Devoret, that implemented and demonstrated the qubit [142].

Chapter 3 is short but pivotal to our exposition. Here we introduce quaternions and demonstrate a simple and natural mapping between the qubit's fiducial representation and quaternions. In this chapter we encounter the von Neumann equation, as well as the legendary *trace formula,* which turns out to be the same as taking the arithmetic mean over the statistical ensemble of the qubit. We learn to manipulate quaternions by the means of commutation relations and discover the sole source of their power: they capture simultaneously in a single formula the cross and the dot products of two vectors. The quaternion formalism is, in a nutshell, the density operator theory. It appears here well before the *wave function* and follows naturally from the qubit's probabilistic description.

Chapter 4 continues the story, beginning with a search for a simplest matrix representation of quaternions, which yields Pauli matrices. We then build the Hilbert space, which the quaternions, represented by Pauli matrices, act on and discover within it the images of the basis states of the qubit we saw in Chapter 2. We discover the notion of state superposition and *derive* the probabilistic interpretation of transition amplitudes. We also look at the transformation properties of spinors, something that will come handy when we get to contemplate Bell inequalities in Chapter 5. We rephrase the properties of the density operator in the unitary language and then seek the unitary equivalent of the quaternion von Neumann equation, which is how we arrive at the Schrödinger equation. We study its general solution and revisit and reinterpret the phenomenon of Larmor precession. We investigate single qubit *gates*, a topic that leads to the discussion of Berry phase [12], which is further illustrated by the beautiful 1988 experiment of Richardson, Kilvington, Green, and Lamoreaux [119].

In Chapter 5 we encounter the simplest bipartite quantum system, the biqubit.

We introduce the reader to the notion of entanglement and then illustrate it with experimental examples. We strike while the iron is hot; otherwise who would believe such weirdness to be possible? We begin by showing a Josephson junction biqubit made by Berkley, Ramos, Gubrud, Strauch, Johnson, Anderson, Dragt, Lobb, and Wellstood in 2003 [11]. Then we show an even more sophisticated Josephson junction biqubit made in 2006 by Steffen, Ansmann, Bialczak, Katz, Lucero, cDermott, Neeley, Weig, Cleland, and Martinis [134]. In case the reader is still not convinced by the functioning of these quantum microelectronic devices, we discuss a very clean example of entanglement between an ion and a photon that was demonstrated by Blinov, Moehring, Duan, and Monroe in 2004 [13]. Having (we hope) convinced the reader that an entangled biqubit is not the stuff of fairy tales, we discuss its representation in a rotated frame and arrive at Bell inequalities. We discuss their philosophical implications and possible ontological solutions to the puzzle at some length before investigating yet another feature of a biqubit—its single qubit expectation values, which are produced by partial traces. This topic is followed by a quite detailed classification of biqubit states, based on Englert and Metwally [39], and discussion of biqubit separability that is based on the Peres-Horodeckis criterion [113, 66].

Mathematics of biqubits is a natural place to discuss nonunitary evolution and to present simple models of important nonunitary phenomena such as depolarization, dephasing, and spontaneous emission. To a future quantum device engineer, these are of fundamental importance, inasmuch as every classical device engineer must have a firm grasp of thermodynamics. One cannot possibly design a working engine, or a working computer, while ignoring the fundamental issue of heat generation and dissipation. Similarly, one cannot possibly contemplate designing working quantum devices while ignoring the inevitable loss of unitarity in every realistic quantum process.

We close this chapter with the discussion of the Schrödinger cat paradox and a beautiful 1996 experiment of Brune, Hagley, Dreyer, Maitre, Maali, Wunderlich, Raimond, and Haroche [18]. This experiment clarifies the muddled notion of what constitutes a quantum measurement and, at the same time, is strikingly "quantum computational" in its concepts and methodology.

The last major chapter of the book, Chapter 6. puts together all the physics and mathematics developed in the previous chapters to strike at the heart of quantum computing: the controlled-NOT gate. We discuss here the notion of quantum gate universality and demonstrate, following Deutsch [29], Khaneja and Glaser [78], and Vidal and Dawson [140], that the controlled-NOT gate is universal for quantum computation. Then we look closely at the Cirac-Zoller idea of 1995 [22] and its elegant 2003 implementation by Schmidt-Kaler, Häffner, Riebe, Gulde, Lancaster, Deuschle, Bechner, Roos, Eschner, and Blatt [125]. On this occasion we also discuss the functioning of the linear Paul trap, electron shelving technique, laser cooling, and side-band transitions, which are all crucial in this experiment. We also look at the 2007 superconducting controlled-NOT gate developed by Plantenberg, de Groot, Harmans, and Mooij [114] and at the 2003 all-optical controlled-NOT gate demonstrated by O'Brien, Pryde, White, Ralph, and Branning [101].

In the closing chapter of the book we outline a roadmap for readers who wish to learn more about quantum computing and, more generally, about quantum information theory. Various quantum computing algorithms as well as error correction procedures are discussed in numerous texts that have been published as far back as 2000, many of them "classic." The device physics background provided by this book should be sufficient to let its readers follow the subject and even read professional publications in technical journals.

But there is another aspect of the story we draw the reader's attention to in this chapter. How "quantum" is quantum computing? Is "quantum" really so unique and different that it cannot be faked at all by classical systems? When comparisons are made between quantum and classical algorithms and statements are made along the lines that "no classical algorithm can possibly do this," the authors, rather narrow-mindedly, restrict themselves to comparisons with classical *digital* algorithms. But the principle of superposition, which makes it possible for quantum algorithms to attain exponential speedup, is not limited to quantum physics only. The famous Grover search algorithm can be implemented on a classical analog computer, as Grover himself demonstrated together with Sengupta in 2002 [57]. It turns out that a great many features of quantum computers can be implemented by using classical analog systems, even entanglement [24, 133, 103, 104, 105]. For a device engineer this is a profound revelation. Classical analog systems are far easier to construct and operate than are quantum systems. If similar computational efficiencies can be attained this way, may not this be an equally profitable endeavor? We don't know the full answer to this question, perhaps because it has not been pursued with as much vigor as has quantum computing itself. But it is an interesting fundamental question in its own right, even from a natural scientist's point of view.

Throughout the whole text and in all quoted examples, I have continuously made the point that everything in quantum physics is about probabilities. A single detection is meaningless and useless, even in those rare situations when theoretical reasoning lets us reduce a problem's solution to such. Experimental realities ensure that we must repeat our detection many times to provide us with classical, not necessarily quantum, error estimates. When a full characterization of a quantum state is needed, the whole statistical ensemble that represents the state must be explored. After all, what is a "quantum state" if not an abstraction that refers to the vector of probabilities that characterize it [60]? And there is but one way to arrive at this characterization. One has to measure and record sometimes hundreds of thousands of detections in order to estimate the probabilities with such error as the context requires.

And don't you ever forget it!

This, of course, does have some bearing on the cost and efficiency of quantum computation, even if we were to overlook quantum computation's energetic inefficiency [49], need for extraordinary cooling and isolation techniques, great complexity and slowness of multiqubit gates, and numerous other problems that all derive from ... physics. This is where quantum computing gets stripped of its magic and dressed in the cloak of reality. But this is not a drab cloak. It has all the coarseness and rich texture of wholemeal bread, and wholemeal bread is good for you.

Acknowledgments

This book owes its existence to many people who contributed to it in various ways, often unknowingly, over the years. But in the first place I would like to thank the editors for their forbearance and encouragement, and to Gail Pieper, who patiently read the whole text herself and through unquestionable magic of her craft made it readable for others.

To Mike McRobbie of Indiana University I owe the very idea that a book could be made of my early lecture notes, and the means and opportunity to do so. Eventually, little of my early notes made it into the book, which is perhaps for the better.

I owe much inspiration, insights and help to my professional colleagues, Zhenghan Wang, Lucien Hardy, Steven Girvin, and Mohammad Amin, whose comments, suggestions, ideas, and questions helped me steer this text into what it has eventually become.

My interest in the foundations of quantum mechanics was awoken many years ago by Asher Peres, who visited my alma mater briefly and talked about the field, and by two of my professors Bogdan Mielnik and Iwo Birula-Białynicki. Although I understood little of it at the time, I learned that the matter was profound and by no means fully resolved. It was also immensely interesting.

To my colleagues and friends at the University of Western Australia, Armenag Nassibian, Laurie Faraone, Paul McCormick, Armando Scolaro, Zig Budrikis, and Yianni Attikiouzel, I owe my electronic engineering background and common sense that, ultimately, helped me navigate through the murky waters of quantum computing and quantum device engineering.

To my many colleagues and friends at the Australian National University, among them Bob Maier, John Slaney, Bob Dewar, Dayal Wickramasighe, and Bob Gingold, I owe my return to physics and deeper interest in computing, as well as countless hours of discussions on completely unrelated topics, though enjoyable nevertheless.

And to my cats, who did all they could to stop me from writing this book, I owe it that they gave up in the end.

Quantum Computing without Magic

"If you want to amount to anything as a witch, Magrat Garlick, you got to learn three things. What's real, what's not real, and what's the difference—"

Terry Pratchett, Witches Abroad

1 Bits and Registers

1.1 Physical Embodiments of a Bit

Bits and bytes

Information technology devices, such as desktop computers, laptops, palmtops, cellular phones, and DVD players, have pervaded our everyday life to such extent that it is difficult to find a person who would not have at least some idea about what bits and bytes are. I shall assume therefore that the reader knows about both, enough to understand that a bit is the "smallest indivisible chunk of information" and that a byte is a string of eight bits.

Discretization of information is a convention.

Yet the concept of a bit as the smallest indivisible chunk of information is a somewhat stifling convention. It is possible to dose information in any quantity, not necessarily in discrete chunks, and this is how many analog devices, including analog computers, work. What's more, it takes a considerable amount of signal processing, and consequently also power and time, to maintain a nice rectangular shape of pulses representing bits in digital circuits. Electronic circuits that can handle information directly, without chopping it to bits and arranging it into bytes, can be orders of magnitude faster and more energy efficient than digital circuits.

Storing and manipulating bits

How are bits and bytes actually stored, moved, and processed inside digital devices? There are many ways to do so. Figure 1.1 shows a logic diagram of one of the simplest memory cells, a flip-flop.

A flip-flop as a 1-bit memory cell

The flip-flop in Figure 1.1 comprises two cross-coupled NAND gates. It is easy to analyze the behavior of the circuit. Let us suppose R is set to 0 and S is set to 1. If R is 0, then regardless of what the second input to the NAND gate at the bottom is, its output must be 1. Therefore the second input to the NAND gate at the top is 1, and so its output Q must be 0. The fact that the roles of R and S in the device are completely symmetric implies that if R is set to 1 and S to 0, we'll get that $Q = 1$ and $\neg Q = 0$. Table 1.1 sums up these simple results.

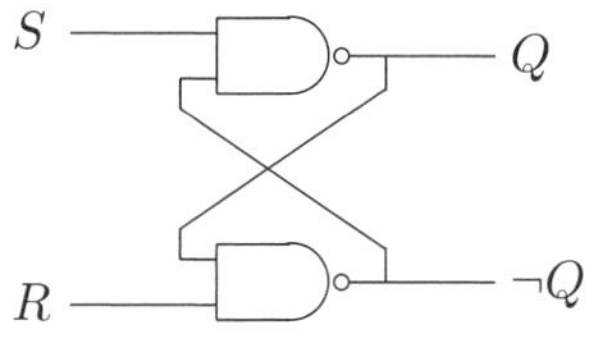

Figure 1.1: A very simple flip-flop comprising two cross-coupled NAND gates.

Table 1.1: Q and $\neg Q$ as functions of R and S for the flip-flop of Figure 1.1.

R	S	Q	$\neg Q$
0	1	0	1
1	0	1	0

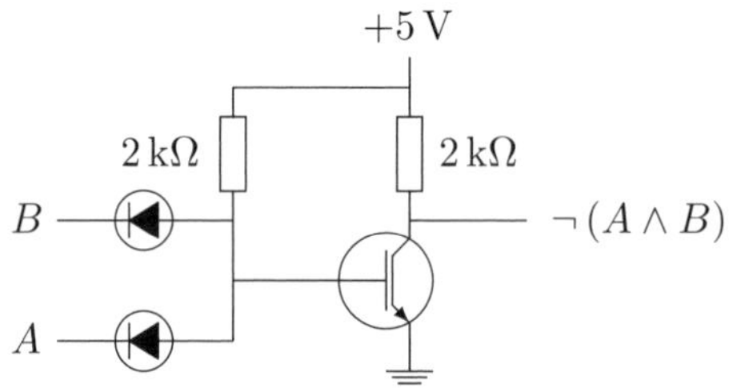

Figure 1.2: A diode-transistor-logic implementation of a NAND gate.

We observe that once the value of Q has been set to either 0 or 1, setting both R and S to 1 retains the preset value of Q. This is easy to see. Let us suppose Q has been set to 1. Therefore $\neg Q$ is 0, and so one of the inputs to the upper NAND gate is 0, which implies that its output must be 1. In order for $\neg Q$ to be 0, both inputs to the lower NAND gate must be 1, and so they are, because $R = 1$.

Now suppose that Q has been preset to 0 instead. In this case the second input to the lower NAND gate is 0, and therefore the output of the gate, $\neg Q$ is 1, which is exactly what is required in order for Q to be 1, on account of $\neg Q$ being the second input to the upper NAND gate.

And so our flip-flop behaves like a simple memory device. By operating on its inputs we can set its output to either 0 or 1, and then by setting both inputs to 1 we can make it remember the preset state.

What is inside the NAND *gate*

It is instructive to have a closer look at what happens inside the NAND gates when the device remembers its preset state. How is this remembering accomplished?

Figure 1.2 shows a simple diode-transistor logic (DTL) implementation of a NAND gate. Each of the diodes on the two input lines A and B conducts when 0 is applied to its corresponding input. The diodes disconnect when 1 is applied to their inputs. The single transistor in the circuit is an n-channel transistor. This means that the channel of the transistor conducts when a positive charge, logical 1, is applied to the gate. Otherwise the channel blocks. Let us consider what is going to happen if

Table 1.2: Truth table of the DTL NAND gate shown in Figure 1.2.

A	B	$\neg(A \wedge B)$
0	0	1
0	1	1
1	0	1
1	1	0

either of the two inputs is set to 0. In this case the corresponding diode conducts, and the positive charge drains from the gate of the transistor. Consequently its channel blocks and the output of the circuit ends up being 1. On the other hand, if both inputs are set to 1, both diodes block. In this case positive charge flows toward the gate of the transistor and accumulates there, and the transistor channel conducts. This sets the potential on the output line to 0. The resulting truth table of the device is shown in Table 1.2. This is indeed the table of a NAND gate.

The important point to observe in the context of our considerations is that it is the presence or the absence of the charge on the transistor gate that determines the value of the output line. If there is no accumulation of positive charge on the gate, the output line is set to 0; if there is a sufficient positive charge on the gate, the output line is set to 1.

The gate charge embodies the bit.

Returning to our flip-flop example, we can now see that the physical embodiment of the bit, which the flip-flop "remembers," is the electric charge stored on the gate of the transistor located in the upper NAND gate of the flip-flop circuit. If there is an accumulation of positive charge on the transistor's gate, the Q line of the flip-flop becomes 0; and if the charge has drained from the gate, the Q line becomes 1. The Q line itself merely provides us with the means of reading the bit.

We could replace the flip-flop simply with a box and a pebble. An empty box would correspond to a drained transistor gate, and this we would then *read* as 1. If we found a pebble in the box, we would read this as 0. The box and the pebble would work very much like the flip-flop in this context.

A pebble in a box

It is convenient to reverse the convention and read a pebble in the box as 1 and its absence as 0. We could do the same, of course, with the flip-flop, simply by renaming Q to $\neg Q$ and vice versa.

Seemingly we have performed an act of conceptual digitization in discussing and then translating the physics of the flip-flop and of the DTL NAND gate to *the box and the pebble* picture.

Continuous transitions between states

A transistor is really an analog amplifier, and one can apply any potential to its gate, which yields a range of continuous values to its channel's resistance. In order for the transistor to behave like a switch and for the circuit presented in Figure 1.2 to behave like a NAND gate, we must condition its input and output voltages: these are usually restricted to $\{0\,\mathrm{V}, +5\,\mathrm{V}\}$ and switched very rapidly between the two values. Additionally, parts of the circuit may be biased at $-5\,\mathrm{V}$ in order to provide adequate polarization. Even then, when looked at with an oscilloscope, pulses representing bits do not have sharp edges. Rather, there are transients, and these must be analyzed rigorously at the circuit design stage in order to eliminate unexpected faulty behavior.

On the other hand, the presence or the absence of the pebble in the box apparently represents two distinct, separate states. There are no transients here. The pebble either is or is not in the box. *Tertium non datur.*

Yet, let us observe that even this is a convention, because, for example, we could place the pebble in such a way that only a half of it would be in the box and the other half would be outside. How should we account for this situation?

Many-valued logic

In binary, digital logic we ignore such states. But other types of logic do allow for the pebble to be halfway or a third of the way or any other portion in the box. Such logic systems fall under the category of *many-valued logics* [44], some of which are even *infinitely valued*. An example of an infinitely valued logic is the popular *fuzzy logic* [81] commonly used in robotics, data bases, image processing, and expert systems.

When we look at quantum logic more closely, these considerations will acquire a new deeper meaning, which will eventually lead to the notion of superposition of quantum states. Quantum logic is one of these systems, where a pebble can be halfway in one box and halfway in another one.

And the boxes don't even have to be adjacent.

1.2 Registers

A 3-bit counter

A row of flip-flops connected with each other in various ways constitutes a register. Depending on how the flip-flops are connected, the register may be used just as a store, or it can be used to perform some arithmetic operations.

Figure 1.3 shows a simple 3-bit modulo-7 counter implemented with three JK flip-flops. A JK flip-flop is a more complex device than the one shown in Figure 1.1; but to understand how the counter works, the reader needs to know only the following two rules:

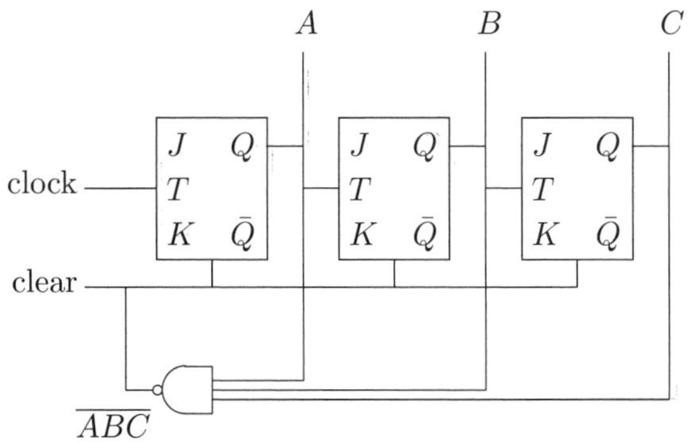

Figure 1.3: A modulo-7 counter made of three JK flip-flops.

1. The state Q of the JK flip-flop toggles on the *trailing* edge of the clock pulse T, that is, when the state of the input T changes from 1 to 0;
2. Applying 0 to *clear* resets Q to 0.

Register states

Let us assume that the whole counter starts in the $\{C = 0, B = 0, A = 0\}$ state. On the first application of the pulse to the *clock* input, A toggles to 1 on the trailing edge of the pulse and stays there. The state of the register becomes $\{C = 0, B = 0, A = 1\}$. On the second application of the clock pulse A toggles back to 0, but this change now toggles B to 1, and so the state of the register becomes $\{C = 0, B = 1, A = 0\}$. On the next trailing edge of the clock pulse A toggles to 1 and the state of the register is now $\{C = 0, B = 1, A = 1\}$. When A toggles back to 0 on the next application of the clock pulse, this triggers the change in B from 1 to 0, but this in turn toggles C, and so the state of the register becomes $\{C = 1, B = 0, A = 0\}$, and so on. Dropping $C =$, $B =$, and $A =$ from our notation describing the state of the register, we can see the following progression:

$$\{000\} \rightarrow \{001\} \rightarrow \{010\} \rightarrow \{011\} \rightarrow \{100\} \rightarrow \ldots . \tag{1.1}$$

We can interpret the strings enclosed in curly brackets as binary numbers; and upon having converted them to decimal notation, we obtain

$$0 \rightarrow 1 \rightarrow 2 \rightarrow 3 \rightarrow 4 \rightarrow \ldots . \tag{1.2}$$

The device counts clock pulses by remembering the previous value and then adding 1 to it on detecting the trailing edge of the clock pulse. When A, B, and C all become 1 at the same time, the NAND gate at the bottom of the circuit applies *clear* to all three flip-flops, and so A, B, and C get reset to 0. This process happens so fast that the counter does not stay in the $\{111\}$ configuration for an appreciable amount of time. It counts from 0 *through* 6 transitioning through seven distinct stable *states* in the process.

State transitions

At first glance we may think that the counter *jumps* between the discrete states. A closer observation of the transitions with an oscilloscope shows that the counter *glides* between the states through a continuum of various configurations, which cannot be interpreted in terms of digital logic. But the configurations in the continuum are unstable, and the gliding takes very little time, so that the notion of *jumps* is a good approximation.[1]

By now we know that a possible physical embodiment of a bit is an accumulation of electric charge on the gate of a transistor inside a flip-flop. We can also think about the presence or the absence of the charge on the gate in the same way we think about the presence or absence of a pebble in a box. And so, instead of working with a row of flip-flops, we can work with a row of boxes and pebbles. Such a system is also a register, albeit a much slower one and more difficult to manipulate.

An "almost quantum" register

The following figure shows an example of a box and a pebble register that displays some features that are reminiscent of quantum physics.

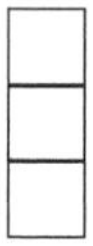

The register contains three boxes stacked vertically. Their position corresponds to the energy of a pebble that may be placed in a box. The higher the location of the box, the higher the energy of the pebble. The pebbles that are used in the register have a peculiar property. When two pebbles meet in a single box, they annihilate, and the energy released in the process creates a higher energy pebble in the box above. Of course, if there is already a pebble there, the newly created pebble and the previously inserted pebble annihilate, too, and an even higher energy pebble is created in the next box up.

[1] An alert reader will perhaps notice that what we call a *jump* in our everyday life is also a gliding transition that takes a jumper (e.g., a cat) through a continuum of unstable configurations that may end eventually with the cat sitting stably on top of a table.

Let us observe what is going to happen if we keep adding pebbles to the box at the bottom of the stack. When we place the first pebble there, the system looks as follows.

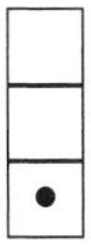

Now we add another pebble to the box at the bottom. The two pebbles annihilate, and a new, higher-energy pebble is created in the middle box.

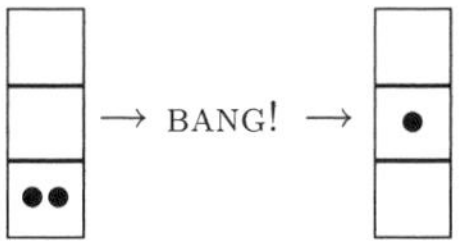

When we add a pebble again to the box at the bottom, nothing much happens, because there is no other pebble in it, and so the state of the register becomes as shown below.

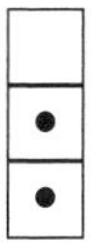

But fireworks fly again when we add a yet another pebble to the box at the bottom of the stack.

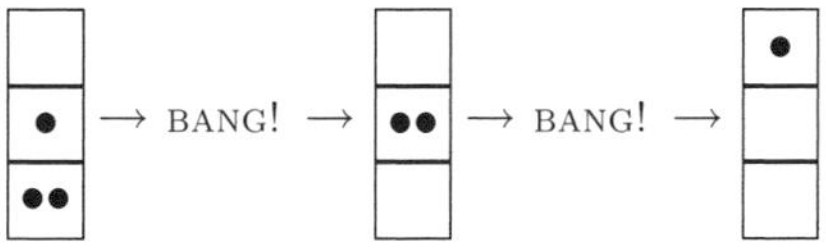

The first bang occurs because there are two pebbles in the first box. The pebbles annihilate, and the energy released creates a higher-energy pebble in the middle box. But a pebble is already there, so the two annihilate in the second bang, and a pebble of even higher energy is now created in the top box.

In summary, the register has transitioned through the following stable states:

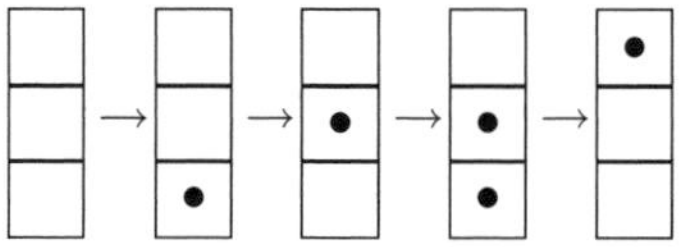

The register, apparently, is also a counter. Eventually we'll end up having pebbles in all three boxes. Adding a yet another pebble to the box at the bottom triggers a chain reaction that will clear all boxes and will eject a very high energy pebble from the register altogether. This, therefore, is a modulo-7 counter.

Creation and annihilation operators

The first reason the register is reminiscent of a quantum system is that its successive states are truly separate, without any in-betweens; that is, pebbles don't move between boxes. Instead they disappear from a box; and if the energy released in the process is high enough, a new pebble reappears from nothingness in a higher-energy box. The model bears some resemblance to quantum field theory, where particle states can be acted on by annihilation and creation operators. We shall see similar formalism applied in the discussion of vibrational states in the Paul trap in Section 6.3, Chapter 6.

The second reason is that here we have a feature that resembles the Pauli exclusion principle, discussed in Section 5.2, Chapter 5, which states that no two fermions can coexist in the same state.

1.3 Fluctuating Registers

The stable states of the register we have seen in the previous sections were all well defined. For example, the counter would go through the sequence of seven *stable* configurations:

$$\{000\} \rightarrow \{001\} \rightarrow \{010\} \rightarrow \{011\} \rightarrow \{100\} \rightarrow \{101\} \rightarrow \{110\}. \qquad (1.3)$$

Once a register would glide into one of these, it would stay there, the values of its bits unchanging, until the next clock pulse would shift it to the next state. In general, a 3-bit register can store numbers from 0 ($\{000\}$) through 7 ($\{111\}$) inclusive. Let us then focus on such a 3-bit register. It does not have to be a counter this time.

A fluctuating register

Also, let us suppose that the register is afflicted by the following malady.[2] When set by some electronic procedure to hold a binary number {101}—there may be some toggle switches to do this on the side of the package—its bits start to fluctuate randomly so that the register spends only 72% of the time in the {101} configuration and 28% of the time in every other configuration, flickering at random between them[3].

Let us assume that the same happens when the register is set to hold other numbers, {000}, {001}, ..., {111} as well; that is, the register ends up flickering between all possible configurations at random but visits its *set* configuration 72% of the time. At first glance a register like this seems rather useless, but we could employ its fluctuations, for example, in Monte Carlo codes.

To make the game more fun, after the register has been set, we are going to cover its toggles with a masking tape, so that its state cannot be ascertained by looking at the toggles. Instead we have to resort to other means. The point of this exercise is to prepare the reader for a description of similar systems that have no toggles at all.

Flactuating register states

The register exists in one of the eight fluctuating *states*. Each state manifests itself by visiting a certain configuration more often than other configurations. This time we can no longer associate the state with a specific configuration as closely as we have done for the register that was not subject to random fluctuations. The state is now something more abstract, something that we can no longer associate with a simple single observation of the register. Instead we have to look at the register for a long time in order to identify its preferred configuration, and thus its state.

Let us introduce the following notation for the states of the fluctuating register:

$\boldsymbol{p}_{\bar{0}}$ is the state that visits {000} most often,
$\boldsymbol{p}_{\bar{1}}$ is the state that visits {001} most often,
$\boldsymbol{p}_{\bar{2}}$ is the state that visits {010} most often,
$\boldsymbol{p}_{\bar{3}}$ is the state that visits {011} most often,
$\boldsymbol{p}_{\bar{4}}$ is the state that visits {100} most often,
$\boldsymbol{p}_{\bar{5}}$ is the state that visits {101} most often,
$\boldsymbol{p}_{\bar{6}}$ is the state that visits {110} most often,
$\boldsymbol{p}_{\bar{7}}$ is the state that visits {111} most often.

[2]The *malady* may have been designed into the register on purpose.
[3]Any similarity to Intel devices is incidental and unintended.

Although we have labeled the states $\boldsymbol{p}_{\bar{0}}$ through $\boldsymbol{p}_{\bar{7}}$, we cannot at this early stage associate the labels with mathematical objects. To endow the states of the fluctuating register with a mathematical structure, we have to figure out how they can be measured and manipulated—and then map this onto mathematics.

State observation

So, how can we ascertain which one of the eight states defined above the register is in, if we are not allowed to peek at the setting of its switches?

To do so, we must observe the register for a long time, writing down its observed configurations perhaps at random time intervals.[4] If the register is in the $\boldsymbol{p}_{\bar{5}}$ state, approximately 72% of the observations should return the $\{101\}$ configuration, with other observations evenly spread over other configurations. If we made n_5 measurements of the register in total, $n^0{}_5$ observations would show the register in the $\{000\}$ configuration, $n^1{}_5$ observations would show it in the $\{001\}$ configuration, and so on for every other configuration, ending with $n^7{}_5$ for the $\{111\}$ configuration.[5] We can now build a column vector for which we would expect the following:

$$\begin{pmatrix} n^0{}_5/n_5 \approx 0.04 \\ n^1{}_5/n_5 \approx 0.04 \\ n^2{}_5/n_5 \approx 0.04 \\ n^3{}_5/n_5 \approx 0.04 \\ n^4{}_5/n_5 \approx 0.04 \\ n^5{}_5/n_5 \approx 0.72 \\ n^6{}_5/n_5 \approx 0.04 \\ n^7{}_5/n_5 \approx 0.04 \end{pmatrix}. \tag{1.4}$$

We do not expect $n^6{}_5/n_5 = 0.04$ exactly, because, after all, the fluctuations of the register are random, but we do expect that we should get very close to 0.04 if n_5 is very large. For $n_5 \rightarrow \infty$ the ratios in the column vector above become *probabilities*.

State description

This lets us identify state $\boldsymbol{p}_{\bar{5}}$ with the column of probabilities of finding the register

[4] An important assumption here is that we can observe the register without affecting its state, which is the case with classical registers but not with quantum ones. One of the ways to deal with it in the quantum case is to discard the observed register and get a new one in the same state for the next observation. Another way is to reset the observed register, if possible, put it in the same state, and repeat the observation.

[5] The superscripts 0 through 7 in $n^0{}_5$ through $n^7{}_5$ and also in $p^0{}_5$ through $p^7{}_5$ further down are *not* exponents. We do not raise n_5 (or p_5) to the powers of 0 through 7. They are just indexes, which say that, for example, $n^4{}_5$ is the number of observations made on a register in state $\boldsymbol{p}_{\bar{5}}$ that found it in configuration $\{100\} \equiv 4$. There is a reason we want this index to be placed in the superscript position rather than in the subscript position. This will be explained in more detail when we talk about forms and vectors in Section 1.7 on page 28. If we ever need to exponentiate an object with a superscript index, for example, $p^3{}_5$, we shall enclose this object in brackets to distinguish between a raised index and an exponent, for example, $\left(p^3{}_5\right)^2$.

in each of its eight possible configurations:

$$\boldsymbol{p}_{\bar{5}} \equiv \begin{pmatrix} {p^0}_5 = 0.04 \\ {p^1}_5 = 0.04 \\ {p^2}_5 = 0.04 \\ {p^3}_5 = 0.04 \\ {p^4}_5 = 0.04 \\ {p^5}_5 = 0.72 \\ {p^6}_5 = 0.04 \\ {p^7}_5 = 0.04 \end{pmatrix}. \tag{1.5}$$

A statistical ensemble of registers

It is often convenient to think of the fluctuating register in terms of a *statistical ensemble.*

Let us assume that instead of a single fluctuating register we have a very large number of static, nonfluctuating registers, of which 4% are in the {000} configuration, 4% are in the {001} configuration, 4% are in the {010} configuration, 4% are in the {011} configuration, 4% are in the {100} configuration, 72% are in the {101} configuration, 4% are in the {110} configuration, and 4% are in the {111} configuration. Now let us put all the registers in a hat, mix them thoroughly, and draw at random n_5 registers from the hat. Of these ${n^0}_5$ will be in the {000} configuration, ${n^1}_5$ in the {001} configuration, ..., ${n^6}_5$ in the {110} configuration and ${n^7}_5$ in the {111} configuration. If the whole *ensemble* has been mixed well, we would expect the following:

$$\begin{aligned} {n^0}_5/n_5 &\approx 0.04, \\ {n^1}_5/n_5 &\approx 0.04, \\ {n^2}_5/n_5 &\approx 0.04, \\ {n^3}_5/n_5 &\approx 0.04, \\ {n^4}_5/n_5 &\approx 0.04, \\ {n^5}_5/n_5 &\approx 0.72, \\ {n^6}_5/n_5 &\approx 0.04, \\ {n^7}_5/n_5 &\approx 0.04. \end{aligned}$$

Logically and arithmetically such an ensemble of static registers from which we *sample* n_5 registers is equivalent to a single randomly fluctuating register at which we *look* (without disturbing its overall condition) n_5 times.

Probabilities

The eight states our fluctuating register can be put in can be characterized by the following column vectors of probabilities:

$$\boldsymbol{p}_{\bar{0}} \equiv \begin{pmatrix} p^0{}_0 = 0.72 \\ p^1{}_0 = 0.04 \\ p^2{}_0 = 0.04 \\ p^3{}_0 = 0.04 \\ p^4{}_0 = 0.04 \\ p^5{}_0 = 0.04 \\ p^6{}_0 = 0.04 \\ p^7{}_0 = 0.04 \end{pmatrix}, \quad \boldsymbol{p}_{\bar{1}} \equiv \begin{pmatrix} p^0{}_1 = 0.04 \\ p^1{}_1 = 0.72 \\ p^2{}_1 = 0.04 \\ p^3{}_1 = 0.04 \\ p^4{}_1 = 0.04 \\ p^5{}_1 = 0.04 \\ p^6{}_1 = 0.04 \\ p^7{}_1 = 0.04 \end{pmatrix}, \quad \boldsymbol{p}_{\bar{2}} \equiv \begin{pmatrix} p^0{}_2 = 0.04 \\ p^1{}_2 = 0.04 \\ p^2{}_2 = 0.72 \\ p^3{}_2 = 0.04 \\ p^4{}_2 = 0.04 \\ p^5{}_2 = 0.04 \\ p^6{}_2 = 0.04 \\ p^7{}_2 = 0.04 \end{pmatrix},$$

$$\boldsymbol{p}_{\bar{3}} \equiv \begin{pmatrix} p^0{}_3 = 0.04 \\ p^1{}_3 = 0.04 \\ p^2{}_3 = 0.04 \\ p^3{}_3 = 0.72 \\ p^4{}_3 = 0.04 \\ p^5{}_3 = 0.04 \\ p^6{}_3 = 0.04 \\ p^7{}_3 = 0.04 \end{pmatrix}, \quad \boldsymbol{p}_{\bar{4}} \equiv \begin{pmatrix} p^0{}_4 = 0.04 \\ p^1{}_4 = 0.04 \\ p^2{}_4 = 0.04 \\ p^3{}_4 = 0.04 \\ p^4{}_4 = 0.72 \\ p^5{}_4 = 0.04 \\ p^6{}_4 = 0.04 \\ p^7{}_4 = 0.04 \end{pmatrix}, \quad \boldsymbol{p}_{\bar{5}} \equiv \begin{pmatrix} p^0{}_5 = 0.04 \\ p^1{}_5 = 0.04 \\ p^2{}_5 = 0.04 \\ p^3{}_5 = 0.04 \\ p^4{}_5 = 0.04 \\ p^5{}_5 = 0.72 \\ p^6{}_5 = 0.04 \\ p^7{}_5 = 0.04 \end{pmatrix},$$

$$\boldsymbol{p}_{\bar{6}} \equiv \begin{pmatrix} p^0{}_6 = 0.04 \\ p^1{}_6 = 0.04 \\ p^2{}_6 = 0.04 \\ p^3{}_6 = 0.04 \\ p^4{}_6 = 0.04 \\ p^5{}_6 = 0.04 \\ p^6{}_6 = 0.72 \\ p^7{}_6 = 0.04 \end{pmatrix}, \quad \boldsymbol{p}_{\bar{7}} \equiv \begin{pmatrix} p^0{}_7 = 0.04 \\ p^1{}_7 = 0.04 \\ p^2{}_7 = 0.04 \\ p^3{}_7 = 0.04 \\ p^4{}_7 = 0.04 \\ p^5{}_7 = 0.04 \\ p^6{}_7 = 0.04 \\ p^7{}_7 = 0.72 \end{pmatrix}.$$

We shall call the probabilities that populate the arrays *fiducial measurements*, and we shall call the arrays of probabilities *fiducial vectors*[6] [60].

For every state $\boldsymbol{p}_{\bar{i}}$, $i = 0, 1, \ldots, 7$, listed above, we have that $p^0{}_i + p^1{}_i + p^2{}_i + p^3{}_i + p^4{}_i + p^5{}_i + p^6{}_i + p^7{}_i = 1$. This means that the probability of finding the register in any one of the configurations from $\{000\}$ through $\{111\}$ is 1. States $\boldsymbol{p}_{\bar{i}}$ that have this property are said to be *normalized*.

Given the collection of normalized states $\boldsymbol{p}_{\bar{i}}$ we can construct statistical ensembles with other values for probabilities p^0 through p^7 by *mixing* states $\boldsymbol{p}_{\bar{i}}$ in various proportions.

---

[6]The word *fiducial* in physics means an object or a system that is used as a standard of reference or measurement. It derives from the Latin word *fiducia*, which means *confidence* or *reliance*. The notion of confidence is closely related to the notion of probabiblity. We often hear meteorologists say they are 80% *confident* it's going to rain in the afteroon. It normally means, it is not going to rain at all.

1.4 Mixtures and Pure States

Let us suppose we have a very large number, N, of fluctuating registers affected by the malady discussed in the previous section. Let us also suppose that N_0 of these have been put in state $\boldsymbol{p}_{\bar{0}}$ and the remaining $N - N_0 = N_3$ have been put in state $\boldsymbol{p}_{\bar{3}}$.

Mixing statistical ensembles

Now let us place all N registers into a hat and mix them thoroughly. We can draw them from the hat at random and look at their configuration but only once per register drawn. What probabilities should we expect for any possible register configuration in the ensemble?

The easiest way to answer the question is to *expand* states $\boldsymbol{p}_{\bar{0}}$ and $\boldsymbol{p}_{\bar{3}}$ into their corresponding statistical ensembles and say that we have N_0 ensembles that correspond to state $\boldsymbol{p}_{\bar{0}}$ and N_3 ensembles that correspond to state $\boldsymbol{p}_{\bar{3}}$. In each $\boldsymbol{p}_{\bar{0}}$ ensemble we have ${n^0}_0$ registers out of n_0 in the $\{000\}$ configuration, and in each $\boldsymbol{p}_{\bar{3}}$ ensemble we have ${n^0}_3$ registers out of n_3 in the $\{000\}$ configuration. Hence, the total number of registers in the $\{000\}$ configuration is

$$N_0{n^0}_0 + N_3{n^0}_3. \tag{1.6}$$

The total number of registers after this expansion of states into ensembles is

$$N_0 n_0 + N_3 n_3. \tag{1.7}$$

Therefore, the probability of drawing a register in the $\{000\}$ configuration is going to be

$$\frac{N_0{n^0}_0 + N_3{n^0}_3}{N_0 n_0 + N_3 n_3}, \tag{1.8}$$

in the limit $N \rightarrow \infty$, $n_0 \rightarrow \infty$, and $n_3 \rightarrow \infty$.

We should also assume at this stage that we have an identical number of registers in the ensembles for $\boldsymbol{p}_{\bar{0}}$ and $\boldsymbol{p}_{\bar{3}}$, that is, that $n_0 = n_3 = n$. The reason is that if the ensembles for $\boldsymbol{p}_{\bar{0}}$ have, say, a markedly smaller number of registers than ensembles for $\boldsymbol{p}_{\bar{0}}$, the latter will weigh more heavily than the former, and so our estimates of probabilities for the whole mixture, based on finite sampling, will be skewed. Then

$$\begin{aligned}
&\frac{N_0{n^0}_0 + N_3{n^0}_3}{N_0 n_0 + N_3 n_3} \\
&\quad = \frac{N_0{n^0}_0 + N_3{n^0}_3}{n\,(N_0 + N_3)} = \frac{N_0{n^0}_0 + N_3{n^0}_3}{nN} = \frac{N_0}{N}\frac{{n^0}_0}{n} + \frac{N_3}{N}\frac{{n^0}_3}{n}.
\end{aligned}$$

In the limit $n \rightarrow \infty$ and $N \rightarrow \infty$ this becomes

$$P_0{p^0}_0 + P_3{p^0}_3, \tag{1.9}$$

where P_0 is the probability of drawing a register in state $\boldsymbol{p}_{\bar{0}}$, P_3 is the probability of drawing a register in state $\boldsymbol{p}_{\bar{3}}$, ${p^0}_0$ is the probability that a register in state $\boldsymbol{p}_{\bar{0}}$ is observed in configuration $\{000\}$, and ${p^0}_3$ is the probability that a register in state $\boldsymbol{p}_{\bar{3}}$ is observed in configuration $\{000\}$.

To get a clearer picture, let us assume that $P_0 = 0.3$ and that $P_3 = 0.7$. At this level (of probabilities pertaining to the mixture) we have that $P_0 + P_3 = 1$. What are the probabilities for each configuration in the mixture?

$$
\begin{aligned}
p^0 = P_0 {p^0}_0 + P_3 {p^0}_3 &= 0.3 \cdot 0.72 + 0.7 \cdot 0.04 = 0.244,\\
p^1 = P_0 {p^1}_0 + P_3 {p^1}_3 &= 0.3 \cdot 0.04 + 0.7 \cdot 0.04 = 0.04,\\
p^2 = P_0 {p^2}_0 + P_3 {p^2}_3 &= 0.3 \cdot 0.04 + 0.7 \cdot 0.04 = 0.04,\\
p^3 = P_0 {p^3}_0 + P_3 {p^3}_3 &= 0.3 \cdot 0.04 + 0.7 \cdot 0.72 = 0.516,\\
p^4 = P_0 {p^4}_0 + P_3 {p^4}_3 &= 0.3 \cdot 0.04 + 0.7 \cdot 0.04 = 0.04,\\
p^5 = P_0 {p^5}_0 + P_3 {p^5}_3 &= 0.3 \cdot 0.04 + 0.7 \cdot 0.04 = 0.04,\\
p^6 = P_0 {p^6}_0 + P_3 {p^6}_3 &= 0.3 \cdot 0.04 + 0.7 \cdot 0.04 = 0.04,\\
p^7 = P_0 {p^7}_0 + P_3 {p^7}_3 &= 0.3 \cdot 0.04 + 0.7 \cdot 0.04 = 0.04.
\end{aligned}
$$

The mixture remains normalized. All probabilities p^i, $i = 1, \ldots, 7$, still add to one.

Fiducial vector

Using symbol $\boldsymbol{p}$ for the array of probabilities p^0 through p^7, we can write the above as follows:

$$\boldsymbol{p} = P_0 \boldsymbol{p}_{\bar{0}} + P_3 \boldsymbol{p}_{\bar{3}}. \tag{1.10}$$

In general, assuming that we use all possible states in the mixture, we would have

$$\boldsymbol{p} = \sum_i P_i \boldsymbol{p}_{\bar{i}}. \tag{1.11}$$

Convexity

The mixture state is a *linear combination* of its constituents. The linearity is restricted by two conditions, namely, that $\sum_i P_i = 1$ and $\forall_i\, 0 \le P_i \le 1$. Linearity so restricted is called *convexity*, but we are going to show in Section 1.6 that it can be extended to full linearity as long as what is on the left-hand side of equation (1.11) is still a physically meaningful state.

One can easily see that $\sum_k p^k = 1$ for any convex linear combination that represents a mixture

$$
\begin{aligned}
\sum_k p^k &= \sum_k \sum_i P_i {p^k}_i = \sum_i P_i \sum_k {p^k}_i = \\
&= \sum_i P_i \cdot 1 = \sum_i P_i = 1.
\end{aligned} \tag{1.12}
$$

States $\boldsymbol{p}_{\bar{0}}$ through $\boldsymbol{p}_{\bar{7}}$ are mixtures, too. For example, state $\boldsymbol{p}_{\bar{3}}$, which is specified by the array *Mixed states*

$$\boldsymbol{p}_{\bar{3}} \equiv \begin{pmatrix} p^0{}_3 = 0.04 \\ p^1{}_3 = 0.04 \\ p^2{}_3 = 0.04 \\ p^3{}_3 = 0.72 \\ p^4{}_3 = 0.04 \\ p^5{}_3 = 0.04 \\ p^6{}_3 = 0.04 \\ p^7{}_3 = 0.04 \end{pmatrix}, \tag{1.13}$$

can be thought of as a mixture of, say, 1,000,000 nonfluctuating registers, of which 720,000 are in the {011} configuration at all times and the remaining 280,000 registers are evenly spread over the remaining configurations, with 40,000 registers in each.

A register that is in a nonfluctuating {000} configuration can still be described in terms of a column vector of probabilities as follows: *Pure states*

$$\begin{pmatrix} p^0 = 1 \\ p^1 = 0 \\ p^2 = 0 \\ p^3 = 0 \\ p^4 = 0 \\ p^5 = 0 \\ p^6 = 0 \\ p^7 = 0 \end{pmatrix}. \tag{1.14}$$

This states that the probability of finding this register in configuration {000} is 1 or, in other words, that the register spends 100% of its time in this configuration. We can use similar array representations for registers in nonfluctuating configurations {001} through {111}. These states, however, cannot be constructed by *mixing* other states, because probabilities cannot be negative, so there is no way that the zeros can be generated in linear combinations of nonzero coefficients, all of which represent some probabilities. States that are not mixtures are called *pure*. Only for pure states $|\,\boldsymbol{p}\rangle$ do we have that

$$\sum_i p^i = 1 \quad \text{and} \quad \sum_i \left(p^i\right)^2 = 1, \tag{1.15}$$

whereas for mixtures

$$\sum_i \left(p_i\right)^2 < 1. \tag{1.16}$$

One can easily see why this should be so. For $0 < p^i < 1$ we have that $0 < \left(p^i\right)^2 < p^i$. It follows that for such p^is $0 < \sum_i \left(p^i\right)^2 < \sum_i p^i = 1$. The equality $\sum_i \left(p^i\right)^2 = 1$ can therefore happen only if at least one $p^i = 1$, but since $\sum_i p^i = 1$ and none of the p_is can be negative, all other p^is must be zero.

Fiducial representation of pure states

Let us introduce the following notation for the pure states:

$$\boldsymbol{e}_0 \equiv \begin{pmatrix} 1\\0\\0\\0\\0\\0\\0\\0 \end{pmatrix}, \quad \boldsymbol{e}_1 \equiv \begin{pmatrix} 0\\1\\0\\0\\0\\0\\0\\0 \end{pmatrix}, \quad \boldsymbol{e}_2 \equiv \begin{pmatrix} 0\\0\\1\\0\\0\\0\\0\\0 \end{pmatrix}, \quad \boldsymbol{e}_3 \equiv \begin{pmatrix} 0\\0\\0\\1\\0\\0\\0\\0 \end{pmatrix},$$

$$\boldsymbol{e}_4 \equiv \begin{pmatrix} 0\\0\\0\\0\\1\\0\\0\\0 \end{pmatrix}, \quad \boldsymbol{e}_5 \equiv \begin{pmatrix} 0\\0\\0\\0\\0\\1\\0\\0 \end{pmatrix}, \quad \boldsymbol{e}_6 \equiv \begin{pmatrix} 0\\0\\0\\0\\0\\0\\1\\0 \end{pmatrix}, \quad \boldsymbol{e}_7 \equiv \begin{pmatrix} 0\\0\\0\\0\\0\\0\\0\\1 \end{pmatrix}.$$

This time we have dropped the bars above the digits to emphasize that these *pure* states do not fluctuate and therefore the digits that represent them are exact and do not represent averages or most often encountered configurations.

The basis of pure states

Let us consider a randomly fluctuating register state specified by a vector of probabilities $\boldsymbol{p}$. Using straightforward array arithmetic, we can express the array $\boldsymbol{p}$ in terms of $\boldsymbol{e}_i$ as follows:

$$\boldsymbol{p} = \sum_i p^i \boldsymbol{e}_i. \tag{1.17}$$

The expression is reminiscent of equation (1.11) on page 14, but here we have replaced fluctuating states on the right-hand side with nonfluctuating *pure* states. And so, we have arrived at the following conclusion:

Every randomly fluctuating register state is a mixture of pure states.

Vector space

The ability to decompose any randomly fluctuating register state $\boldsymbol{p}$ into a linear (to be exact, a convex) combination of other fluctuating or pure states suggests that we can think of the fluctuating states as belonging to a vector space in which

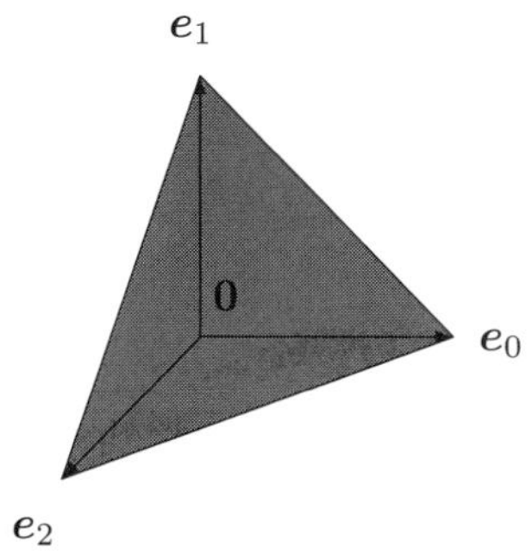

Figure 1.4: The set S in the three-dimensional vector space that corresponds to a 2-bit modulo-3 fluctuating register is the gray triangle spanned by the ends of the three basis vectors of the space.

the natural choice for the basis are the pure states.[7] But fluctuating states do not fill the space entirely, because only vectors for which

$$\sum_i p^i = 1 \quad \text{and} \quad \forall_i \, 0 \le p^i \le 1, \tag{1.18}$$

are physical. Let us call the set of physically meaningful vectors in this space S.

A set of physically meaningful states

Figure 1.4 shows set S for a three-dimensional vector space that corresponds to a 2-bit modulo-3 fluctuating register, that is, a register, for which the $\{11\}$ configuration is unstable and flips the register back to $\{00\}$.

In this case S is the triangle spanned by the tips of the pure states.

$$\boldsymbol{e}_0 \equiv \begin{pmatrix} 1 \\ 0 \\ 0 \end{pmatrix}, \quad \boldsymbol{e}_1 \equiv \begin{pmatrix} 0 \\ 1 \\ 0 \end{pmatrix}, \quad \boldsymbol{e}_2 \equiv \begin{pmatrix} 0 \\ 0 \\ 1 \end{pmatrix}. \tag{1.19}$$

The null state

One can fill the space between the triangle in Figure 1.4 and the zero of the vector space by admitting the *null* state as a possible participant in the mixtures.

[7]We are going to *load* the term *basis* with an additional meaning soon, but what we have just called *the basis states* will remain such in the classical physics context even with this additional loading put on them.

A 3-bit register null state, for which we are going to use symbol $\mathbf{0}$, corresponds to the array of probabilities

$$\mathbf{0} \equiv \begin{pmatrix} 0 \\ 0 \\ 0 \\ 0 \\ 0 \\ 0 \\ 0 \end{pmatrix}. \tag{1.20}$$

The meaning of the null state is that a register in this state does not return any *acceptable* reading at all: we can say that it's broken, that all its LEDs are off.

Let us consider the following mixture: 30% of the 3-bit registers are in the null state $\mathbf{0}$, 40% of the 3-bit registers are in the $\boldsymbol{e}_2$ state, and the remaining 30% are in the $\boldsymbol{e}_4$ state. What are the coefficients p^i of the mixture? Let us expand the mixture into its statistical ensemble, assuming for simplicity that the total number of registers in the ensemble is 100. Then 40 registers will be in the $\{010\}$ configuration, 30 registers will be in the $\{100\}$ configuration, and the remaining 30 registers will be in no readable configuration at all. The probability of drawing a register in the $\{010\}$ configuration from the ensemble is therefore $p^2 = 40/100 = .4$. The probability of drawing a register in the $\{100\}$ configuration from the ensemble is $p^4 = 30/100 = .3$. The probability of drawing a register in any other acceptable configuration is 0. The resulting state of the register is therefore as follows:

$$\boldsymbol{p} \equiv \begin{pmatrix} p^0 = 0.0 \\ p^1 = 0.0 \\ p^2 = 0.4 \\ p^3 = 0.0 \\ p^4 = 0.3 \\ p^5 = 0.0 \\ p^6 = 0.0 \\ p^7 = 0.0 \end{pmatrix}. \tag{1.21}$$

The sum of all p^is is now equal to 0.7, which is less than 1.

Diluted mixtures

A state vector for which $\sum_i p^i < 1$ is said to be *unnormalized* as opposed to a state vector for which $\sum_i p^i = 1$, which, as we have already remarked, is said to be *normalized*. The addition of the null state $\mathbf{0}$ to the mixture has the effect of "diluting" it. Because every unnormalized state vector must have some admixture of the null state, all unnormalized states clearly are mixtures. And conversely, states that are not mixtures and are not null either—that is, pure states—must be normalized. And then we also have the body of states that are not pure but are not diluted either: these states are normalized, and they are also mixtures.

If null states are allowed, then the corresponding set S is no longer restricted to the surface of the triangle shown in Figure 1.4. Instead, the states fill the whole volume of the tetrahedron between the triangle and the zero of the vector space, including both the zero and the triangle.

Pure states as extremal points of S

The pure states and the null state are the *extremal* points of S. This observation lets us arrive at the following *definition* of pure states:

> Pure states correspond to extremal points of $S - \{\mathbf{0}\}$.

The definition will come handy in more complex situations and in richer theories, in which we may not be able to draw a simple picture or recognize that a given state is pure by merely looking at its corresponding column of probabilities.

1.5 Basis States

In Section 1.4 we stated that pure states were the natural choice for the basis of the vector space in which physical states of fluctuating registers filled set S. This assertion was based on elementary algebra, which we used to decompose an arbitrary mixture into a linear combination of pure states,

$$\boldsymbol{p} = \sum_{i=0}^{7} p^i \boldsymbol{e}_i. \tag{1.22}$$

The decomposition, in turn, was based on the representation of pure states by arrays of zeros and ones—arrays, which in the world of elementary algebra are commonly associated with basis vectors.

A basis state

Here we are going to give a quite special physical meaning to what we are going to call the *basis state* throughout the remainder of this book and to what is also called the basis state in quantum mechanics.

> A *basis state* is the configuration of a randomly fluctuating register that can be ascertained by glancing at it momentarily.

Physical and canonical basis states

If we have a fluctuating register and "glance at it momentarily," we are not going to see it fluctuate. Instead, we are going to see this register in a quite specific frozen (albeit momentarily only) configuration, for example, $\{010\}$. If we glance at the register again a moment later, the register may be in the $\{101\}$ configuration.[8] We

[8] Here again we make the assumption that glancing at the register does not affect its state.

are going to call the nonfluctuating states that correspond to these configurations (such as $\boldsymbol{e}_2$ and $\boldsymbol{e}_5$) the physical basis states. Altogether we are going to have eight such basis states for the classical 3-bit randomly fluctuating register, assuming that all bits are allowed to fluctuate freely. The basis states, as defined here, are $\boldsymbol{e}_0$ through $\boldsymbol{e}_7$, which is also what we have called "pure states" and what is also a *canonical* basis in the fiducial vector space.

We are going to use a special notation for such momentarily glanced basis vectors to distinguish between them and the canonical basis. Borrowing from the traditions of quantum mechanics, we'll denote them by $|\,\boldsymbol{e}_0\rangle$ through $|\,\boldsymbol{e}_7\rangle$.

Dimensionality and degrees of freedom

The number of vectors in the physical basis of the classical randomly fluctuating register—this number is also called the *dimensionality* of the system—is the same as the number of probabilities needed to describe the state. The number of probabilities is also called the number of degrees of freedom of the system. This follows clearly from how the probabilities have been defined: they are probabilities of finding the register in one of its specific configurations, which here we have identified with the basis states, because these are the only configurations (and not the fluctuating states) that we can actually see when we give the register a brief glance.

Denoting the dimensionality of the system by N and the numbers of degrees of freedom by K, we can state that for the classical randomly fluctuating register

$$K = N. \tag{1.23}$$

Why should we distinguish at all between canonical and physical basis, especially since they appear to be exactly the same for the case considered so far? The answer is that they will not be the same in the quantum register case. What's more, we shall find there that

$$K = N^2. \tag{1.24}$$

Transitional configurations

A perspicacious reader may be tempted to ask the following question: What if I happen to catch the register in one of the *unstable* states it goes through when it *glides* between the configurations described by symbols $\{000\}$ through $\{111\}$? Such a configuration does not correspond to any of the "basis vectors" we have defined in this section.

This is indeed the case. Our probabilistic description does not cover the configurations of the continuum through which the register glides between its *basis states* at all. Instead we have focused entirely on the stable discrete configurations.

The meaning of "momentarily"

One of the ways to deal with the problem is to define more precisely what we mean by "momentarily." Assuming that the transition between the stable configurations of the register, the gliding phase, lasts δt, we want "momentarily" to be much longer than δt. But "momentarily" must not be too long, because then we'll see the register switch during observation. If the register stays in any given configuration for Δt on average, then we want "momentarily" to be much shorter than Δt. And so we arrive at a somewhat more precise definition of "momentarily":

$$\delta t \ll \text{"momentarily"} \ll \Delta t. \tag{1.25}$$

We shall see that this problem is not limited to classical registers. A similar condition is imposed on quantum observations, although the dynamics of quantum observations is quite different. But it still takes a certain amount of time and effort to force a quantum system into its basis state. If the act of observation is too lightweight and too fast, the quantum system will not "collapse" to the basis state, and the measurement will end up being *insufficiently resolved.*

Inclusion of transitional states

A quite different way to deal with the problem is to include the continuum of configurations the register glides through between its stable states into the model and to allow register configurations such as $\{0.76\ 0.34\ 0.18\}$. We would then have to add probabilities (or probability densities) of finding the register in such a state to our measurements and our theory. The number of dimensions of the system would skyrocket, but this does not necessarily imply that the system would become intractable.

This solution also has its equivalent in the world of quantum physics. Detailed investigations of the spectrum of hydrogen atom revealed that its spectral lines were split into very fine structure and that additional splitting occurred in presence of electric and magnetic fields. To account for every observed feature of the spectrum, physicists had to significantly enlarge the initially simple theory of hydrogen atom so as to incorporate various quantum electrodynamic corrections.

1.6 Functions and Measurements on Mixtures

How can we define a function on a randomly fluctuating register *state*?

Averages

There are various ways to do so. For example, we could associate certain values, $f_i \in \mathbb{R}$, with specific configurations of the register, that is, $\{000\}, \ldots, \{111\}$, and then we could associate an average of f_i over the statistical ensemble that corresponds to $\boldsymbol{p}$ with $f(\boldsymbol{p})$. This is a very physical way of doing things, since all that we can see as we glance at the fluctuating register every now and then are its various basis states, its momentarily frozen configurations. If every one of the configurations is associated with some value f_i, what we're going to perceive in terms of f_i over a longer time, as the register keeps fluctuating, is an average value of f_i.

Arbitrary functions

Another way to define a function f on a fluctuating register state would be to construct an arbitrary mapping of the form

$$S \ni \boldsymbol{p} \mapsto f\left(p^0, p^1, p^2, p^3, p^4, p^5, p^6, p^7\right) \in \mathbb{R}, \tag{1.26}$$

where $\mathbb{R}$ stands for real numbers.

The second way is very general and could be used to define quite complicated nonlinear functions on coefficients p_i, for example,

$$S \ni \boldsymbol{p} \mapsto \left(p^0\right)^2 + p^1 p^3 - \sin\left(p^2 p^3 p^4\right) + e^{p^5 p^6 p^7} \in \mathbb{R}. \tag{1.27}$$

A function like this could be implemented by an electronic procedure. For example, the procedure could observe the register for some time collecting statistics and building a fiducial vector for it. Once the vector is sufficiently well defined, the above operation would be performed on its content and the value of f delivered on output. Knowing value of f on the basis states would not in general help us evaluate f on an arbitrary state $\boldsymbol{p}$. Similarly, knowing values of f on the components of a mixture would not in general help us evaluate f on the mixture itself. In every case we would have to carry out full fiducial measurements for the whole mixture first, and only evaluate the function afterward.

On the other hand, the strategy outlined at the beginning of the section leads to functions that can be evaluated on the go and have rather nice and simple properties. Also, they cover an important special case: that of the fiducial vector itself.

Let us consider a register that has a tiny tunable laser linked to its circuitry and the coupling between the laser and the configuration of the register is such that when the register is in configuration $\{000\}$, the laser emits red light; when the register is in configuration $\{111\}$, the laser emits blue light; and when the register is in any of the intermediate configurations, the laser emits light of some color between red and blue. Let the frequency of light emitted by the laser when the register is in state $\boldsymbol{e}_i$ (which corresponds directly to a specific configuration) be f_i. Let us then define the frequency function f on the basis states[9] as

$$f\left(\boldsymbol{e}_i\right) \doteq f_i. \tag{1.28}$$

We are going to extend the definition to an arbitrary mixture $\boldsymbol{p} = \sum_i p^i \boldsymbol{e}_i$ by calculating the average value of f_i over the ensemble that corresponds to $\boldsymbol{p}$. Let us call the average value $\bar{f}$ and let us use the arithmetic mean formula to calculate

Arithmetic mean over the ensemble

[9]These are, in fact, physical basis states, because we want to associate the definition with register configurations that can be observed by glancing at the register momentarily. So we should really write here $|\,\boldsymbol{e}_i\rangle$. But let us recall that for the classical register they are really the same, so we'll avoid excessive notational complexity by using the canonical basis instead.

it. If the ensemble comprises n registers, then np^0 registers are in state $\boldsymbol{e}_0$, np^1 registers are in state $\boldsymbol{e}_1$, ..., and np^7 registers are in state $\boldsymbol{e}_7$. The arithmetic mean of f_i over the ensemble is

$$\begin{aligned} \bar{f} &= \frac{1}{n}\left(np^0 f_0 + np^1 f_1 + \cdots + np^7 f_7\right) \\ &= \frac{n}{n}\sum_{i=0}^{7} p^i f_i = \sum_i p^i f_i = \sum_i p^i f\left(\boldsymbol{e}_i\right). \end{aligned} \tag{1.29}$$

By defining that $f\left(\boldsymbol{p}\right) \doteq \bar{f}$, we obtain the following formula:

$$f\left(\boldsymbol{p}\right) = f\left(\sum_i p^i \boldsymbol{e}_i\right) = \sum_i p^i f\left(\boldsymbol{e}_i\right). \tag{1.30}$$

Now let us take two arbitrary states[10]

$$\boldsymbol{p}_1 = \sum_i p^i{}_1 \boldsymbol{e}_i \quad \text{and} \quad \boldsymbol{p}_2 = \sum_i p^i{}_2 \boldsymbol{e}_i, \tag{1.31}$$

and evaluate f on their mixture:

$$\begin{aligned} f\left(P_1\boldsymbol{p}_1 + P_2\boldsymbol{p}_2\right) &= f\left(P_1\sum_i p^i{}_1\boldsymbol{e}_i + P_2\sum_i p^i{}_2\boldsymbol{e}_i\right) \\ &= f\left(\sum_i \left(P_1 p^i{}_1 + P_2 p^i{}_2\right)\boldsymbol{e}_i\right) = \sum_i \left(P_1 p^i{}_1 + P_2 p^i{}_2\right) f\left(\boldsymbol{e}_i\right) \\ &= P_1\sum_i p^i{}_1 f\left(\boldsymbol{e}_i\right) + P_2\sum_i p^i{}_2 f\left(\boldsymbol{e}_i\right) \\ &= P_1 f\left(\sum_i p^i{}_1\boldsymbol{e}_i\right) + P_2 f\left(\sum_i p^i{}_2\boldsymbol{e}_i\right) \\ &= P_1 f\left(\boldsymbol{p}_1\right) + P_2 f\left(\boldsymbol{p}_2\right). \end{aligned}$$

In summary,

$$f\left(P_1\boldsymbol{p}_1 + P_2\boldsymbol{p}_2\right) = P_1 f\left(\boldsymbol{p}_1\right) + P_2 f\left(\boldsymbol{p}_2\right). \tag{1.32}$$

We find that f is *convex* on mixtures, or, in other words, it is linear on expressions of the form $P_1\boldsymbol{p}_1 + P_2\boldsymbol{p}_2$, where $\boldsymbol{p}_1$ and $\boldsymbol{p}_2$ belong to S, $0 \le P_1 \le 1$ and $0 \le P_2 \le 1$ and $P_1 + P_2 = 1$.

Convexity of arithmetic mean

[10] These are no longer the same states as our previously defined states $\boldsymbol{p}_{\bar{1}}$ and $\boldsymbol{p}_{\bar{2}}$: the bars above the subscripts are absent. They are simply two general mixture states.

We can extend the definition of f to diluted mixtures by adding the following condition:

$$f(\mathbf{0}) = 0. \tag{1.33}$$

Returning to our model where f_i is a frequency of light emitted by a tunable laser linked to a register configuration that corresponds to state $\boldsymbol{e}_i$, $f(\boldsymbol{p})$ is the combined color of light emitted by the laser as the register in state $\boldsymbol{p}$ fluctuates randomly through various configurations. The color may be different from all colors of frequencies f_i that are observed by glancing at the register momentarily. Frequency of light emitted by a broken register, $f(\mathbf{0})$, is zero. In this case the laser does not emit anything.

Probability as a convex function on mixtures

As we have already remarked, a special class of functions that belong in this category comprises probabilities that characterize a state. Let us recall equation (1.11) on page 14, which we rewrite here as follows:

$$\boldsymbol{p} = P_1\boldsymbol{p}_1 + P_2\boldsymbol{p}_2. \tag{1.34}$$

Let us define a function on $\boldsymbol{p}$ that returns the i-th probability p^i. Let us call this function $\boldsymbol{\omega}^i$. Using this function and the above equation for the probability of a mixture, we can write the following expression for p^i:

$$p^i = \boldsymbol{\omega}^i(\boldsymbol{p}) = P_1\boldsymbol{\omega}^i(\boldsymbol{p}_1) + P_2\boldsymbol{\omega}^i(\boldsymbol{p}_2) = P_1 p^i{}_1 + P_2 p^i{}_2. \tag{1.35}$$

If states $\boldsymbol{p}_1$ and $\boldsymbol{p}_2$ happen to be the canonical basis states, and if we have all of them in the mixture, we get

$$p^i = \boldsymbol{\omega}^i(\boldsymbol{p}) = \boldsymbol{\omega}^i\left(\sum_k p^k \boldsymbol{e}_k\right) = \sum_k p^k \boldsymbol{\omega}^i(\boldsymbol{e}_k). \tag{1.36}$$

For this to be consistent we must have that

$$\boldsymbol{\omega}^i(\boldsymbol{e}_k) = \delta^i{}_k, \tag{1.37}$$

Kronecker delta

where $\delta^i{}_k = 0$ for $i \neq k$ and $\delta^i{}_k = 1$ for $i = k$.

Interpreting $\boldsymbol{\omega}^i$ as an average value of something over the statistical ensemble that corresponds to $\boldsymbol{p}$, we can say that $p^i = \boldsymbol{\omega}^i(\boldsymbol{p})$ is the average frequency with which the basis state $\boldsymbol{e}_i$ is observed as we keep an eye on the randomly fluctuating register, for example, 30 times out of 100 (on average) for $p^i = 30\%$.

The linear combinations of various states discussed so far have been restricted by the conditions

$$\forall_i\, 0 \le p^i \le 1 \quad \text{and} \quad \sum_i p^i = 1, \tag{1.38}$$

or, in the case of diluted states, $0 \leq \sum_i p^i < 1$. When we have observed that function f, defined as the arithmetic mean over statistical ensembles corresponding to states $\boldsymbol{p}$, was linear, the observation has been restricted to coefficients p^i or P_i (for mixtures of mixtures) satisfying the same conditions. So this *partial* linearity, the *convexity*, is not full linearity, which should work also for $p^i > 1$ and for $p^i < 0$, unless we can demonstrate that the former implies the latter.

Convexity implies linearity

So here we are going to demonstrate just this,[11] namely, that function f defined as above, which has the property that

$$f\left(P_1\boldsymbol{p}_1 + P_2\boldsymbol{p}_2\right) = P_1 f\left(\boldsymbol{p}_1\right) + P_2 f\left(\boldsymbol{p}_2\right), \tag{1.39}$$

where

$$0 \leq P_1 \leq 1, \quad \text{and} \quad 0 \leq P_2 \leq 1, \quad \text{and} \quad P_1 + P_2 = 1, \tag{1.40}$$

is fully linear on S, meaning that

$$f\left(\sum_i a_i \boldsymbol{p}_i\right) = \sum_i a_i f\left(\boldsymbol{p}_i\right), \tag{1.41}$$

where

$$\boldsymbol{p}_i \in S, \quad \text{and} \quad \sum_i a_i \boldsymbol{p}_i \in S, \quad \text{and} \quad a_i \in \mathbb{R}. \tag{1.42}$$

In other words, a_i can be greater than 1, and they can be negative, too.

Since we allow for the presence of the null state $\mathbf{0}$, let us assume that $\boldsymbol{p}_2 = \mathbf{0}$. The fiducial vector for $\mathbf{0}$ comprises zeros only and $f(\mathbf{0}) = 0$, hence equation (1.39) implies that in this case

$$f\left(P_1\boldsymbol{p}_1\right) = P_1 f\left(\boldsymbol{p}_1\right). \tag{1.43}$$

Now let us replace P_1 with $1/\nu$ and $P_1\boldsymbol{p}_1$ with $\boldsymbol{p}$. We obtain that

$$f\left(\boldsymbol{p}\right) = \frac{1}{\nu} f\left(\nu\boldsymbol{p}\right), \tag{1.44}$$

or more succinctly

$$\nu f\left(\boldsymbol{p}\right) = f\left(\nu\boldsymbol{p}\right). \tag{1.45}$$

In this new equation state $\boldsymbol{p}$ is unnormalized (or diluted) and $1 < \nu$. Combining equations (1.43) and (1.45) yields

$$f\left(a\boldsymbol{p}\right) = a f\left(\boldsymbol{p}\right) \tag{1.46}$$

[11] This proof follows [60] with minor alterations.

for $0 \le a$ (including $1 < a$) and as long as $a\boldsymbol{p} \in S$, because the expression lacks physical meaning otherwise. But we can extend the expression beyond S from a purely algebraic point of view, and this will come in handy below.

Now let us consider an *arbitrary* linear combination of states that still delivers a state in S,

$$\boldsymbol{p} = \sum_i a_i \boldsymbol{p}_i, \tag{1.47}$$

where, as above, we no longer restrict a_i: they can be negative and/or greater than 1, too. Let us divide coefficients a_i into negative and positive ones. Let us call the list of indexes i, that yield $a_i < 0$, A_-, and the list of indexes i, that yield $a_i > 0$, A_+. This lets us rewrite the equation above as follows:

$$\boldsymbol{p} + \sum_{i \in A_-} |a_i| \boldsymbol{p}_i = \sum_{i \in A_+} a_i \boldsymbol{p}_i. \tag{1.48}$$

Let us introduce

$$\nu = 1 + \sum_{i \in A_-} |a_i|, \tag{1.49}$$

and let us divide both sides of equation (1.48) by ν,

$$\frac{1}{\nu}\boldsymbol{p} + \sum_{i \in A_-} \frac{|a_i|}{\nu} \boldsymbol{p}_i = \sum_{i \in A_+} \frac{a_i}{\nu} \boldsymbol{p}_i. \tag{1.50}$$

This time all coefficients on the left-hand side of the equation; that is,

$$\frac{1}{\nu}, \quad \frac{|a_i|}{\nu}, \quad \text{for} \quad i \in A_-, \tag{1.51}$$

are positive and add up to 1 (because we have defined ν so that they would). Let us define

$$\mu = \sum_{i \in A_+} \frac{a_i}{\nu}. \tag{1.52}$$

Using μ, we can rewrite equation (1.50) as follows:

$$\frac{1}{\nu}\boldsymbol{p} + \sum_{i \in A_-} \frac{|a_i|}{\nu} \boldsymbol{p}_i = \mu \sum_{i \in A_+} \frac{a_i}{\mu\nu} \boldsymbol{p}_i. \tag{1.53}$$

Here all coefficients on the right-hand side of this equation,

$$\frac{a_i}{\mu\nu} \quad \text{for} \quad i \in A_+, \tag{1.54}$$

are positive and add up to 1 (because we have defined μ so that they would). We can now apply function f to both sides of the equation. Let us begin with the left-hand side. Here we have a regular mixture; therefore

$$f\left(\frac{1}{\nu}\boldsymbol{p} + \sum_{i\in A_-} \frac{|a_i|}{\nu}\boldsymbol{p}_i\right) = \frac{1}{\nu} f(\boldsymbol{p}) + \sum_{i\in A_-} \frac{|a_i|}{\nu} f(\boldsymbol{p}_i). \tag{1.55}$$

On the right-hand side we first make use of equation (1.46) on page 25, which yields

$$f\left(\mu \sum_{i\in A_+} \frac{a_i}{\mu\nu}\boldsymbol{p}_i\right) = \mu f\left(\sum_{i\in A_+} \frac{a_i}{\mu\nu}\boldsymbol{p}_i\right) = \cdots. \tag{1.56}$$

Now we simply make use of the fact that what f acts on is a regular mixture and obtain

$$\cdots = \mu \sum_{i\in A_+} \frac{a_i}{\mu\nu} f(\boldsymbol{p}_i). \tag{1.57}$$

Combining both sides yields

$$\frac{1}{\nu} f(\boldsymbol{p}) + \sum_{i\in A_-} \frac{|a_i|}{\nu} f(\boldsymbol{p}_i) = \mu \sum_{i\in A_+} \frac{a_i}{\mu\nu} f(\boldsymbol{p}_i). \tag{1.58}$$

Let us finally multiply both sides of the equation by ν, and let us cancel μ/μ on the right-hand side to get

$$f(\boldsymbol{p}) + \sum_{i\in A_-} |a_i| f(\boldsymbol{p}_i) = \sum_{i\in A_+} a_i f(\boldsymbol{p}_i), \tag{1.59}$$

which is the same as

$$f(\boldsymbol{p}) = \sum_i a_i f(\boldsymbol{p}_i). \tag{1.60}$$

In summary, we have just demonstrated that f is indeed fully linear on S, even for linear combinations that are not convex.

Beyond arithmetic mean

The arithmetic mean is not the only way in which function f can be extended from its definition on the basis states $\boldsymbol{e}_i$ to an arbitrary mixture $\boldsymbol{p}$. Instead of using the arithmetic mean, $\langle \boldsymbol{f}, \boldsymbol{p}\rangle$, we could use the generalized mean, which is defined by the following formula:

$$\bar{f}_t = \left(\sum_i p^i (f_i)^t\right)^{1/t}. \tag{1.61}$$

The generalized mean $\bar{f}_t$ becomes the arithmetic mean for $t = 1$. It becomes the harmonic mean for $t = -1$ and the geometric mean for $t \to 0$. For very large values of t, $\bar{f}_t \approx \max_i f_i$; and for very large negative values of t, $\bar{f}_t \approx \min_i f_i$.

There may be situations in electronics and physics when, for example, the harmonic mean or the geometric mean is a more appropriate way of extending f to mixtures. But f so defined would not be linear and therefore would not mix in direct proportion to the abundances of mixture components.

1.7 Forms and Vectors

Linear functions on a vector space are called *forms*. Such functions can be thought of as mirror images of vectors they operate on,[12] and they form a vector space of their own.

Whenever a *form* $\boldsymbol{f}$, which represents a measurement on a mixture, encounters a vector $\boldsymbol{p}$, which represents a mixture, they get together in an explosive union $\langle \boldsymbol{f}, \boldsymbol{p} \rangle$, which delivers a number $f(\boldsymbol{p})$,

$$\langle \boldsymbol{f}, \boldsymbol{p} \rangle \doteq f(\boldsymbol{p}) . \tag{1.62}$$

Using this notation, we can restate the linearity of function f, equation (1.41), as follows:

$$\langle \boldsymbol{f}, \sum_i a_i \boldsymbol{p}_i \rangle = \sum_i a_i \langle \boldsymbol{f}, \boldsymbol{p}_i \rangle . \tag{1.63}$$

In practical computations we often identify a state vector $\boldsymbol{p}$ with a column of numbers (probabilities), although much can be said and even proven about state vectors without using this particular representation. Similarly, a form $\boldsymbol{f}$ can be identified with a *row* of numbers. For example, a form that returns the first component of a vector would have the following row-of-numbers representation:

$$\boldsymbol{\omega}^0 \equiv (1, 0, 0, 0, 0, 0, 0, 0) . \tag{1.64}$$

[12] An excellent and easy to follow introduction to vectors, forms, and tensors—especially for a physicist or an electronic engineer—can be found in [94], chapters 2–5, pages 47–162. It is not necessary to read anything before or after that, unless a reader is so inclined.

When a form like this is placed to the left of vector $\boldsymbol{p}$, and standard matrix multiplication rules are activated, we have the following result:

$$\langle \boldsymbol{\omega}^0, \boldsymbol{p} \rangle = (1,0,0,0,0,0,0,0) \cdot \begin{pmatrix} p^0 \\ p^1 \\ p^2 \\ p^3 \\ p^4 \\ p^5 \\ p^6 \\ p^7 \end{pmatrix} = p^0, \tag{1.65}$$

where the dot between the row and the column stands for matrix multiplication. The form in this example represents a measurement of probability that the system is in configuration $\{000\}$.

The following listing introduces a canonical basis in the space of forms:

$$\begin{aligned}
\boldsymbol{\omega}^0 &\equiv (1,0,0,0,0,0,0,0)\,, \\
\boldsymbol{\omega}^1 &\equiv (0,1,0,0,0,0,0,0)\,, \\
\boldsymbol{\omega}^2 &\equiv (0,0,1,0,0,0,0,0)\,, \\
\boldsymbol{\omega}^3 &\equiv (0,0,0,1,0,0,0,0)\,, \\
\boldsymbol{\omega}^4 &\equiv (0,0,0,0,1,0,0,0)\,, \\
\boldsymbol{\omega}^5 &\equiv (0,0,0,0,0,1,0,0)\,, \\
\boldsymbol{\omega}^6 &\equiv (0,0,0,0,0,0,1,0)\,, \\
\boldsymbol{\omega}^7 &\equiv (0,0,0,0,0,0,0,1)\,.
\end{aligned}$$

Fiducial forms

As we did with columns of probabilities, we shall call the rows and symbols such as $\boldsymbol{\omega}^i$ *fiducial* forms or *fiducial measurements*, because they act on fiducial vectors.

The basis forms $\boldsymbol{\omega}^i$ satisfy the following relation:

$$\langle \boldsymbol{\omega}^i, \boldsymbol{e}_j \rangle = \delta^i{}_j. \tag{1.66}$$

As we can express mixture $\boldsymbol{p}$ in terms of the basis states

$$\boldsymbol{p} = \sum_i p^i \boldsymbol{e}_i, \tag{1.67}$$

form $\boldsymbol{f}$ can be expressed in terms of the basis forms $\boldsymbol{\omega}^i$, defined above, as follows:

$$\boldsymbol{f} = \sum_i f_i \boldsymbol{\omega}^i. \tag{1.68}$$

The action of $\boldsymbol{f}$ on $\boldsymbol{p}$ now becomes

$$\begin{aligned}\langle \boldsymbol{f}, \boldsymbol{p}\rangle &= \left\langle \sum_i f_i \boldsymbol{\omega}^i, \sum_j p^j \boldsymbol{e}_j \right\rangle \\ &= \sum_i \sum_j f_i p^j \langle \boldsymbol{\omega}^i, \boldsymbol{e}_j\rangle = \sum_i \sum_j f_i p^j \delta^i{}_j = \sum_i f_i p^i. \end{aligned} \tag{1.69}$$

The physical meaning of $\langle \boldsymbol{f}, \boldsymbol{p}\rangle$ is the arithmetic mean of f_i over the statistical ensemble represented by the mixture coefficients p^i, where f_i are the values f assumes on the basis states $\boldsymbol{e}_i$.

Subscripts and superscripts

It is useful to adhere to the following typographic convention. Vectors and forms are typeset in bold font, whereas vector and form coefficients are typeset in light font. Whenever possible we reserve small Latin letters (such as $\boldsymbol{v}$ or $\boldsymbol{e}$) for vectors and small Greek letters (such as $\boldsymbol{\eta}$ or $\boldsymbol{\omega}$) for forms—though sometimes, as we have just done with form $\boldsymbol{f}$, we shall break this convention. Basis vectors (such as $\boldsymbol{e}_i$) are numbered with subscripts. Basis forms (such as $\boldsymbol{\omega}^i$) are numbered with superscripts. On the other hand, vector coefficients (such as p^i) are numbered with superscripts, and form coefficients (such as f_i) are numbered with subscripts. Thus, whenever there is a summation in expressions such as

$$\boldsymbol{p} = \sum_i p^i \boldsymbol{e}_i, \tag{1.70}$$

$$\boldsymbol{f} = \sum_i f_i \boldsymbol{\omega}^i, \tag{1.71}$$

$$\langle \boldsymbol{f}, \boldsymbol{p}\rangle = \sum_i f_i p^i, \tag{1.72}$$

the summation runs over indexes of which one is always down and the other one is always up. A summation like this is called *contraction*. One speaks, for example, about *contraction on index i*.

The convention is useful because it constantly reminds us about what various objects we work with are. It helps debug form and vector expressions, too. For example, an expression such as $\sum_i f_i \boldsymbol{e}_i$ should attract our suspicion because it suggests that we are trying to use form coefficients in order to construct a vector. In some contexts we may wish to do just this, and may even get away with it, but not without giving it some thought in the first place.

Summation convention

The above leads to the *summation convention*, which states that whenever there exists an expression with two identical indexes, of which one is up and the other one down, summation should be assumed. For example,

$$f_i p^i \equiv \sum_i f_i p^i,$$
$$p^i \boldsymbol{e}_i \equiv \sum_i p^i \boldsymbol{e}_i,$$
$$f_i \boldsymbol{\omega}^i \equiv \sum_i f_i \boldsymbol{\omega}^i.$$

The summation convention is handy in advanced tensor calculus, where geometric and dynamic objects may be endowed with several subscripts and superscripts. We shall not use it in this text, though, because we shall seldom work with complicated tensor expressions.

Vector and form transformations

The placement of indexes on form and vector coefficients is not just a matter of esthetics, convenience, and debugging. It reflects transformation properties of these objects, too.

Let us suppose that instead of decomposing vector $\boldsymbol{v}$ in basis $\boldsymbol{e}_i$, $\boldsymbol{v} = \sum_i v^i \boldsymbol{e}_i$, we were to decompose it in another basis, say, $\boldsymbol{e}_{i'}$. The basis vectors $\boldsymbol{e}_{i'}$ are not the same as $\boldsymbol{e}_i$, the prime on the index i' matters, but they are all linearly independent as basis vectors should be. Also, let us suppose that we find another basis in the form space, $\boldsymbol{\omega}^{i'}$, such that $\langle \boldsymbol{\omega}^{i'}, \boldsymbol{e}_{j'} \rangle = \delta^{i'}{}_{j'}$. Vector coefficients in the new basis $\boldsymbol{e}_{i'}$ can be found by using the basis in the form space, namely,

$$v^{i'} = \langle \boldsymbol{\omega}^{i'}, \boldsymbol{v} \rangle. \tag{1.73}$$

Since both $\boldsymbol{e}_i$ and $\boldsymbol{e}_{i'}$ are the bases of linearly independent vectors, there must be a linear transformation that converts one basis onto the other one. Let us call the coefficients of this transformation $\Lambda_{i'}{}^j$. The transformation rule for the basis vectors is then

$$\boldsymbol{e}_{i'} = \sum_j \Lambda_{i'}{}^j \boldsymbol{e}_j. \tag{1.74}$$

We should expect a similar transformation for the forms

$$\boldsymbol{\omega}^{i'} = \sum_j \boldsymbol{\omega}^j \Lambda_j{}^{i'}. \tag{1.75}$$

We do not assume that $\Lambda_{i'}{}^j$ and $\Lambda_j{}^{i'}$ are the same: the typographic placement of primed and unprimed indexes warns us that they may be different. But they are related. The relationship is easy to see by invoking the rules $\langle \boldsymbol{\omega}^i, \boldsymbol{e}_j \rangle = \delta^i{}_j$ and $\langle \boldsymbol{\omega}^{i'}, \boldsymbol{e}_{j'} \rangle = \delta^{i'}{}_{j'}$:

$$\begin{aligned}
\delta^{i'}{}_{j'} &= \langle \boldsymbol{\omega}^{i'}, \boldsymbol{e}_{j'} \rangle = \left\langle \sum_k \boldsymbol{\omega}^k \Lambda_k{}^{i'}, \sum_l \Lambda_{j'}{}^l \boldsymbol{e}_l \right\rangle \\
&= \sum_k \sum_l \Lambda_k{}^{i'} \Lambda_{j'}{}^l \langle \boldsymbol{\omega}^k, \boldsymbol{e}_l \rangle = \sum_k \sum_l \Lambda_k{}^{i'} \Lambda_{j'}{}^l \delta^k{}_l \\
&= \sum_k \Lambda_k{}^{i'} \Lambda_{j'}{}^k. \qquad (1.76)
\end{aligned}$$

This tells us that matrices $\left\| \Lambda_k{}^{i'} \right\|$ and $\left\| \Lambda_{j'}{}^k \right\|$ are inverses of each other.

Now we can turn back to transformation properties of vector and form coefficients. We can easily see that for vectors we have the following:

$$v^{i'} = \left\langle \boldsymbol{\omega}^{i'}, \boldsymbol{v} \right\rangle = \left\langle \sum_j \boldsymbol{\omega}^j \Lambda_j{}^{i'}, \boldsymbol{v} \right\rangle = \sum_j \Lambda_j{}^{i'} \langle \boldsymbol{\omega}^j, \boldsymbol{v} \rangle = \sum_j v^j \Lambda_j{}^{i'}. \qquad (1.77)$$

On the other hand, we get a different relation for forms:

$$\eta_{i'} = \langle \boldsymbol{\eta}, \boldsymbol{e}_{i'} \rangle = \left\langle \boldsymbol{\eta}, \sum_j \Lambda_{i'}{}^j \boldsymbol{e}_j \right\rangle = \sum_j \Lambda_{i'}{}^j \langle \boldsymbol{\eta}, \boldsymbol{e}_j \rangle = \sum_j \Lambda_{i'}{}^j \eta_j. \qquad (1.78)$$

We see that form and vector coefficients transform in opposite directions. Vector coefficients (index is up) transform like basis forms (their index is up, too), and form coefficients (index down) transform like basis vectors (their index is down, too). This is good news because it means that expressions such as $\sum_i \eta_i v^i$ don't transform at all. Transformations of η_i and v^i cancel each other, so that the resulting scalar $\langle \boldsymbol{\eta}, \boldsymbol{v} \rangle$ is *independent* of the choice of vector and form bases.

1.8 Transformations of Mixtures

Let us suppose we have a hat full of randomly mixed 3-bit registers in various *static* configurations. Such a statistical ensemble is equivalent to a single randomly fluctuating register in some *state* $\boldsymbol{p}$. If there is a total of N registers in the hat, the abundances of registers in configurations $\boldsymbol{e}_0, \boldsymbol{e}_1, \ldots, \boldsymbol{e}_7$ are $Np^0, Np^1, \ldots, Np^7$.

Cinderella transformation

Drawing on the popular fairytale of Cinderella, we are going to burden the poor girl with the following ungrateful task. She should draw the registers out of the hat one by one. Whenever she draws a register in state $\boldsymbol{e}_0$ she should tweak its toggles so as to change its state to $\boldsymbol{p}_0$ and then place it in another hat. Whenever she draws a register in state $\boldsymbol{e}_1$ she should tweak it so as to change its state to $\boldsymbol{p}_1$ and then place it in the other hat, too, and so on for the remaining states. In other words, she should perform the following transformation on the *whole* ensemble:

$$\begin{aligned}
\boldsymbol{e}_0 &\rightarrow \boldsymbol{p}_0, \\
\boldsymbol{e}_1 &\rightarrow \boldsymbol{p}_1, \\
\boldsymbol{e}_2 &\rightarrow \boldsymbol{p}_2, \\
\boldsymbol{e}_3 &\rightarrow \boldsymbol{p}_3, \\
\boldsymbol{e}_4 &\rightarrow \boldsymbol{p}_4, \\
\boldsymbol{e}_5 &\rightarrow \boldsymbol{p}_5, \\
\boldsymbol{e}_6 &\rightarrow \boldsymbol{p}_6, \\
\boldsymbol{e}_7 &\rightarrow \boldsymbol{p}_7,
\end{aligned}$$

where $\boldsymbol{p}_i = \sum_j p^j{}_i \boldsymbol{e}_j$.

What Cinderella is going to end up with in the second hat, after the whole operation is finished, is another mixture. Let us now draw a register from the second hat—this time it is going to be a fluctuating one—and glance at it momentarily. What are the probabilities of seeing $\boldsymbol{e}_0, \boldsymbol{e}_1, \ldots, \boldsymbol{e}_7$?

To answer this question, we expand, as we did on previous occasions, states $\boldsymbol{p}_0$ through $\boldsymbol{p}_7$ into their statistical ensembles, remembering that we are going to have

Np^0	ensembles that correspond to	$\boldsymbol{p}_0$,
Np^1	ensembles that correspond to	$\boldsymbol{p}_1$,
Np^2	ensembles that correspond to	$\boldsymbol{p}_2$,
Np^3	ensembles that correspond to	$\boldsymbol{p}_3$,
Np^4	ensembles that correspond to	$\boldsymbol{p}_4$,
Np^5	ensembles that correspond to	$\boldsymbol{p}_5$,
Np^6	ensembles that correspond to	$\boldsymbol{p}_6$,
Np^7	ensembles that correspond to	$\boldsymbol{p}_7$.

Without loss of generality we can assume that each ensemble that corresponds to $\boldsymbol{p}_0$ through $\boldsymbol{p}_7$ comprises the same number of non-fluctuating registers. Let's call this number n. Consequently, in an ensemble that corresponds to $\boldsymbol{p}_0$ we are going to have

$np^0{}_0$	registers in state	$\boldsymbol{e}_0$,

$$\begin{array}{lll} np^1{}_0 & \text{registers in state} & \boldsymbol{e}_1, \\ np^2{}_0 & \text{registers in state} & \boldsymbol{e}_2, \\ np^3{}_0 & \text{registers in state} & \boldsymbol{e}_3, \\ np^4{}_0 & \text{registers in state} & \boldsymbol{e}_4, \\ np^5{}_0 & \text{registers in state} & \boldsymbol{e}_5, \\ np^6{}_0 & \text{registers in state} & \boldsymbol{e}_6, \\ np^7{}_0 & \text{registers in state} & \boldsymbol{e}_7, \end{array}$$

and similarly for the other ensembles. The total number of registers in state $\boldsymbol{e}_0$ in all the ensembles is going to be

$$\begin{aligned} & Np^0 np^0{}_0 + Np^1 np^0{}_1 + Np^2 np^0{}_2 + Np^3 np^0{}_3 \\ & \quad + Np^4 np^0{}_4 + Np^5 np^0{}_5 + Np^6 np^0{}_6 + Np^7 np^0{}_7 \\ & \quad = Nn \sum_{i=0}^{7} p^i p^0{}_i. \end{aligned} \tag{1.79}$$

The total number of all registers in all the ensembles is going to be Nn, because we had N registers in the first hat and, after tweaking the toggles, each register got "expanded" into an ensemble of n registers. Let the state of the mixture in the second hat be called $\boldsymbol{q}$ with coefficients $q^0, q^1, \ldots, q^7$. Using the above formula, we have for q^0

$$q^0 = \frac{1}{Nn} Nn \sum_{i=0}^{7} p^i p^0{}_i = \sum_{i=0}^{7} p^0{}_i p^i. \tag{1.80}$$

We have reversed here the order of p_i and p_{0i} for a purely cosmetic reason. Similarly we can write the following equations for the remaining coefficients:

$$\begin{aligned} q^1 &= \sum_{i=0}^{7} p^1{}_i p^i, \quad q^2 = \sum_{i=0}^{7} p^2{}_i p^i, \quad q^3 = \sum_{i=0}^{7} p^3{}_i p^i, \\ q^4 &= \sum_{i=0}^{7} p^4{}_i p^i, \quad q^5 = \sum_{i=0}^{7} p^5{}_i p^i, \quad q^6 = \sum_{i=0}^{7} p^6{}_i p^i, \\ q^7 &= \sum_{i=0}^{7} p^7{}_i p^i. \end{aligned}$$

All of this can be written more concisely as follows:

$$q^j = \sum_i p^j{}_i p^i. \tag{1.81}$$

We can rewrite the formula by using the vector and form notation and obtain

$$\begin{aligned} \boldsymbol{q} &= \sum_j q^j \boldsymbol{e}_j = \sum_j \sum_i p^j{}_i \boldsymbol{e}_j p^i = \sum_j \sum_i p^j{}_i \boldsymbol{e}_j \langle \boldsymbol{\omega}^i, \boldsymbol{p} \rangle \\ &= \left\langle \sum_j \sum_i p^j{}_i \boldsymbol{e}_j \otimes \boldsymbol{\omega}^i, \boldsymbol{p} \right\rangle = \langle \boldsymbol{P}, \boldsymbol{p} \rangle , \end{aligned} \tag{1.82}$$

where we have defined

$$\boldsymbol{P} \doteq \sum_j \sum_i p^j{}_i \boldsymbol{e}_j \otimes \boldsymbol{\omega}^i .$$

Tensor product

The symbol $\otimes$ is called a *tensor product*. All that it means here is that we put $\boldsymbol{e}_j$ and $\boldsymbol{\omega}^i$ together next to each other *typographically*. If form $\boldsymbol{\omega}^i$ finds a vector to prey on, for example, $\boldsymbol{p}$, it vanishes together with the vector, leaving a number $\langle \boldsymbol{\omega}^i, \boldsymbol{p} \rangle$ behind. All that is left on the paper then is $\boldsymbol{e}_j$ multiplied by that number.

Symbol $\boldsymbol{P}$ denotes an *operator* that describes the transformation of the ensemble performed by Cinderella. Let us observe that every jth term of this operator, $\sum_i p^j{}_i \boldsymbol{\omega}^i$, is a form. The operator can therefore be thought of as eight forms arranged so that they together transform one state vector, $\boldsymbol{p}$, into another one, $\boldsymbol{q}$. Each of the forms is a convex function, but, as we have already seen, it is also fully linear on S. Consequently $\boldsymbol{P}$, the collection of the forms, is linear, too.

Typographic conventions

We shall usually adhere throughout this text to a typographic convention that uses capital bold letters, like $\boldsymbol{P}$, for operators and other complex objects that have one or more tensor products inside them, although we shall deviate from this convention in some cases where tradition dictates that, for example, a metric tensor should be denoted by $\boldsymbol{g}$.

Another typographic convention drops brackets $\langle$ and $\rangle$ when describing an action of an operator on a vector. And so, instead of writing $\langle \boldsymbol{P}, \boldsymbol{p} \rangle$ we can write simply $\boldsymbol{Pp}$. This has the additional benefit of translating naturally into a matrix (representing the $\boldsymbol{P}$) times a column (representing the $\boldsymbol{p}$) expression.

Convexity of Cinderella transformations

Although every Cinderella transformation like the one discussed here is going to be linear, not every linear transformation that we can apply to S is going to be a valid Cinderella transformation. Rotations and reflections, for example, are linear, but here they would rotate or reflect some of the states out of S. Cinderella transformations, on the other hand, keep everything within S because of the following three conditions, which together imply the convexity of $\boldsymbol{P}$:

$$\forall_{ji} 0 \leq p^j{}_i \leq 1, \tag{1.83}$$

$$\forall_j \sum_i p^j{}_i = 1, \tag{1.84}$$

$$\forall_i \sum_j p^j{}_i = 1. \tag{1.85}$$

Reversibility of Cinderella transformations

Can Cinderella transformations be reversed? The answer to this question is interesting: the only reversible Cinderella transformations are permutations of pure states; in other words, if a reversible Cinderella transformation is applied to a pure state, another pure state must come out. Since reversibility also implies that no two different pure states may be converted to the same output state, the only possibility we are left with is a permutation of pure states.

The way to see this is as follows. Let the transformation in question be called $\boldsymbol{C}$. If it is possible for $\boldsymbol{C}$ to convert a pure state to a mixture, then we would have the following:

$$\boldsymbol{C}\boldsymbol{e}_i = \sum_k c^k \boldsymbol{e}_k. \tag{1.86}$$

Since $\boldsymbol{C}$ is reversible, $\boldsymbol{C}^{-1}$ exists, and we can apply it to both sides of the equation above, which yields

$$\boldsymbol{C}^{-1}\boldsymbol{C}\boldsymbol{e}_i = \boldsymbol{e}_i = \sum_k c^k \boldsymbol{C}^{-1}\boldsymbol{e}_k. \tag{1.87}$$

But this says that $\boldsymbol{e}_i$ is a mixture,[13] which it is not. Hence we must conclude that $\boldsymbol{C}$ cannot convert $\boldsymbol{e}_k$ to a mixture.

Beyond linearity

We can always think of a quite general transformation on a mixture defined by the following formula:

$$\boldsymbol{q} = \sum_i q^i\left(p^0, p^1, \ldots, p^7\right)\boldsymbol{e}_i, \tag{1.88}$$

where $q^i\left(p^0, p^1, \ldots, p^7\right)$ are some arbitrary, possibly nonlinear, real-valued functions that convert coefficients p^i $(i = 0, 1, \ldots, 7)$ into real numbers between 0 and 1. The only additional condition we would impose on functions q^i would be that

$$\forall_{\boldsymbol{p}\in S} \sum_i q^i(\boldsymbol{p}) \leq 1, \tag{1.89}$$

where the ≤ 1 condition would cover the option of generating diluted mixtures. How could we implement such a general transformation? It would not be sufficient to provide Cinderella with a prescription such as before. This, as we have seen, would result in a linear transformation. In order to generate $\boldsymbol{q}$ Cinderella would have to empty the first hat entirely, counting abundances for each configuration.

[13] $\boldsymbol{C}^{-1}\boldsymbol{e}_k$ must be linearly independent states because otherwise $\boldsymbol{C}$ wouldn't be reversible.

Only then, having collected sufficient fiducial statistics to ascertain the state of the *whole* mixture $\boldsymbol{p}$, could she sit down and calculate q^i for $i = 0, 1, \ldots, 7$. Having done so, she could then set switches on each register to generate $\boldsymbol{q}$, and only then would she place the register in the second hat.

This transformation would be quite different physically from what we call a Cinderella transformation. A Cinderella transformation can be implemented on the go: we don't have to have the complete knowledge of the mixture in order to begin processing the registers. The prescription allows us to perform the transformation on each register separately and have things still add up to $\boldsymbol{q} = \boldsymbol{P}\boldsymbol{p}$.

1.9 Composite Systems

In this section we have a closer look at what happens when we combine smaller randomly fluctuating registers into a larger one.

Let us consider two 2-bit randomly fluctuating registers. A 2-bit randomly fluctuating register is described by states that belong to the $2^2 = 4$-dimensional vector space. The canonical basis vectors in this space are $\boldsymbol{e}_0 \equiv \{00\}$, $\boldsymbol{e}_1 \equiv \{01\}$, $\boldsymbol{e}_2 \equiv \{10\}$ and $\boldsymbol{e}_3 \equiv \{11\}$. A vector space that contains the register obtained by combining the two 2-bit registers is $4 \times 4 = 16$ dimensional. Let us call the two 2-bit registers A and B, and let us label the basis states that refer to the two registers with indexes A and B, too. We have the following basis states for register A:

$$\{00\}_A, \{01\}_A, \{10\}_A, \{11\}_A, \tag{1.90}$$

and similarly for register B:

$$\{00\}_B, \{01\}_B, \{10\}_B, \{11\}_B. \tag{1.91}$$

What will the basis states look like for a system made by placing registers A and B next to each other, so that they form a 4-bit register? Since we are not going to do anything special to these two registers, other than just place them next to each other, whenever we give them a brief glance as they keep fluctuating, we're going to see one of the following:

Basis of the combined register system

$$\begin{array}{ll}
\{00\}_A\{00\}_B, & \{00\}_A\{01\}_B, \\
\{00\}_A\{10\}_B, & \{00\}_A\{11\}_B, \\
\{01\}_A\{00\}_B, & \{01\}_A\{01\}_B, \\
\{01\}_A\{10\}_B, & \{01\}_A\{11\}_B, \\
\{10\}_A\{00\}_B, & \{10\}_A\{01\}_B,
\end{array}$$

$$\begin{array}{ll} \{10\}_A\{10\}_B, & \{10\}_A\{11\}_B, \\ \{11\}_A\{00\}_B, & \{11\}_A\{01\}_B, \\ \{11\}_A\{10\}_B, & \{11\}_A\{11\}_B. \end{array}$$

The pairs must be then the basis state vectors of the combined system. We can replace binary digits in the curly brackets with our symbolic notation for basis vectors, and write

$$\{01\}_A\{10\}_B \rightarrow \boldsymbol{e}_{1A}\boldsymbol{e}_{2B}, \tag{1.92}$$

where we have simply placed $\boldsymbol{e}_{1A}$ and $\boldsymbol{e}_{2B}$ next to each other on the sheet of paper. But we have already seen something similar when we defined an operator in terms of the tensor product $\otimes$. To avoid possible confusion and to emphasize that we do not really multiply these vectors by each other but merely write them next to each other, let us use the same symbol here:

$$\{01\}_A\{10\}_B \rightarrow \boldsymbol{e}_{1A} \otimes \boldsymbol{e}_{2B}. \tag{1.93}$$

The basis of the combined register system can now be described as listed below:

$$\begin{array}{ll} \boldsymbol{e}_{0A} \otimes \boldsymbol{e}_{0B}, & \boldsymbol{e}_{0A} \otimes \boldsymbol{e}_{1B}, \\ \boldsymbol{e}_{0A} \otimes \boldsymbol{e}_{2B}, & \boldsymbol{e}_{0A} \otimes \boldsymbol{e}_{3B}, \\ \boldsymbol{e}_{1A} \otimes \boldsymbol{e}_{0B}, & \boldsymbol{e}_{1A} \otimes \boldsymbol{e}_{1B}, \\ \boldsymbol{e}_{1A} \otimes \boldsymbol{e}_{2B}, & \boldsymbol{e}_{1A} \otimes \boldsymbol{e}_{3B}, \\ \boldsymbol{e}_{2A} \otimes \boldsymbol{e}_{0B}, & \boldsymbol{e}_{2A} \otimes \boldsymbol{e}_{1B}, \\ \boldsymbol{e}_{2A} \otimes \boldsymbol{e}_{2B}, & \boldsymbol{e}_{2A} \otimes \boldsymbol{e}_{3B}, \\ \boldsymbol{e}_{3A} \otimes \boldsymbol{e}_{0B}, & \boldsymbol{e}_{3A} \otimes \boldsymbol{e}_{1B}, \\ \boldsymbol{e}_{3A} \otimes \boldsymbol{e}_{2B}, & \boldsymbol{e}_{3A} \otimes \boldsymbol{e}_{3B}. \end{array}$$

Mixing combined register systems

Using the pairs, we can construct various mixtures in the same way we did with a single register. For example, we could take 30% of $\boldsymbol{e}_{1A} \otimes \boldsymbol{e}_{3B}$, 25% of $\boldsymbol{e}_{3A} \otimes \boldsymbol{e}_{0B}$, 25% of $\boldsymbol{e}_{2A} \otimes \boldsymbol{e}_{1B}$, and 20% of $\boldsymbol{e}_{0A} \otimes \boldsymbol{e}_{2B}$, and obtain

$$\begin{aligned} \boldsymbol{p} = {} & 0.3\,\boldsymbol{e}_{1A} \otimes \boldsymbol{e}_{3B} + 0.25\,\boldsymbol{e}_{3A} \otimes \boldsymbol{e}_{0B} \\ & + 0.25\,\boldsymbol{e}_{2A} \otimes \boldsymbol{e}_{1B} + 0.2\,\boldsymbol{e}_{0A} \otimes \boldsymbol{e}_{2B}, \end{aligned} \tag{1.94}$$

and this mixture would still leave 12 other basis states unused.

Forms on combined register states

If $\boldsymbol{\omega}_A$ is a form (or a measurement) that acts on states of register A and $\boldsymbol{\eta}_B$ is a form (or a measurement) that acts on states of register B, then the two can be combined into a form $\boldsymbol{\omega}_A \otimes \boldsymbol{\eta}_B$ that acts on states of the combined register as shown in the following example:

$$\begin{aligned}
&\langle \boldsymbol{\omega}_A \otimes \boldsymbol{\eta}_B, \boldsymbol{p}\rangle \\
&\quad = \langle \boldsymbol{\omega}_A \otimes \boldsymbol{\eta}_B, 0.3\, \boldsymbol{e}_{1A} \otimes \boldsymbol{e}_{3B} + 0.25\, \boldsymbol{e}_{3A} \otimes \boldsymbol{e}_{0B} \\
&\qquad + 0.25\, \boldsymbol{e}_{2A} \otimes \boldsymbol{e}_{1B} + 0.2\, \boldsymbol{e}_{0A} \otimes \boldsymbol{e}_{2B}\rangle \\
&\quad = 0.3\langle \boldsymbol{\omega}_A \otimes \boldsymbol{\eta}_B, \boldsymbol{e}_{1A} \otimes \boldsymbol{e}_{3B}\rangle + 0.25\langle \boldsymbol{\omega}_A \otimes \boldsymbol{\eta}_B, \boldsymbol{e}_{3A} \otimes \boldsymbol{e}_{0B}\rangle \\
&\qquad + 0.25\langle \boldsymbol{\omega}_A \otimes \boldsymbol{\eta}_B, \boldsymbol{e}_{2A} \otimes \boldsymbol{e}_{1B}\rangle + 0.2\langle \boldsymbol{\omega}_A \otimes \boldsymbol{\eta}_B, \boldsymbol{e}_{0A} \otimes \boldsymbol{e}_{2B}\rangle \\
&\quad = \ldots
\end{aligned}$$

Now we proceed exactly as we did in the definition of the operator. Let us first unite A forms with A vectors. This step will produce numbers that will get thrown out in front of the $\langle$ and $\rangle$ brackets leaving B forms and B vectors to do the same. So, in the final account we get

$$\begin{aligned}
\ldots \quad = \quad & 0.3\langle \boldsymbol{\omega}_A, \boldsymbol{e}_{1A}\rangle\langle \boldsymbol{\eta}_B, \boldsymbol{e}_{3B}\rangle + 0.25\langle \boldsymbol{\omega}_A, \boldsymbol{e}_{3A}\rangle\langle \boldsymbol{\eta}_B, \boldsymbol{e}_{0B}\rangle \\
& + 0.25\langle \boldsymbol{\omega}_A, \boldsymbol{e}_{2A}\rangle\langle \boldsymbol{\eta}_B, \boldsymbol{e}_{1B}\rangle + 0.2\langle \boldsymbol{\omega}_A, \boldsymbol{e}_{0A}\rangle\langle \boldsymbol{\eta}_B, \boldsymbol{e}_{2B}\rangle. \qquad (1.95)
\end{aligned}$$

A function so defined is clearly linear on states $\boldsymbol{p} \in \mathbb{R}^{16}$, and hence it can be interpreted as an average over the statistical ensemble that corresponds to $\boldsymbol{p}$.

Tensor product

Because under the action of $\boldsymbol{\omega}_A \otimes \boldsymbol{\eta}_B$ a pair such as $\boldsymbol{e}_{0A} \otimes \boldsymbol{e}_{3B}$ gets converted into a *product* of two reals, $\langle \boldsymbol{\omega}_A, \boldsymbol{e}_{0A}\rangle\langle \boldsymbol{\eta}_B, \boldsymbol{e}_{3B}\rangle$, one can think of the tensor product $\otimes$ as a *product in waiting*. It turns into a real product of two numbers eventually, but only after its two component vectors have been devoured by forms.

Another way to think of a tensor product is as a logical *and*. Instead of saying that we have observed the pair of registers A and B in state

$$\boldsymbol{e}_{0A} \otimes \boldsymbol{e}_{3B},$$

we can say that we have observed the pair components in states

$$\boldsymbol{e}_{0A} \quad \text{and} \quad \boldsymbol{e}_{3B}. \qquad (1.96)$$

2 The Qubit

2.1 The Evil Quanta

Deceptive simplicity of macroscopic physics

Macroscopic matter, meaning things that surround us in our everyday life—like cups, saucers, telephones and frog-infested ponds—are subject to well-known and well-understood laws of macroscopic physics. The laws are pretty simple, although when applied to almost any realistic system they tend to yield complex equations that are almost impossible to understand, seldom admit analytical solutions and frequently display chaotic behavior. Often, we have to help ourselves with common-sense understanding of the macroscopic world in order to construct, analyze, and solve equations that describe it.

Nineteenth-century physicists expected that the laws of macroscopic physics, which they distilled from their macroscopic observations and refined with their macroscopic brains should extend to the microscopic domain as well. They imagined that atoms (and they did not suspect at the time that atoms could be made of even smaller constituents) would be subject to the same the laws of Newtonian dynamics that worked so well on cannon balls and anvils sliding off rotating wedges with rough surfaces.

According to classical physics we should not be here.

It is just as well that they were wrong because, otherwise, nothing would work and we could not and would not be here to discuss these issues. Matter based on the principles of classical physics would cease to exist almost instantaneously. The Rutherford model predicted that a typical lifetime of a hydrogen atom should be about 10^{-10} s. Even if we were to ignore this little difficulty, classical stars should run out of puff in a mere 100 million years, as Kelvin and Helmholtz discovered. Needless to say, Rutherford knew well that hydrogen atoms did not decay after only 10^{-10} s, and Kelvin and Helmholtz were well aware of the fact that Mother Earth was several billion years old.

Microscopic physics averages away in thermodynamic limit.

Physics of the microscopic domain is then quite different, and one should expect this. A typical macroscopic chunk of matter contains about the Avogadro number of molecules. It may be 1/100 or 1/1000 or perhaps a thousand times more than the Avogadro number. It does not make much difference, because the Avogadro number is so huge, 6.023×10^{23}/mole. The laws of microscopic physics are not only different, but they also appear richer than the laws of macroscopic physics. But when one puts 6.023×10^{23} quantum objects together, couples them to other equally voluminous lumps of macroscopic matter, and immerses the whole lot in a thermal bath of an also huge number of photons,[1] the spectrum of which corresponds to

[1] The number of photons per unit volume in the room temperature black-body radiation is $4.125 \times 10^{14}\,\mathrm{m}^{-3}$. The formula is $16\pi k^3\zeta(3)T^3/(c^3h^3)$, where $\zeta(3) \approx 1.202$ is Apéry's constant.

the room temperature, most of this different and rich microscopic physics averages away, and all we are left with in the macroscopic world is our well-known and intuitively sound macroscopic physics ... which predicts that we should not exist.

One should expect that macroscopic physics ought to be derivable from microscopic physics. How would one go about this? One would derive mathematical formulas describing the behavior of a system of N quantum objects interacting with one another. Because of the complexity involved, it may not be possible to do this exactly; but one could make some simplifying assumptions on the way. One would then take a limit $N \rightarrow \infty$ (but in such a way that $N/V = \text{density} = \text{constant}$), since the Avogadro number is large enough to be replaced by infinity, and one would also assume the ambient temperature to be sufficiently high for quantum statistics to be replaced by the Boltzmann statistics, and in this limit we would expect the laws of macroscopic physics to emerge. This procedure is called *taking the thermodynamic limit of the quantum theory*. It is indeed the case that the laws of microscopic physics, as we know and understand them today, yield the laws of macroscopic physics in the thermodynamic limit. Something would be seriously wrong if they did not.

Microscopic physics cannot be derived from macroscopic physics.

But one should not expect that microscopic physics could be derived from macroscopic physics. Because of the averaging away of quantum effects on taking the thermodynamic limit, various theories of microscopic physics, some of them blatantly at odds with each other and with experimental phenomenology as well, may yield the same macroscopic physics in the thermodynamic limit. A simple example of this is the Boltzmann distribution, which can be derived both from the Fermi-Dirac statistics and from the Bose-Einstein statistics, though these are quite different. There is no way then that, say, the Fermi-Dirac statistics can be derived from the Boltzmann statistics. Additional assumptions that pertain only to the microscopic domain must be involved.

Smuggling macroscopic concepts into the domain of microscopic physics

Neither should one expect that macroscopic concepts such as the space-time continuum or differential manifolds would be applicable to the description of microscopic systems. There is no reason to expect that microscopic "space" should even be Hausdorff.[2]

Let us consider an example of neutron decay. Free neutrons live 885.7 ± 0.8 seconds on average [73]. Their lifetime is measured with macroscopic clocks, and

[2] A space is called "Hausdorff" if every two points of the space can be surrounded with open sets that do not intersect with one another. It is not easy to conceive of a space that is not Hausdorff, because almost all spaces in normal mathematical analysis are Hausdorff. Examples of non-Hausdorff spaces that cannot be trivially made into Hausdorff ones are Zariski topologies on algebraic varieties and Heyting algebras.

we see some neutrons living less than 885.7 seconds and some more. The decay rate is exponential, and the process of decay is probabilistic. But what is the macroscopic time that is used in the measurement? What if every neutron has its own microscopic clock and all these clocks work at different speeds, the macroscopic time being the average of all the little microscopic times? The neutron's lifetime measured by its own clock could very well be fixed; but because their clocks work at different speeds, we would perceive the neutron decay process as random when measured against the macroscopic clock.[3]

Yet one has to start somewhere, and the devil we know is better than the one we don't. So, in effect, physicists smuggled a lot of macroscopic conceptual framework into their description of the microscopic world. And whereas it seems to work in general, one is tempted to wonder sometimes whether the use of such concepts in microscopic physics is not abuse.

Scientific method and microscopic physics

How, then, can we arrive at correct theories of the microscopic world? The answer is straightforward, if disappointing: educated guesswork combined with laboratory verification. The so-called quantization procedures are just educated guesswork. They are not real derivations, and they seldom yield correct quantum theories without the need for additional manhandling. Their real purpose is to ensure that whatever is eventually concocted on the microscopic level yields an expected macroscopic theory in the thermodynamic limit, and this is fair enough. It does not imply, however, that microscopic theories cannot be constructed in other ways, while still yielding the correct thermodynamic limit. At present there are many different ways to choose from, the most popular being canonical quantization, Feynman path integrals, spin networks, and, more recently, topological field theories.[4]

Microscopic measurement is extremely difficult.

But at the end of the day all that matters is the microscopic description itself, which may as well have been guessed, its predictions, and its laboratory verification. And verification turns out to be a real can of worms because, as yet, nobody has come up with a way to observe a microscopic system without entangling it at the same time with a macroscopic (or at best a mesoscopic) measuring apparatus. So, in effect, we don't really know what microscopic systems do when they are left on their own and how they interact with each other in the absence of the macroscopic

[3]Neutron's own clocks run with different speeds, even by reckoning of present day physics, because of relativistic time dilation. And this is a measurable effect. But here we want to go further and equip them with clocks that would be more like our human aging clocks.

[4]One of my colleagues so believed in the physical reality of quantization procedures that she thought a device for manipulating quantum objects could be made based on the procedure she worked on. She was disappointed to learn that the device could not be made because her quantization procedure, like any other, did not and could not correspond to any real physical process.

measuring apparatus—although we may have some vague ideas. The presence of the measuring apparatus in our investigations of the microscopic world is so important that we have been forced to acknowledge that certain measured quantities and perhaps even measured microscopic objects themselves are *made* by the act of the measurement and do not exist in the same form prior to the measurement.

Measurement as a physical process

Until fairly recently the measurement process used to be captured by an axiom of quantum mechanics and was thought to stand apart from other quantum processes. But today we view it as a lopsided, dynamic, physical process that results from an interaction of a microscopic object with a system comprising a large number of other microscopic objects. It can be understood, analyzed, and verified even within the existing framework of microscopic physics. This important change in how we understand the measurement is among the greatest accomplishments of modern physics [17, 153]. We shall discuss the related Haroche-Ramsey experiment in Section 5.12.1.

The difficulties of the microscopic measurement can be understood better by pondering what it means to "observe."

We observe behavior of macroscopic objects by looking at them or looking at instruments that, in turn, look at the objects. "Looking" at something implies that we have a light source emitting photons, which bounce off the observed object, then enter our eyes and get absorbed by the retina, which converts light into chemical energy. The chemical energy activates nerves that transmit the signals to the brain, which interprets them. The crucial link in this chain is where photons bounce off the observed object. The amount of momentum transmitted from the photons to the object is so small, compared to the momentum of the object, that the act of illumination does not affect the object and its behavior appreciably. Consequently we can ignore the effect that the light source has on the object. Planets do not change their orbits and rotation because the sun shines on them (on the other hand, comets do—because they are light and fluffy and lose material when heated). An anvil sliding off a rotating wedge is not going to stop suddenly and jump back to the top because a laboratory lamp is turned on (on the other hand, a startled mouse will scurry away, leaving a crumbled cookie behind—because the mouse has photosensitive eyes, whereas the anvil does not).

Fragility of microscopic systems

But what if instead of using a benign light source in the form of a candle or a light bulb, we were to use an exploding bomb? This would certainly be overkill, and we would end up none the wiser, because the explosion would obliterate all we wanted to look at. Yet this is quite like what happens in the microscopic world. Microscopic systems are so delicate, so fragile, that even bouncing as little as a single photon against them can change their state dramatically. The situation is exacerbated further by the fact that the wavelength of a photon is inversely proportional to its momentum. A photon of low momentum has large wavelength, so it is not going to be a precise enough instrument with which to observe microscopic systems. But if we attempt to select a photon with wavelength sufficiently short to give us a well-resolved picture of a microscopic object, its momentum will be so large that it will destroy the object we are trying to observe. Because the same relation affects all other elementary particles, we can't get around this difficulty by choosing, for example, neutrons to observe microscopic systems—although it is sometimes possible to get just a little further by observing with particles other than photons.

The Heisenberg uncertainty principle

But yet another complication is perhaps the weirdest. Quantum objects such as photons and electrons appear to be nonlocal. They turn into pointlike energy discharges only when they get snatched away from their free-range spread-out status by a macroscopic measuring apparatus. It is as if the apparatus sucked them into a point. All other free-ranging quantum systems the measured particle "overlapped" with prior to this act of kidnapping and localization detect the sudden absence of their companion and react to it in various ways. So, in quantum physics it is not just the act of shining light onto an observed object that affects it. The act of kidnapping bounced-off photons affects the object, too. It is as if the anvil sliding off the rotating wedge stopped suddenly and jumped back to the top because the observer's eyes have absorbed the scattered photons.

Nonlocality of microscopic objects

Einstein, the great physicist of the twentieth century, could not stomach this. Yet stomach this he had to, because this was what had clearly transpired from laboratory experiments. He responded by writing about the "evil quanta" a year before his death [63]. But quanta aren't evil. They are what they are, and the best way to make them likable is to understand their weirdness and seek to explore it.

The Einstein curse

So, what can we say about microscopic systems? After all we have been investigating them for some 100 years or so. Numerous Nobel prizes were awarded for successful predictions pertaining to and then discoveries and exploitations of various microscopic phenomena. Devices such as lasers and semiconductor switches, the functioning of which is based on principles of quantum physics, are incorporated in common household appliances.

The first point to observe is that microscopic systems are essentially unpredictable. It is usually impossible to predict exactly what a given single electron or a photon is going to do in various experimental situations, although on rare occasions such predictions can be made. When a single microscopic system is subjected to a blasting force of a measurement with a macroscopic apparatus, we end up with random read-outs, which, to make matters worse, may not always tell us what the microscopic system's properties were prior to the measurement. As we

Unpredictability of microscopic systems

have already pointed out, some properties and perhaps even the objects themselves appear to be made by the measurement.

Probabilistic description

But this random read-out is not necessarily white-random. If we repeat the measurement over and over on microscopic systems that have been prepared in exactly the same way, we'll discover that there may be certain probability distributions associated with experimental results. And these distributions, it turns out, can be predicted with great accuracy.

Quantum physics

> Quantum physics is a discipline that tells us how to describe, manipulate, and predict evolution of probability distributions associated with measurements made on microscopic systems with macroscopic measuring devices.

Universality

Much had been said about universality of quantum physics (see, for example, [111] and references therein). At first glance quantum physics cannot be universal, because it describes the interaction between microscopic systems and macroscopic devices (a measuring apparatus or just a macroscopic environment—as far as the microscopic system is concerned, there is no difference). But microscopic systems are *always* in the presence of some macroscopic environment,[5] so the question about what microscopic systems may possibly do "on their own" is perhaps as silly as the question about the number of angels that can dance on the head of a pin.[6]

Frontiers

And then, again, once physicists have in hand a theory, such as quantum physics, they always try to extend it and apply to phenomena that at first may seem beyond the theory's original area of competence. This is a worthwhile endeavor because it is through such activities that our knowledge and understanding of nature expand, too. So, one can try to observe microscopic systems with mesoscopic devices—the present-day technology allows, at last, for such measurements to be carried out [99] [62] [117]—and one can then test whether predictions of quantum physics still agree with what such mesoscopic measurements return. What we find is that certain older formulations of quantum mechanics may have to be ever so subtly revised and enriched in order to account for what the new experiments tell us.

Quantum computing

These activities are also of great importance to quantum computing because by

[5] According to some physicists the universal force of gravity constantly measures every quantum object within its reach—and it reaches everywhere, which is why we call it "universal"—and by doing so brings everything, including human thoughts, into existence [109].

[6] The medieval argument was over how many angels could stand on the *point* of a pin [65]. Ironically, this question has recently received some attention of modern science, because aspiring quantum theories of gravity let us reform it in mathematical terms and seek answer to it within the confines of the theory [123], which yet again demonstrates that there are no "stupid questions." Whether a question is "stupid" depends on the framework within which it is asked.

these means we learn how to manipulate microscopic systems to our advantage. And it is exactly here that the frontiers of present-day quantum physics and computer science meet.

This is going to be our battlefield.

2.2 The Fiducial Vector of a Qubit

What is a qubit?

A qubit is a quantum relative of a classical randomly fluctuating one-bit register.[7] It is the simplest nontrivial quantum system.

As is the case with randomly fluctuating 1-bit registers, there are many possible physical embodiments of a qubit. Qubits can be "natural"; for example, a neutron placed in a very strong uniform magnetic field of about 12 T is a natural qubit—and we'll work with this example occasionally. Qubits can be engineered, too; for example, the *quantronium* circuit [142] discussed toward the end of this chapter makes an excellent qubit.

Qubit's fiducial vector

Yet, regardless of the details of their engineering, which can be sometimes quite complex, qubits' fiducial mathematics and dynamics are always the same. Qubits can be described in terms of fiducial vectors indeed. Each entry in the vector specifies a probability of finding a qubit in the corresponding configuration.

Basis states

A qubit is a two-dimensional system ($N = 2$) in the sense that has been explained in Section 1.5 on page 19. Hence, if we were to give a qubit a quick glance—speaking figuratively, of course, since glancing at quantum objects is far from trivial—we would find it in one of two possible states. These states are often referred to as "up" and "down" and are denoted by symbols

$$|\uparrow\rangle \quad \text{and} \quad |\downarrow\rangle.$$

Qubit computing

We may associate the binary number 0 with $|\uparrow\rangle$ (because $|\uparrow\rangle$ is often the lower energy state of the two) and the binary number 1 with $|\downarrow\rangle$. We end up with an object that, like a bit, can be used for counting. This leads to the following notation that has been adopted in quantum computing:

$$|\uparrow\rangle \equiv |\,0\rangle, \tag{2.1}$$

$$|\downarrow\rangle \equiv |\,1\rangle. \tag{2.2}$$

Having glanced at a qubit and having found it in either of its two basis states, we can no longer repeat the observation, because the original state of the qubit usually,

[7] It can even be faked faithfully by a special combination of a classical randomly fluctuating display driven by a pair of coupled oscillators [103, 104, 105]. We talk more about it in Section 7.5.

though not always, is destroyed by the observation. But we can prepare another qubit, or even the same one sometimes, in the same way, and repeat the experiment. By doing so a sufficiently large number of times, we can estimate probabilities of finding it either in the $|\uparrow\rangle$ state or in the $|\downarrow\rangle$ state. Because in this experiment the qubit cannot be found in any other state, the probabilities must add to 1. Let us call these p^0 and p^1, respectively. Then

$$p^0 + p^1 = 1. \tag{2.3}$$

Parameterizing probabilities

The equation lets us parameterize p^0 and p^1 with a single number r^z such that

$$p^0 = \frac{1}{2}(1 + r^z) \quad \text{and} \tag{2.4}$$

$$p^1 = \frac{1}{2}(1 - r^z). \tag{2.5}$$

The parameterization ensures that equation (2.3) is always satisfied, and restricting r^z to $-1 \le r^z \le 1$ (or, in other words, $r^z \in [-1, 1]$[8]) ensures that p^0 and p^1 stay within $[0, 1]$.

But p^0 and p^1 are not enough to fully describe the state of a qubit. A qubit's fiducial vector has two more probabilities, which remain hidden to this experiment.[9] We can find them by performing additional observations in a different setup. How this is done exactly we'll discuss in the next section. For the time being, it suffices to say that the other two parameters are probabilities. They are not complementary with respect to each other and with respect to p^0 and p^1. But because they are probabilities, we can parameterize them similarly:

$$p^2 = \frac{1}{2}(1 + r^x), \quad \text{and} \tag{2.6}$$

$$p^3 = \frac{1}{2}(1 + r^y), \tag{2.7}$$

where r^x and r^y are real numbers between -1 and 1, or, in other words, $r^x \in [-1, 1]$ and $r^z \in [-1, 1]$.

[8]The notation means "all real numbers between -1 and $+1$ inclusive." It should not be confused with $\{-1, +1\}$, which is a two-element set comprising -1 and $+1$.

[9]These *are* "hidden parameters" with respect to this measurement. But they are not of the kind that were sought by Einstein, Podolsky, and Rosen [38], because they are not deterministic.

In summary, the full fiducial vector of a qubit is four-dimensional ($K = 4$) and can be parameterized as follows:

$$\boldsymbol{p} = \frac{1}{2}\begin{pmatrix} 1 + r^z \\ 1 - r^z \\ 1 + r^x \\ 1 + r^y \end{pmatrix}, \tag{2.8}$$

where r^x, r^y, and r^z are all real numbers between -1 and 1.

The vector illustrates fundamental differences between classical and quantum systems. First, even though the system is two dimensional, its fiducial vector is four dimensional ($K = 4 = N^2$) [60]. We shall no longer have the trivial mapping between basis states and fiducial states that characterized classical randomly fluctuating registers. In the quantum world the fiducial-level description is quite different from the basis state-level description.

The second point to observe is that

$$\sum_{i=0}^{3} p^i = 2 + \frac{r^x + r^y}{2}. \tag{2.9}$$

Vector $\boldsymbol{p}$ is normalized, but not in the classical sense. The normalization is restricted to its first two components only, as given by equation (2.3).

2.3 The Stern-Gerlach Experiment

To develop a better understanding of the qubit's fiducial vector, we shall discuss in this section how to measure the fiducial vector with its four probabilities for a neutron encoded qubit.

Magnetic properties of neutrons

Although neutrons don't have an electric charge, they are known to have a magnetic moment [143]

$$\mu_n = -1.9130427 \pm 0.0000005\, \mu_N, \tag{2.10}$$

where [95]

$$\mu_N = \frac{q_e \hbar}{2m_p} = (5.05078343 \pm 0.00000043) \times 10^{-27}\ \mathrm{Am}^2 \tag{2.11}$$

is the nuclear magneton and where q_e is the elementary charge, m_p is the mass of the proton, and $\hbar$ is the Planck constant divided by 2π.

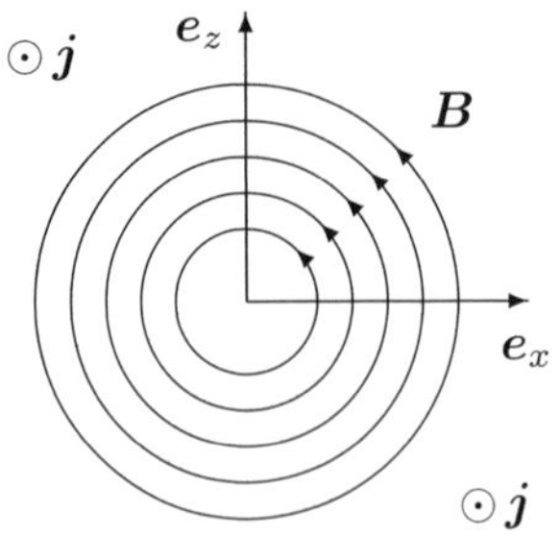

Figure 2.1: Magnetic field configuration for $\boldsymbol{j} = -\epsilon_0 c^2 \alpha \boldsymbol{e}_y$. The symbol $\odot$ to the left of $\boldsymbol{j}$ stands for the tip of the arrow that points at the reader. We find that $\boldsymbol{B} = \alpha x \boldsymbol{e}_z$ along the x axis.

According to classical physics the mechanical energy of the neutron immersed in magnetic field $\boldsymbol{B}$ is[10]

$$E = -\boldsymbol{\mu}_n \cdot \boldsymbol{B}. \tag{2.12}$$

Spatially varying magnetic field

Because the magnetic moment of a neutron is negative, neutrons that are counteraligned to the magnetic field have lower energy than neutrons that are aligned with it. It is the other way round with protons.

According to classical reasoning, if the magnetic field varies in space, then a force is exerted on the neutron by the gradient of the magnetic field $\boldsymbol{\nabla B}$,

$$\boldsymbol{F} = -\boldsymbol{\nabla} E = \boldsymbol{\nabla} \left(\boldsymbol{\mu}_n \cdot \boldsymbol{B} \right). \tag{2.13}$$

Let us suppose that $\boldsymbol{B} = \alpha x \boldsymbol{e}_z$. For the magnetic field so defined, $\boldsymbol{\nabla} \cdot \boldsymbol{B} = 0$ and $\boldsymbol{\nabla} \times \boldsymbol{B} = -\alpha \boldsymbol{e}_y$. $\boldsymbol{B}$ has vanishing divergence, as it should, and it can be generated by current density $\boldsymbol{j} = -\epsilon_0 c^2 \alpha \boldsymbol{e}_y$.[11] To be more precise, the Maxwell equations tell us that there ought to be a B^x component varying with z in this situation, too. The field's configuration is shown in Figure 2.1.

The solution $\boldsymbol{B} = \alpha x \boldsymbol{e}_z$ corresponds to $\boldsymbol{B}$ along the x axis. We shall confine ourselves to the narrow neighborhood of the x axis, then, in order to have $\boldsymbol{B}$ described by the formula.

[10] To brush up on the dynamics of a current loop in the magnetic field, see, for example, [43], Section 15-1, "The Forces on a Current Loop; Energy of a Dipole."

[11] To brush up on the Maxwell equations see, for example, [43], Chapter 18, "The Maxwell Equations."

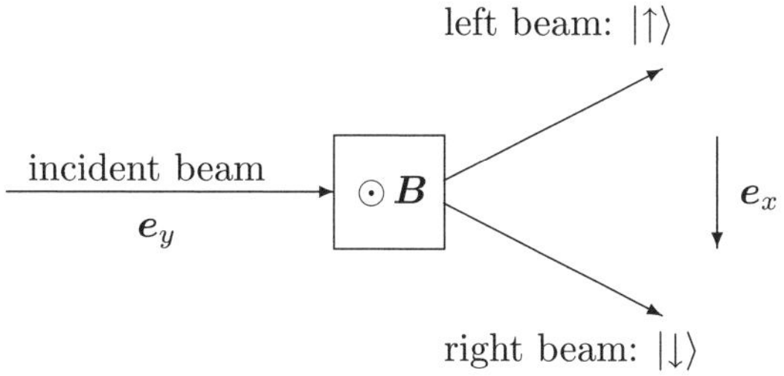

Figure 2.2: Splitting of the incident neutron beam by the chamber filled with $\boldsymbol{B} = \alpha x \boldsymbol{e}_z$. The symbol $\odot$ inside the box representing the chamber stands for the tip of the $\boldsymbol{B}$ arrow that points at the reader.

The mechanical energy of the neutron immersed in $\boldsymbol{B}$ is $-\mu_n^z \alpha x$, and the force acting on it becomes

$$\boldsymbol{F} = \mu_n^z \alpha \boldsymbol{e}_x. \tag{2.14}$$

We assume that the neutron's magnetic moment can point in every direction, $-\mu_n \leq \mu_n^z \leq \mu_n$, which, in turn, yields force $\boldsymbol{F} = F^x \boldsymbol{e}_x$, where

$$-|\mu_n \alpha| \leq F^x \leq |\mu_n \alpha|. \tag{2.15}$$

Let us suppose that a well-collimated monochromatic and unpolarized[12] beam of neutrons is fired in the y direction as shown in Figure 2.2. Then the beam enters a chamber filled with $\boldsymbol{B} = \alpha x \boldsymbol{e}_z$. On leaving the chamber the beam should fan out in the x direction.

This is what classical physics says, but this is not what happens.

The beam splits instead of fanning out.

Unless the beam has been specially prepared, about which more below, it *splits*, as shown in Figure 2.2, into two well-collimated beams, one of which corresponds to $F^x = -|\mu_n \alpha|$ and the other one to $F^x = |\mu_n \alpha|$, and the whole middle that corresponds to $-|\mu_n \alpha| < F^x < |\mu_n \alpha|$ is missing. This result tells us that as the neutrons encounter $\boldsymbol{B} = \alpha x \boldsymbol{e}_z$ they either fully align or fully counteralign with the direction of the ambient magnetic field. Furthermore, the alignment appears to happen instantaneously and in such a way that no energy is released in the process. Nobody has ever managed to capture a neutron that would be inclined under some angle to the direction of the ambient magnetic field and that would then gradually align with it, releasing the excess of $-\boldsymbol{\mu}_n \cdot \boldsymbol{B}$ in the process.

Similarity to birefringence

A classical physics phenomenon that resembles closely what we observe here is a

[12] *Collimated* means well focused, *monochromatic* means that all neutrons have the same energy and momentum, *unpolarized*—well, this will be explained later; just read on.

passage of a light beam through a birefringent crystal. The crystal splits the beam into two components, which become physically separated from each other. Each component is linearly polarized in a direction perpendicular to that of the other component.[13]

Drawing on this similarity, we refer to the two neutron beams that emerge from our apparatus as *fully polarized*. The beam that corresponds to $F^x = |\mu_n\alpha|$ will be deflected to the right,[14] and the beam that corresponds to $F^x = -|\mu_n\alpha|$ will be deflected to the left. Assuming that α is positive, and remembering that the neutron magnetic moment is negative, all neutrons in the right beam have their $\boldsymbol{\mu}_n$ counteraligned to $\boldsymbol{B}$, and all neutrons in the left beam have their $\boldsymbol{\mu}_n$ aligned with $\boldsymbol{B}$. Let us call the state of the ones in the right beam $|\downarrow\rangle$ and the state of the ones in the left beam $|\uparrow\rangle$.

Normalization of the fiducial vector

Now we can ask questions about neutrons constituting the incident beam. We can ask, for example, about the probability that a neutron entering the chamber will emerge from it in the $|\uparrow\rangle$ state (the left beam); the probability is p^0. We can then ask about the probability that a neutron entering the chamber will emerge from it in the $|\downarrow\rangle$ state (the right beam); the probability is p^1. Because the chamber does not swallow neutrons, every neutron should emerge from it in either of the two possible states. Consequently the probabilities p^0 and p^1 must add to 1:

$$p^0 + p^1 = 1. \tag{2.16}$$

Rotating the chamber by 90° lets us measure p^2.

If we were to rotate the whole set up about the y axis by 90° clockwise, we would end up with the chamber filled with $\boldsymbol{B} = \beta z \boldsymbol{e}_x$. If we try the classical description again, the energy of a neutron entering the chamber would be $E = -\mu_n^x \beta z$ and the force exerted on the neutron would be $\boldsymbol{F} = -\boldsymbol{\nabla} E = \mu_n^x \beta \boldsymbol{e}_z$. According to classical reasoning, the beam of incident neutrons should fan out in the z direction. And this is again wrong, because the beam instead *splits* in the z direction, the two new beams corresponding to $F^z = |\mu_n\beta|$ and $F^z = -|\mu_n\beta|$ with the whole middle $-|\mu_n\beta| < F^z < |\mu_n\beta|$ missing. This is shown in Figure 2.3.

Neutrons in the upper beam have their magnetic moments $\boldsymbol{\mu}_n$ counteraligned to the direction of the magnetic field $\boldsymbol{B} = \beta z \boldsymbol{e}_x$, whereas neutrons in the lower beam have their magnetic moments aligned with the direction of $\boldsymbol{B}$. Let us call the states of the neutrons in the upper beam $|\leftarrow\rangle$ and the states of the neutrons in the lower

[13] Birefringence is discussed in [41], Section 33-3, "Birefringence."

[14] Here we assume that the three vectors $\boldsymbol{e}_x$, $\boldsymbol{e}_y$, and $\boldsymbol{e}_z$ have the right-hand screw orientation; that is, as one rotates from $\boldsymbol{e}_x$ to $\boldsymbol{e}_y$, one moves up in the direction of $\boldsymbol{e}_z$.

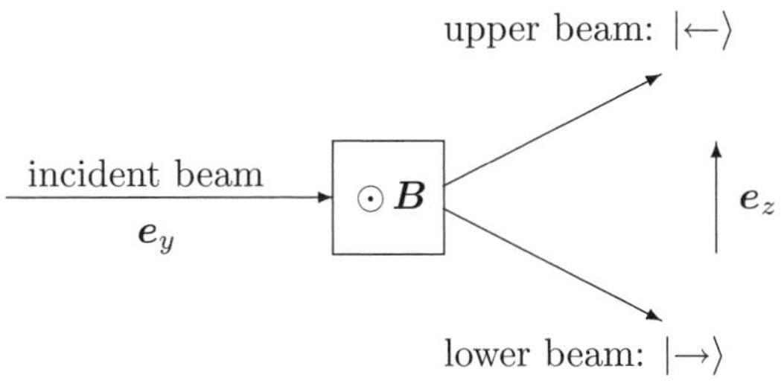

Figure 2.3: Splitting of the incident neutron beam by the chamber filled with $\boldsymbol{B} = \beta z \boldsymbol{e}_x$. As before, the symbol $\odot$ inside the box representing the chamber stands for the tip of the arrow that points at the reader.

beam $|\rightarrow\rangle$. We shall call the probability that a neutron incident on the chamber emerges from it in the lower beam p^2, and we'll parameterize it by

$$p^2 = \frac{1}{2}\left(1 + r^x\right), \tag{2.17}$$

where $-1 \leq r^x \leq 1$. The probability that the neutron emerges from the chamber in the upper beam is then $1 - p^2 = \frac{1}{2}\left(1 - r^x\right)$, because every neutron that enters the chamber *must* leave it either in the upper or in the lower beam.

Finally, let us consider shooting the beam in the y direction pervaded by $\boldsymbol{B} = \gamma x \boldsymbol{e}_y$. Neutrons entering the chamber acquire energy $E = -\boldsymbol{\mu}_n \cdot \boldsymbol{B} = -\mu_n^y \gamma x$ and are subjected to force $\boldsymbol{F} = -\boldsymbol{\nabla} E = \mu_n^y \gamma \boldsymbol{e}_x$. Classically, on leaving the chamber the beam should fan out in the x direction, but again this is not what happens. The beam *splits* in the x direction, producing two new beams that correspond to $\boldsymbol{\mu}_n$ aligned with $\boldsymbol{B} = \gamma x \boldsymbol{e}_y$ (the left beam) and aligned against the direction of $\boldsymbol{B}$ (the right beam). This is shown in Figure 2.4

Let us call the state of neutrons in the left beam (i.e., the ones aligned with $\boldsymbol{B}$) $|\otimes\rangle$ and the state of neutrons in the right beam (i.e., the ones aligned against the direction of $\boldsymbol{B}$) $|\odot\rangle$. We have used this notation already in Figures 2.2 and 2.3. The symbol $\otimes$ is suggestive of an arrow flying away from the reader—in this case the arrow points in the direction of the beam itself. Symbol $\odot$ is suggestive of an arrow flying toward the reader—in this case the arrow points against the direction of the beam.

We can ask about the probability that a neutron entering the chamber will leave it in the $|\otimes\rangle$ state. The probability is p^3, and we can parameterize it by

$$p^3 = \frac{1}{2}\left(1 + r^y\right), \tag{2.18}$$

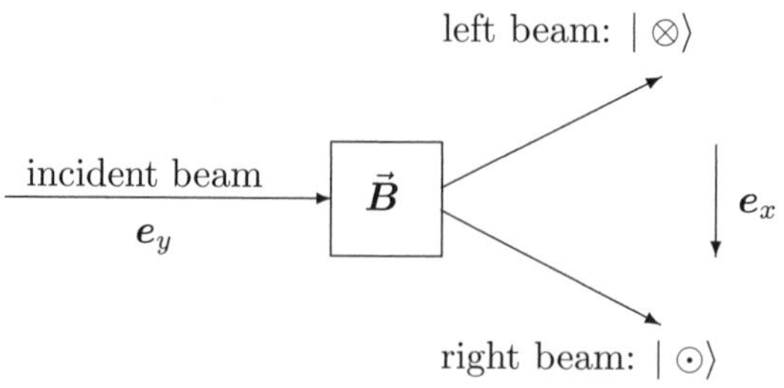

Figure 2.4: Splitting of the incident neutron beam by the chamber filled with $\boldsymbol{B} = \gamma x \boldsymbol{e}_y$. The arrow above $\boldsymbol{B}$ indicates that the magnetic field inside the chamber points to the right.

where $-1 \le r^y \le 1$. The probability of finding the neutron in the $|\odot\rangle$ state is $1 - p^3 = \frac{1}{2}(1 - r^y)$, because every neutron that enters the chamber *must* exit it either in the $|\otimes\rangle$ or in the $|\odot\rangle$ state.

Probabilities p^0, p^2, and p^3 cannot be measured at the same time.

We observe that we cannot measure simultaneously p^0, p^2, and p^3. The apparatus needed to measure p^0 is oriented differently from the apparatus needed to measure p^2 or p^3. To measure all three quantities, we have to subject the beam of identically prepared neutrons to all three measurements separately.

The measurements do not represent mutually exclusive alternatives either. If we were to produce a polarized beam in the $|\uparrow\rangle$ state and then pass it through a device that should split it between $|\rightarrow\rangle$ and $|\leftarrow\rangle$ states, the beam would indeed split—in this case, evenly. This is the reason the components of the fiducial vector $\boldsymbol{p}$ add to more than 1. Yet, to fully characterize the state of a qubit, we have to measure all three probabilities: p^0, p^2, and p^3.

The gedanken experiment discussed in this section follows closely a similar experiment carried out by Otto Stern and Walther Gerlach in 1922 [51]. Instead of neutron beams they used ionized atoms of silver, and their configuration of magnetic field was more complicated and produced by a wedge-shaped magnet. We have oversimplified the experiment to flesh out the significant aspects of it. Choosing a neutron beam eliminates electrical interactions that silver ions are subject to, and choosing a linearly varying magnetic field—which is not easy to produce in practice—simplifies the calculations.

2.4 Polarized States

Now that we know how to measure neutron beam qubit probabilities[15] that form its fiducial vector $\boldsymbol{p}$, let us consider certain special situations.

Beams leaving the apparatus are fully polarized.

Let us suppose the beam is fired in the y direction, whereupon it enters a chamber filled with $\boldsymbol{B} = \alpha x \boldsymbol{e}_z$. The beam splits in the x direction so that neutrons in the $|\uparrow\rangle$ state shoot to the left and neutrons in the $|\downarrow\rangle$ state shoot to the right as shown in Figure 2.2. The fiducial vector that describes neutrons in the left beam looks as follows:

$$|\uparrow\rangle \equiv \begin{pmatrix} 1 \\ 0 \\ 0.5 \\ 0.5 \end{pmatrix}. \tag{2.19}$$

Of the three probability parameters, r^x, r^y, and r^z, only the third one does not vanish and equals 1. We can combine the three parameters into another column vector, $\boldsymbol{r}$, given by

$$\boldsymbol{r} = \begin{pmatrix} 0 \\ 0 \\ 1 \end{pmatrix} = \boldsymbol{e}_z. \tag{2.20}$$

The fiducial vector that describes neutrons in the right beam is

$$|\downarrow\rangle \equiv \begin{pmatrix} 0 \\ 1 \\ 0.5 \\ 0.5 \end{pmatrix}. \tag{2.21}$$

This time, the corresponding vector $\boldsymbol{r}$ looks as follows:

$$\boldsymbol{r} = \begin{pmatrix} 0 \\ 0 \\ -1 \end{pmatrix} = -\boldsymbol{e}_z. \tag{2.22}$$

We observe that p^2 and p^3 in both fiducial vectors are not zero. The reason is that in both cases $r^x = r^y = 0$, but this merely leaves $p^2 = p^3 = \frac{1}{2}(1+0) = \frac{1}{2}$.

Before we discuss what this means, we observe one more fact. Vector $\boldsymbol{r}$, which is called the qubit *polarization vector,* is a real three-dimensional vector, not only

[15] Although we do not normally say so, it is the whole neutron beam that corresponds to the notion of a qubit here, not individual neutrons in it. We cannot use a single neutron for any meaningful measurements in this context, especially none that would yield probabilities.

because it has three components, but also because it transforms the right way on transformations of the coordinate system, (x, y, z).

Let us go back to Figure 2.2. Let us rotate the coordinate system about the y axis by 90°, so that the z axis becomes the x axis and vice versa. Rotating the frame against which the measurements are made does nothing to the physics of the neutron beam and the beam splitting apparatus, which are oblivious to our conventions. Yet, in the new frame the beam splits in the z direction, and the field points in the x direction. Consequently what was r^z before becomes r^x now, while the value of the parameter has not changed otherwise. The resulting transformation is

$$\boldsymbol{e}_z \rightarrow \boldsymbol{e}_x \Longrightarrow r^z \rightarrow r^x. \tag{2.23}$$

Similar reasoning repeated for other directions will show that the components of $\boldsymbol{r}$ indeed transform the right way for a three-dimensional vector.

Now, let us get back to our discussion of nonzero values of p^2 and p^3.

A fully polarized beam is split again.

Let us take the left beam from Figure 2.2 (neutrons in the $|\uparrow\rangle$ state), and let us direct it into another chamber filled with $\boldsymbol{B} = \beta z \boldsymbol{e}_x$ as shown in Figure 2.3. The fiducial vector that corresponds to the $|\uparrow\rangle$ state tells us that the beam is going to split in this chamber evenly, that is, approximately half of all neutrons (let us recall that probabilities become exact measures of what is going to happen only when the number of neutrons in the beam becomes infinite) that enter it will swing upwards and the other half will swing downwards. The prediction pertains to the statistical ensemble only. We have no means of predicting what any given neutron is going to do. It may just as well swing upwards as downwards, much the same as a well-thrown coin may land just as well heads or tails. In any case, half of all neutrons, *on average*, will emerge from the chamber in the $|\rightarrow\rangle$ state, and the other half will emerge in the $|\leftarrow\rangle$ state. Their corresponding fiducial vectors will be

$$|\rightarrow\rangle \equiv \begin{pmatrix} 0.5 \\ 0.5 \\ 1 \\ 0.5 \end{pmatrix}, \quad \boldsymbol{r} = \boldsymbol{e}_x, \tag{2.24}$$

and

$$|\leftarrow\rangle \equiv \begin{pmatrix} 0.5 \\ 0.5 \\ 0 \\ 0.5 \end{pmatrix}, \quad \boldsymbol{r} = -\boldsymbol{e}_x. \tag{2.25}$$

At first we might think that the $|\uparrow\rangle$ state is a mixture of $|\rightarrow\rangle$ and $|\leftarrow\rangle$. But it is not, as the following calculation shows:[16]

$$0.5\begin{pmatrix}0.5\\0.5\\1\\0.5\end{pmatrix}+0.5\begin{pmatrix}0.5\\0.5\\0\\0.5\end{pmatrix}=0.5\begin{pmatrix}1\\1\\1\\1\end{pmatrix}. \qquad (2.26)$$

The fiducial vector corresponds to $\boldsymbol{r}=\boldsymbol{0}$, not to $\boldsymbol{r}=\boldsymbol{e}_z$. The state of the beam has been changed by interaction with the second chamber, and whatever information was stored in the input state has been irretrievably lost. Whereas the input state was fully polarized, the output state, if we were to merge the two beams, contains no useful information: it is all white noise.

Incident state is replaced with a mixture.

We shall learn in Chapter 4 that $|\uparrow\rangle$ is indeed related to $|\rightarrow\rangle$ and $|\leftarrow\rangle$, but not as a mixture of the two states. Instead we shall find that it is a superposition of the two states. We shall be able to write

Superposition versus mixture

$$|\uparrow\rangle=\frac{1}{\sqrt{2}}\left(|\rightarrow\rangle+|\leftarrow\rangle\right), \qquad (2.27)$$

but this will *not* translate into

$$\boldsymbol{p}_{|\uparrow\rangle}=\frac{1}{2}\left(\boldsymbol{p}_{|\rightarrow\rangle}+\boldsymbol{p}_{|\leftarrow\rangle}\right). \qquad (2.28)$$

As we remarked earlier, the relationship between the basis states and the fiducial vectors in quantum physics is nontrivial.

Let us go back to the left beam that emerges from the first chamber (still Figure 2.2), the beam that is in the $|\uparrow\rangle$ state. If instead of directing the beam into the chamber filled with $\boldsymbol{B}=\beta z\boldsymbol{e}_x$ we were to direct it into a chamber filled with $\boldsymbol{B}=\alpha x\boldsymbol{e}_z$, meaning a chamber that works exactly the same way as the first chamber, we would find that *all* neutrons entering the chamber swing to the left. Hence, this time the chamber would not change their state but would merely confirm the state of the incident neutrons and preserve it intact. This is one of the rare circumstances in quantum physics when we can predict *with certainty* what is going to happen to *every* individual neutron.

Confirming the state does not destroy it.

This behavior is similar to the behavior of photons passing through a series of polarizing plates. Let us assume that the first plate polarizes incident light in the z direction. If we insert a second plate behind the first plate and rotate its

Neutron and photon polarizations

[16] We're getting a little ahead here. Mixtures of qubit states will be discussed in Section 2.5, page 59.

polarization axis by 90° relative to the first plate, no photons will pass through the two plates. If we rotate the polarization axis of the second plate by 90° again, both plates will have their axes aligned, and every photon that passes through the first plate is guaranteed to pass through the second plate as well. We can make this prediction *with certainty* about every photon incident on the second plate. But if the second plate has been rotated by 45° relative to the first plate, only about a half of the photons incident on the second plate will be transmitted. Yet, we have no means to make exact predictions about any individual photon incident on the second plate in this case. We can make statements only about the statistical ensemble that, in this case, corresponds to the beam of the incident photons.

Neutron beam polarized in an arbitrary direction

So far we have contemplated polarized states aligned with one of the three principal directions, x, y, and z only. What about a state described by vector $\boldsymbol{r}$ of length 1 but tilted arbitrarily? We can describe components of such a vector by using spherical coordinates θ and ϕ (see Figure 2.5),

$$\boldsymbol{r} = \begin{pmatrix} \sin\theta\cos\phi \\ \sin\theta\sin\phi \\ \cos\theta \end{pmatrix}, \tag{2.29}$$

where θ is the angle between $\boldsymbol{e}_z$ and $\boldsymbol{r}$, and ϕ is the angle between $\boldsymbol{e}_x$ and the projection of $\boldsymbol{r}$ on the equatorial plane, $z = 0$. The corresponding fiducial vector is

$$\boldsymbol{p} = \frac{1}{2} \begin{pmatrix} 1 + \cos\theta \\ 1 - \cos\theta \\ 1 + \sin\theta\cos\phi \\ 1 + \sin\theta\sin\phi \end{pmatrix}. \tag{2.30}$$

Is the beam fully polarized?

If we were to shoot a beam of neutrons so polarized in the y direction and through the chamber filled with magnetic field parallel to $\boldsymbol{r}$ and varying linearly in the direction perpendicular to $\boldsymbol{r}$, *all* incident neutrons would swing to the same side in the direction perpendicular to $\boldsymbol{r}$. If we can find such a direction by trial and error and confirm that all neutrons swing to the same side, we'll know that the incident beam is fully polarized.

Polarized photons

It is the same as with the polarized light. If we have a fully polarized incident light beam and a polarizer, we shall always be able to find an angle of the polarizer that lets all incident photons through. If, on the other hand, some photons get absorbed for *every* angle of the polarizer, the incident beam is not fully polarized. It is a *mixture* of various polarization states.

If we combine the requirement that each component of $\boldsymbol{r}$ is confined to $[-1, 1]$, in order to ensure that all components of $\boldsymbol{p}$ are confined to $[0, 1]$, with the observation

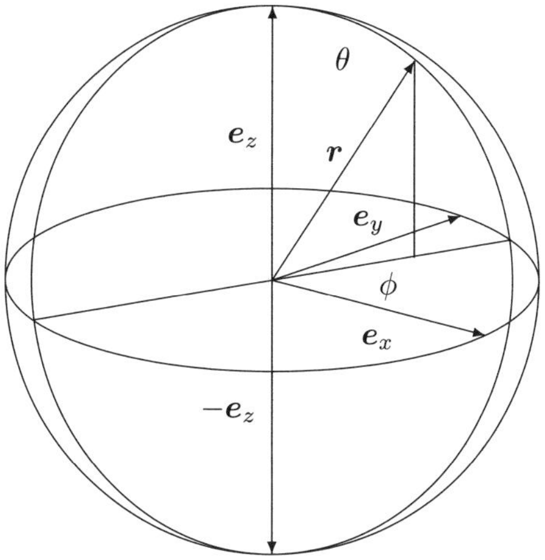

Figure 2.5: Bloch sphere. The radius of the Bloch sphere is 1. All vectors shown in the figure touch the surface of the Bloch sphere with their tips. θ, the zenith angle, is the angle between $\boldsymbol{e}_z$ and $\boldsymbol{r}$, and ϕ, the azimuth angle, is the angle between $\boldsymbol{e}_x$ and the projection of $\boldsymbol{r}$ on the equatorial plane.

that $\boldsymbol{r}$ is a real vector in the three-dimensional Euclidean space, we arrive at the notion that fully polarized states of a neutron beam are described by vector $\boldsymbol{r}$ of length 1 pointing in some arbitrary direction. Such states are *pure*. They cannot be produced by mixing neutrons of various polarizations. We have seen what happened when we tried to mix neutrons in the $|\rightarrow\rangle$ and $|\leftarrow\rangle$ states.

Bloch sphere

The set of all *pure* states of the neutron beam corresponds to the surface traced by vector $\boldsymbol{r}$ of length 1 as it points in all possible directions. It is the surface of a sphere of radius 1. The sphere, shown in Figure 2.5, is called the *Bloch sphere*.

2.5 Mixtures of Qubit States

Let us consider two qubit states. One is given by

$$\boldsymbol{p}_1 = \frac{1}{2}\begin{pmatrix} 1 + r_1^z \\ 1 - r_1^z \\ 1 + r_1^x \\ 1 + r_1^y \end{pmatrix}, \tag{2.31}$$

and the other one by

$$p_2 = \frac{1}{2}\begin{pmatrix} 1 + r_2^z \\ 1 - r_2^z \\ 1 + r_2^x \\ 1 + r_2^y \end{pmatrix}. \tag{2.32}$$

Constructing mixtures

We can construct a mixture of the two states in the same way we constructed mixtures for classical, randomly fluctuating registers. The state of the mixture is given by

$$\boldsymbol{p} = P_1 \boldsymbol{p}_1 + P_2 \boldsymbol{p}_2, \tag{2.33}$$

where $P_1 + P_2 = 1$. What is going to be the resulting vector $\boldsymbol{r}$ for the mixture? The answer is

$$\boldsymbol{p} = \frac{P_1}{2}\begin{pmatrix} 1 + r_1^z \\ 1 - r_1^z \\ 1 + r_1^x \\ 1 + r_1^y \end{pmatrix} + \frac{P_2}{2}\begin{pmatrix} 1 + r_1^z \\ 1 - r_1^z \\ 1 + r_1^x \\ 1 + r_1^y \end{pmatrix} = \frac{1}{2}\begin{pmatrix} P_1 + P_2 + P_1 r_1^z + P_2 r_2^z \\ P_1 + P_2 - P_1 r_1^z - P_2 r_2^z \\ P_1 + P_2 + P_1 r_1^x + P_2 r_2^x \\ P_1 + P_2 + P_1 r_1^y + P_2 r_2^y \end{pmatrix}.$$

We can make use of $P_1 + P_2 = 1$ to collect this into

$$\boldsymbol{p} = \frac{1}{2}\begin{pmatrix} 1 + r^z \\ 1 - r^z \\ 1 + r^x \\ 1 + r^y \end{pmatrix}, \tag{2.34}$$

where

$$\boldsymbol{r} = P_1 \boldsymbol{r}_1 + P_2 \boldsymbol{r}_2. \tag{2.35}$$

Let us suppose that both states that constitute the mixture, that is, $\boldsymbol{p}_1$ and $\boldsymbol{p}_2$, correspond to fully polarized states, meaning states for which $\boldsymbol{r}_i \cdot \boldsymbol{r}_i = 1$, where $i = 1$ or $i = 2$. Let us see how mixing affects the length of $\boldsymbol{r}$ by evaluating

$$\begin{aligned} \boldsymbol{r} \cdot \boldsymbol{r} &= (P_1 \boldsymbol{r}_1 + P_2 \boldsymbol{r}_2) \cdot (P_1 \boldsymbol{r}_1 + P_2 \boldsymbol{r}_2) \\ &= \left(P_1^2 r_1^2 + 2 P_1 P_2 \boldsymbol{r}_1 \cdot \boldsymbol{r}_2 + P_2^2 r_2^2\right) \\ &= \left(P_1^2 + 2 P_1 P_2 \cos\theta + P_2^2\right), \end{aligned} \tag{2.36}$$

where θ is the angle between $\boldsymbol{r}_1$ and $\boldsymbol{r}_2$. When $\theta = 0$, we get

$$\boldsymbol{r} \cdot \boldsymbol{r} = \left(P_1^2 + 2 P_1 P_2 + P_2^2\right) = (P_1 + P_2)^2 = 1^2 = 1, \tag{2.37}$$

but this is not a mixture, because $\theta = 0$ means that $\boldsymbol{p}_1 = \boldsymbol{p}_2$. For a real mixture, $\theta \neq 0$, which means that $\cos\theta < 1$. In this case we get

$$\boldsymbol{r} \cdot \boldsymbol{r} = \left(P_1^2 + 2 P_1 P_2 \cos\theta + P_2^2\right) < \left(P_1^2 + 2 P_1 P_2 + P_2^2\right) = 1. \tag{2.38}$$

We see that mixing two fully polarized (and *different*) states results in a state with a shorter $\boldsymbol{r}$.

This implies that states that fill the interior of the Bloch sphere are all mixtures. It also confirms that the fully polarized, pure states on the Bloch sphere cannot be produced by mixing other states, because mixing always delivers a state that is located somewhere inside the Bloch sphere. The pure states on the Bloch sphere are extremal, in agreement with our definition of pure states given on page 19. Set S of all possible physical mixtures and pure states of a qubit[17] forms a ball of radius 1. This ball is called the *Bloch ball*.

Bloch ball

Set S of all possible qubit states is quite different from the set of all possible states of a randomly fluctuating classical register. The latter is edgy, with pure states well separated from each other. On the other hand, pure, that is, fully polarized states of a qubit form a smooth continuous surface, the Bloch sphere. The only reversible Cinderella transformations for a randomly fluctuating classical register were permutations of its pure states and the resulting transformations of mixtures. None of them were continuous. On the other hand, we have an infinite number of continuous reversible Cinderella transformations of a qubit, because a sphere can be mapped onto itself in an infinite number of rotations.

Let us consider a mixture state given by $\boldsymbol{r}$ where $\|\boldsymbol{r}\| < 1$. We can always rotate our system of coordinates to align z with $\boldsymbol{r}$, and we can always rotate the chamber that splits the incident neutron beam accordingly. Without loss of generality we can therefore assume that $\boldsymbol{r} = r\boldsymbol{e}_z$ and $0 \leq r < 1$. The fiducial vector that describes the mixture is

$$\boldsymbol{p} = \frac{1}{2}\begin{pmatrix} 1+r \\ 1-r \\ 1 \\ 1 \end{pmatrix}. \tag{2.39}$$

Because $0 \leq r < 1$, we find that $0.5 \leq p^0 < 1$ and $0 < p^1 \leq 0.5$. This result tells us that for the mixture state the incident beam is *always* going to split, even if we happen to adjust the magnetic field inside the chamber so that it is parallel to $\boldsymbol{r}$. There will be always some neutrons that swing to the left, even if most (or at worst a half for $\boldsymbol{r} = \boldsymbol{0}$) swing to the right. The beam is not fully polarized. On the other hand, each of the two beams that leave the beam splitter is fully polarized. The

A mixed beam always splits.

[17] Compare with set S for a classical, randomly fluctuating register, Figure 1.4 on page 17.

one that swings to the left is described by

$$\boldsymbol{p}_{|\uparrow\rangle} = \frac{1}{2} \begin{pmatrix} 1 \\ 0 \\ 0.5 \\ 0.5 \end{pmatrix}, \tag{2.40}$$

and the one that swings to the right is described by

$$\boldsymbol{p}_{|\downarrow\rangle} = \frac{1}{2} \begin{pmatrix} 0 \\ 1 \\ 0.5 \\ 0.5 \end{pmatrix}. \tag{2.41}$$

2.6 The Measurement

Let us consider a general fully polarized state given by

$$\boldsymbol{p} = \frac{1}{2} \begin{pmatrix} 1 + \cos\theta \\ 1 - \cos\theta \\ 1 + \sin\theta\cos\phi \\ 1 + \sin\theta\sin\phi \end{pmatrix}. \tag{2.42}$$

Let us dwell a moment on what happens when the state is subjected to a full set of measurements as described in Section 2.2 on page 47.

Polarized beam splitting

Let us suppose the first measurement splits the beam in the field $\boldsymbol{B} = \alpha x \boldsymbol{e}_z$. The measurement is going to align or counteralign each incident neutron with $\boldsymbol{e}_z$. Approximately $(1 + \cos\theta)/2$ of all incident neutrons will swing to the left and will emerge from the apparatus in state

$$\boldsymbol{p}_{|\uparrow\rangle} = \begin{pmatrix} 1 \\ 0 \\ 0.5 \\ 0.5 \end{pmatrix}. \tag{2.43}$$

The remaining neutrons, $(1 - \cos\theta)/2$ of the incident neutrons, will swing to the right and emerge from the apparatus in state

$$\boldsymbol{p}_{|\downarrow\rangle} = \begin{pmatrix} 0 \\ 1 \\ 0.5 \\ 0.5 \end{pmatrix}. \tag{2.44}$$

We can now merge both beams, creating a new state,

$$\begin{aligned} \boldsymbol{p} &= \frac{1}{2}(1+\cos\theta)\,\boldsymbol{p}_{|\uparrow\rangle} + \frac{1}{2}(1-\cos\theta)\,\boldsymbol{p}_{|\downarrow\rangle} \\ &= \frac{1}{2}\begin{pmatrix} 1+\cos\theta \\ 1-\cos\theta \\ 1 \\ 1 \end{pmatrix}. \end{aligned} \tag{2.45}$$

The corresponding vector $\boldsymbol{r}$ is given by

$$\boldsymbol{r} = \begin{pmatrix} \cos\theta \\ 0 \\ 0 \end{pmatrix}. \tag{2.46}$$

Original state is lost.

The length of the vector is 1 only for $\theta = 0$, which would mean that we have managed to align $\boldsymbol{B}$ inside the splitting apparatus with the original direction of $\boldsymbol{r}$. Otherwise $r^2 < 1$. The state created by merging beams that have left the beam splitter is a mixture. We see here that the measurement destroys the original state, which cannot be reconstructed by merely merging the two issuing beams together. The measurement destroys information about the other angle, ϕ.

Wouldn't we be luckier, then, trying a measurement against $\boldsymbol{B} = \beta z \boldsymbol{e}_x$?

Let us try this option. The measuring apparatus in this case is going to split the beam so that $(1+\sin\theta\cos\phi)/2$ of the incident neutrons swing down. The neutrons leave the apparatus in the $|{\rightarrow}\rangle$ state described by

$$\boldsymbol{p}_{|\rightarrow\rangle} = \begin{pmatrix} 0.5 \\ 0.5 \\ 1 \\ 0.5 \end{pmatrix}. \tag{2.47}$$

The remaining $(1-\sin\theta\cos\phi)/2$ of the incident neutrons swing down, leaving the apparatus in the $|{\leftarrow}\rangle$ state described by

$$\boldsymbol{p}_{|\leftarrow\rangle} = \begin{pmatrix} 0.5 \\ 0.5 \\ 0 \\ 0.5 \end{pmatrix}. \tag{2.48}$$

Merging the two beams together results in the state described by

$$\boldsymbol{p} = \frac{1}{2}(1+\sin\theta\cos\phi)\,\boldsymbol{p}_{|\rightarrow\rangle} + \frac{1}{2}(1-\sin\theta\cos\phi)\,\boldsymbol{p}_{|\leftarrow\rangle}$$

$$= \frac{1}{2}\begin{pmatrix} 1 \\ 1 \\ 1+\sin\theta\cos\phi \\ 1 \end{pmatrix}. \tag{2.49}$$

The corresponding vector $\boldsymbol{r}$ is given by

$$\boldsymbol{r} = \begin{pmatrix} 0 \\ \sin\theta\cos\phi \\ 0 \end{pmatrix}. \tag{2.50}$$

Only for $\theta = 90°$ and $\phi = 0$ is $r^2 = 1$. For all other angles $r^2 < 1$, and so we end up with the mixture again.

Three measurements are required to determine ϕ and θ uniquely.

Carrying out this measurement alone is not going to tell us much about either θ or ϕ. All we are going to see is some $r < 1$ and $\boldsymbol{r}$ pointing in the direction of $\boldsymbol{e}_x$. A great many combinations of θ and ϕ can produce such $\boldsymbol{r} = r\boldsymbol{e}_x$. But if we were to perform the measurement *after* the previous measurement on an identically prepared neutron beam, we would already have θ, and this should give us $\cos\phi$. Yet, knowing $\cos\phi$ still does not let us determine $\boldsymbol{r}$ uniquely, because $\cos\phi = \cos(360° - \phi)$. We would have to find about the sign of $\sin\phi$ in order to determine ϕ uniquely. It is here that the third measurement against $\boldsymbol{B} = \gamma x \boldsymbol{e}_y$ comes in.

Every measurement we have discussed in this section converts a pure state into a mixture unless it happens to *confirm* the pure state. The resulting mixture contains less information than the original state. One has to carry out three independent measurements on the incident beam, ensuring that the state of the beam does not change between the measurements, in order to reconstruct the original state of the beam.

Measuring a mixture

How would we go about measuring a mixture? The procedure discussed above refers to fully polarized states, so we assume from the beginning that $r = 1$. We would still measure the three probabilities that would yield $r^x = r\sin\theta\cos\phi$, $r^y = r\sin\theta\sin\phi$, and $r^z = r\cos\theta$. Squaring and adding the three components yield $(r^x)^2 + (r^y)^2 + (r^z)^2 = r^2$ and hence r. Now we can divide each of the three components by r, and we end up with the same problem we had for the fully polarized state; that is, we get θ from r^z/r, then $\cos\phi$ from r^x/r, and finally the sign of $\sin\phi$ from r^y/r in order to determine ϕ uniquely.

2.7 Pauli Vectors and Pauli Forms

The generic form of the qubit fiducial vector

$$\boldsymbol{p} = \frac{1}{2}\begin{pmatrix} 1 + r^z \\ 1 - r^z \\ 1 + r^x \\ 1 + r^y \end{pmatrix}, \tag{2.51}$$

can be rewritten in the following way:

$$\boldsymbol{p} = \frac{1}{2}\left(\boldsymbol{\varsigma}_1 + r^x \boldsymbol{\varsigma}_x + r^y \boldsymbol{\varsigma}_y + r^z \boldsymbol{\varsigma}_z\right), \tag{2.52}$$

where we are going to call

$$\boldsymbol{\varsigma}_1 = \begin{pmatrix} 1 \\ 1 \\ 1 \\ 1 \end{pmatrix}, \quad \boldsymbol{\varsigma}_x = \begin{pmatrix} 0 \\ 0 \\ 1 \\ 0 \end{pmatrix}, \quad \boldsymbol{\varsigma}_y = \begin{pmatrix} 0 \\ 0 \\ 0 \\ 1 \end{pmatrix}, \quad \boldsymbol{\varsigma}_z = \begin{pmatrix} 1 \\ -1 \\ 0 \\ 0 \end{pmatrix} \tag{2.53}$$

Pauli vectors and Pauli matrices

Pauli vectors. Although Pauli didn't invent these four vectors, they are closely related to Pauli matrices, as we shall see in Chapters 3 and 4, and fulfill a similar role to Pauli matrices within the framework of the fiducial formalism. We are going to use the wiggly symbol $\boldsymbol{\varsigma}$ (pronounced "varsigma") for Pauli vectors, bowing to tradition, because its close relative, the Greek letter $\boldsymbol{\sigma}$ (pronounced "sigma"), is used commonly to denote Pauli matrices.

Pauli vectors can be represented in terms of canonical basis vectors of the fiducial space,

$$\boldsymbol{e}_0 = \begin{pmatrix} 1 \\ 0 \\ 0 \\ 0 \end{pmatrix}, \boldsymbol{e}_1 = \begin{pmatrix} 0 \\ 1 \\ 0 \\ 0 \end{pmatrix}, \boldsymbol{e}_2 = \begin{pmatrix} 0 \\ 0 \\ 1 \\ 0 \end{pmatrix}, \boldsymbol{e}_3 = \begin{pmatrix} 0 \\ 0 \\ 0 \\ 1 \end{pmatrix}, \tag{2.54}$$

as follows:

$$\boldsymbol{\varsigma}_1 = \boldsymbol{e}_0 + \boldsymbol{e}_1 + \boldsymbol{e}_2 + \boldsymbol{e}_3, \tag{2.55}$$

$$\boldsymbol{\varsigma}_x = \boldsymbol{e}_2, \tag{2.56}$$

$$\boldsymbol{\varsigma}_y = \boldsymbol{e}_3, \tag{2.57}$$

$$\boldsymbol{\varsigma}_z = \boldsymbol{e}_0 - \boldsymbol{e}_1. \tag{2.58}$$

This can be easily inverted to yield $\boldsymbol{e}_i$ in terms of $\boldsymbol{\varsigma}_i$:

$$\boldsymbol{e}_0 = \frac{1}{2}\left(\boldsymbol{\varsigma}_1 + \boldsymbol{\varsigma}_z - \boldsymbol{\varsigma}_x - \boldsymbol{\varsigma}_y\right), \tag{2.59}$$

$$
\begin{aligned}
\boldsymbol{e}_1 &= \frac{1}{2}\left(\varsigma_1 - \varsigma_z - \varsigma_x - \varsigma_y\right), & (2.60)\\
\boldsymbol{e}_2 &= \varsigma_x, & (2.61)\\
\boldsymbol{e}_3 &= \varsigma_y. & (2.62)
\end{aligned}
$$

Pauli forms

We are also going to introduce four Pauli forms defined by

$$
\begin{aligned}
\varsigma^1 &= (1, 1, 0, 0), & (2.63)\\
\varsigma^x &= (-1, -1, 2, 0), & (2.64)\\
\varsigma^y &= (-1, -1, 0, 2), & (2.65)\\
\varsigma^z &= (1, -1, 0, 0). & (2.66)
\end{aligned}
$$

Duality

Pauli forms are dual to Pauli vectors, meaning that

$$\langle \varsigma^i, \varsigma_j \rangle = 2\delta^i{}_j, \quad \text{for} \quad i, j = 1, x, y, z. \tag{2.67}$$

From this we can easily derive that

$$
\begin{aligned}
\langle \varsigma^1, \boldsymbol{p} \rangle &= 1, & (2.68)\\
\langle \varsigma^x, \boldsymbol{p} \rangle &= r^x, & (2.69)\\
\langle \varsigma^y, \boldsymbol{p} \rangle &= r^y, & (2.70)\\
\langle \varsigma^z, \boldsymbol{p} \rangle &= r^z. & (2.71)
\end{aligned}
$$

Extracting $\boldsymbol{r}$ from $\boldsymbol{p}$

We can therefore use Pauli forms as devices for extracting r^x, r^y, and r^z from arbitrary fiducial vectors $\boldsymbol{p}$.

It is easy to express canonical forms,

$$
\begin{aligned}
\boldsymbol{\omega}^0 &= (1, 0, 0, 0), & (2.72)\\
\boldsymbol{\omega}^1 &= (0, 1, 0, 0), & (2.73)\\
\boldsymbol{\omega}^2 &= (0, 0, 1, 0), & (2.74)\\
\boldsymbol{\omega}^3 &= (0, 0, 0, 1), & (2.75)
\end{aligned}
$$

in terms of Pauli forms and vice versa:

$$
\begin{aligned}
\boldsymbol{\omega}^0 &= \frac{1}{2}\left(\varsigma^1 + \varsigma^z\right), & (2.76)\\
\boldsymbol{\omega}^1 &= \frac{1}{2}\left(\varsigma^1 - \varsigma^z\right), & (2.77)\\
\boldsymbol{\omega}^2 &= \frac{1}{2}\left(\varsigma^1 + \varsigma^x\right), & (2.78)
\end{aligned}
$$

$$\omega^3 = \frac{1}{2}\left(\varsigma^1 + \varsigma^y\right), \tag{2.79}$$

and

$$\varsigma^1 = \omega^0 + \omega^1, \tag{2.80}$$
$$\varsigma^x = -\omega^0 - \omega^1 + 2\omega^2, \tag{2.81}$$
$$\varsigma^y = -\omega^0 - \omega^1 + 2\omega^3, \tag{2.82}$$
$$\varsigma^z = \omega^0 - \omega^1. \tag{2.83}$$

The bases of Pauli vectors and forms

But we can do more with Pauli forms and vectors, since they form natural bases in the fiducial vector and form spaces of the qubit.

Metric tensor

The first thing we are going to do with them is to form the metric tensors in the fiducial vector and form spaces. The metric tensor $\boldsymbol{g}$ in the fiducial vector space is defined by

$$\boldsymbol{g} = \frac{1}{2}\left(\varsigma_1 \otimes \varsigma_1 + \varsigma_x \otimes \varsigma_x + \varsigma_y \otimes \varsigma_y + \varsigma_z \otimes \varsigma_z\right). \tag{2.84}$$

Its counterpart in the fiducial form space, $\tilde{\boldsymbol{g}}$, is defined by

$$\tilde{\boldsymbol{g}} = \frac{1}{2}\left(\varsigma^1 \otimes \varsigma^1 + \varsigma^x \otimes \varsigma^x + \varsigma^y \otimes \varsigma^y + \varsigma^z \otimes \varsigma^z\right). \tag{2.85}$$

Metric tensors $\boldsymbol{g}$ and $\tilde{\boldsymbol{g}}$ can be used to convert vectors to forms and vice versa. Let us observe the following:

$$\langle \tilde{\boldsymbol{g}}, \varsigma_j \rangle = \langle \frac{1}{2} \sum_{i=1,x,y,z} \varsigma^i \otimes \varsigma^i, \varsigma_j \rangle = \frac{1}{2} \sum_{i=1,x,y,z} \varsigma^i 2\delta^i{}_j = \varsigma^j. \tag{2.86}$$

Similarly

$$\langle \varsigma^i, \boldsymbol{g} \rangle = \langle \varsigma^i, \frac{1}{2} \sum_{j=1,x,y,z} \varsigma_j \otimes \varsigma_j \rangle = \frac{1}{2} \sum_{j=1,x,y,z} 2\delta^i{}_j \varsigma_j = \varsigma_i. \tag{2.87}$$

Converting vectors to forms

Without much ado we can use the above to convert the fiducial vector $\boldsymbol{p}$ to its dual fiducial form $\tilde{\boldsymbol{p}}$ [18]:

$$\tilde{\boldsymbol{p}} = \langle \tilde{\boldsymbol{g}}, \boldsymbol{p} \rangle$$

[18] The notation $r^x\varsigma^x + r^y\varsigma^y + r^z\varsigma^z$ does not look quite as elegant as $r^x\varsigma_x + r^y\varsigma_y + r^z\varsigma_z$ because indexes x, y, and z are on the same level, instead of being placed on the alternate levels, as we have emphasized in Section 1.7. The reason is that form $\tilde{\boldsymbol{p}}$ has been generated by conversion from vector $\boldsymbol{p}$. Still, we can rescue the situation by lowering indexes on $\boldsymbol{r}$, that is, by rewriting the $\boldsymbol{r}$ dependent part of the form as $r_x\varsigma^x + r_y\varsigma^y + r_z\varsigma^z$. We can do so because in orthonormal (Cartesian) coordinates $r^i = r_i$ for $i = x, y, z$.

$$\begin{aligned} &= \frac{1}{2}\left(\varsigma^1 + r^x\varsigma^x + r^y\varsigma^y + r^z\varsigma^z\right) \\ &= \frac{1}{2}\left(1 - r^x - r^y + r^z, 1 - r^x - r^y - r^z, 2r^x, 2r^y\right). \end{aligned} \tag{2.88}$$

It works in the other direction, too:

$$\boldsymbol{p} = \langle \tilde{\boldsymbol{p}}, \boldsymbol{g} \rangle. \tag{2.89}$$

Length of $\boldsymbol{p}$

The metric tensor $\tilde{\boldsymbol{g}}$ can be used to evaluate the length of vectors in the fiducial space, hence its name, "the metric." To evaluate the length of $\boldsymbol{p}$, we proceed as follows:

$$\begin{aligned} \langle \tilde{\boldsymbol{g}}, \boldsymbol{p} \otimes \boldsymbol{p} \rangle &= \langle \frac{1}{2} \sum_{i=1,x,y,z} \varsigma^i \otimes \varsigma^i, \boldsymbol{p} \otimes \boldsymbol{p} \rangle \\ &= \frac{1}{2}\left(1 \cdot 1 + r^x r^x + r^y r^y + r^z r^z\right) \\ &= \frac{1}{2}\left(1 + \boldsymbol{r} \cdot \boldsymbol{r}\right). \end{aligned} \tag{2.90}$$

The same can be obtained by evaluating $\langle \tilde{\boldsymbol{p}}, \boldsymbol{p} \rangle$ or

$$\frac{1}{2}\left(1 - r^x - r^y + r^z, 1 - r^x - r^y - r^z, 2r^x, 2r^y\right) \cdot \frac{1}{2} \begin{pmatrix} 1 + r^z \\ 1 - r^z \\ 1 + r^x \\ 1 + r^y \end{pmatrix}. \tag{2.91}$$

For fully polarized states, the pure states, we have that $\boldsymbol{r} \cdot \boldsymbol{r} = 1$ and so

$$\langle \tilde{\boldsymbol{p}}, \boldsymbol{p} \rangle = 1. \tag{2.92}$$

For completely chaotic states, that is, states for which $\boldsymbol{r} = \boldsymbol{0}$

$$\langle \tilde{\boldsymbol{p}}, \boldsymbol{p} \rangle = \frac{1}{2}. \tag{2.93}$$

Tensors $\boldsymbol{g}$ and $\tilde{\boldsymbol{g}}$ have matrix representations, which can be derived from fiducial representations of Pauli vectors and forms. The procedure is somewhat laborious, but eventually two symmetric matrices come out:

$$\boldsymbol{g} \equiv \frac{1}{2} \begin{pmatrix} 2 & 0 & 1 & 1 \\ 0 & 2 & 1 & 1 \\ 1 & 1 & 2 & 1 \\ 1 & 1 & 1 & 2 \end{pmatrix} \tag{2.94}$$

and

$$\tilde{g} \equiv \frac{1}{2} \begin{pmatrix} 4 & 2 & -2 & -2 \\ 2 & 4 & -2 & -2 \\ -2 & -2 & 4 & 0 \\ -2 & -2 & 0 & 4 \end{pmatrix}. \tag{2.95}$$

One can easily check that $\|g\| = \|\tilde{g}\|^{-1}$, that is, that the matrix that corresponds to g is the inverse of the matrix that corresponds to $\tilde{g}$.

But there is a better way to see that g and $\tilde{g}$ are each other's inverses. Let us consider $\langle \tilde{g}, g \rangle$ not contracted fully, but instead contracted on one vector and one form only. The result is

$$\boldsymbol{\delta} = \frac{1}{2} \left(\varsigma^1 \otimes \varsigma_1 + \varsigma^x \otimes \varsigma_x + \varsigma^y \otimes \varsigma_y + \varsigma^z \otimes \varsigma_z \right). \tag{2.96}$$

The reason there is only one $\frac{1}{2}$ in front is that the other one cancels with the factor of 2 thrown out by the contractions. It is now easy to see that this new object is simply the Kronecker delta or, in other words, the identity, because it converts an arbitrary fiducial vector $\boldsymbol{p}$ into itself:

$$\langle \boldsymbol{\delta}, \boldsymbol{p} \rangle = \frac{1}{2} \sum_{i=1,x,y,z} \langle \varsigma^i, \boldsymbol{p} \rangle \varsigma_i = \frac{1}{2} \sum_{i=1,x,y,z} 2p^i \varsigma_i = \boldsymbol{p}. \tag{2.97}$$

Extracting $\boldsymbol{r}$ from $\tilde{\boldsymbol{p}}$

As we used Pauli forms ς^1 through ς^z to extract vector $\boldsymbol{r}$ from $\boldsymbol{p}$, we can use Pauli vectors ς_1 through ς_z to extract $\boldsymbol{r}$ from $\tilde{\boldsymbol{p}}$:

$$\langle \tilde{\boldsymbol{p}}, \varsigma_1 \rangle = 1, \tag{2.98}$$
$$\langle \tilde{\boldsymbol{p}}, \varsigma_x \rangle = r_x, \tag{2.99}$$
$$\langle \tilde{\boldsymbol{p}}, \varsigma_y \rangle = r_y, \tag{2.100}$$
$$\langle \tilde{\boldsymbol{p}}, \varsigma_z \rangle = r_z. \tag{2.101}$$

2.8 The Hamiltonian Form

The Hamiltonian form for a qubit is given by[19] [20]

$$\boldsymbol{\eta} = -\mu \left(B_x \varsigma^x + B_y \varsigma^y + B_z \varsigma^z \right). \tag{2.102}$$

[19]Here we keep indexes x, y, and z on $\boldsymbol{B}$ on the lower level because we really want to use $\boldsymbol{B}$ as a form, not as a vector. This makes no difference computationally, because $B_i = B^i, i = x, y, z$, in orthonormal (Cartesian) coordinates. See footnote 18 on page 67.

[20] There is a reason we have chosen $\boldsymbol{\eta}$ for the Hamiltonian form. The usual letter used for the Hamiltonian operator in quantum mechanics is $\boldsymbol{H}$. But we need a lower-case Greek letter here. Since H happens to be the Greek capital letter version of "eta," the lower case of which is η, we end up with $\boldsymbol{\eta}$ for the Hamiltonian *form*. The Greek word from which the word "energy" derives is $\varepsilon\nu\varepsilon\rho\gamma\varepsilon\iota\alpha$, which means "activity" or "effect." Its first letter is ε, not η.

If the qubit is implemented as a beam of neutrons, then μ is the magnetic moment of the neutron and $\boldsymbol{B} = B_x \boldsymbol{e}_x + B_y \boldsymbol{e}_y + B_z \boldsymbol{e}_z$ is the magnetic field. But a qubit may be implemented in other ways too, and then the physical meaning of μ and $\boldsymbol{B}$ is bound to be different. Yet the Hamiltonian form still looks the same.

Extracting $\boldsymbol{B}$ from $\boldsymbol{\eta}$

Coefficients B_x, B_y and B_z can be extracted from the Hamiltonian form $\boldsymbol{\eta}$ by the application of Pauli vectors to it:

$$\begin{aligned} \langle \boldsymbol{\eta}, \varsigma_1 \rangle &= 0, & (2.103) \\ \langle \boldsymbol{\eta}, \varsigma_x \rangle &= -2\mu B_x, & (2.104) \\ \langle \boldsymbol{\eta}, \varsigma_y \rangle &= -2\mu B_y, & (2.105) \\ \langle \boldsymbol{\eta}, \varsigma_z \rangle &= -2\mu B_z. & (2.106) \end{aligned}$$

The reason we get $-2\mu B_x$ instead of just $-\mu B_x$ is that there is no $1/2$ in front of $\boldsymbol{\eta}$, and let us remember that $\langle \varsigma^i, \varsigma_i \rangle = 2$ for $i = x, y, z$.

The average energy

The Hamiltonian form is used to calculate the average value of the energy over the statistical ensemble of a qubit described by the fiducial vector $\boldsymbol{p}$. The formula looks the same as formulas we have derived for classical randomly fluctuating registers in Section 1.6, page 21:

$$\begin{aligned} \langle E \rangle &= \langle \boldsymbol{\eta}, \boldsymbol{p} \rangle \\ &= \left\langle -\mu \left(B_x \varsigma^x + B_y \varsigma^y + B_z \varsigma^z \right), \frac{1}{2} \left(\varsigma_1 + r^x \varsigma_x + r^y \varsigma_y + r^z \varsigma_z \right) \right\rangle \\ &= -\mu \left(B_x r^x + B_y r^y + B_z r^z \right). \end{aligned} \quad (2.107)$$

Similarity to the classical formula

The expression looks exactly like the classical expression that describes the energy of the magnetic dipole $\boldsymbol{\mu}$ in the magnetic field $\boldsymbol{B}$

$$E = -\boldsymbol{\mu} \cdot \boldsymbol{B} \quad (2.108)$$

if we identify

$$\boldsymbol{\mu} = -\mu \begin{pmatrix} r^x \\ r^y \\ r^z \end{pmatrix}. \quad (2.109)$$

But we must remember that in the world of quantum mechanics the qubit polarization vector $\boldsymbol{r}$ parameterizes probability measurements made on the statistical ensemble of, for example, neutrons and does not represent the space orientation of the individual neutron—about which we know nothing. The giveaway that distinguishes the quantum formula from its classical cousin is the angular brackets around the E.

As we could convert the fiducial vector $\boldsymbol{p}$ to the fiducial form $\tilde{\boldsymbol{p}}$ by contracting $\boldsymbol{p}$ with metric $\tilde{\boldsymbol{g}}$, similarly we can convert the Hamiltonian form to the Hamiltonian vector by contracting the form with metric $\boldsymbol{g}$,

Hamiltonian vector

$$\tilde{\boldsymbol{\eta}} = \langle \boldsymbol{\eta}, \boldsymbol{g} \rangle = -\mu \left(B_x \varsigma_x + B_y \varsigma_y + B_z \varsigma_z \right). \tag{2.110}$$

The measurement of the average energy on the statistical ensemble that corresponds to $\boldsymbol{p}$ can be then expressed also in this way:

$$\langle E \rangle = \langle \tilde{\boldsymbol{p}}, \tilde{\boldsymbol{\eta}} \rangle. \tag{2.111}$$

Forms and vectors are interchangeable—as long as we remember to interchange both at the same time!

It is instructive to evaluate the Hamiltonian form on the basis states $|\uparrow\rangle$ and $|\downarrow\rangle$. Let us recall that neutrons in the beam always align or counteralign with the direction of the magnetic field. The basis states $|\uparrow\rangle$ and $|\downarrow\rangle$ can therefore be observed if a neutron is placed in $\boldsymbol{B} = B\boldsymbol{e}_z$. For the $|\uparrow\rangle$ state we have $\boldsymbol{r} = \boldsymbol{e}_z$, and for the $|\downarrow\rangle$ state we have $\boldsymbol{r} = -\boldsymbol{e}_z$. Hence

Energy of the basis states

$$\langle \boldsymbol{\eta}, \uparrow \rangle = -\mu B = |\mu|\, B, \tag{2.112}$$

and

$$\langle \boldsymbol{\eta}, \downarrow \rangle = \mu B = -\,|\mu|\, B, \tag{2.113}$$

where we have used the absolute value of μ and the explicit sign, to remind the reader that the neutron's μ is negative. The aligned state is the higher energy state of the two.

The energy difference between the two states is

$$\Delta E = E_{|\uparrow\rangle} - E_{|\downarrow\rangle} = 2\,|\mu|\, B. \tag{2.114}$$

Let us suppose we take a neutron in the $|\uparrow\rangle$ state and stick it in the chamber filled with $\boldsymbol{B} = B\boldsymbol{e}_z$. Initially the neutron is going to maintain its original orientation of the magnetic moment. But this orientation is not stable, because there is another orientation possible with lower energy, namely, $|\downarrow\rangle$. Consequently the neutron will eventually flip to the lower energy state, releasing a photon of energy $2\,|\mu|\, B$ in the process.

The photon release that results from the spin flip is called the energy *dissipation* event. It is quite unpredictable, although the *half-life* of the higher energy state can be measured for various circumstances. The energy dissipation process is among the main causes of errors in quantum computing, so a better understanding of this

Energy dissipation

phenomenon is clearly called for. We shall present a simplified model of dissipation in Chapter 5.

The frequency of light emitted in the dissipation process is given by

$$\Delta E = 2\,|\mu|\,B = \hbar\omega, \tag{2.115}$$

which yields

$$\omega = \frac{2\,|\mu|\,B}{\hbar}. \tag{2.116}$$

The important point to note here is that we never observe neutrons emitting smaller portions of energy in this process (which would result in $\omega < 2\,|\mu|\,B/\hbar$). In other words, the flip does not proceed by going through some intermediate states, which we could interpret as various angles of tilt between "up" and "down." The neutron switches instantaneously between the two opposite and discrete configurations *without* going through intermediate states, because ... there aren't any to go through. If there were, we would have seen them in the beam-splitting experiment, and our vector $\boldsymbol{p}$ would have to have a larger number of slots.

2.9 Qubit Evolution

Deterministic evolution of probabilities

Among the most astonishing discoveries of quantum physics is that whereas the behavior of any individual microscopic system is chaotic and in general unpredictable, yet probability distributions that pertain to microscopic systems can be described with great precision and their evolution predicted accurately by the means of deterministic differential equations.

This is not an entirely new situation in physics. For example, the classical diffusion equation is deterministic, yet it can be derived from a microscopic picture that assumes completely random molecular motion.

On the other hand, in quantum mechanics the question "where do quantum probabilities come from" remains unanswered,[21] and, while an orthodox pronouncement states that quantum probabilities are fundamental and unexplainable, most physicists and chemists, who work with quantum mechanics daily, have a vague picture in their mind that associates the wave function of a particle with a sort of tension traveling through space—wherever the tension is greater, the particle is more likely

[21] Although the question remains unanswered, various answers have been proposed. These range from the de Broglie-Bohm pilot wave theory [14] through the Everett's many-worlds interpretation of quantum mechanics [31] and more. The reason for our statement that this remains an unsolved problem is that none of the proposed solutions, which are all mathematically sound, has been verified experimentally.

to materialize. In the absence of macroscopic objects the particle is dissolved in the "tension wave" and nonlocal, but as soon as it encounters a "measuring apparatus" (it can as well be a large lump of matter without any dials—the particle doesn't care), the whole shebang shrinks and, pronto, we get a pointlike energy transfer: the particle has been registered.

Lightning strike analogy

This is somewhat similar to a lightning strike, which, as we know, results from large-scale distribution of charge in the storm cloud and in the ground. Yet when the lightning strikes, the discharge is pointlike compared to the size of the storm cloud. An elementary particle in its pointlike manifestation can be thought of as a lightning strike between its "probability cloud" and a measuring apparatus or some other macroscopic system it interacts with. Once the charge in the lightning has been transferred along the lightning channel, it's gone. It has been neutralized and can no longer be retrieved. So is an elementary particle, once it has been registered. We can no longer pick it up in tweezers and put it back into its probability cloud. A pointlike particle, like the lightning, is a transient phenomenon. Its probability cloud, like the storm cloud, is not.

Professional physicists and chemists are normally careful not to divulge, especially to their colleagues and students, their feelings on this matter [8], and just stick to the lore. Only in the most intimate moments of marital bliss may a spouse overhear the physicist or the chemist uttering an illicit thought about quantum ontology during sleep or in the shower.

Limited applicability of the Schrödinger equation

The equation we are going to introduce in this section does not describe every aspect of microscopic behavior. For example, it does not describe the measurement process or the dissipation process. To analyze these processes, we will have to advance to multiqubit systems, because both derive from interactions of a qubit with its environment. The environment can be modeled by other qubits our selected qubit interacts with. In simple, though quite revealing, models it is enough to add just a few more qubits to the description.

The evolution equation of a qubit can be guessed as follows. We seek the simplest possible equation that describes the evolution of the polarization vector $\boldsymbol{r}$ in the presence of an externally applied field $\boldsymbol{B}$, such that the polarization state of the qubit, meaning the length of vector $\boldsymbol{r}$, does not change. If the qubit is fully polarized initially, it remains fully polarized.

We shall see toward the end of this chapter that this is an idealization. Real qubits do not behave like this. They normally lose their polarization. But this idealization is no different from equally unrealistic Newton equations that neglect friction or from the Lorentz force equation. No classical charges can possibly move following the Lorentz equation, because accelerated charges radiate and thus lose energy. But the Lorentz equation idealization is useful, and radiative effects are normally added to it as corrections, although an exact classical equation does exist, due to Lorentz, Abraham, and Dirac, that includes radiative corrections to a motion of a charged point particle automatically [122].

The simplest evolution of vector $\boldsymbol{r}$ in the presence of vector $\boldsymbol{B}$, so that the length of $\boldsymbol{r}$ does not change, is rotation or precession of $\boldsymbol{r}$ about $\boldsymbol{B}$. This, in turn, is accomplished if the infinitesimal change of $\boldsymbol{r}$ in time $\mathrm{d}t$ is perpendicular to $\boldsymbol{r}$. One vector made of a combination of $\boldsymbol{r}$ and $\boldsymbol{B}$ that's always perpendicular to $\boldsymbol{r}$ is

$$\boldsymbol{r} \times \boldsymbol{B}. \tag{2.117}$$

We can therefore postulate that

$$\frac{\mathrm{d}}{\mathrm{d}t}\boldsymbol{r} \propto \boldsymbol{r} \times \boldsymbol{B}. \tag{2.118}$$

The coefficient of proportionality should have μ in it because, after all, this is how the qubit couples to the magnetic field. But μB is energy, measured in joules, whereas what we have on the left-hand side of equation (2.118) is meters per second. We have meters on the right-hand side, too, in the form of $\boldsymbol{r}$. To get s^{-1} that is still missing, we should divide μB (joules) by something with the dimensions of joules $\cdot$ second. The obvious candidate is the Planck constant, because it has just the right dimension, due to $E = \hbar\omega$. In other words,

$$\frac{\mathrm{d}}{\mathrm{d}t}\boldsymbol{r} \propto \frac{\mu}{\hbar}\boldsymbol{r} \times \boldsymbol{B}. \tag{2.119}$$

The units on both sides of the equation match now. Whatever other proportionality coefficient enters the equation, it is dimensionless. The factor must be such that the precession frequency, which we are going to figure out in the next section, corresponds to the frequency of light emitted by the qubit when it flips between its two quantum energy states $|\uparrow\rangle$ and $|\downarrow\rangle$. The energy difference between them, as we have seen in equation (2.114), is $2|\mu|B$, and so the missing numerical factor is 2, and our evolutionary equation for the qubit becomes

$$\frac{\mathrm{d}}{\mathrm{d}t}\boldsymbol{r} = \frac{2\mu}{\hbar}\boldsymbol{r} \times \boldsymbol{B}. \tag{2.120}$$

Some people sneer at this equation, calling it "semiclassical," which is considered a great insult in quantum physics. But, as we'll see later, this equation is entirely equivalent to the much celebrated Schrödinger equation and at the same time is easier to understand and verify. After all, it describes in direct terms the evolution of neutron beam polarization in the presence of the external magnetic field—polarization being what fully characterizes the qubit and what is measured in the Stern-Gerlach experiment.

It is easy to observe that equation (2.120) does not change the length of vector $\boldsymbol{r}$, namely,

$\boldsymbol{r} \cdot \boldsymbol{r}$ is preserved.

$$\frac{\mathrm{d}}{\mathrm{d}t}(\boldsymbol{r} \cdot \boldsymbol{r}) = 2\boldsymbol{r} \cdot \frac{\mathrm{d}}{\mathrm{d}t}\boldsymbol{r} = \frac{4\mu}{\hbar}\boldsymbol{r} \cdot (\boldsymbol{r} \times \boldsymbol{B}) = 0. \tag{2.121}$$

States that are pure in the beginning remain pure. Because the measurement process normally converts pure states into mixtures, equation (2.120) clearly cannot describe it in its present form.

In the next two sections we shall see that what equation (2.120) *does* describe can still be quite complex. We shall encounter a fair amount of interesting physics that is directly applicable to quantum computing.

Quantum gates

2.10 Larmor Precession

Let us consider a qubit described by the fiducial vector $\boldsymbol{p}$ in the presence of the static (unchanging in time) and uniform (unchanging in space) field $\boldsymbol{B} = B\boldsymbol{e}_z$. Substituting $B\boldsymbol{e}_z$ for $\boldsymbol{B}$ in equation (2.120) results in

$$\frac{\mathrm{d}r^x}{\mathrm{d}t} = \frac{2\mu}{\hbar} r^y B, \tag{2.122}$$

$$\frac{\mathrm{d}r^y}{\mathrm{d}t} = -\frac{2\mu}{\hbar} r^x B, \tag{2.123}$$

$$\frac{\mathrm{d}r^z}{\mathrm{d}t} = 0. \tag{2.124}$$

We assume that the initial state of the qubit is fully polarized and so, as equation (2.121) tells us, the qubit remains fully polarized. Since the tip of the arrow touches the surface of the Bloch sphere at all times, we can parameterize the $\boldsymbol{r}$ vector by the two angles θ and ϕ shown in Figure 2.5:

$$\boldsymbol{r} = \begin{pmatrix} \cos\phi(t)\sin\theta(t) \\ \pm\sin\phi(t)\sin\theta(t) \\ \cos\theta(t) \end{pmatrix}. \tag{2.125}$$

The sign of the second component depends on whether the qubit rotates clockwise or counterclockwise.

Equation (2.124) is the easiest to solve. It says that $r^z = \cos\theta$ is constant and hence θ is constant, which implies that the probabilities of finding the qubit in states $|\uparrow\rangle$ and $|\downarrow\rangle$ are constant as well.

p^0 and p^1 remain unaffected.

Harmonic oscillator

Equations (2.122) and (2.123) are coupled in the same way that the position and the velocity are coupled in the equations of the harmonic oscillator. To show this, we take the second derivative of r^x with respect to time:

$$\frac{\mathrm{d}^2 r^x}{\mathrm{d}t^2} = \frac{\mathrm{d}}{\mathrm{d}t}\frac{\mathrm{d}r^x}{\mathrm{d}t} = \frac{2\mu B}{\hbar}\frac{\mathrm{d}r^y}{\mathrm{d}t} = -\left(\frac{2\mu B}{\hbar}\right)^2 r^x. \tag{2.126}$$

Similarly, for r^y

$$\frac{\mathrm{d}^2 r^y}{\mathrm{d}t^2} = -\left(\frac{2\mu B}{\hbar}\right)^2 r^y. \tag{2.127}$$

Larmor frequency

Equation (2.126) can be satisfied by the following ansatz:

$$r^x(t) = r_0^x \cos\omega_L t. \tag{2.128}$$

In other words, we say that $\phi(t) = \omega_L t$. The ansatz assumes implicitly that at $t = 0$ $r^x = r_0^x = \sin\theta$ and $\mathrm{d}r^x/\mathrm{d}t = 0$. The constant ω_L is given by

$$\omega_L = \frac{2\mu B}{\hbar} \tag{2.129}$$

and is called the *Larmor frequency* after the Irish physicist Sir Joseph Larmor (1857–1942), who was the first to explain the splitting of spectral lines by the magnetic field.

We can obtain r^y from equation (2.122), which states that

$$r^y = \frac{1}{\omega_L}\frac{\mathrm{d}}{\mathrm{d}t}r^x = -\frac{1}{\omega_L}\omega_L r_0^x \sin\omega_L t = -r_0^x \sin\omega_L t. \tag{2.130}$$

In summary, this is our solution:

$$r^x = \sin\theta\cos\omega_L t, \tag{2.131}$$

$$r^y = -\sin\theta\sin\omega_L t, \tag{2.132}$$

$$r^z = \cos\theta. \tag{2.133}$$

Larmor precession

The solution describes vector $\boldsymbol{r}$ that rotates about the direction of $\boldsymbol{B}$ in such a way that its projection on the direction of $\boldsymbol{B}$ remains constant; see Figure 2.6. Such a motion is called a precession; hence this phenomenon is referred to as the Larmor precession.

Although the probabilities of finding the qubit in $|\uparrow\rangle$ and $|\downarrow\rangle$ are constant, the probabilities of finding the qubit in states $|\rightarrow\rangle$ and $|\otimes\rangle$ change all the time. The state remains fully polarized, that is, pure, but the direction of its polarization precesses around $\boldsymbol{B}$.

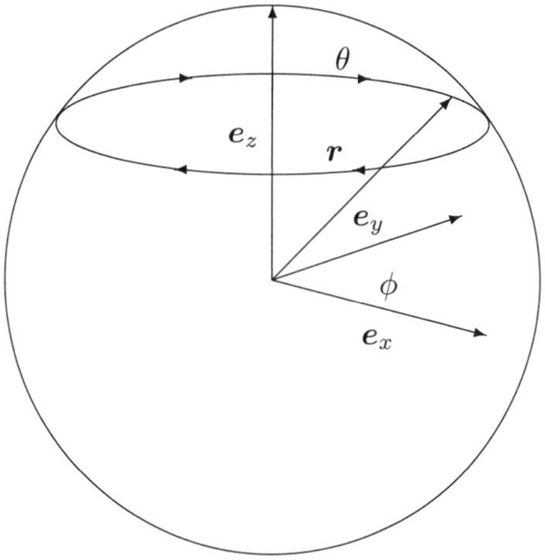

Figure 2.6: Larmor precession

The Larmor precession occurs with the angular frequency of $\omega_L = 2\mu B/\hbar$, which is the same as the frequency of the photon emitted as the result of the dissipative transition from $|\uparrow\rangle$ to $|\downarrow\rangle$ (cf. equation (2.116) on page 72). Well, we have, after all, concocted our evolutionary equation (2.120) to make it so.

Probabilities

To evaluate probabilities of finding the qubit in various basis states, we have to substitute solutions (2.131), (2.132), and (2.133) into the fiducial vector $\boldsymbol{p}$.

The probability of finding the qubit in state $|\uparrow\rangle$ is

$1 + \cos\alpha = 2\cos^2\frac{\alpha}{2}$

$$p^0 = \frac{1}{2}\left(1 + r^z\right) = \frac{1}{2}\left(1 + \cos\theta\right) = \cos^2\frac{\theta}{2}. \tag{2.134}$$

The probability of finding the qubit in state $|\downarrow\rangle$ is

$1 - \cos\alpha = 2\sin^2\frac{\alpha}{2}$

$$p^1 = \frac{1}{2}\left(1 - r^z\right) = \frac{1}{2}\left(1 - \cos\theta\right) = \sin^2\frac{\theta}{2}. \tag{2.135}$$

The probability of finding the qubit in state $|\rightarrow\rangle$ is

$$p^2 = \frac{1}{2}\left(1 + r^x\right) = \frac{1}{2}\left(1 + \sin\theta\cos\omega_L t\right), \tag{2.136}$$

and the probability of finding it in state $|\otimes\rangle$ is

$$p^3 = \frac{1}{2}\left(1 + r^y\right) = \frac{1}{2}\left(1 - \sin\theta\sin\omega_L t\right). \tag{2.137}$$

Larmor precession does not measure the qubit.

We note that just placing the qubit in a static and uniform magnetic field $\boldsymbol{B} = B\boldsymbol{e}_z$ does not in itself result in the measurement, which is why the qubit remains in a pure state, although the state itself varies with time. To measure the qubit, we have to pass it through a chamber where magnetic field has a nonvanishing gradient. Another way to measure the qubit is to measure its own magnetic field with highly sensitive Helmholtz coils, which is how the results of computations are read out in nuclear magnetic resonance experiments.

If the qubit is implemented by other means, that is, not as a microscopic magnet, the physical meaning of $\boldsymbol{B}$ and μ are different, and then appropriate measurement methods have to be devised.

Larmor precession as a parasitic effect

The Larmor precession can be a parasitic effect in the context of quantum computing. Usually we want our qubits to stay just as they are. It is sometimes possible to just switch $\boldsymbol{B}$ off between applications of various quantum gates. This, again, depends on how qubits are implemented. In nuclear magnetic resonance experiments $\boldsymbol{B}$ cannot be switched off. In this case an elaborate procedure called refocusing must be deployed to cancel the effects of Larmor precession in the final account. We shall discuss this procedure in Section 7.2.

Refocusing

2.11 Rabi Oscillations

The Larmor precession does not change the proportions of $|\uparrow\rangle$ to $|\downarrow\rangle$ in a qubit state described by $\boldsymbol{p}$. If we were to make the associations $|\uparrow\rangle \equiv |\, 0\rangle$ and $|\downarrow\rangle \equiv |\, 1\rangle$, then the precessing qubit would stay put, as far as its computational value is concerned, even though something would keep going inside it, so to speak. But the Larmor precession may have an effect if the qubit is subjected to manipulations that depend on the exact position of $\boldsymbol{r}$.

How to flip a qubit

The question for this section is how we can change the proportion of $|\uparrow\rangle$ to $|\downarrow\rangle$ in a qubit, and, in particular, how we can flip $|\uparrow\rangle$ into $|\downarrow\rangle$ and vice versa in a controlled way, that is, without waiting for a dissipation event to occur.

This can be accomplished in various ways. The simplest and the most commonly practiced way is to apply a small magnetic oscillation in the plane perpendicular to the background magnetic field $\boldsymbol{B}$. The result of this oscillation will be a slow—very slow, in fact, compared with the Larmor precession—latitudinal movement of the tip of vector $\boldsymbol{r}$. The latitudinal drift when combined with the Larmor precession results in drawing a spiral curve on the surface of the Bloch sphere that connects its two poles.

To analyze this effect in more detail, we are going to solve equation (2.120) yet again, taking all components of $\boldsymbol{B}$ into account, although we are still going to make some simplifying assumptions. Expanding $\mathrm{d}\boldsymbol{r}/\mathrm{d}t = (2\mu/\hbar)\,\boldsymbol{r}\times\boldsymbol{B}$ in the x, y, z coordinates explicitly gives us the following three equations:

$$\frac{\mathrm{d}r^x}{\mathrm{d}t} = \frac{2\mu}{\hbar}\left(r^y B^z - r^z B^y\right), \tag{2.138}$$

$$\frac{\mathrm{d}r^y}{\mathrm{d}t} = \frac{2\mu}{\hbar}\left(r^z B^x - r^x B^z\right), \tag{2.139}$$

$$\frac{\mathrm{d}r^z}{\mathrm{d}t} = \frac{2\mu}{\hbar}\left(r^x B^y - r^y B^x\right). \tag{2.140}$$

We are going to assume that

Approximations

$$|B^x| \ll |B^z|, \tag{2.141}$$

$$|B^y| \ll |B^z|, \tag{2.142}$$

$$B^z = \text{const.} \tag{2.143}$$

We are also going to assume that the initial state is fully polarized, which implies that the solution for all values of t must be fully polarized, too.

Since B^x and B^y are small compared to B^z, we can try the following form of the solution:

General form of the solution

$$r^x = \sin\theta(t)\cos\omega_L t, \tag{2.144}$$

$$r^y = -\sin\theta(t)\sin\omega_L t, \tag{2.145}$$

$$r^z = \cos\theta(t). \tag{2.146}$$

In other words, we assume that the qubit keeps precessing as in Section 2.10, but this time the angle θ is no longer constant. Instead, it is a slowly varying function of time—slowly, compared to $\omega_L t$. The assumption is equivalent to saying that we are going to ignore $r^z B^y$ compared to $r^y B^z$ in equation (2.138) and $r^z B^x$ compared to $r^x B^z$ in equation (2.139). With these simplifications, equations (2.138) and (2.139) are the same as equations (2.122) and (2.123) on page 75, which, as we know, describe the Larmor precession. This leaves us with equation (2.140).

LHS worked on

First let us substitute (2.146) in the left-hand side of equation (2.140). The result is

$$\frac{\mathrm{d}}{\mathrm{d}t}r^z = \frac{\mathrm{d}}{\mathrm{d}t}\cos\theta(t) = -\sin\theta(t)\frac{\mathrm{d}}{\mathrm{d}t}\theta(t). \tag{2.147}$$

RHS worked on

Substituting (2.144) and (2.145) in the right-hand side of equation (2.140) yields

$$\frac{2\mu}{\hbar}\left(r^x B^y - r^y B^x\right)$$

$$= \frac{2\mu}{\hbar}\left(B^y \sin\theta(t)\cos\omega_L t + B^x \sin\theta(t)\sin\omega_L t\right). \tag{2.148}$$

LHS = RHS

Comparing equations (2.147) and (2.148) tells us that we can cancel $\sin\theta(t)$ that occurs on both sides. We obtain a simpler differential equation for $\theta(t)$, namely,

$$\frac{\mathrm{d}}{\mathrm{d}t}\theta(t) = \frac{2\mu}{\hbar}\left(B^x \sin\omega_L t + B^y \cos\omega_L t\right). \tag{2.149}$$

Specifications for B^x and B^y

Until now we have not specified B^x and B^y other than to say that they're much smaller than B^z. We are now going to specify both, in order to make equation (2.149) easier to solve. Let us assume the following:

$$B^x = B_\perp \sin\omega t, \tag{2.150}$$

$$B^y = B_\perp \cos\omega t. \tag{2.151}$$

On this occasion we also rename B^z to $B_\parallel$. Plugging (2.150) and (2.151) into (2.149) results in

$$\frac{\mathrm{d}}{\mathrm{d}t}\theta(t) = \frac{2\mu B_\perp}{\hbar}\left(\sin\omega_L t \sin\omega t + \cos\omega_L t\cos\omega t\right). \tag{2.152}$$

$\cos\alpha\cos\beta + \sin\alpha\sin\beta = \cos(\alpha-\beta)$

Now we can invoke the well-known high-school trigonometric formula (shown in the margin note) to wrap this into

$$\frac{\mathrm{d}}{\mathrm{d}t}\theta(t) = \frac{2\mu B_\perp}{\hbar}\cos(\omega_L - \omega)t, \tag{2.153}$$

which at long last can be solved easily for $\theta(t)$,

$$\theta(t) = \frac{2\mu B_\perp}{\hbar}\frac{\sin(\omega_L-\omega)t}{\omega_L-\omega}, \tag{2.154}$$

assuming that $\theta(0) = 0$, that is, assuming that at $t = 0$ the qubit is in the $|\uparrow\rangle$ state.

2.11.1 Solution at Resonance

Resonance condition

We shall first consider solutions at frequencies ω that are very close to ω_L or right on the spot, that is, $\omega = \omega_L$. This corresponds to buzzing the qubit with a small transverse magnetic field $\boldsymbol{B}_\perp = (B_\perp \sin\omega_L t)\,\boldsymbol{e}_x + (B_\perp\cos\omega_L t)\,\boldsymbol{e}_y$ that rotates with the Larmor frequency of the qubit itself.

$\lim_{x\to 0}\sin x = x$

For very small values of $(\omega_L - \omega)$ we can replace the sine function with

$$\sin(\omega_L-\omega)t \approx (\omega_L-\omega)t. \tag{2.155}$$

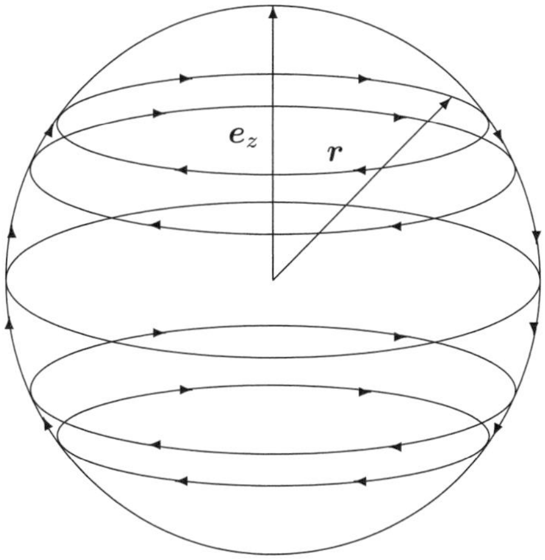

Figure 2.7: Rabi oscillations. As vector $\boldsymbol{r}$ continues to precess rapidly about $\boldsymbol{e}_z$, its latitude θ (measured from the north pole) increases slowly, until $\boldsymbol{r}$ flips to the southern hemisphere and approaches the south pole, whereupon the process reverses.

Solution at resonance

This way the $(\omega_L - \omega)$ factor cancels out, and equation (2.154) becomes

$$\theta(t) = \frac{2\mu B_\perp}{\hbar} t. \tag{2.156}$$

Rabi oscillations

The angle θ grows linearly with time t. Although θ is restricted to $0 \le \theta \le \pi$ in principle, the solution doesn't care about this and continues to grow beyond π. What does this mean? First, let us note that $\cos\theta = \cos(2\pi - \theta)$. This implies that as far as r^z is concerned, once θ has exceeded π, we can replace it with $\theta' = 2\pi - \theta$ and reduce the angle to the $0 \ldots \pi$ range. In other words, once θ has reached π, it begins to swing back, until it returns to $\theta = 0$, whereupon it is going to resume its journey south.

So, what we have here are oscillations of vector $\boldsymbol{r}$ between north and south interposed on Larmor precession. The north-south oscillations of $\boldsymbol{r}$ produced by this process, and illustrated in Figure 2.7, are called Rabi oscillations, after U.S. physicist and Nobel prize winner (1944) Isidor Isaac Rabi (1898–1988), and occur with a frequency much lower than the Larmor frequency.

Rabi frequency

The period of these oscillations is derived from

$$\theta(T_R) = 2\pi = \frac{2\mu B_\perp}{\hbar} T_R, \tag{2.157}$$

which yields

$$T_R = \frac{\pi\hbar}{\mu B_\perp}. \tag{2.158}$$

The angular frequency of Rabi oscillations, ω_R, is

$$\omega_R = \frac{2\pi}{T_R} = \frac{2\mu B_\perp}{\hbar} \ll \frac{2\mu B_\parallel}{\hbar} = \omega_L, \tag{2.159}$$

because

$$|B_\perp| \ll |B_\parallel|.$$

How long does it take to flip a qubit?

If we were to begin the process with a qubit in the basis state $|\uparrow\rangle \equiv |\,0\rangle$, aligned with $\boldsymbol{B}_\parallel$, it would take $\Delta t = T_R/2 = \pi\hbar/(2\mu B_\perp)$ of buzzing the qubit with $\boldsymbol{B}_\perp = (B_\perp \sin\omega t)\,\boldsymbol{e}_x + (B_\perp \cos\omega t)\,\boldsymbol{e}_y$ to flip it to $|\downarrow\rangle \equiv |\,1\rangle$. And similarly, to flip the qubit from $|\downarrow\rangle \equiv |\,1\rangle$ to $|\uparrow\rangle \equiv |\,0\rangle$ would take the same amount of time.

The full form of the solution for $\boldsymbol{r}(t)$ is given by

$$r^x = \sin\omega_R t \cos\omega_L t, \tag{2.160}$$

$$r^y = -\sin\omega_R t \sin\omega_L t, \tag{2.161}$$

$$r^z = \cos\omega_R t. \tag{2.162}$$

More about $\boldsymbol{r}$ going back north

Let us have another close look at what is going to happen when θ grows above π. On the way back from the south pole $\pi < \theta < 2\pi$ and in this region, $\sin\theta$ is negative. Replacing θ with $\theta' = 2\pi - \theta$, which has the effect of reducing θ back to $[0,\pi]$, and writing the minus in front of $\sin\theta$ explicitly, we find that the solution becomes

$$r^x = -\sin\theta' \cos\omega_L t = \sin\theta' \cos(\omega_L t + \pi), \tag{2.163}$$

$$r^y = \sin\theta' \sin\omega_L t = -\sin\theta' \sin(\omega_L t + \pi), \tag{2.164}$$

$$r^z = \cos\theta', \tag{2.165}$$

where we have made use of the fact that $-\sin\alpha = \sin(\alpha+\pi)$ and $-\cos\alpha = \cos(\alpha+\pi)$ in order to absorb the changed sign into the sin and cos of the Larmor precession.

The result tells us that on its way back north vector $\boldsymbol{r}$ is going to visit a point of the Bloch sphere, which, for a given angle θ, is on the other side, compared to the point it had visited for the same angle θ on the way south.

As vector $\boldsymbol{r}$ nears the north pole, its returning trajectory on the Bloch sphere is going to align itself with the starting trajectory, because it is its reflection with respect to the $\boldsymbol{e}_z$ axis, so that after having reached the north pole vector $\boldsymbol{r}$ will go south on exactly the same trajectory it traced originally.

As we go south and north, and south again and north again, we're going to follow the same spiral all the time, crossing the equator at exactly the same two points, the point on the return voyage being on the other side, with respect to the point at which we cross the equator on the way south.

This property can be used in the following operation, which is due to U. S. physicist and Nobel prize winner (1989) Norman Foster Ramsey (born in 1915).

Ramsey experiment

Let us suppose we use the Rabi oscillations to tilt $\boldsymbol{r}$ by 90° south. We can now switch off the buzzing field $\boldsymbol{B}_\perp$ and just wait a while, allowing for the Larmor precession to rotate $\boldsymbol{r}$ about the z axis. Then we can switch $\boldsymbol{B}_\perp$ on again and continue tilting the qubit. If the Larmor precession has rotated the qubit by a multiple of 2π (meaning, 360°) in the meantime, the qubit will come back to the point from which it left and on receiving the buzzing signal will resume its journey to the south pole. But if the Larmor precession has rotated the qubit by an odd multiple of π (meaning, 180°), the qubit will resume its journey along the Rabi trajectory at the point that's on the other side of the Bloch sphere, and so it'll come back north instead.

What will happen if the Larmor precession leaves vector $\boldsymbol{r}$ stranded at some point that is between the two points where the Rabi spiral crosses the equator? When the buzzing signal kicks in, it is going to be out of phase now with respect to the Larmor precession. We can describe this by altering equation (2.153) on page 80 and adding a fixed angle ϕ to the phase of the qubit:

$$\frac{\mathrm{d}}{\mathrm{d}t}\theta(t) = \omega_R \cos\left(\left(\omega_L - \omega\right) t + \phi\right). \tag{2.166}$$

Making use again of the school formula shown in the margin, we can rewrite this as

$\cos(\alpha + \beta) = \cos\alpha\cos\beta - \sin\alpha\sin\beta$

$$\frac{\mathrm{d}}{\mathrm{d}t}\theta(t) = \omega_R \left(\cos\left(\omega_L - \omega\right) t \cos\phi - \sin\left(\omega_L - \omega\right) t \sin\phi\right). \tag{2.167}$$

The resonance condition $\omega = \omega_L$ kills the $\sin(\omega_L - \omega)t$ term and converts $\cos(\omega_L - \omega)t$ to 1, which yields

$$\frac{\mathrm{d}}{\mathrm{d}t}\theta(t) = \omega_R \cos\phi, \tag{2.168}$$

the solution of which is

$$\theta(t) = \omega_R t \cos\phi + \theta_0. \tag{2.169}$$

Let us assume this time that at $t = 0$ $\theta = \pi/2$, in other words, that the tip of vector $\boldsymbol{r}$ is on the equator and its longitude is ϕ. Then

$$\theta(t) = \omega_R t \cos\phi + \frac{\pi}{2}. \tag{2.170}$$

First, let us observe that for $\phi = 0$ we get exactly what we had before: θ is going to increase or, in other words, $\boldsymbol{r}$ will continue on its way south at the rate of ω_R per second. On the other hand, if $\phi = \pi$, $\cos\phi = -1$, then θ is going to *decrease* at the rate of ω_R per second; that is, $\boldsymbol{r}$ will turn north. So we have now reproduced the basic characteristics of our Rabi spiral. For any other angle ϕ between 0 and π, the progress of $\boldsymbol{r}$ is going to be slowed. For $0 < \phi < \pi/2$ $\boldsymbol{r}$ will move south. For $\pi/2 < \phi < \pi$ $\boldsymbol{r}$ will move north instead. But for $\phi = \pi/2$ $\boldsymbol{r}$ will get stuck on the equator.

Rabi oscillations do not constitute a measurement.

As was the case with the Larmor precession, wagging $\boldsymbol{r}$ to and fro on the Bloch sphere with transverse oscillating magnetic fields still does not constitute a measurement. If we start with a fully polarized qubit state, we end with a fully polarized qubit state, too.

Let us have a look at how probability of finding the qubit in state $|\uparrow\rangle$ varies with time:

$$p^0 = \frac{1}{2}\left(1 + r^z\right) = \frac{1}{2}\left(1 + \cos\omega_R t\right). \tag{2.171}$$

$\frac{1}{2}\left(1 + \cos 2\alpha\right) = \cos^2\alpha$

We are again going to make use of the high-school trigonometric formula to wrap the above into

$$p^0 = \cos^2\frac{\omega_R t}{2}. \tag{2.172}$$

At $t = 0$ we have that $p^0 = 1$. Then the probability begins to diminish, and at $\omega_R t/2 = \pi/2$ we get $p^0 = 0$, which means that $|\uparrow\rangle$ has flipped completely to $|\downarrow\rangle$. This is the same picture as given by the evolution of $\boldsymbol{r}$, this time expressed in terms of probabilities.

Exact predictions

This is also another rare circumstance when an outcome of a quantum mechanical experiment can be predicted exactly and with certainty for every individual quantum system. If one takes a neutron from an $|\uparrow\rangle$ beam and sends it into a chamber with $\boldsymbol{B}_{\|}$ and $\boldsymbol{B}_{\perp}$ as specified in this section, after $\pi\hbar/(2\mu B_{\perp})$ seconds spent in the chamber (but not a fraction of a second longer!) the neutron is guaranteed to be in the $|\downarrow\rangle$ state.[22]

[22] In practice we cannot make this prediction with certainty because the $\boldsymbol{B}_{\perp}$ field is usually highly nonuniform and because we may not be able to switch it on and off exactly on time. But we want to distinguish here between the fundamental quantum probabilities and probabilities that arise from imperfections of the experimental procedure.

2.11.2 Solution off Resonance

Solution (2.154) on page 80 is valid for all values of ω regardless of whether ω is close to ω_L. It is valid as long as the approximation we have made, $B_\perp \ll B_\parallel$, is valid. For $\omega = \omega_L$ we found that θ varied linearly with time, eventually swinging onto the other side, namely, from $\theta = 0°$ to $\theta = 180°$.

Far from the resonance the qubit does not absorb energy.

One can easily see that for ω far away from ω_L, something quite different is going to happen. Let us recall equation (2.154):

$$\theta(t) = \frac{2\mu B_\perp}{\hbar} \frac{\sin(\omega_L - \omega)t}{\omega_L - \omega}. \tag{2.173}$$

When $|\omega_L - \omega| >> 0$, the denominator becomes very large.[23] Therefore the amplitude of oscillations described by (2.173) becomes very small, and the oscillations themselves become very fast. The result is that vector $\boldsymbol{r}$ keeps pointing up and just vibrates very quickly around $\theta = 0$. The qubit does not absorb energy from the incident radiation.

Near the resonance

What if $\omega \neq \omega_L$ but they are not so far apart that $\boldsymbol{r}$ gets stuck on $\theta = 0$? We observe that only for $\omega = \omega_L$ can θ wander all over the place. Otherwise $-1 \leq \sin(\omega_L - \omega)t \leq 1$, and θ is restricted to

$$-\frac{2\mu B_\perp}{\hbar(\omega_L - \omega)} \leq \theta \leq \frac{2\mu B_\perp}{\hbar(\omega_L - \omega)}. \tag{2.174}$$

We can therefore ask the simple question: How far away can we move ω from ω_L so that θ can still reach π—in however circuitous a manner? The answer to this question is

$$\pi = \frac{2\mu B_\perp}{\hbar(\omega_L - \omega)}, \tag{2.175}$$

which yields

$$\omega_L - \omega = \frac{2\mu B_\perp}{\hbar\pi}. \tag{2.176}$$

A somewhat better measure here would be $(\omega_L - \omega)/\omega_L$, because this quantity is nondimensional and it will let us eliminate $\hbar$ and μ from the equation. Dividing (2.176) by Larmor frequency (given by equation (2.129) on page 76) results in

Energy absorption condition

$$\frac{\omega_L - \omega}{\omega_L} = \frac{2\mu B_\perp}{\hbar\pi} \frac{\hbar}{2\mu B_\parallel} = \frac{B_\perp}{\pi B_\parallel}. \tag{2.177}$$

[23] We point out that the Planck constant, $\hbar$, in the denominator does not squash $|\omega_L - \omega|$ at all, because there is another $\hbar$ hidden inside μ and they cancel out.

This gives us a measure of how precisely we have to tune the frequency of the buzzing field $\boldsymbol{B}_\perp$ in order to eventuate a qubit flip. On the one hand, the smaller $B_\perp/B_\|$, the more accurate is the solution given by equation (2.173); on the other hand, the smaller the $B_\perp$ field, the closer the ω has to approach ω_L in order to still eventuate a qubit flip.

In a typical NMR experiment $B_\|$ may be on the order of 12 T. The amplitude of the buzzing field, which is generated by Helmholtz coils, is tiny, usually on the order of one Gauss, where 1 Gauss $= 10^{-4}$ T.[24] This tells us that we must ensure at least

$$\frac{\omega_L - \omega}{\omega_L} < \frac{10^{-5}}{\pi}. \tag{2.178}$$

2.12 The Quantronium

A quantum device

The quantronium [142] is a quantum electronic circuit realization of a qubit. It is not a classical electronic device that simulates a quantum system. It is a true man-made quantum system in its own right, even though it does not incorporate obvious microscopic elements such as, to pick just one, an individual atom of phosphorus embedded in silicon lattice in a precisely defined location and surrounded with controlling electrodes [75] [76]. Instead, the quantronium relies on one of the few macroscopic manifestations of quantum physics, superconductivity. This has the advantage that the circuit can be made by using a fairly standard, though not necessarily "industry standard," microelectronic technology.[25]

The quantronium circuit together with other auxiliary circuitry is shown in Figure 2.8. We cannot analyze the functioning of this circuit in great detail because this would call for quantum physics background that goes way beyond the scope of this text. But we shall explain enough of it to illustrate how the circuit is used to observe the Rabi oscillations and the Larmor precession of the qubit.

Where is the qubit?

The qubit itself is contained in the large black dot to the left of letter N. The dot symbolizes a low-capacitance superconducting electrode, which in this configuration is called the Cooper pair box. The Cooper pair box is connected to the rest of the circuit through two Josephson tunnel junctions represented in Figure 2.8 by the two square boxes with $\frac{1}{2}E_J$ written inside them. The Josephson junctions and the Cooper pair box are biased across the gate capacitance C_g by the voltage source U.

[24]The parameters provided are based on the Varian Inova 500 MHz specifications.

[25]The quantronium was patterned by using electron beam lithography and aluminum evaporation. Electron beam lithography is a very precise laboratory technique that allows for nanolevel pattern definition. Because of its slowness it cannot be used in mass-produced devices.

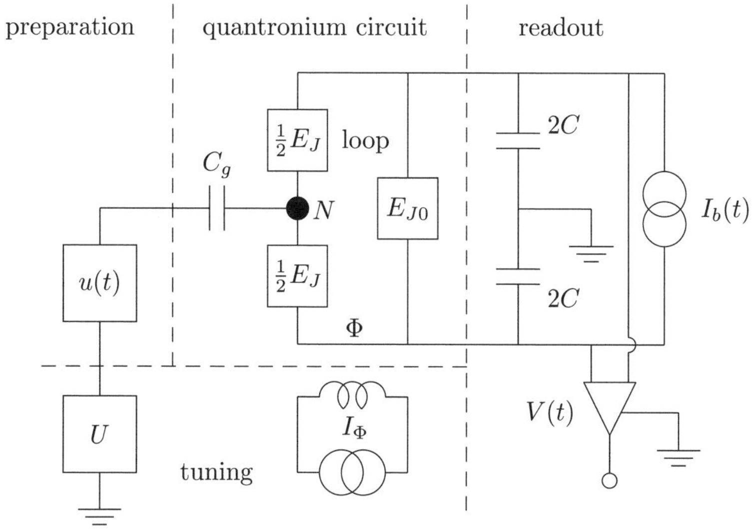

Figure 2.8: The quantronium and its auxiliary electronic circuitry. Redrawn after [142]. With permission from AAAS.

A Cooper pair is a pair of electrons, coupled to each other and to the lattice of the crystal they live in, so that they form a quasiparticle, meaning a composite and *nonlocal* quantum object with elementary particle characteristics. The total number of such pairs in the Cooper pair box is N. The energy of the box is quantized, meaning that it can assume several discrete values, which depend on the bias voltage U and the DC current I_Φ that flows in the coil adjacent to the quantronium circuit (shown just below the circuit in Figure 2.8) and generates the magnetic field flux Φ in the quantronium circuit loop. The ground energy level and the first excited energy level of the Cooper pair box form a two-state quantum system, that is, a qubit. We associate the ground state with $| \, 0\rangle$ and the first excited energy level with $| \, 1\rangle$. For the operational parameters of the circuit both states are characterized by the same average number of Cooper pairs in the box in order to make the qubit insensitive to fluctuations of the gate charge.

Tuning the qubit: $\omega_L = \omega_L(U, I_\Phi)$

The bias voltage U and the current I_Φ are used to tune the properties of the qubit with the effect that its Larmor frequency, $\omega_L = 2\mu B_{\|}/\hbar$, is a function of U and I_Φ. But there is no simple formula that we can use to separate μ from $B_{\|}$. For a given pair of U and I_Φ we get $\mu B_{\|}$ bundled together, although, of course, we can separate them experimentally after we have carried out sufficient measurements on the circuit.

Preparing the qubit with the $u(t)$ pulse

The buzzing field $\boldsymbol{B}_\perp$ is represented by the variable voltage $u(t)$ that is superimposed on U. In order to drive the qubit, the frequency of the pulse must match the qubit's Larmor frequency $f_L = \omega_L/(2\pi)$, which, for the circuit drawn in Figure 2.8, was $16.4635 \pm 0.0008\,\mathrm{GHz}$. By choosing the amplitude and the duration of the pulse, we can swing vector $\boldsymbol{r}$ up and down, thus affecting p^0 and p^1 of the qubit. Having done so we can commence the measurement, which is implemented by the part of the circuit drawn on the right-hand side of Figure 2.8.

Measuring the qubit

States $|\,0\rangle$ and $|\,1\rangle$ are differentiated by the supercurrent in the loop that develops as the result of the trapezoidal readout pulse $I_b(t)$, which can be sent into the circuit. The current flows through both Josephson junctions labeled by $\frac{1}{2}E_J$ and through the third Josephson junction on the right hand side of the loop, labeled by E_{J0}. The two capacitors in the readout part of the circuit are meant to reduce phase fluctuations in the loop. The supercurrent in combination with the bias current in the E_{J0} junction can switch the junction to a measurable voltage state $V(t)$. This switching is probabilistic, too. There is a high probability that the junction will switch if the qubit is in state $|\,1\rangle$ and a low probability that it will switch if the qubit is in state $|\,0\rangle$. So we have to play with two probabilistic processes here: we have quantum probabilities associated with the qubit itself, and we have another layer of probabilities associated with the readout circuitry. The efficiency of the readout is 60%; that is, in 60% of cases, the readout circuit will correctly discriminate between $|\,0\rangle$ and $|\,1\rangle$. While not perfect, this is sufficient to let us observe Rabi oscillations.

Isolating the qubit

The qubit is isolated from the environment and from the readout circuitry by a variety of means. First, the whole circuit is cooled to 15 mK. Then additional protection is provided by large ratios of E_{J0}/E_J and C/C_J, where C_J is the capacitance of the Josephson junction. Parameters U and I_Φ are chosen so as to eliminate charge fluctuation noise and flux and bias current noise.

Sequential exploration of the statistical ensemble

The observed Rabi fluctuations of the quantronium are shown in Figure 2.9 (A). Every point in this graph is the result of 50,000 identical qubit preparations and measurements, which were carried out in order to collect sufficient statistics for the estimate of the switching probability. The standard deviation on 50,000 measurements of this type is $\sqrt{50,000} \approx 224$, which means that our probability estimates are loaded with no more than 0.5% error.[26]

[26] See, for example, Chapter 6 in [41].

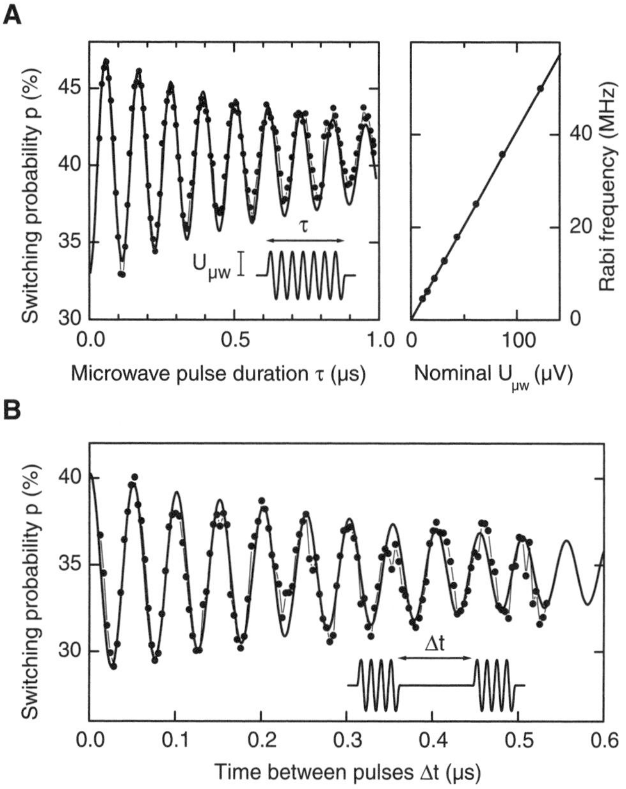

Figure 2.9: Rabi oscillations (A) and Ramsey fringes (B) in quantronium. From [142]. Reprinted with permission from AAAS.

The buzzing signal $u(t)$ of amplitude $B_\perp = 22\,\mu\mathrm{V}$ and frequency $f_L = 16.4635\,\mathrm{GHz}$ was used for all points. Flux Φ was set to zero.

In order to prepare the qubit for a given point on the graph, the qubit was first allowed to thermalize and align in $U \equiv B_\|$, dropping to its ground state $|\,0\rangle$. Then the qubit was buzzed with $u(t)$ for a specific duration (up to $1\,\mu$s). As soon as the buzzing had stopped, the trapezoidal pulse $I_b(t)$ was sent into the circuit in order to trigger the switching of the large Josephson junction E_{J0}. If the junction had switched, the resulting pulse $V(t)$ was observed; otherwise there was no pulse. The ratio of "switched" to "not switched" for 50,000 shots/point is what has been plotted in Figure 2.9 (A).

Rabi oscillations

We can clearly see oscillations in the graph that are the function of the pulse duration. Because the switching probability in the large junction E_{J0} is proportional to p^0, the observed oscillations are indeed Rabi oscillations. For the Larmor frequency in the tens of GHz range, the Rabi frequency is in the tens of MHz range. The Rabi oscillations for this qubit are therefore about a thousand times slower than the Larmor precession.

Qubit depolarization

The amplitude of the oscillations is clearly damped. The reason is that the qubit becomes depolarized as it interacts with the environment, however much the circuit's designers had tried to reduce such interaction. This effect is not described by our somewhat simplistic model that does not capture qubit depolarization. Nevertheless we can fit the data with an exponentially damped sinusoid and extract the Rabi frequency from it. The fitted curve is overlaid on the data points and shown here in paler shade.

Dependence of ω_R on $B_\perp$

Having done so, we can repeat the whole experiment for different values of the buzzing signal $u(t)$ amplitude, $B_\perp$, and verify that the Rabi frequency ω_R obtained from the measurements increases linearly with $B_\perp$ as equation (2.159) on page 82 ($\omega_R = 2\mu B_\perp/\hbar$) asserts.

The results of these measurements, shown in the right panel of Figure 2.9 (A), fully confirm equation (2.159). The Rabi frequency ω_R is, as expected, directly proportional to $B_\perp$.

Ramsey experiment

Figure 2.9 (B) shows the result of the Ramsey operation on the quantronium qubit. The operation is carried out as follows. First the qubit is thermalized and brought to state $|\,0\rangle$. Then it is buzzed with $\boldsymbol{B}_\perp$ for the time required to rotate it by $\pi/2$, that is, by 90°. When the buzzing stops, the qubit is allowed to precess around the equator for Δt microseconds, whereupon the buzzing is resumed. We buzz the qubit again for the time required to rotate it by the further 90° *in normal circumstances*. But let us recall our discussion of the Ramsey experiment on page 83. Only if the qubit has precessed by a multiple of 2π in time Δt will it resume its journey south with the same angular polar velocity ω_R. If the qubit has precessed by an odd multiple of π in the time Δt, it will instead return back north with the same angular polar velocity ω_R. But for any other angle the qubit has precessed in the time Δt its march south or north will be slowed down by the cosine of the angle. Consequently, when we apply the second buzz to the qubit, it won't be enough to make it go all the way to the top or all the way to the bottom.

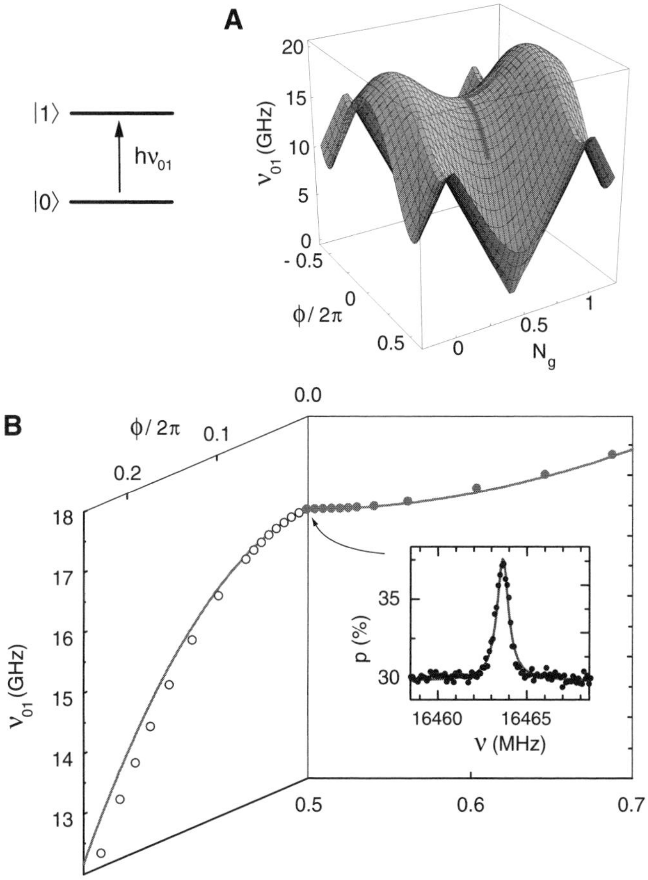

Figure 2.10: Quantronium's transition frequency in function of its various parameters (A). Probability of transition between the $|\,0\rangle$ and $|\,1\rangle$ states of the quantronium for circuit parameters that correspond to the saddle point (B). From [142]. Reprinted with permission from AAAS.

When the measurement is made, the qubit will have some probability of being found in $\mid 1\rangle$ and some probability of being found in $\mid 0\rangle$. The probability will fluctuate with Δt, as shown in Figure 2.9 (B). The fluctuations, in essence, show us that the qubit precesses with the Larmor frequency around the z direction, even though we have no means in this circuit to measure p^2 and p^3. The amplitude of the fluctuations diminishes exponentially, as was the case with the Rabi oscillations, even though the qubit is disconnected from the driving force during the time Δt when it is expected to precess freely. The depolarization observed in the measurement can be used to estimate the so-called coherence time of the qubit. Any quantum computations we want to carry out using the qubit must be completed well before the qubit's quantum state decoheres, which manifests in its depolarization.

Resonant absorption

We have argued in Section 2.11.2 on page 85 that the buzzing signal could force qubit to flip its basis state only if the buzzing frequency was very close to the Larmor frequency of the qubit. Equation (2.177) quantified this by stating that the qubit could not flip its state at all if $|\omega - \omega_L|/\omega_L > B_\perp/(\pi B_\parallel)$. The insert in Figure 2.10 (B) shows how the qubit flip probability for the quantronium depends on the frequency of the buzzing signal $u(t)$ for circuit parameters that correspond to the saddle point of the diagram shown in Figure 2.10 (A). The saddle point was then used in the Rabi and Ramsey experiments because of the parametric stability of the circuit in its vicinity.

Every point in the graph in Figure 2.10 (B) is the result of 50,000 measurements, too. For each measurement the quantronium qubit was first thermalized and allowed to align with $B_\parallel$ in its ground state $\mid 0\rangle$. The qubit was then irradiated with microwaves of a given frequency, emitted by a small antenna inserted into the cryostat together with the circuit, for up to 100 ns (T_R in Figure 2.9 is about 100 ns), and then the measurement was activated by sending the trapezoidal pulse $I_b(t)$ into the loop of the quantronium circuit. The big Josephson junction E_{J0} would switch sometimes, which would be detected by observing the pulse $V(t)$, and sometimes it wouldn't. Whether it switched or not would depend on the state of the qubit, as we have pointed out already, so the switching probability here is related to whether the qubit itself flipped to $\mid 1\rangle$. After 50,000 of such trials sufficient statistics were collected to give us an estimate of the switching probability for a given frequency $f = \omega/(2\pi)$. The measurements were repeated while the frequency of the microwave signal was varied so that the whole neighborhood of the resonance point $f = f_L = \omega_L/(2\pi)$ was covered. This is shown in Figure 2.10 (B). We can clearly see a well-defined peak. The peak can be fitted with the Lorentz absorption curve yielding the exact position of the resonance at $f_L = 16.4635\,\mathrm{GHz}$ and the peak width $\Delta f = 0.8\mathrm{MHz}$.

The quantronium is more than just a yet another attempt to fabricate a qubit as a solid state device—and perhaps the first one that gives us more than just a glimmer of its quantum nature. It is an elucidating example of a simple quantum system that brings up two of the most important features of quantum physics: its probabilistic nature and its extreme sensitivity to the environment. Just as important, the quantronium illustrates how quantum experiments are carried out in general and how the qubit's statistical ensemble is explored in particular.

Quantum systems are probabilistic and fragile.

3 Quaternions

3.1 Continuing Development of Quantum Mechanics

The fiducial formalism presented in the preceding chapters is inherent in the physics of a qubit: it is complete and entirely physical without any admixture of unnecessary metaphysics. It operates on directly measurable entities such as probabilities and various quantites averaged over a statistical ensemble of a qubit. It accounts for mixtures, fully polarized (pure) states, discrete energy spectrum of a qubit, qubit precession, qubit flipping, and other effects—all within a simple algebra of the conventional probability calculus, albeit enhanced by the addition of "hidden variables"—and without our ever having to resort to the use of complex numbers, Hilbert spaces, probability amplitudes, and other esoteric weapons from the armory of the traditional quantum mechanics. It would make Ernst Mach exuberant.

This may come as a surprise to some physicists, because statements were sometimes made in the past about the impossibility of such a description.[1] Unfortunately quantum mechanics is littered with various statements of this nature, often uttered by famous people, that—in time—were demonstrated to be blatantly false.[2]

Quantum mechanics continues to develop.

In spite of the great maturity of quantum mechanics, new and surprising results continue to crop up all the time. It was only in 1983 that the Berry phase was discovered [12], yet it is such a fundamental quantum effect. It was only in 1996 that a bi-qubit separability criterion was discovered by Peres and Horodeckis [66] [113]. And it was only in 2004 that Durt published his remarkable theorem about entanglement and interaction [35]. The observation that every quantum system and its evolution can be described entirely in terms of probabilities and various physical parameters averaged over the statistical ensemble of the system (such parameters are called *expectation values*) was published as late as January 2000 [145] by Stefan Weigert of the Université de Neuchâtel in Switzerland. The specific observation that such a description can be formulated in terms of a slightly generalized probability calculus is due to Lucien Hardy of the Clarendon Laboratory in Oxford [60] and is even more recent.

In this chapter we will see that the single qubit probability calculus maps naturally onto an even simpler type of calculus, for which we no longer need to bother about vectors and forms because the state of a qubit as well as its Hamiltonian form map onto numbers—albeit of a rather special type: quaternions.

[1] See, for example, [112] or [42].

[2] See, for example, [8] and [14].

3.2 Hamilton Quaternions

To map a vector or a form with four entries onto a number without the loss of information, we must have numbers with four slots. Such numbers were invented by Irish mathematician, physicist, and astronomer Sir William Rowan Hamilton (1805–1865), whose accomplishments and insights were so great that the most important mathematical device of quantum physics, the Hamiltonian (we called it the Hamiltonian form), was named after him. Hamilton was so clever that if he had lived long enough, he would have invented both special relativity and quantum mechanics himself. He nearly did so. He certainly can be credited with the invention of spacetime, when he said the following [52]:

> Time is said to have only one dimension, and space to have three dimensions... The mathematical quaternion partakes of both these elements; in technical language it may be said to be "time plus space," or "space plus time": and in this sense it has, or at least involves a reference to, four dimensions.
>
> And how the One of Time, of Space the Three,
> Might in the Chain of Symbols girdled be.

Similarity of quaternions to complex numbers

Quaternions are similar to complex numbers, which have two slots. The slots of a complex number are called *real* and *imaginary*. The imaginary slot is marked by writing the letter "i" in front of it:[3]

$$z = a + \mathrm{i}b. \tag{3.1}$$

Definition of a quaternion

Quaternions have four slots, of which one is real and the remaining three are marked by letters $\boldsymbol{i}$, $\boldsymbol{j}$ and $\boldsymbol{k}$:

$$\boldsymbol{q} = a + b\boldsymbol{i} + c\boldsymbol{j} + d\boldsymbol{k}. \tag{3.2}$$

The coefficients a, b, c, and d are normally real, but complex numbers may be used in their place. We will see that even though complex numbers are deployed in the mapping between fiducial vectors and forms and quaternions, their use is cosmetic and dictated by tradition rather than necessity. Purely real-number-based mapping can be used, too, with almost identical results.

Quaternion commutation relations

The three *imaginary* units of the quaternion world have similar properties to the imaginary unit of complex numbers:

$$\boldsymbol{ii} = \boldsymbol{jj} = \boldsymbol{kk} = -1. \tag{3.3}$$

[3] Electrical engineers prefer "j."

But there is one additional rule,

$$\boldsymbol{ijk} = -1, \tag{3.4}$$

from which the following can be derived:

$$\boldsymbol{ij} = -\boldsymbol{ji} = \boldsymbol{k}, \tag{3.5}$$
$$\boldsymbol{jk} = -\boldsymbol{kj} = \boldsymbol{i}, \tag{3.6}$$
$$\boldsymbol{ki} = -\boldsymbol{ik} = \boldsymbol{j}. \tag{3.7}$$

The derivation is quite simple. Let us consider, for example,

$$\boldsymbol{ij} = -\boldsymbol{ij}(\boldsymbol{kk}) = -(\boldsymbol{ijk})\boldsymbol{k} = \boldsymbol{k}. \tag{3.8}$$

Now let us take

$$\boldsymbol{j}(\boldsymbol{ii})\boldsymbol{j} = -\boldsymbol{jj} = 1 = (\boldsymbol{ji})(\boldsymbol{ij}) = (\boldsymbol{ji})\boldsymbol{k}, \tag{3.9}$$

which implies that

$$\boldsymbol{ji} = -\boldsymbol{k} = -\boldsymbol{ij}. \tag{3.10}$$

In other words, quaternion imaginary units do not commute. They anticommute instead:

$$\boldsymbol{ij} + \boldsymbol{ji} \doteq \{\boldsymbol{i}, \boldsymbol{j}\} = 0, \tag{3.11}$$

and similarly for other pairs. The symbol $\{\boldsymbol{i}, \boldsymbol{j}\}$ is called the anti-commutator.

Quaternion imaginary units anti-commute.

3.3 Pauli Quaternions

For historic and some technical reasons present-day physicists do not use symbols $\boldsymbol{i}$, $\boldsymbol{j}$, and $\boldsymbol{k}$. Instead they use different symbols,[4] namely,

Quaternions and Pauli matrices

$$\boldsymbol{\sigma}_x = \mathrm{i}\boldsymbol{i}, \tag{3.12}$$
$$\boldsymbol{\sigma}_y = \mathrm{i}\boldsymbol{j}, \tag{3.13}$$
$$\boldsymbol{\sigma}_z = \mathrm{i}\boldsymbol{k}, \tag{3.14}$$

where $\mathrm{i} = \sqrt{-1}$. The resulting properties of Pauli quaternions are

Pauli matrices commutation and anti-commutation relations

$$\boldsymbol{\sigma}_x\boldsymbol{\sigma}_x = \boldsymbol{\sigma}_y\boldsymbol{\sigma}_y = \boldsymbol{\sigma}_z\boldsymbol{\sigma}_z = \mathbf{1}, \tag{3.15}$$

[4]... in effect, they sometimes do not realize they work with quaternions.

and

$$\boldsymbol{\sigma}_x\boldsymbol{\sigma}_y = -\boldsymbol{\sigma}_y\boldsymbol{\sigma}_x = (\mathrm{i}\boldsymbol{i})(\mathrm{i}\boldsymbol{j}) = -\boldsymbol{k} = \mathrm{i}\mathrm{i}\boldsymbol{k} = \mathrm{i}\boldsymbol{\sigma}_z, \tag{3.16}$$

$$\boldsymbol{\sigma}_y\boldsymbol{\sigma}_z = -\boldsymbol{\sigma}_z\boldsymbol{\sigma}_y = \mathrm{i}\boldsymbol{\sigma}_x, \tag{3.17}$$

$$\boldsymbol{\sigma}_z\boldsymbol{\sigma}_x = -\boldsymbol{\sigma}_x\boldsymbol{\sigma}_z = \mathrm{i}\boldsymbol{\sigma}_y. \tag{3.18}$$

To avoid a complete disconnect from the way everybody else does things in physics, we're going to use this notation, too, even though it is clumsier than the Hamilton quaternion notation used in mathematics.

A couple of typographic notes are in order here. An astute reader will have already observed one such convention, namely, that a roman "i" is used as $\sqrt{-1}$, and an italicized bold $\boldsymbol{i}$ is used for a quaternion unit. Also, a bold **1** is used instead of just 1 in equation (3.15). This is also due to tradition. People who use $\boldsymbol{\sigma}_x$, $\boldsymbol{\sigma}_y$, and $\boldsymbol{\sigma}_z$ instead of $\boldsymbol{i}$, $\boldsymbol{j}$, and $\boldsymbol{k}$ like to think of **1** as an identity matrix rather than a number. But we don't have to do so, nor do we need to look inside $\boldsymbol{\sigma}_x$, $\boldsymbol{\sigma}_y$ and $\boldsymbol{\sigma}_z$, even though these can be represented by matrices as well. To us they will be just 1, $\boldsymbol{i}$, $\boldsymbol{j}$, and $\boldsymbol{k}$ in disguise.

Pauli quaternion commutation rules can be usefully encapsulated in a single expression:

$$\boldsymbol{\sigma}_i\boldsymbol{\sigma}_j = \delta_{ij}\mathbf{1} + \mathrm{i}\sum_k \epsilon_{ijk}\boldsymbol{\sigma}_k, \tag{3.19}$$

where i, j, and k run through x, y, and z; δ_{ij} is the Kronecker delta; and ϵ_{ijk} is the Levi Civita tensor, sometimes also called the totally antisymmetric symbol, which is $+1$ for every even permuation of $\{x, y, z\}$, -1 for every odd permutation, and zero otherwise. It is easy to see that for $i = j$ we get $\delta_{ii} = 1$ but $\epsilon_{iik} = 0$ for every k, and then

$$\boldsymbol{\sigma}_i\boldsymbol{\sigma}_i = \mathbf{1}. \tag{3.20}$$

On the other hand, for $i \neq j$ we get $\delta_{ij} = 0$ but $\epsilon_{ijk} = \pm 1$ as long as k is different from both i and j, in which case the sign depends on the ordering of whatever i, j, and k stand for. In particular, if $i = x$ and $j = y$, we get

$$\boldsymbol{\sigma}_x\boldsymbol{\sigma}_y = \mathrm{i}\epsilon_{xyz}\boldsymbol{\sigma}_z = \mathrm{i}\boldsymbol{\sigma}_z, \tag{3.21}$$

because ϵ_{xyz} is the only nonvanishing ϵ_{xyk}, and it is equal to 1.

3.4 From Fiducial Vectors to Quaternions

The mapping between fiducial vectors and quaternions is as simple as it can possibly get. It is going to be linear, and we are going to map

$$\varsigma_1 \rightarrow \mathbf{1}, \tag{3.22}$$
$$\varsigma_x \rightarrow \boldsymbol{\sigma}_x, \tag{3.23}$$
$$\varsigma_y \rightarrow \boldsymbol{\sigma}_y, \tag{3.24}$$
$$\varsigma_z \rightarrow \boldsymbol{\sigma}_z, \tag{3.25}$$

and, since there does not have to be a distinction between vectors and forms in the quaternion world—after all, they are all just numbers—we're going to map fiducial forms similarly:

$$\varsigma^1 \rightarrow \mathbf{1}, \tag{3.26}$$
$$\varsigma^x \rightarrow \boldsymbol{\sigma}_x, \tag{3.27}$$
$$\varsigma^y \rightarrow \boldsymbol{\sigma}_y, \tag{3.28}$$
$$\varsigma^z \rightarrow \boldsymbol{\sigma}_z. \tag{3.29}$$

This way a fiducial vector of a qubit,

$$\boldsymbol{p} = \frac{1}{2}\left(\varsigma_1 + r^x\varsigma_x + r^y\varsigma_y + r^z\varsigma_z\right), \tag{3.30}$$

becomes

$$\boldsymbol{\rho} = \frac{1}{2}\left(\mathbf{1} + r^x\boldsymbol{\sigma}_x + r^y\boldsymbol{\sigma}_y + r^z\boldsymbol{\sigma}_z\right), \tag{3.31}$$

and the Hamilton form,

$$\boldsymbol{\eta} = -\mu\left(B_x\varsigma^x + B_y\varsigma^y + B_z\varsigma^z\right), \tag{3.32}$$

becomes

$$\boldsymbol{H} = -\mu\left(B_x\boldsymbol{\sigma}_x + B_y\boldsymbol{\sigma}_y + B_z\boldsymbol{\sigma}_z\right). \tag{3.33}$$

Density and Hamiltonian operators

We are going to call the two quaternions defined this way $\boldsymbol{\rho}$ and $\boldsymbol{H}$, a *density quaternion* and a *Hamiltonian quaternion,* respectively—and temporarily. Traditional physics terminology is a little different because most physicists don't think of these objects as quaternions. They think of them as *operators* instead and call them a *density operator* and a *Hamiltonian operator,* or a *Hamiltonian* for short.

3.5 Expectation Values

But there remains the question of how we should map the operation that yields the average energy of the ensemble, namely,

$$\langle \boldsymbol{\eta}, \boldsymbol{p} \rangle = -\mu \left(B_x r^x + B_y r^y + B_z r^z \right). \tag{3.34}$$

The simplest approach is to multiply the two quaternions and see what comes out:

$$\begin{aligned} \boldsymbol{H}\boldsymbol{\rho} &= -\mu \left(B_x \boldsymbol{\sigma}_x + B_y \boldsymbol{\sigma}_y + B_z \boldsymbol{\sigma}_z \right) \frac{1}{2} \left(\mathbf{1} + r^x \boldsymbol{\sigma}_x + r^y \boldsymbol{\sigma}_y + r^z \boldsymbol{\sigma}_z \right) \\ &= -\frac{\mu}{2} \Big[B_x \boldsymbol{\sigma}_x + B_y \boldsymbol{\sigma}_y + B_z \boldsymbol{\sigma}_z \\ &\qquad + r^x \left(B_x \boldsymbol{\sigma}_x + B_y \boldsymbol{\sigma}_y + B_z \boldsymbol{\sigma}_z \right) \boldsymbol{\sigma}_x \\ &\qquad + r^y \left(B_x \boldsymbol{\sigma}_x + B_y \boldsymbol{\sigma}_y + B_z \boldsymbol{\sigma}_z \right) \boldsymbol{\sigma}_y \\ &\qquad + r^z \left(B_x \boldsymbol{\sigma}_x + B_y \boldsymbol{\sigma}_y + B_z \boldsymbol{\sigma}_z \right) \boldsymbol{\sigma}_z \Big] \\ &= \ldots . \end{aligned}$$

Let us recall that a square of each sigma is $\mathbf{1}$, whereas $\boldsymbol{\sigma}_i \boldsymbol{\sigma}_j$ is some other sigma for $i \neq j$ multiplied by "i." This makes it easy to collect terms that are proportional to $\mathbf{1}$:

$$\begin{aligned} \ldots &= -\frac{\mu}{2} \Big[\left(r^x B_x + r^y B_y + r^z B_z \right) \mathbf{1} \\ &\qquad\qquad + \text{various terms multiplied by sigmas} \Big]. \end{aligned} \tag{3.35}$$

$\Re$ extracts the real part of a quaternion.

We find our solution $\langle E \rangle = -\mu \left(\boldsymbol{r} \cdot \boldsymbol{B} \right)$ standing right next to $\mathbf{1}$. All we need to do is to get rid of the remaining sigma terms. This is easy: we introduce a projection operation that extracts a *real* part from the quaternion, similarly to how we extract a real part from a complex number. Let us define

$$\Re \left(\boldsymbol{q} \right) = \Re \left(a \mathbf{1} + b \boldsymbol{\sigma}_x + c \boldsymbol{\sigma}_y + d \boldsymbol{\sigma}_z \right) \doteq a. \tag{3.36}$$

Hence,

$$2 \Re \left(\boldsymbol{H} \boldsymbol{\rho} \right) = -\mu \left(\boldsymbol{r} \cdot \boldsymbol{B} \right) = \langle \boldsymbol{\eta}, \boldsymbol{p} \rangle = \langle E \rangle. \tag{3.37}$$

Extracting fiducial vectors from quaternions

In a similar fashion we can obtain probabilities for finding a qubit in a specific state, expressions that are expectation values of a sort, too, from the density quaternion $\boldsymbol{\rho}$. Let us consider the following four quaternions:

$$\boldsymbol{P}^0 = \frac{1}{2} \left(\mathbf{1} + \boldsymbol{\sigma}_z \right), \tag{3.38}$$

$$\begin{aligned}
\boldsymbol{P}^1 &= \frac{1}{2}\left(\mathbf{1}-\boldsymbol{\sigma}_z\right), &(3.39)\\
\boldsymbol{P}^2 &= \frac{1}{2}\left(\mathbf{1}+\boldsymbol{\sigma}_x\right), &(3.40)\\
\boldsymbol{P}^3 &= \frac{1}{2}\left(\mathbf{1}+\boldsymbol{\sigma}_y\right). &(3.41)
\end{aligned}$$

First, let us observe that they are direct images of the canonical forms in the fiducial space, when expressed in terms of Pauli forms (cf. equations (2.76) on page 66):

$$\begin{aligned}
\boldsymbol{\omega}^0 &= \frac{1}{2}\left(\varsigma^1+\varsigma^z\right), &(3.42)\\
\boldsymbol{\omega}^1 &= \frac{1}{2}\left(\varsigma^1-\varsigma^z\right), &(3.43)\\
\boldsymbol{\omega}^2 &= \frac{1}{2}\left(\varsigma^1+\varsigma^x\right), &(3.44)\\
\boldsymbol{\omega}^3 &= \frac{1}{2}\left(\varsigma^1+\varsigma^y\right). &(3.45)
\end{aligned}$$

Let us apply $\boldsymbol{P}^2$ to $\boldsymbol{\rho}$ and then take $2\Re$ of the result:

$$\begin{aligned}
\boldsymbol{P}^2\boldsymbol{\rho} &= \frac{1}{2}\left(\mathbf{1}+\boldsymbol{\sigma}_x\right)\frac{1}{2}\left(\mathbf{1}+r^x\boldsymbol{\sigma}_x+r^y\boldsymbol{\sigma}_y+r^z\boldsymbol{\sigma}_z\right)\\
&= \frac{1}{4}\left(\mathbf{1}+r^x\boldsymbol{\sigma}_x+r^y\boldsymbol{\sigma}_y+r^z\boldsymbol{\sigma}_z+\boldsymbol{\sigma}_x\left(\mathbf{1}+r^x\boldsymbol{\sigma}_x+r^y\boldsymbol{\sigma}_y+r^z\boldsymbol{\sigma}_z\right)\right)\\
&= \ldots .
\end{aligned}$$

Now we are again going to collect terms that are proportional to $\mathbf{1}$. There is the single $\mathbf{1}$ in front, and then $\boldsymbol{\sigma}_x r^x \boldsymbol{\sigma}_x$ is going to produce another $\mathbf{1}$. All other terms will be multiplied by sigmas. Thus, we have the following.

$$\begin{aligned}
\ldots &= \frac{1}{4}\left(\mathbf{1}+r^x\mathbf{1}\right) &(3.46)\\
&\quad + \text{various terms multiplied by sigmas}
\end{aligned}$$

Taking $2\Re$ of it yields

$$2\Re\left(\boldsymbol{P}^2\boldsymbol{\rho}\right) = \frac{1}{2}\left(1+r^x\right). \tag{3.47}$$

In the same way one can easily see that

$$\begin{aligned}
2\Re\left(\boldsymbol{P}^0\boldsymbol{\rho}\right) &= \frac{1}{2}\left(1+r^z\right), &(3.48)\\
2\Re\left(\boldsymbol{P}^1\boldsymbol{\rho}\right) &= \frac{1}{2}\left(1-r^z\right), &(3.49)
\end{aligned}$$

$$2\Re\left(\boldsymbol{P}^3\boldsymbol{\rho}\right) = \frac{1}{2}\left(1 + r^y\right). \tag{3.50}$$

We have arrived at the following formula that extracts *probabilities* from the density quaternion of a qubit:

$$p^i = 2\Re\left(\boldsymbol{P}^i\boldsymbol{\rho}\right). \tag{3.51}$$

Probabilities so obtained are consistent with our original mapping (3.22)–(3.25) and show that the mapping can be reversed. To extract probabilities from the density quaternion, we can use the $\boldsymbol{P}^i$ quaternions and the $2\Re$ rule, or we can replace sigmas with varsigmas in the density quaternion and read the probabilities this way. The latter is, of course, easier.

Generally speaking, as we would represent any measurement on the quantum system by a fiducial form, upon switching to the quaternion representation of the system, we represent any measurement Q on the quantum system by a quaternion $\boldsymbol{Q}$ and the result of this measurement in terms of Q averaged over the ensemble by

$$\langle Q\rangle = 2\Re\left(\boldsymbol{Q}\boldsymbol{\rho}\right). \tag{3.52}$$

The formula $p^i = 2\Re\left(\boldsymbol{P}^i\boldsymbol{\rho}\right)$ tells us how to reconstruct the fiducial vector $\boldsymbol{p}$ from the density quaternion $\boldsymbol{\rho}$. Is there an analogous formula that would generate the Hamiltonian form $\boldsymbol{\eta}$ from the Hamilton quaternion $\boldsymbol{H}$, without resorting to replacing sigmas with the corresponding form-varsigmas?

Extracting fiducial forms from quaternions

Such a formula can be read out from $\langle\boldsymbol{\eta},\boldsymbol{p}\rangle$ as follows:

$$\langle E\rangle = \langle\boldsymbol{\eta},\boldsymbol{p}\rangle = \sum_i \eta_i p^i = \sum_i \eta_i 2\Re\left(\boldsymbol{P}^i\boldsymbol{\rho}\right) = 2\Re\left(\sum_i \eta_i\boldsymbol{P}^i\boldsymbol{\rho}\right) = 2\Re\left(\boldsymbol{H}\boldsymbol{\rho}\right), \tag{3.53}$$

where we have made use of the fact that $2\Re$ is a linear operation. From this it is now clear that

$$\boldsymbol{H} = \sum_i \eta_i\boldsymbol{P}^i. \tag{3.54}$$

So the way to read coefficients η_i is to express $\boldsymbol{H}$ *not* in terms of sigmas, but in terms of $\boldsymbol{P}^i$ instead!

Let us try it:

$$\mathbf{1} = \boldsymbol{P}^0 + \boldsymbol{P}^1, \tag{3.55}$$

$$\boldsymbol{\sigma}_z = \boldsymbol{P}^0 - \boldsymbol{P}^1, \tag{3.56}$$

$$\boldsymbol{\sigma}_x = 2\boldsymbol{P}^2 - \mathbf{1} = 2\boldsymbol{P}^2 - \boldsymbol{P}^0 - \boldsymbol{P}^1, \tag{3.57}$$

$$\boldsymbol{\sigma}_y = 2\boldsymbol{P}^3 - \mathbf{1} = 2\boldsymbol{P}^3 - \boldsymbol{P}^0 - \boldsymbol{P}^1. \tag{3.58}$$

Substituting this in place of sigmas in

$$\boldsymbol{H} = -\mu(B_x\boldsymbol{\sigma}_x + B_y\boldsymbol{\sigma}_y + B_z\boldsymbol{\sigma}_z) \tag{3.59}$$

yields

$$\begin{aligned}\boldsymbol{H} &= -\mu\left(B_x\left(2\boldsymbol{P}^2 - \boldsymbol{P}^0 - \boldsymbol{P}^1\right) + B_y\left(2\boldsymbol{P}^3 - \boldsymbol{P}^0 - \boldsymbol{P}^1\right) + B_z\left(\boldsymbol{P}^0 - \boldsymbol{P}^1\right)\right) \\ &= -\mu\left(\left(B_z - B_x - B_y\right)\boldsymbol{P}^0 - \left(B_z + B_x + B_y\right)\boldsymbol{P}^1 + 2B_x\boldsymbol{P}^2 + 2B_y\boldsymbol{P}^3\right),\end{aligned} \tag{3.60}$$

and so we get

$$\eta_0 = B_z - B_x - B_y, \tag{3.61}$$
$$\eta_1 = -B_z - B_x - B_y, \tag{3.62}$$
$$\eta_2 = 2B_x, \tag{3.63}$$
$$\eta_3 = 2B_y, \tag{3.64}$$

which is the same as $\boldsymbol{\eta} = B_x\boldsymbol{\varsigma}^x + B_y\boldsymbol{\varsigma}^y + B_z\boldsymbol{\varsigma}^z$, where

$$\boldsymbol{\varsigma}^x \equiv (-1, -1, 2, 0), \tag{3.65}$$
$$\boldsymbol{\varsigma}^y \equiv (-1, -1, 0, 2), \tag{3.66}$$
$$\boldsymbol{\varsigma}^z \equiv (1, -1, 0, 0). \tag{3.67}$$

3.6 Mixtures

Now that we know how to switch between quaternions, on the one hand, and vectors and forms of the qubit fiducial space, on the other, let us try expressing some ideas we explored in the preceding chapter in the quaternion language.

The first of these is going to be mixing two pure states. A pure state is described by a vector $\boldsymbol{r}$ with length 1. Within the fiducial formalism we used to construct a mixture by taking a convex linear combination of two or more pure states represented by probability vectors such as $\boldsymbol{p}_1$ and $\boldsymbol{p}_2$:

$$\boldsymbol{p} = a_1\boldsymbol{p}_1 + a_2\boldsymbol{p}_2, \tag{3.68}$$

where $a_1 + a_2 = 1$. Since the mapping between fiducial vectors and quaternions is linear, the same should hold for the density quaternion $\boldsymbol{\rho}$:

Mixing density quaternions

$$\begin{aligned}
\boldsymbol{\rho} &= a_1\boldsymbol{\rho_1} + a_2\boldsymbol{\rho_2} \\
&= a_1\frac{1}{2}\left(\mathbf{1} + r_1^x\boldsymbol{\sigma}_x + r_1^y\boldsymbol{\sigma}_y + r_1^z\boldsymbol{\sigma}_z\right) + a_2\frac{1}{2}\left(\mathbf{1} + r_2^x\boldsymbol{\sigma}_x + r_2^y\boldsymbol{\sigma}_y + r_2^z\boldsymbol{\sigma}_z\right) \\
&= \frac{1}{2}\left((a_1+a_2)\mathbf{1} + (a_1 r_1^x + a_2 r_2^x)\boldsymbol{\sigma}_x + (a_1 r_1^y + a_2 r_2^y)\boldsymbol{\sigma}_y + (a_1 r_1^z + a_2 r_2^z)\boldsymbol{\sigma}_z\right).
\end{aligned} \tag{3.69}$$

Since $a_1 + a_2 = 1$, the first term in the large brackets, $(a_1 + a_2)\mathbf{1}$, becomes $\mathbf{1}$. The remaining three sigma terms become $r^x\boldsymbol{\sigma}_x + r^y\boldsymbol{\sigma}_y + r^z\boldsymbol{\sigma}_z$, where

$$\boldsymbol{r} = a_1\boldsymbol{r}_1 + a_2\boldsymbol{r}_2, \tag{3.70}$$

and

$$\begin{aligned}
\boldsymbol{r}\cdot\boldsymbol{r} &= (a_1\boldsymbol{r}_1 + a_2\boldsymbol{r}_2)\cdot(a_1\boldsymbol{r}_1 + a_2\boldsymbol{r}_2) \\
&= a_1^2 + a_2^2 + 2a_1a_2\cos(\boldsymbol{r}_1, \boldsymbol{r}_2) \\
&\le a_1^2 + a_2^2 + 2a_1a_2 = (a_1+a_2)^2 = 1.
\end{aligned} \tag{3.71}$$

This is the same result we had obtained for the mixture of two fiducial vectors.

3.7 Qubit Evolution

The qubit evolution equation that looked like

$$\frac{\mathrm{d}}{\mathrm{d}t}\boldsymbol{r} = \frac{2\mu}{\hbar}\boldsymbol{r}\times\boldsymbol{B} \tag{3.72}$$

Von Neumann equation

has the following quaternion formulation:

$$\frac{\mathrm{d}}{\mathrm{d}t}\boldsymbol{\rho} = -\frac{\mathrm{i}}{\hbar}\left[\boldsymbol{H}, \boldsymbol{\rho}\right], \tag{3.73}$$

where

$$[\boldsymbol{H}, \boldsymbol{\rho}] \doteq \boldsymbol{H}\boldsymbol{\rho} - \boldsymbol{\rho}\boldsymbol{H} \tag{3.74}$$

is called the *commutator* of $\boldsymbol{H}$ and $\boldsymbol{\rho}$. Written in this form, the qubit evolution equation is called the von Neumann equation, after John von Neumann (1903–1957), one of the greatest scientists of the twentieth century, renowned for his extravagant parties, bad driving, unfailingly elegant attire, and short life—he died of cancer after having been exposed to high doses of radiation while working on nuclear weapons. Back then many scientists, beginning with Marie Curie, were quite careless about these things. We've learned from their misfortune.

To transform the equation back into (3.72) is straightforward.

First, let us consider the left-hand side of the von Neumann equation:

$$\begin{aligned}\frac{\mathrm{d}}{\mathrm{d}t}\boldsymbol{\rho} &= \frac{1}{2}\frac{\mathrm{d}}{\mathrm{d}t}\left(\mathbf{1} + r^x\boldsymbol{\sigma}_x + r^y\boldsymbol{\sigma}_y + r^z\boldsymbol{\sigma}_z\right) \\ &= \frac{1}{2}\left(\left(\frac{\mathrm{d}}{\mathrm{d}t}r^x\right)\boldsymbol{\sigma}_x + \left(\frac{\mathrm{d}}{\mathrm{d}t}r^y\right)\boldsymbol{\sigma}_y + \left(\frac{\mathrm{d}}{\mathrm{d}t}r^z\right)\boldsymbol{\sigma}_z\right) \\ &= \frac{1}{2}\left(\frac{\mathrm{d}}{\mathrm{d}t}\boldsymbol{r}\right)\cdot\vec{\boldsymbol{\sigma}}, \end{aligned} \tag{3.75}$$

The meaning of $\vec{\boldsymbol{\sigma}}$

because $\mathbf{1}$, $\boldsymbol{\sigma}_x$, $\boldsymbol{\sigma}_y$ and $\boldsymbol{\sigma}_z$ are all quaternion constants. The symbol $\vec{\boldsymbol{\sigma}}$ is a tri-vector of the three sigmas: $\boldsymbol{\sigma}_x$, $\boldsymbol{\sigma}_y$ and $\boldsymbol{\sigma}_z$. Expressions such as $v^x\boldsymbol{\sigma}_x + v^y\boldsymbol{\sigma}_y + v^z\boldsymbol{\sigma}_z$ occur so often that a typographic shortcut $\boldsymbol{v}\cdot\vec{\boldsymbol{\sigma}}$ was invented. But this should *not* be understood as a real scalar product of two vectors in which infomation is lost. Rather, it should be understood as something similar to $v^x\boldsymbol{e}_x + v^y\boldsymbol{e}_y + v^z\boldsymbol{e}_z \doteq \boldsymbol{v}\cdot\vec{\boldsymbol{e}}$. Here no information is lost as the operation is performed. There are three fully extractable components of $\boldsymbol{v}$ on both sides of the equation.

Now, let us turn to the commutator:

$$\begin{aligned}&\boldsymbol{H\rho} - \boldsymbol{\rho H} \\ &\quad = -\frac{\mu}{2}\Big[\left(B_x\boldsymbol{\sigma}_x + B_y\boldsymbol{\sigma}_y + B_z\boldsymbol{\sigma}_z\right)\left(\mathbf{1} + r^x\boldsymbol{\sigma}_x + r^y\boldsymbol{\sigma}_y + r^z\boldsymbol{\sigma}_z\right) \\ &\qquad - \left(\mathbf{1} + r^x\boldsymbol{\sigma}_x + r^y\boldsymbol{\sigma}_y + r^z\boldsymbol{\sigma}_z\right)\left(B_x\boldsymbol{\sigma}_x + B_y\boldsymbol{\sigma}_y + B_z\boldsymbol{\sigma}_z\right)\Big] \\ &\quad \ldots .\end{aligned}$$

Before we plunge into the fury of computational rage, let us have a sanguine look at the equation. To begin with, we are going to have terms resulting from multiplication of the Hamilton quaternion by the $\mathbf{1}$ of the density quaternion. These will produce $\boldsymbol{B}\cdot\vec{\boldsymbol{\sigma}}$ from $\boldsymbol{H\rho}$ and $-\boldsymbol{B}\cdot\vec{\boldsymbol{\sigma}}$ from $-\boldsymbol{\rho H}$, so they'll cancel out. Next, we're going to have terms resulting from $\boldsymbol{\sigma}_i$, for $i = x, y, z$, multiplying themselves, for example, $\boldsymbol{\sigma}_x\boldsymbol{\sigma}_x$. For each i, $\boldsymbol{\sigma}_i\boldsymbol{\sigma}_i = \mathbf{1}$, so these terms are going to produce $\boldsymbol{B}\cdot\boldsymbol{r}\,\mathbf{1}$ from $\boldsymbol{H\rho}$, which is exactly what we had in $2\Re(\boldsymbol{H\rho})$, and $-\boldsymbol{B}\cdot\boldsymbol{r}\,\mathbf{1}$ from $\boldsymbol{\rho H}$, so they'll cancel out, too.

The only terms that are going to survive, then, will be the asymmetric terms:

$$\begin{aligned}\ldots = -\frac{\mu}{2}\Big[&\big(B_x r^y\boldsymbol{\sigma}_x\boldsymbol{\sigma}_y + B_x r^z\boldsymbol{\sigma}_x\boldsymbol{\sigma}_z \\ &+B_y r^x\boldsymbol{\sigma}_y\boldsymbol{\sigma}_x + B_y r^z\boldsymbol{\sigma}_y\boldsymbol{\sigma}_z \\ &+B_z r^x\boldsymbol{\sigma}_z\boldsymbol{\sigma}_x + B_z r^y\boldsymbol{\sigma}_z\boldsymbol{\sigma}_y\big) \\ &-\big(B_x r^y\boldsymbol{\sigma}_y\boldsymbol{\sigma}_x + B_x r^z\boldsymbol{\sigma}_z\boldsymbol{\sigma}_x\end{aligned}$$

$$
\begin{aligned}
&+B_y r^x \boldsymbol{\sigma}_x \boldsymbol{\sigma}_y + B_y r^z \boldsymbol{\sigma}_z \boldsymbol{\sigma}_y \\
&+B_z r^x \boldsymbol{\sigma}_x \boldsymbol{\sigma}_z + B_z r^y \boldsymbol{\sigma}_y \boldsymbol{\sigma}_z)\Big] \\
= \;& \ldots .
\end{aligned}
$$

Let us observe that the expression in the second round bracket is the same as the expression in the first round bracket but with sigmas ordered the other way. Switching them around produces minus, which cancels with the minus of the commutator, so that we end up with the following:

$$
\begin{aligned}
\ldots \;=\; & -\frac{\mu}{2} 2 \left(B_x r^y \boldsymbol{\sigma}_x \boldsymbol{\sigma}_y + B_x r^z \boldsymbol{\sigma}_x \boldsymbol{\sigma}_z\right. \\
& +B_y r^x \boldsymbol{\sigma}_y \boldsymbol{\sigma}_x + B_y r^z \boldsymbol{\sigma}_y \boldsymbol{\sigma}_z \\
& \left.+B_z r^x \boldsymbol{\sigma}_z \boldsymbol{\sigma}_x + B_z r^y \boldsymbol{\sigma}_z \boldsymbol{\sigma}_y\right) \\
=\;& \ldots .
\end{aligned}
$$

But here we also have sigma pairs that are just switched around. Reordering them the "right way" produces minuses, so that we end up with

$$
\begin{aligned}
\ldots \;=\; & -\mu\Big[\left(B_x r^y - B_y r^x\right) \boldsymbol{\sigma}_x \boldsymbol{\sigma}_y \\
& + \left(B_y r^z - B_z r^y\right) \boldsymbol{\sigma}_y \boldsymbol{\sigma}_z \\
& + \left(B_z r^x - B_x r^z\right) \boldsymbol{\sigma}_z \boldsymbol{\sigma}_x \Big] \\
=\;& \ldots .
\end{aligned}
$$

Finally, let us replace $\boldsymbol{\sigma}_x \boldsymbol{\sigma}_y$ with $i\boldsymbol{\sigma}_z$, and similarly for the other two pairs, to get

$$
\begin{aligned}
\ldots \;=\; & -i\mu\Big[\left(B_x r^y - B_y r^x\right) \boldsymbol{\sigma}_z + \left(B_y r^z - B_z r^y\right) \boldsymbol{\sigma}_x + \left(B_z r^x - B_x r^z\right) \boldsymbol{\sigma}_y \Big] \\
=\;& \ldots . \qquad (3.76)
\end{aligned}
$$

Expressions in the round brackets are components of a vector product $\boldsymbol{B} \times \boldsymbol{r}$. We can therefore rewrite our result in a more compact form as follows:

$$
[\boldsymbol{H}, \boldsymbol{\rho}] = -\mathrm{i}\mu \left(\boldsymbol{B} \times \boldsymbol{r}\right) \cdot \vec{\boldsymbol{\sigma}}. \qquad (3.77)
$$

The von Neumann equation for the quaternion-described qubit turns into

$$
\frac{1}{2}\left(\frac{\mathrm{d}}{\mathrm{d}t}\boldsymbol{r}\right) \cdot \vec{\boldsymbol{\sigma}} = -\frac{\mathrm{i}}{\hbar}\left(-\mathrm{i}\mu\right)\left(\boldsymbol{B} \times \boldsymbol{r}\right) \cdot \vec{\boldsymbol{\sigma}}, \qquad (3.78)
$$

which then yields

$$
\frac{\mathrm{d}}{\mathrm{d}t}\boldsymbol{r} = \frac{2\mu}{\hbar}\boldsymbol{r} \times \boldsymbol{B}, \qquad (3.79)
$$

quod erat demonstrandum.

3.8 Why Does It Work?

Before we attempt to answer this question of why the quaternion trick works, let us first explain that all that we've done with Pauli quaternions, the ones preferred by the physicists, we could have done just as easily with the original Hamilton quaternions. The imaginary unit "i" that pops up in the von-Neumann equation is there only because of the way we had defined the sigmas as "i" times Hamilton quaternions $\boldsymbol{i}$, $\boldsymbol{j}$, or $\boldsymbol{k}$. It does not really represent anything deep or fundamental. It is merely ornamental and somewhat misleading.

If we were to use the original Hamilton quaternions, our equations would look similar, with only a sign different here or there, and without any imaginary units. For example, if we were to map

There is no need for "i" when using Hamilton quaternions.

$$\varsigma_1 \leftrightarrow 1, \tag{3.80}$$

$$\varsigma_x \leftrightarrow \boldsymbol{i}, \tag{3.81}$$

$$\varsigma_y \leftrightarrow \boldsymbol{j}, \tag{3.82}$$

$$\varsigma_z \leftrightarrow \boldsymbol{k}, \tag{3.83}$$

we would get

$$\langle \boldsymbol{\eta}, \boldsymbol{p} \rangle = -2\Re\left(\boldsymbol{H}\boldsymbol{\rho}\right). \tag{3.84}$$

But we could be a little fancier here and could, for example, map forms onto *conjugate* quaternions, meaning

Forms could be mapped on conjugate quaternions

$$\boldsymbol{\eta} = B_x\boldsymbol{\varsigma}^x + B_y\boldsymbol{\varsigma}^y + B_z\boldsymbol{\varsigma}^z \quad \rightarrow \quad -B_x\boldsymbol{i} - B_y\boldsymbol{j} - B_z\boldsymbol{k}. \tag{3.85}$$

This would yield

$$\langle E \rangle = 2\Re\left(\boldsymbol{H}\boldsymbol{\rho}\right). \tag{3.86}$$

The von Neumann equation in the Hamilton quaternion formalism and with forms mapped onto normal quaternions, not the conjugate ones, is

$$\frac{\mathrm{d}}{\mathrm{d}t}\boldsymbol{\rho} = \frac{1}{\hbar}\left[\boldsymbol{H}, \boldsymbol{\rho}\right]. \tag{3.87}$$

There is no "i" here and no minus either. The minus could be restored by mapping fiducial forms onto conjugate quaternions instead.

So, whether we use sigmas or $\boldsymbol{i}$, $\boldsymbol{j}$, and $\boldsymbol{k}$, we get much the same picture—and we still don't have to do quantum mechanics with complex numbers as quaternion coefficients.

Now, why does the quaternion trick work? The reason is that their peculiar commutation properties encode both a scalar product (or a *dot* product as some call it) and a vector product. In general

$$\begin{aligned} \boldsymbol{ab} &= (a^x\boldsymbol{i} + a^y\boldsymbol{j} + a^z\boldsymbol{k})\,(b^x\boldsymbol{i} + b^y\boldsymbol{j} + b^z\boldsymbol{k}) \\ &= (a^x b^x + a^y b^y + a^z b^z)\,(-1) \\ &\quad + [(a^y b^z - a^z b^y)\,\boldsymbol{jk} + (a^z b^x - a^x b^z)\,\boldsymbol{ki} + (a^x b^y - a^y b^x)\,\boldsymbol{ij}] \\ &= -\vec{a}\cdot\vec{b} + \left(\vec{a}\times\vec{b}\right)^x \boldsymbol{i} + \left(\vec{a}\times\vec{b}\right)^y \boldsymbol{j} + \left(\vec{a}\times\vec{b}\right)^z \boldsymbol{k}. \end{aligned} \tag{3.88}$$

Quaternions encode a dot and a cross product of two vectors

This is where all the magic comes from. To extract the scalar product from $\boldsymbol{ab}$, we could just take the real part of it, the $\Re$, and multiply by whatever coefficient is needed to get the right answer, or we could take $\boldsymbol{ab} + \boldsymbol{ba}$ and then the $(\boldsymbol{i}, \boldsymbol{j}, \boldsymbol{k})$ part of the quaternion would cancel out. To extract the vector product from $\boldsymbol{ab}$ we could extract the coefficients that multiply $\boldsymbol{i}$, $\boldsymbol{j}$, and $\boldsymbol{k}$, or we could take $\boldsymbol{ab} - \boldsymbol{ba}$, and then the real part of the quaternion, the scalar product part, would cancel out.

To extract the coefficients that multiply $\boldsymbol{i}$, $\boldsymbol{j}$, and $\boldsymbol{k}$, we can use operators similar to $\Re$. A complex number operator that extracts the imaginary component of a complex number is called $\Im$. But in the case of the quaternions we need three such operators, and so we're going to call them $\Im_i$, $\Im_j$, and $\Im_k$.

Mapping theories onto quaternions

In summary, any theory that contains a combination of scalar and vector products can be mapped onto quaternions and special quaternion rules devised to extract whatever equations of the original theory are wanted.

Quaternions encode elementary three-dimensional vector algebra in the form of "numbers" with four slots each.

Describing rotations with quaternions

For this reason Hamilton was able to encode rotations and their combinations by using quaternions. This is, in fact, what he invented them for. Various equations of special and even general relativity can be mapped onto quaternion algebra, too.

Our theory of qubits, which describes their statistical ensembles in terms of probability vectors, eventually resolves to scalar and vector products, namely, $\langle E\rangle = -\mu\,(\boldsymbol{r}\cdot\boldsymbol{B})$ and $\mathrm{d}\boldsymbol{r}/\mathrm{d}t = 2\mu\,(\boldsymbol{r}\times\boldsymbol{B})\,/\hbar$. Hence, it is only natural that a mapping exists that lets us express it in terms of quaternions.

That a mapping comes out to be so simple, with Pauli vectors and forms mapping directly onto 1 (or $\mathbf{1}$) and $(\boldsymbol{i}, \boldsymbol{j}, \boldsymbol{k})$ (or $(\boldsymbol{\sigma}_x, \boldsymbol{\sigma}_y, \boldsymbol{\sigma}_z)$), derives from the fact that we had defined ς_1, ς_x, ς_y, and ς_z so that this would be the case, knowing in advance the result we wanted to achieve.

But we didn't cheat. Indeed, $\boldsymbol{p} = \frac{1}{2}\,(\varsigma_1 + r^x\varsigma_x + r^y\varsigma_y + r^z\varsigma_z)$ fully describes a qubit state, meaning a state of a statistical ensemble that represents it.

A more profound question is why the following works for qubits:

$$\boldsymbol{p} = \frac{1}{2}\begin{pmatrix} 1 + r^z \\ 1 - r^z \\ 1 + r^x \\ 1 + r^y \end{pmatrix}, \tag{3.89}$$

where $\boldsymbol{r}^2 \leq 1$. All else, after all, follows from it.

The reason is that it is so general.

The first two entries in $\boldsymbol{p}$ represent the fact that a beam of qubits splits in two in the presence of a magnetic field that has a nonvanishing gradient in the z direction. Individual qubits must go either up or down. *Tertium non datur.* This exhausts all possibilities. Consequently $p^0 + p^1 = 1$. To express p^0 and p^1 as $(1 + r^z)/2$ and $(1 - r^z)/2$ with $r^z \in [0, 1]$ merely parameterizes this observation. But the experiment also tells us that the description of a qubit in terms of r^z alone is incomplete. Knowing p^0 and p^1 is *not* enough to fully describe the state of a qubit. To fully describe the qubit, we must know how it is going to behave in magnetic fields whose gradients are in the x and y directions, too. Hence the additional two terms, which are similarly parameterized by r^x and r^y, and the transformation rules that make $\boldsymbol{r} = (r^x, r^y, r^z)$ a vector.

Contrary to what some physicists think and say, a spin qubit can be polarized in *any* direction in the three-dimensional space. This can be always *confirmed* by rotating the magnet arrangement in all possible directions until the beam of qubits no longer splits and all qubits in it are deflected in the same direction, which corresponds to $\boldsymbol{r}$. Such a direction can be always found if a beam is fully polarized. If such a direction does not exist, the beam is not fully polarized; it is a mixture.

So we come to a surprising conclusion: The reason $\boldsymbol{p}$ describes a qubit is that it has enough slots to do so. The reason quaternions describe a qubit is that they capture any theory that has scalar and vector products in it.

An astute reader may stop here and ask: "OK, I'll accept this explanation about $\boldsymbol{p}$ and quaternions, but what about $\mathrm{d}\boldsymbol{r}/\mathrm{d}t = 2\mu\left(\boldsymbol{r} \times \boldsymbol{B}\right)/\hbar$ and $\langle E \rangle = -\mu\,\boldsymbol{r} \cdot \boldsymbol{B}$? Why do these work?"

These are the simplest possible equations that combine two vectors, $\boldsymbol{r}$ and $\boldsymbol{B}$, to produce a scalar (energy) or a vector (torque).

But this is also where the real physics is. After all, it is Nature that makes this choice, to be simple. What the equations say is that a qubit behaves *on average* like a classical magnetic dipole. This on-average behavior represents the classical thermodynamic limit of the theory. If we were to immerse a macroscopic sample full of qubits in a magnetic field $\boldsymbol{B}$, we would observe precession of the average

A qubit behaves like a classical magnetic dipole on average.

magnetic field of the sample exactly as described by $\mathrm{d}\boldsymbol{r}/\mathrm{d}t = 2\mu\left(\boldsymbol{r} \times \boldsymbol{B}\right)/\hbar$. If we were to buzz the sample with the $\boldsymbol{B}_\perp$ field, we'd see Rabi oscillations.

This, in quick summary, is *all* that quantum mechanics can say about a qubit: that a beam (or a sample) of qubits splits in two in the presence of varying magnetic fields (or some other drivers that are described by $\boldsymbol{B}$) and that an ensemble of identically prepared fully polarized qubits behaves, mathematically, like a classical magnetic dipole *on average*.

Quantum mechanics is a probability theory. All its pronouncements refer only to statistical ensembles of quantum objects. Quantum mechanics has nothing whatsoever to say about a particular individual quantum object—unless in very special and rare circumstances when the behavior of the whole ensemble can be predicted with 100% certainty. Only in such cases can we be certain about what an individual quantum object is going to do.

4 The Unitary Formalism

4.1 Unpacking Pauli Quaternions

The *unitary* description of qubits arises when the $\boldsymbol{\sigma}_x$, $\boldsymbol{\sigma}_y$, and $\boldsymbol{\sigma}_z$ quaternions are "unpacked" into matrices.

Why should we bother about unpacking them in the first place, if we can obtain all the information we need directly from the quaternion picture or by using fiducial vectors and forms? This is a good question. It translates into another even more profound question. Since the fiducial formalism and its mapping onto quaternions—otherwise known as the density operator formalism—already cover all qubit physics, what new physics can we possibly arrive at by unpacking the sigmas? Are we merely going to delude ourselves with unnecessary metaphysics? This question latches directly onto the business of quantum computing, because the whole idea derives from the *notation* of the unitary calculus.

Matrix representation of quaternions

What do we mean by "unpacking the sigmas"? We are going to look for matrix representations of sigmas that have the same commutation and anticommutation properties, namely,

$$\boldsymbol{\sigma}_x^2 = \boldsymbol{\sigma}_y^2 = \boldsymbol{\sigma}_z^2 = \mathbf{1}, \tag{4.1}$$

$$\boldsymbol{\sigma}_x \boldsymbol{\sigma}_y \boldsymbol{\sigma}_z = \mathrm{i}\mathbf{1}. \tag{4.2}$$

The other sigma properties, for example, $\boldsymbol{\sigma}_x \boldsymbol{\sigma}_y = -\boldsymbol{\sigma}_y \boldsymbol{\sigma}_x = \mathrm{i}\boldsymbol{\sigma}_z$, can be derived from the two—as we have done with the Hamilton quaternions $\boldsymbol{i}$, $\boldsymbol{j}$, and $\boldsymbol{k}$.

4.2 Pauli Matrices

It is easy to see that we cannot represent the sigmas by numbers alone, real or complex, because the sigmas anticommute and neither real nor complex numbers do. But matrices do not commute in general either, and so the simplest representation of sigmas can be sought in the form of 2×2 matrices. Because we have the "i," the imaginary unit, in our commutation relations for sigmas, we will have to consider 2×2 complex matrices. Pure real matrices will no longer do, because how would we generate the "i." Also, it will become clear, after we will have completed the exercise, that if we were to use Hamilton's original $\boldsymbol{i}$, $\boldsymbol{j}$, and $\boldsymbol{k}$ symbols instead, we would still end up with complex valued matrices. At this level, there is no escape from complex numbers.

Let us begin with the following general parameterization of 2×2 matrices:

$$\boldsymbol{\sigma}_x = \begin{pmatrix} a_{11} & a_{12} \\ a_{21} & a_{22} \end{pmatrix}, \tag{4.3}$$

$$\boldsymbol{\sigma}_y = \begin{pmatrix} b_{11} & b_{12} \\ b_{21} & b_{22} \end{pmatrix}, \tag{4.4}$$

$$\boldsymbol{\sigma}_z = \begin{pmatrix} c_{11} & c_{12} \\ c_{21} & c_{22} \end{pmatrix}, \tag{4.5}$$

where a_{ij}, b_{ij}, and c_{ij} are complex numbers.

The first rule that applies equally to all sigmas is that they square to $\mathbf{1}$. Let us perform the corresponding computation on $\boldsymbol{\sigma}_z$:

$$\begin{aligned} \boldsymbol{\sigma}_z\boldsymbol{\sigma}_z &= \begin{pmatrix} c_{11} & c_{12} \\ c_{21} & c_{22} \end{pmatrix}\begin{pmatrix} c_{11} & c_{12} \\ c_{21} & c_{22} \end{pmatrix} \\ &= \begin{pmatrix} c_{11}^2 + c_{12}c_{21} & c_{12}(c_{11}+c_{22}) \\ c_{21}(c_{11}+c_{22}) & c_{22}^2 + c_{12}c_{21} \end{pmatrix} \\ &= \begin{pmatrix} 1 & 0 \\ 0 & 1 \end{pmatrix}. \end{aligned} \tag{4.6}$$

This yields the following two groups of equations:

$$c_{12}(c_{11}+c_{22}) = 0, \tag{4.7}$$

$$c_{21}(c_{11}+c_{22}) = 0, \tag{4.8}$$

and

$$c_{11}^2 + c_{12}c_{21} = 1, \tag{4.9}$$

$$c_{22}^2 + c_{12}c_{21} = 1. \tag{4.10}$$

The first two equations can be satisfied by either

$$c_{11} = -c_{22}, \tag{4.11}$$

or

$$c_{12} = c_{21} = 0. \tag{4.12}$$

Let us suppose the latter holds. This means that the matrix is diagonal, and we also get that

$$c_{11}^2 = c_{22}^2 = 1, \tag{4.13}$$

which implies that

$$c_{11} = \pm 1 \quad \text{and} \quad c_{22} = \pm 1. \tag{4.14}$$

We end up with two possibilities:

$$\boldsymbol{\sigma}_z = \pm\begin{pmatrix} 1 & 0 \\ 0 & 1 \end{pmatrix}, \quad \text{or} \tag{4.15}$$

$$\boldsymbol{\sigma}_z = \pm \begin{pmatrix} 1 & 0 \\ 0 & -1 \end{pmatrix}. \tag{4.16}$$

The first solution is just **1**. This is not a good solution here because we want $\boldsymbol{\sigma}_z$ to be something other than **1** (we already have a **1** in the quaternion quartet), so we're going to take the second one, and we're going to choose the plus sign for it:

$$\boldsymbol{\sigma}_z = \begin{pmatrix} 1 & 0 \\ 0 & -1 \end{pmatrix}. \tag{4.17}$$

In effect we have satisfied both equations, (4.11) and (4.12).

For the remaining two matrices, $\boldsymbol{\sigma}_x$ and $\boldsymbol{\sigma}_y$, we have to choose the remaining option, that is,

$$a_{12} \neq 0 \quad \text{and} \tag{4.18}$$

$$a_{21} \neq 0, \tag{4.19}$$

and similarly for $\boldsymbol{\sigma}_y$ and the bs, because otherwise we'd end up with either **1** or $\begin{pmatrix} 1 & 0 \\ 0 & -1 \end{pmatrix}$ again. But if $a_{12} \neq 0$ and $a_{21} \neq 0$, then we must have

$$a_{11} = -a_{22} \doteq a \quad \text{and} \tag{4.20}$$

$$b_{11} = -b_{22} \doteq b, \tag{4.21}$$

where a and b are such that

$$a^2 + a_{12}a_{21} = 1 \quad \text{and} \tag{4.22}$$

$$b^2 + b_{12}b_{21} = 1. \tag{4.23}$$

Now, let us make use of the anticommutation rule $\boldsymbol{\sigma}_x\boldsymbol{\sigma}_z + \boldsymbol{\sigma}_z\boldsymbol{\sigma}_x = \mathbf{0}$. First we have

$$\begin{aligned} \boldsymbol{\sigma}_x\boldsymbol{\sigma}_z &= \begin{pmatrix} a & a_{12} \\ a_{21} & -a \end{pmatrix}\begin{pmatrix} 1 & 0 \\ 0 & -1 \end{pmatrix} \\ &= \begin{pmatrix} a & -a_{12} \\ a_{21} & a \end{pmatrix}, \end{aligned} \tag{4.24}$$

but

$$\begin{aligned} \boldsymbol{\sigma}_z\boldsymbol{\sigma}_x &= \begin{pmatrix} 1 & 0 \\ 0 & -1 \end{pmatrix}\begin{pmatrix} a & a_{12} \\ a_{21} & -a \end{pmatrix} \\ &= \begin{pmatrix} a & a_{12} \\ -a_{21} & a \end{pmatrix}. \end{aligned} \tag{4.25}$$

Adding these two yields

$$\boldsymbol{\sigma}_x\boldsymbol{\sigma}_z + \boldsymbol{\sigma}_z\boldsymbol{\sigma}_x = 2\begin{pmatrix} a & 0 \\ 0 & a \end{pmatrix}. \tag{4.26}$$

For the anticommutation rule to hold, a must be zero. Since the same argument is going to work for $\boldsymbol{\sigma}_y$ as well, we get

$$a = 0 \quad \text{and} \tag{4.27}$$

$$b = 0. \tag{4.28}$$

In summary,

$$\boldsymbol{\sigma}_x = \begin{pmatrix} 0 & a_{12} \\ a_{21} & 0 \end{pmatrix}, \tag{4.29}$$

$$\boldsymbol{\sigma}_y = \begin{pmatrix} 0 & b_{12} \\ b_{21} & 0 \end{pmatrix}. \tag{4.30}$$

But let us recall that as a and b vanish, we are left with

$$a_{12}a_{21} = 1 \quad \text{and} \tag{4.31}$$

$$b_{12}b_{21} = 1. \tag{4.32}$$

We can satisfy these equations by setting $a_{12} = x$ and $a_{21} = 1/x$ and similarly $b_{12} = y$ and $b_{21} = 1/y$, so that

$$\boldsymbol{\sigma}_x = \begin{pmatrix} 0 & x \\ x^{-1} & 0 \end{pmatrix}, \tag{4.33}$$

$$\boldsymbol{\sigma}_y = \begin{pmatrix} 0 & y \\ y^{-1} & 0 \end{pmatrix}. \tag{4.34}$$

Now, we are ready to make use of the rule stating that

$$\boldsymbol{\sigma}_x\boldsymbol{\sigma}_y\boldsymbol{\sigma}_z = \mathrm{i}\mathbf{1}. \tag{4.35}$$

Substituting our matrix expressions for $\boldsymbol{\sigma}_x$, $\boldsymbol{\sigma}_y$, and $\boldsymbol{\sigma}_z$ yields

$$\begin{pmatrix} 0 & x \\ x^{-1} & 0 \end{pmatrix}\begin{pmatrix} 0 & y \\ y^{-1} & 0 \end{pmatrix}\begin{pmatrix} 1 & 0 \\ 0 & -1 \end{pmatrix} = \begin{pmatrix} x/y & 0 \\ 0 & -y/x \end{pmatrix} = \begin{pmatrix} \mathrm{i} & 0 \\ 0 & \mathrm{i} \end{pmatrix}. \tag{4.36}$$

This yields the following two equations in combination with what we have arrived at already:

$$x/y = \mathrm{i} \quad \text{and} \tag{4.37}$$

$$-y/x = \mathrm{i}. \tag{4.38}$$

Both solve to

$$y = -\mathrm{i}x. \tag{4.39}$$

Since $1/y = 1/(-\mathrm{i}x) = \mathrm{i}x^{-1}$, we end up with

$$\boldsymbol{\sigma}_x = \begin{pmatrix} 0 & x \\ x^{-1} & 0 \end{pmatrix}, \tag{4.40}$$

$$\boldsymbol{\sigma}_y = \begin{pmatrix} 0 & -\mathrm{i}x \\ \mathrm{i}x^{-1} & 0 \end{pmatrix}, \tag{4.41}$$

$$\boldsymbol{\sigma}_z = \begin{pmatrix} 1 & 0 \\ 0 & -1 \end{pmatrix}. \tag{4.42}$$

However surprising this may be to some physicists on account of the x factor, it is easy to check, for example, with Maple, or Mathematica, or manually even, that these matrices indeed satisfy all quaternion relations expected of $\boldsymbol{\sigma}_x$, $\boldsymbol{\sigma}_y$, and $\boldsymbol{\sigma}_z$, which is sufficient to get the right expressions for $\langle E \rangle = -\mu\left(\boldsymbol{r} \cdot \boldsymbol{B}\right)$ via $\langle E \rangle = 2\Re\left(\boldsymbol{H}\boldsymbol{\rho}\right)$ and for $\mathrm{d}\boldsymbol{r}/\mathrm{dt} = 2\mu\left(\mathrm{r} \times \mathrm{B}\right)/\hbar$ via $\mathrm{d}\boldsymbol{\rho}/\mathrm{dt} = -\mathrm{i}\left[\boldsymbol{H}, \boldsymbol{\rho}\right]/\hbar$.

The choice of $x = 1$ is a natural one in this context, though not strictly necessary at this stage. But it makes $\boldsymbol{\sigma}_x$ pleasingly symmetric and $\boldsymbol{\sigma}_y$ pleasingly Hermitian, which means that $A_{ij} = A^*_{ji}$, and both properties will prove useful as we go along.

Hermitian matrices

Hermitian matrices are of special importance in quantum physics. We will see on page 129 that x will be further restricted to $e^{\mathrm{i}\phi_x}$, so that $\boldsymbol{\sigma}_y$, even with this remaining degree of freedom left, will end up being Hermitian anyway, as will $\boldsymbol{\sigma}_x$. The condition $A_{ij} = A^*_{ji}$ says that if we transpose a Hermitian matrix $\boldsymbol{A}$ and then complex conjugate it, the resulting matrix is the same as the original one,

$$\left(\boldsymbol{A}^T\right)^* = \boldsymbol{A}. \tag{4.43}$$

Hermitian adjoint

A special symbol, † (a dagger, not a cross), represents this *Hermitian adjoint* of $\boldsymbol{A}$:

$$\boldsymbol{A}^\dagger \doteq \left(\boldsymbol{A}^T\right)^*. \tag{4.44}$$

Using this symbol, we can say that operator $\boldsymbol{A}$ is Hermitian when

$$\boldsymbol{A}^\dagger = \boldsymbol{A}. \tag{4.45}$$

Pauli matrices

Having made the choice of $x = 1$—and we are *not* going to forget about it, instead working on its further justification as we develop the unitary formalism—we arrive at the quaternion representation in terms of Pauli matrices:

$$\mathbf{1} = \begin{pmatrix} 1 & 0 \\ 0 & 1 \end{pmatrix}, \tag{4.46}$$

$$\boldsymbol{\sigma}_x = \begin{pmatrix} 0 & 1 \\ 1 & 0 \end{pmatrix}, \tag{4.47}$$

$$\boldsymbol{\sigma}_y = \begin{pmatrix} 0 & -\mathrm{i} \\ \mathrm{i} & 0 \end{pmatrix}, \tag{4.48}$$

$$\boldsymbol{\sigma}_z = \begin{pmatrix} 1 & 0 \\ 0 & -1 \end{pmatrix}. \tag{4.49}$$

It is useful to express the canonical basis in the space of 2×2 matrices in terms of Pauli matrices, a procedure similar to what we did earlier with canonical forms in the fiducial space and Pauli forms (cf. equations 2.76 on page 66). So,

$$\boldsymbol{M}_0 = \begin{pmatrix} 1 & 0 \\ 0 & 0 \end{pmatrix} = \frac{1}{2}\left(\mathbf{1} + \boldsymbol{\sigma}_z\right), \tag{4.50}$$

$$\boldsymbol{M}_1 = \begin{pmatrix} 0 & 1 \\ 0 & 0 \end{pmatrix} = \frac{1}{2}\left(\boldsymbol{\sigma}_x + \mathrm{i}\boldsymbol{\sigma}_y\right), \tag{4.51}$$

$$\boldsymbol{M}_2 = \begin{pmatrix} 0 & 0 \\ 1 & 0 \end{pmatrix} = \frac{1}{2}\left(\boldsymbol{\sigma}_x - \mathrm{i}\boldsymbol{\sigma}_y\right), \tag{4.52}$$

$$\boldsymbol{M}_3 = \begin{pmatrix} 0 & 0 \\ 0 & 1 \end{pmatrix} = \frac{1}{2}\left(\mathbf{1} - \boldsymbol{\sigma}_z\right). \tag{4.53}$$

Let us observe that only $\boldsymbol{M}_0$ and $\boldsymbol{M}_3$ bear similarity to $\boldsymbol{P}^0$ and $\boldsymbol{P}^1$ of equations 3.41 on page 101. Matrix representations of $\boldsymbol{P}^2$ and $\boldsymbol{P}^3$ are

$$\boldsymbol{P}^2 = \frac{1}{2}\begin{pmatrix} 1 & 1 \\ 1 & 1 \end{pmatrix}, \tag{4.54}$$

$$\boldsymbol{P}^3 = \frac{1}{2}\begin{pmatrix} 1 & -\mathrm{i} \\ \mathrm{i} & 1 \end{pmatrix}. \tag{4.55}$$

Pauli matrices are traceless.

With the exception of $\mathbf{1}$, all other Pauli matrices are *traceless*, which means that the sum of their diagonal elements is zero. $\mathbf{1}$ itself has trace of 2. We can therefore use the matrix operation of taking trace, Tr, in place of the quaternion operation $\Re$. Also, let us note that since $\mathrm{Tr}\left(\mathbf{1}\right) = 2$, we can drop the 2 factor that appeared in front of $\Re$, that is,

$$2\Re\left(\boldsymbol{H}\boldsymbol{\rho}\right) = \mathrm{Tr}\left(\boldsymbol{H}\boldsymbol{\rho}\right). \tag{4.56}$$

This $\mathrm{Tr}(\mathbf{1}) = 2$, in fact, is related to some of the 1/2s and 2's that appeared in definitions of the qubit's fiducial vector and its close cousin, the density quaternion, as well as in the definitions of Pauli forms and of the metric tensor of the fiducial space.

Of the three Pauli matrices one, $\boldsymbol{\sigma}_y$, is imaginary. Could we do away with imaginary numbers if we were to carry out the procedure for Hamilton quaternions $\boldsymbol{i}$, $\boldsymbol{j}$, and $\boldsymbol{k}$? The answer is "no"; we wouldn't do better. Let us recall that

$$\boldsymbol{i} = -\mathrm{i}\boldsymbol{\sigma}_x, \tag{4.57}$$

$$\boldsymbol{j} = -\mathrm{i}\boldsymbol{\sigma}_y, \tag{4.58}$$

$$\boldsymbol{k} = -\mathrm{i}\boldsymbol{\sigma}_z. \tag{4.59}$$

This transformation would make $\boldsymbol{j}$ real, but then we'd end up with imaginary $\boldsymbol{i}$ and $\boldsymbol{k}$.

If we were to substitute a_{ij}, b_{ij}, and c_{ij}, as we did previously, for $\boldsymbol{i}$, $\boldsymbol{j}$, and $\boldsymbol{k}$, equation $\boldsymbol{kk} = -\mathbf{1}$ would yield

$$c_{12}\left(c_{11} + c_{22}\right) = 0, \tag{4.60}$$

$$c_{21}\left(c_{11} + c_{22}\right) = 0, \tag{4.61}$$

as before, but this time

$$c_{11}^2 + c_{12}c_{21} = -1, \tag{4.62}$$

$$c_{22}^2 + c_{12}c_{21} = -1. \tag{4.63}$$

In the case of $c_{12} = c_{21} = 0$ we get that $c_{11}^2 = -1$ and $c_{22}^2 = -1$, which implies $c_{11} = \pm\mathrm{i} = c_{22}$ right away.

There is no real 2×2 representation of quaternions.

In summary, whereas we can describe qubit dynamics and kinematics *completely* in terms of real numbers and measurable probabilities alone, as long as we work either within the fiducial formalism or within the density quaternion framework, the moment we "unpack" the quaternions, we have to let the imaginary numbers in. Like Alice falling into the rabbit hole, we will encounter strange creatures and notions, some of which may well belong in the fantasy world.

4.3 The Basis Vectors and the Hilbert Space

Density matrix

A fully unpacked density quaternion of a qubit becomes a 2×2 complex matrix that looks as follows:

$$\boldsymbol{\rho} = \frac{1}{2}\left(\mathbf{1} + r^x \boldsymbol{\sigma}_x + r^y \boldsymbol{\sigma}_y + r^z \boldsymbol{\sigma}_z\right) = \frac{1}{2}\begin{pmatrix} 1 + r^z & r^x - \mathrm{i} r^y \\ r^x + \mathrm{i} r^y & 1 - r^z \end{pmatrix}, \tag{4.64}$$

and in this form it is called a *density operator* or a *density matrix* of a qubit.

We had seen an operator $\boldsymbol{P}$ in Section 1.8 (page 32) that talked about "transformations of mixtures" represented as a sum of tensor products of basis vectors and forms, where the vectors and forms were ordered "the other way round" so that they wouldn't eat each other:

$$\boldsymbol{P} \doteq \sum_j \sum_i p^j{}_i \boldsymbol{e}_j \otimes \boldsymbol{\omega}^i. \tag{4.65}$$

The trick can be applied to every linear operator that is represented by a matrix, including $\boldsymbol{\rho}$, in which case we can write

$$\begin{aligned} \boldsymbol{\rho} &= \frac{1}{2}\Big((1 + r^z)\, \boldsymbol{e}_0 \otimes \boldsymbol{\omega}^0 + (r^x - \mathrm{i} r^y)\, \boldsymbol{e}_0 \otimes \boldsymbol{\omega}^1 \\ &\qquad + (r^x + \mathrm{i} r^y)\, \boldsymbol{e}_1 \otimes \boldsymbol{\omega}^0 + (1 - r^z)\, \boldsymbol{e}_1 \otimes \boldsymbol{\omega}^1 \Big), \end{aligned} \tag{4.66}$$

where $\boldsymbol{e}_i$ and $\boldsymbol{\omega}^i$ are canonical basis vectors and forms in this new two-dimensional complex vector space into which we have unpacked our quaternions. What they are will transpire when we have a closer look at some specific qubit states we know and understand well by now.

Let us consider a state that is described by

$$\boldsymbol{r} = \begin{pmatrix} 0 \\ 0 \\ 1 \end{pmatrix}, \tag{4.67}$$

or, in other words, by $\boldsymbol{r} = \boldsymbol{e}_z$, where $\boldsymbol{e}_z$ is a unit-length vector that points in the z direction, that is, vertically up. This is a fully polarized (pure) state. Its fiducial representation is $\boldsymbol{p} = \frac{1}{2}\left(\varsigma_1 + \varsigma_z\right)$, its quaternion representation is $\boldsymbol{\rho} = \frac{1}{2}\left(\mathbf{1} + \boldsymbol{\sigma}_z\right)$, and its density matrix representation is

$$\boldsymbol{\rho} = \begin{pmatrix} 1 & 0 \\ 0 & 0 \end{pmatrix} = \boldsymbol{e}_0 \otimes \boldsymbol{\omega}^0. \tag{4.68}$$

The pair $\boldsymbol{e}_0$ and $\boldsymbol{\omega}^0$ represent the $\boldsymbol{r} = \boldsymbol{e}_z$ state.

This tells us that the pair $\boldsymbol{e}_0$ and $\boldsymbol{\omega}^0$ can be employed to represent the $\boldsymbol{r} = \boldsymbol{e}_z$ state. We had encountered this state in Section 2.4 (page 55) that talked about polarized states, and on that occasion we called it $|\uparrow\rangle$. We are going to adopt the same notation here, including the complementary notation for the $\boldsymbol{\omega}^0$ form:

$$\boldsymbol{e}_0 \doteq |\uparrow\rangle, \tag{4.69}$$

$$\boldsymbol{\omega}^0 \doteq \langle\uparrow| \,. \tag{4.70}$$

Let us call the corresponding density operator $\boldsymbol{\rho}_\uparrow$, to distinguish it from density operators that will describe other states. We can write

$$\boldsymbol{\rho}_\uparrow = |\uparrow\rangle \otimes \langle\uparrow|, \tag{4.71}$$

or

$$\boldsymbol{\rho}_\uparrow = |\uparrow\rangle\langle\uparrow| \tag{4.72}$$

for short. Physicists normally drop the tensor product symbol $\otimes$ in this context.[1]

$\boldsymbol{\rho}_\uparrow = |\uparrow\rangle\langle\uparrow|$

Vector $|\uparrow\rangle$ and its dual form $\langle\uparrow|$ can be described in terms of columns and rows of numbers:

$$\boldsymbol{e}_0 \doteq |\uparrow\rangle \doteq \begin{pmatrix} 1 \\ 0 \end{pmatrix}, \tag{4.73}$$

$$\boldsymbol{\omega}^0 \doteq \langle\uparrow| \doteq (1, 0)\,, \tag{4.74}$$

$$\boldsymbol{e}_0 \otimes \boldsymbol{\omega}^0 = \begin{pmatrix} 1 \\ 0 \end{pmatrix} (1,0) = \begin{pmatrix} 1 & 0 \\ 0 & 0 \end{pmatrix}. \tag{4.75}$$

The last equation is a genuine matrix multiplication; that is, multiplying a column vector by a row one, with the column vector on the left side of the row, and applying the usual matrix multiplication rules, builds a 2×2 matrix that is in this case $\boldsymbol{\rho}_\uparrow$. Also, let us observe that

$\langle\uparrow|\uparrow\rangle = 1$

$$\langle\uparrow|\uparrow\rangle = \langle\boldsymbol{\omega}^0, \boldsymbol{e}_0\rangle = (1,0)\begin{pmatrix} 1 \\ 0 \end{pmatrix} = 1, \tag{4.76}$$

which is as it should be, because $\boldsymbol{\omega}^0$ is dual to $\boldsymbol{e}_0$.

Forms, bras and rows versus vectors, kets and columns

The terminology of *a form* and *a vector* adopted in this text is relatively new to mathematics and newer still to physics, if we were to measure time in centuries—it goes back about a century to Élie Cartan (1869–1951). When Dirac invented his angular bracket notation used in quantum mechanics today, he called *kets* what we call vectors here, and what we call forms, he called *bras*. Thus a conjugation of a form and a vector, for example, $\langle\uparrow|\downarrow\rangle$, becomes a *bra-ket*. Einstein and his differential geometry colleagues, on the other hand, called forms *covariant vectors* and vectors *contravariant vectors*, because form coefficients transform like basis vectors and vector coefficients transform the other way round. Finally, people who work with computers prefer to call forms *row vectors* and to call vectors *column*

[1]Some aren't even aware of its existence!

vectors. All these terminologies are still in use today, depending on who we talk to, and sometimes even depending on a context.

Now, let us consider a state that is given by

$$\boldsymbol{r} = \begin{pmatrix} 0 \\ 0 \\ -1 \end{pmatrix}, \tag{4.77}$$

or, in other words, by $\boldsymbol{r} = -\boldsymbol{e}_z$. This is also a fully polarized state whose fiducial representation is $\boldsymbol{p} = \frac{1}{2}(\varsigma_1 - \varsigma_z)$ and whose quaternion representation is $\boldsymbol{\rho} = \frac{1}{2}(\mathbf{1} - \boldsymbol{\sigma}_z)$. Its density matrix representation is

$$\boldsymbol{\rho} = \begin{pmatrix} 0 & 0 \\ 0 & 1 \end{pmatrix} = \boldsymbol{e}_1 \otimes \boldsymbol{\omega}^1. \tag{4.78}$$

This tells us that the pair $\boldsymbol{e}_1$ and $\boldsymbol{\omega}^1$ can be used to represent the $\boldsymbol{r} = -\boldsymbol{e}_z$ state. We encountered this state in Section 2.4, too, and called it $|\downarrow\rangle$ back then. So, we are going to adopt this notation here, as well, together with the complementary notation for its dual form:

$$\begin{aligned} \boldsymbol{e}_1 &\doteq |\downarrow\rangle, && (4.79)\\ \boldsymbol{\omega}^1 &\doteq \langle\downarrow| \,. && (4.80) \end{aligned}$$

$\boldsymbol{\rho}_\downarrow = |\downarrow\rangle\langle\downarrow|$

Let us call the corresponding density operator $\boldsymbol{\rho}_\downarrow$. We can now write

$$\boldsymbol{\rho}_\downarrow = |\downarrow\rangle \otimes \langle\downarrow|, \quad \text{or} \quad \boldsymbol{\rho}_\downarrow = |\downarrow\rangle\langle\downarrow| \,. \tag{4.81}$$

Vector $|\downarrow\rangle$ and its dual form $\langle\downarrow|$ can be described in terms of columns and rows of numbers as follows:

$$\begin{aligned} \boldsymbol{e}_1 &\doteq |\downarrow\rangle \doteq \begin{pmatrix} 0 \\ 1 \end{pmatrix}, && (4.82)\\ \boldsymbol{\omega}_1 &\doteq \langle\downarrow| \doteq (0,1)\,, && (4.83)\\ \boldsymbol{e}_1 \otimes \boldsymbol{\omega}^1 &= \begin{pmatrix} 0 \\ 1 \end{pmatrix}(0,1) = \begin{pmatrix} 0 & 0 \\ 0 & 1 \end{pmatrix}, && (4.84) \end{aligned}$$

$\langle\downarrow|\downarrow\rangle = 1$

where the last, as before, is a genuine matrix multiplication. Since $|\downarrow\rangle$ and $\langle\downarrow|$ are dual, we have that

$$\langle\downarrow|\downarrow\rangle = 1. \tag{4.85}$$

But we have

$$\langle\downarrow|\uparrow\rangle = (0,1)\begin{pmatrix} 1 \\ 0 \end{pmatrix} = 0, \quad \text{and} \tag{4.86}$$

$$\langle\uparrow|\downarrow\rangle = (1,0)\begin{pmatrix}0\\1\end{pmatrix} = 0, \quad (4.87)$$

as well.

And so we find that the two vectors, $|\uparrow\rangle$ and $|\downarrow\rangle$, comprise a basis of the two-dimensional complex vector space in which $\boldsymbol{\rho}_\uparrow$ and $\boldsymbol{\rho}_\downarrow$ operate—once we have replaced the quaternion symbols $\boldsymbol{\sigma}_x$, $\boldsymbol{\sigma}_y$ and $\boldsymbol{\sigma}_z$ with Pauli matrices.[2]

Hilbert space of a qubit

The vector space spanned by $|\downarrow\rangle$ and $|\uparrow\rangle$ is called the *Hilbert space* of a qubit. Apart from being a complex vector space and, in this case, two dimensional, it has some other properties that make it *Hilbert.* We're going to discover them one by one as we explore it.

Vectors $|\uparrow\rangle$ and $|\downarrow\rangle$ correspond to the *physical basis* states of the qubit, in the sense that was discussed in Sections 1.5 (page 19) and 2.2 (page 47).

4.4 The Superstition of Superposition

What about qubit states such as $|\rightarrow\rangle$ and $|\otimes\rangle$, states that correspond to polarization directions that are perpendicular to $\boldsymbol{e}_z$? They are, after all, perfectly normal qubit beam states that can be *confirmed* by orienting the beam-splitting magnet appropriately. Have they been excluded from the unitary formalism?

Since they can be described in terms of fiducial vectors, and therefore in terms of quaternions, too, it should be possible to describe them within the framework of the unitary formalism, even though the only basis vectors that we can build our vector space from correspond to $|\uparrow\rangle$ and $|\downarrow\rangle$.

Let us consider a density matrix that corresponds to $\boldsymbol{r} = \boldsymbol{e}_x$. The state is described by $\boldsymbol{p} = \frac{1}{2}(\varsigma_1 + \varsigma_x)$ or by $\boldsymbol{\rho} = \frac{1}{2}(\mathbf{1} + \boldsymbol{\sigma}_x)$. Upon having unpacked $\boldsymbol{\sigma}_x$ we find

$$\begin{aligned}\boldsymbol{\rho}_\rightarrow &= \frac{1}{2}\begin{pmatrix}1 & 1\\1 & 1\end{pmatrix}\\ &= \frac{1}{2}\left(|\uparrow\rangle\otimes\langle\uparrow| + |\uparrow\rangle\otimes\langle\downarrow| + |\downarrow\rangle\otimes\langle\uparrow| + |\downarrow\rangle\otimes\langle\downarrow|\right)\\ &= \frac{1}{\sqrt{2}}\left(|\uparrow\rangle + |\downarrow\rangle\right)\otimes\frac{1}{\sqrt{2}}\left(\langle\uparrow| + \langle\downarrow|\right).\end{aligned} \quad (4.88)$$

[2]These are also called $\boldsymbol{\sigma}_x$, $\boldsymbol{\sigma}_y$, and $\boldsymbol{\sigma}_z$. Whenever there is a possibility of confusion, we will attempt to clarify whether the sigmas employed in various formulas should be thought of as quaternion symbols or matrices. The general rule is that if we operate on sigmas using their commutation and anticommutation properties only, they are quaternions. If we unpack them and use their matrix properties, they are Pauli matrices. Physicists call them Pauli matrices in all contexts.

$|\rightarrow\rangle$ is a superposition of $|\uparrow\rangle$ and $|\downarrow\rangle$.

This lets us identify

$$|\rightarrow\rangle = \frac{1}{\sqrt{2}}\left(|\uparrow\rangle + |\downarrow\rangle\right). \tag{4.89}$$

Similarly, for the qubit state that corresponds to $\boldsymbol{r} = -\boldsymbol{e}_x$ we get

$$\begin{aligned} \boldsymbol{\rho}_{\leftarrow} &= \frac{1}{2}\begin{pmatrix} 1 & -1 \\ -1 & 1 \end{pmatrix} \\ &= \frac{1}{2}\left(|\uparrow\rangle \otimes \langle\uparrow| - |\uparrow\rangle \otimes \langle\downarrow| - |\downarrow\rangle \otimes \langle\uparrow| + |\downarrow\rangle \otimes \langle\downarrow|\right) \\ &= \frac{1}{\sqrt{2}}\left(|\uparrow\rangle - |\downarrow\rangle\right) \otimes \frac{1}{\sqrt{2}}\left(\langle\uparrow| - \langle\downarrow|\right). \end{aligned} \tag{4.90}$$

$|\leftarrow\rangle$ is a superposition of $|\uparrow\rangle$ and $|\downarrow\rangle$, too.

Hence

$$|\leftarrow\rangle = \frac{1}{\sqrt{2}}\left(|\uparrow\rangle - |\downarrow\rangle\right). \tag{4.91}$$

States $|\otimes\rangle$ and $|\odot\rangle$ that correspond do $\boldsymbol{r} = \boldsymbol{e}_y$ and $\boldsymbol{r} = -\boldsymbol{e}_y$ have similar representation in the unitary space of a qubit but with one subtle difference. Let us start with $|\otimes\rangle = \frac{1}{2}\left(\varsigma_1 + \varsigma_y\right)$. The density matrix equivalent is

$$\begin{aligned} \boldsymbol{\rho}_{\otimes} &= \frac{1}{2}\begin{pmatrix} 1 & -\mathrm{i} \\ \mathrm{i} & 1 \end{pmatrix} \\ &= \frac{1}{\sqrt{2}}\left(|\uparrow\rangle + \mathrm{i}\,|\downarrow\rangle\right) \otimes \frac{1}{\sqrt{2}}\left(\langle\uparrow| - \mathrm{i}\,\langle\downarrow|\right). \end{aligned} \tag{4.92}$$

A form dual to $\left(|\uparrow\rangle + \mathrm{i}\,|\downarrow\rangle\right)/\sqrt{2}$ is $\left(\langle\uparrow| - \mathrm{i}\,\langle\downarrow|\right)/\sqrt{2}$.

At first, it may look like we have a vector here and a form that are not related, or not *dual*. But let us evaluate the following:

$$\begin{aligned} &\frac{1}{\sqrt{2}}\left(\langle\uparrow| - \mathrm{i}\,\langle\downarrow|\right)\frac{1}{\sqrt{2}}\left(|\uparrow\rangle + \mathrm{i}\,|\downarrow\rangle\right) \\ &\quad = \frac{1}{2}\left(\langle\uparrow|\uparrow\rangle + \mathrm{i}\,\langle\uparrow|\downarrow\rangle - \mathrm{i}\,\langle\downarrow|\uparrow\rangle + \langle\downarrow|\downarrow\rangle\right) \\ &\quad = \frac{1}{2}\left(1 + \mathrm{i}\,0 - \mathrm{i}\,0 + 1\right) = 1. \end{aligned} \tag{4.93}$$

The result tells us that in this particular two-dimensional vector space with complex coefficients, whenever we want to make a form out of a vector, we need to convert all occurrences of "i" to "-i."

Making a dual in the Hilbert space

This is another feature of what's called a Hilbert space. If a vector space wants to be one, it is not enough for the space to be just a complex vector space; it has to have this property as well. Furthermore if we have an operator $\boldsymbol{A}$ acting on a vector $|\Psi\rangle$, namely $\boldsymbol{A}\,|\Psi\rangle$, then the image of this operation in the form world is

$\langle \Psi \mid \boldsymbol{A}^\dagger$, where $\dagger$ represents *Hermitian adjoint* defined by equation (4.44) on page 115. Only if the operator is Hermitian, that is, such that $\boldsymbol{A} = \boldsymbol{A}^\dagger$, we find that the dual of $\boldsymbol{A} \mid \Psi\rangle$ is $\langle \Psi \mid \boldsymbol{A}$, without the dagger.

In summary:

$$| \otimes \rangle = \frac{1}{\sqrt{2}} \left(|\uparrow\rangle + \mathrm{i}\, |\downarrow\rangle \right) \quad \text{and} \tag{4.94}$$

$$\langle \otimes | = \frac{1}{\sqrt{2}} \left(\langle\uparrow| - \mathrm{i} \langle\downarrow| \right). \tag{4.95}$$

Without further calculation we can easily guess that

$$| \odot \rangle = \frac{1}{\sqrt{2}} \left(|\uparrow\rangle - \mathrm{i}\, |\downarrow\rangle \right) \quad \text{and} \tag{4.96}$$

$$\langle \odot | = \frac{1}{\sqrt{2}} \left(\langle\uparrow| + \mathrm{i} \langle\downarrow| \right). \tag{4.97}$$

What is the meaning of $|{\rightarrow}\rangle = \left(|\uparrow\rangle + |\downarrow\rangle\right)/\sqrt{2}$?

A plus in the superposition does not translate into a mixture of states.

First, the plus operator, $+$, used in this context is *not* the same plus we had used in adding probability vectors or their corresponding quaternions. Back then, adding two probability vectors of pure states (similarly for quaternions) would *always* have resulted in a mixed state, unless the constituents were one and the same state. But here we add two pure states, which are not identical at all, and we end up with another pure state. The addition of two vectors in the qubit's Hilbert space, which is called a *superposition*, does *not* map onto addition of the two probability vectors, or quaternions. The transition

$$\boldsymbol{\rho}_{\rightarrow} \rightarrow |{\rightarrow}\rangle \tag{4.98}$$

is clearly nonlinear. But this one

$$\boldsymbol{\rho}_{\rightarrow} \rightarrow |{\rightarrow}\rangle\langle{\rightarrow}| \tag{4.99}$$

is. We can think of the unitary formalism, in which states are represented by Hilbert space vectors, as something akin to a square root of the density operator formalism.

Hiding information

The second aspect to ponder is that this feature of the qubit's Hilbert space not so much reveals new physics as hides some of what is transparent in the fiducial or quaternion formalisms, namely, polarization states that are perpendicular to $\boldsymbol{e}_z$.

The measurable physics of a qubit in the $|{\rightarrow}\rangle$ state is described by its fiducial vector of probabilities

$$\boldsymbol{p}_{\rightarrow} = \frac{1}{2}(\varsigma_1 + \varsigma_x) = \begin{pmatrix} 1/2 \\ 1/2 \\ 1 \\ 1/2 \end{pmatrix}. \tag{4.100}$$

The vector says that if a beam of qubits in this state is sent through a magnetic beam splitter with the magnetic field gradient pointing in the $\boldsymbol{e}_z$ direction, then the beam is going to split in half: one half of all incident qubits will swing upwards, the other half downwards. If the beam splitter is rotated so that its magnetic field gradient points in the $\boldsymbol{e}_y$ direction, again, the beam is going to split in two equal halves. But if the beam splitter is rotated so that its magnetic field gradient points in the $\boldsymbol{e}_x$ direction, then the beam is not going to split at all. Instead, all qubits in the beam will be deflected in the same direction. The last measurement therefore *confirms* that the beam is polarized in the $\boldsymbol{e}_x$ direction.

Unitary interpretation of superposition

But people who take the notation of the unitary formalism too literally have a different interpretation of this state. They say that $|{\rightarrow}\rangle = \frac{1}{\sqrt{2}}(|{\uparrow}\rangle + |{\downarrow}\rangle)$ means that *every* qubit in the beam is *simultaneously* polarized in the $\boldsymbol{e}_z$ direction *and* in the $-\boldsymbol{e}_z$ direction, and they argue that this is what the experiment shows: If we send a beam of qubits through the beam splitter oriented in the $\boldsymbol{e}_z$ direction, the beam splits in half—some qubits get deflected upwards, some downwards.

One may have the following argument with this viewpoint.

First, its proponents attribute to every individual qubit what is clearly a property *not* of an individual qubit but of a statistical ensemble of qubits. This is what every theory of probability is all about, and quantum mechanics of a qubit is merely another theory of probability. The problem derives from thinking of probability as a measure of tendency, attributable to every individual member of an ensemble, as opposed to thinking of probability in terms of how it is measured: by counting frequencies within the ensemble. There is a certain probability that I may find a \$100 bill on the floor of the supermarket tomorrow. Does this mean that I half-have and half-not-have this bill in my hand today? Does this mean that I have a certain proclivity toward finding \$100 bills on supermarket floors?

Second, they tend to forget about the option of rotating the beam splitter so as to find a direction for which the beam no longer splits. The reason is that this possibility cannot be clearly read from the unitary representation of the state. It is there, but it is hidden.

Fiducial interpretation of superposition

On the other hand, the interpretation that is read from the fiducial vector $\boldsymbol{p}$ or from its equivalent quaternion $\boldsymbol{\rho}$ is that the qubits in the beam are polarized *neither* in the $\boldsymbol{e}_z$ direction *nor* in the $-\boldsymbol{e}_z$ direction. They are polarized in the $\boldsymbol{e}_x$ direction instead, which is perpendicular to both $\boldsymbol{e}_z$ and $-\boldsymbol{e}_z$. They are going to be flipped onto either the $\boldsymbol{e}_z$ or the $-\boldsymbol{e}_z$ directions only when they are measured by an $\boldsymbol{e}_z$ oriented apparatus, and with such probability as can be read from $\boldsymbol{p}$ or $\boldsymbol{\rho}$.

It is not always easy to find a clear, physical interpretation of a quantum state that is described by a linear combination of some basis vectors in the Hilbert space. For example, what does $|\rightarrow\rangle$ mean for the quantronium? But rather than saying that the quantum object has *all* the constituent properties of a superposition, it is probably better to say that it has *neither* of them, that its property is altogether different. What it is exactly we're not going to see until we switch back to the density operator formalism—through the $|\Psi\rangle \rightarrow |\Psi\rangle\langle\Psi|$ operation—or, even better, to the fiducial formalism—through the $\boldsymbol{\sigma}_i \rightarrow \varsigma_i$ mapping—and evaluate probabilities of every experimental measurement needed to fully characterize the state.

> The full vector of these probabilities *is* the state. $|\rightarrow\rangle$ is its mathematical abstraction.

Sometimes an investigated system, such as quantronium, is not equipped to carry out a full set of measurements, and all it can deliver are p^0 and p^1. But, at least in principle, p^2 and p^3 are there, too, and they can be detected with appropriately improved measurement setup, or by more elaborate experimentation. For example, Ramsey fringes generated by the quantronium demonstrated to us the presence of Larmor precession, even though the quantronium does not have a circuitry needed to measure p^2 and p^3.

N-dimensional quantum systems, where N is the number of dimensions of their corresponding Hilbert space map onto N^2-dimensional fiducial systems [60]. Hence, the number of probabilities needed to fully characterize a quantum system grows rapidly with the dimension of the system. This is a practical reason why we often prefer to work within the confines of the unitary formalism, especially for more complex systems. Whereas it is easy to talk about p^0, p^1, p^2, and p^3 for a single qubit, if we were to consider an 8-qubit register, the number of dimensions of the corresponding Hilbert space would be $2^8 = 256$, but the number of dimensions of the corresponding fiducial space would be 65,536. Consequently, even if the fiducial viewpoint is right, the unitary viewpoint is more practical.

Fiducial vector of a general state in the qubit's Hilbert space

Let us consider a general superposition of two Hilbert space basis vectors $|\uparrow\rangle$ and $|\downarrow\rangle$:

$$|\Psi\rangle = a\,|\uparrow\rangle + b\,|\downarrow\rangle, \tag{4.101}$$

where a and b are two complex numbers. We can convert it to its corresponding density matrix, first by constructing the form

$$\langle \Psi \mid = a^* \langle \uparrow \mid + b^* \langle \downarrow \mid, \tag{4.102}$$

where the asterisk denotes complex conjugation, that is, an operation that swaps "i" for "-i", and then by building an operator out of the two:

$$\mid \Psi \rangle \langle \Psi \mid = aa^* \; |\uparrow\rangle\langle\uparrow| + ab^* \; |\uparrow\rangle\langle\downarrow| + ba^* \; |\downarrow\rangle\langle\uparrow| + bb^* \; |\downarrow\rangle\langle\downarrow| \, . \tag{4.103}$$

Let us compare this to the general form of the density matrix as obtained from the quaternion representation in order to figure out how the full information about the state of the polarization of the qubit beam is hidden in the unitary formalism's superposition:

$$aa^* = \frac{1}{2}\left(1 + r^z\right), \tag{4.104}$$

$$bb^* = \frac{1}{2}\left(1 - r^z\right), \tag{4.105}$$

$$ab^* = \frac{1}{2}\left(r^x - \mathrm{i}r^y\right), \tag{4.106}$$

$$ba^* = \frac{1}{2}\left(r^x + \mathrm{i}r^y\right). \tag{4.107}$$

Adding the first two equations yields

$$aa^* + bb^* = 1. \tag{4.108}$$

This tells us that we cannot construct just any linear combinations of $|\uparrow\rangle$ and $|\downarrow\rangle$. Combinations that are physically meaningful are restricted by the condition $aa^* + bb^* = 1$. This is called the normalization condition. All Hilbert space vectors that correspond to physical states of a qubit must satisfy it.

Physically meaningful Hilbert space vectors must be normalized.

Also, let us observe that $aa^* = \frac{1}{2}\left(1 + r^z\right) = p^0$ is the probability of registering the qubit with its spin up and $bb^* = \frac{1}{2}\left(1 - r^z\right) = p^1$ is the probability of registering the qubit with its spin down. The normalization condition is therefore the same as $p^0 + p^1 = 1$.

Subtracting the second equation from the first one yields

$$aa^* - bb^* = r^z. \tag{4.109}$$

We can extract r^x and r^y similarly from the third and the fourth equations:

$$ab^* + ba^* = r^x \quad \text{and} \tag{4.110}$$

$$\mathrm{i}\left(ab^* - ba^*\right) \quad = \quad r^y. \tag{4.111}$$

Now, let us see if we can fill the whole Bloch *ball* with superpositions of $|\uparrow\rangle$ and $|\downarrow\rangle$:

Only the Bloch sphere, not the full ball, can be covered with Hilbert space states.

$$\begin{aligned}
& r^x r^x + r^y r^y + r^z r^z \\
& = \left(ab^* + ba^*\right)^2 + \mathrm{i}^2\left(ab^* - ba^*\right)^2 + \left(aa^* - bb^*\right)^2 \\
& = \left(ab^*\right)^2 + 2ab^*ba^* + \left(ba^*\right)^2 \\
& \quad - \left(ab^*\right)^2 + 2ab^*ba^* - \left(ba^*\right)^2 \\
& \quad + \left(aa^*\right)^2 - 2aa^*bb^* + \left(bb^*\right)^2 \\
& = \left(aa^*\right)^2 + 2aa^*bb^* + \left(bb^*\right)^2 \\
& = \left(aa^* + bb^*\right)^2 \\
& = 1.
\end{aligned} \tag{4.112}$$

We can fill the Bloch sphere, but *not* the Bloch ball. In other words, the unitary formalism presented so far is restricted to fully polarized beams only. We have no means to describe mixtures within this framework. It looks like what we have here represents only a subset of the quantum probability theory, and as such it is incomplete.

Mixtures in unitary formalism

This would be a high price to pay. If we cannot describe mixtures, we cannot describe the measurement process, since, as we had seen in Chapter 2, Section 2.6 (page 62), the measurement process converted a pure state to a mixture.

Measurement

Within the framework of the traditionally formulated unitary theory the measurement process has been handled by introducing an auxiliary unphysical *axiom*. The axiom is unphysical because it elevates the measurement process above the theory, making it a special act rather than a result of a physical interaction between a measuring apparatus and an observed quantum system.

Depolarization

A similar problem would seem to affect our ability to describe the effects of the interaction between an observed quantum system and the environment within the framework of the unitary formalism. We saw in Section 2.12 (page 86), which talked about the quantronium circuit, that the qubit depolarized gradually. This manifested in the diminishing amplitude of Rabi and Ramsey oscillations. The initially pure state of the qubit converted into a mixture.

But it is not so bad.

The situation here is similar to the situation we encounter in investigating classical dissipative systems. There, energy is obviously lost to the environment, yet it does not mean that energy on the whole is not conserved. The energy may leak from an observed subsystem, for example a damped oscillator, but we can recover it eventually by including air and its expansion, friction, and heat, as well as temperature and lengthening of the spring in the model.

Dissipative quantum systems

It is similar with the unitary formalism. We will show in the next chapter that a "dissipative," meaning a nonunitary, quantum system can be always embedded in a larger "nondissipative" unitary system. In this way, unitarity, like energy, is "conserved" globally, even if it appears to leak out of the portion of the system under observation. Careful analysis of what happens when a unitary quantum system interacts with the environment lets us derive dissipative quantum equations, their classical analog being Newton equations with friction.

Is the universe unitary?

This mathematical trick leads some physicists to proclaim that the universe itself must be a quantum unitary system, but this is just a far-fetched extrapolation well beyond the domain of the theory. We cannot even be certain that the theory of gravity, as it is commonly accepted today, is applicable to objects of galactic size and beyond—some claim, with sound justification, that it is not [16]. Although we know a great deal about the universe, there is apparently much that we do not know either.

Further specification of Pauli matrices

Let us return to the choice of $x = 1$ in equations (4.40) and (4.41) in Section 4.2 (page 115) and explain it some more.

Let us focus on equation (4.88) on page 121, but this time let us use

$$\boldsymbol{\sigma}_x = \begin{pmatrix} 0 & x \\ x^{-1} & 0 \end{pmatrix}. \tag{4.113}$$

This would yield

$$\begin{aligned} \boldsymbol{\sigma}_{\rightarrow} &= \frac{1}{2}\begin{pmatrix} 1 & x \\ x^{-1} & 1 \end{pmatrix} \\ &= \frac{1}{2}\left(|\uparrow\rangle \otimes \langle\uparrow| + x\,|\uparrow\rangle \otimes \langle\downarrow| + x^{-1}\,|\downarrow\rangle \otimes \langle\uparrow| + |\downarrow\rangle \otimes \langle\downarrow|\right). \end{aligned} \tag{4.114}$$

To find a Hilbert space vector and its dual form that correspond to this operator, we can try to match it against a tensor product of a general superposition and its dual form,

$$\begin{aligned} &(a\,|\uparrow\rangle + b\,|\downarrow\rangle) \otimes (a^*\,\langle\uparrow| + b^*\,\langle\downarrow|) \\ &\quad = aa^*\,|\uparrow\rangle \otimes \langle\uparrow| + ab^*\,|\uparrow\rangle \otimes \langle\downarrow| + ba^*\,|\downarrow\rangle \otimes \langle\uparrow| + bb^*\,|\downarrow\rangle \otimes \langle\downarrow|. \end{aligned} \tag{4.115}$$

This yields the following equations:

$$aa^* = |a|^2 = \frac{1}{2}, \quad \text{hence} \quad |a| = \frac{1}{\sqrt{2}}, \tag{4.116}$$

$$\begin{aligned} bb^* &= |b|^2 = \frac{1}{2}, \quad \text{hence} \quad |b| = \frac{1}{\sqrt{2}}, && (4.117) \\ ab^* &= \frac{1}{2}x, && (4.118) \\ ba^* &= \frac{1}{2}x^{-1}. && (4.119) \end{aligned}$$

Let us use Euler notation for a and b, namely,

$$\begin{aligned} a &= |a|e^{\mathrm{i}\phi_a} \quad \text{and} && (4.120) \\ b &= |b|e^{\mathrm{i}\phi_b}. && (4.121) \end{aligned}$$

Then equations (4.118) and (4.119) become

$$\begin{aligned} ab^* &= |a||b|e^{\mathrm{i}(\phi_a-\phi_b)} = \frac{1}{2}e^{\mathrm{i}(\phi_a-\phi_b)} = \frac{1}{2}x, && (4.122) \\ ba^* &= |b||a|e^{\mathrm{i}(\phi_b-\phi_a)} = \frac{1}{2}e^{-\mathrm{i}(\phi_a-\phi_b)} = \frac{1}{2}x^{-1}, && (4.123) \end{aligned}$$

which yield

$$x = e^{\mathrm{i}(\phi_a-\phi_b)}. \qquad (4.124)$$

If we were to restrict ourselves to a real x, the only choice for us would be $x = \pm 1$, and this would land us exactly where we are already, with the only freedom left as to the sign in front of the $\boldsymbol{\sigma}_x$ matrix. We could also make a purely imaginary choice $x = \pm\mathrm{i}$, and this would merely swap $\boldsymbol{\sigma}_x$ and $\boldsymbol{\sigma}_y$. A nontrivial choice as to x is possible, too. In this case x would have to be a complex number of length 1 given by

$$x = e^{\mathrm{i}\phi_x}. \qquad (4.125)$$

Any other choice of x would make it impossible for us to recover the $|\rightarrow\rangle$ state within the resulting formalism; that is, we could not identify a single Hilbert space vector that would correspond to it.

Choosing $x = e^{\mathrm{i}\phi_x}$, which is the only choice we're ultimately left with, results in a Hermitian representation of $\boldsymbol{\sigma}_x$ and $\boldsymbol{\sigma}_y$ namely,

$$\begin{aligned} \boldsymbol{\sigma}_x &= \begin{pmatrix} 0 & e^{\mathrm{i}\phi_x} \\ e^{-\mathrm{i}\phi_x} & 0 \end{pmatrix} && (4.126) \\ \boldsymbol{\sigma}_y &= \begin{pmatrix} 0 & -\mathrm{i}e^{i\phi_x} \\ ie^{-\mathrm{i}\phi_x} & 0 \end{pmatrix} = \begin{pmatrix} 0 & e^{\mathrm{i}(\phi_x-\pi/2)} \\ e^{-\mathrm{i}(\phi_x-\pi/2)} & 0 \end{pmatrix}. && (4.127) \end{aligned}$$

If we were to choose $x = e^{\mathrm{i}\phi_x}$ with $\phi_x \neq 0$, the coefficients a and b in

$$|\rightarrow\rangle = a\,|\uparrow\rangle + b\,|\downarrow\rangle$$

would have to be such that $|a| = |b| = 1/\sqrt{2}$, and

$$\phi_a - \phi_b = \phi_x. \tag{4.128}$$

Phase factor invariance

In particular, for $\phi_x = 0$, which is equivalent to our choice of $x = 1$, we end up with $\phi_a = \phi_b$, but we are not forced to take $\phi_a = \phi_b = 0$. This means that we can multiply the superposition by $e^{i\phi_a} = e^{i\phi_b} = e^{i\phi}$ and this is still going to yield the same physical state.

4.5 Probability Amplitudes

A probability that a qubit in a *pure* state defined by $\boldsymbol{r}_1$, where $\boldsymbol{r}_1 \cdot \boldsymbol{r}_1 = 1$, is going to be filtered onto an "up" beam in some measuring apparatus is given by $p^0 = (1 + r_1^z)/2$. What is a probability that the qubit is going to be filtered onto an "up" beam in a differently oriented apparatus?

Let us assume that the orientation of the apparatus is $\boldsymbol{r}_2$, where $\boldsymbol{r}_2 \cdot \boldsymbol{r}_2 = 1$. We can answer the question by rotating our system of coordinates so that $\boldsymbol{r}_2 = \boldsymbol{e}_{z'}$, and now we have that $p^{0'} = \left(1 + r_1^{z'}\right)/2$. But what is $r_1^{z'}$? Let us recall that $r^z = \langle \boldsymbol{\omega}^z, \boldsymbol{r} \rangle = \boldsymbol{e}_z \cdot \boldsymbol{r}$. This holds for every orthonormal basis $\boldsymbol{e}_i$, including the new basis defined by $\boldsymbol{r}_2$. Consequently, $r_1^{z'} = \boldsymbol{e}_{z'} \cdot \boldsymbol{r}_1 = \boldsymbol{r}_2 \cdot \boldsymbol{r}_1$. The probability of a *transition* from a *pure* state that corresponds to $\boldsymbol{r}_1$ to a *pure* state that corresponds to $\boldsymbol{r}_2$ is therefore

Probability of a transition from $\boldsymbol{r}_1$ to $\boldsymbol{r}_2$

$$p_{\boldsymbol{r}_2 \leftarrow \boldsymbol{r}_1} = \frac{1}{2}\left(1 + \boldsymbol{r}_2 \cdot \boldsymbol{r}_1\right). \tag{4.129}$$

Let us digress here for a moment. The above formula is expressed in terms of fiducial vector parameterizations. Can we express it instead in terms of fiducial vectors themselves?

Let the two states be described by

$$\boldsymbol{p}_1 = \frac{1}{2}\left(\varsigma_1 + r_1^x \varsigma_x + r_1^y \varsigma_y + r_1^z \varsigma_z\right) \quad \text{and} \tag{4.130}$$

$$\boldsymbol{p}_2 = \frac{1}{2}\left(\varsigma_1 + r_2^x \varsigma_x + r_2^y \varsigma_y + r_2^z \varsigma_z\right). \tag{4.131}$$

Let us now evaluate $\langle \tilde{\boldsymbol{p}}_2, \boldsymbol{p}_1 \rangle$:

$$\begin{aligned}\langle \tilde{\boldsymbol{p}}_2, \boldsymbol{p}_1 \rangle &= \frac{1}{4}\left\langle \varsigma^1 + r_2^x \varsigma^x + r_2^y \varsigma^y + r_2^z \varsigma^z, \varsigma_1 + r_1^x \varsigma_x + r_1^y \varsigma_y + r_1^z \varsigma_z \right\rangle \\ &= \frac{1}{4}\left(\langle \varsigma^1, \varsigma_1 \rangle + r_1^x r_2^x \langle \varsigma^x, \varsigma_x \rangle + r_1^y r_2^y \langle \varsigma^y, \varsigma_y \rangle + r_1^z r_2^z \langle \varsigma^z, \varsigma_z \rangle\right)\end{aligned}$$

$$= \frac{1}{2}\left(1 + \boldsymbol{r}_1 \cdot \boldsymbol{r}_2\right), \tag{4.132}$$

because $\langle \varsigma^i, \varsigma_j \rangle = 2\delta^i{}_j$. We will soon see that there is a nice typographic correspondence between

$$p_{2\leftarrow 1} = \langle \tilde{\boldsymbol{p}}_2, \boldsymbol{p}_1 \rangle \tag{4.133}$$

and its unitary equivalent.

Although we have derived this formula for *pure* states only, it may be extended to a situation in which $\boldsymbol{r}_1$ is *any* state, possibly a mixed one, and $\boldsymbol{r}_2$ is pure. The latter is required because in our derivation we really thought of $\boldsymbol{r}_2$ as defining a direction and we made use of the fact that its length is 1 (when we equated $\boldsymbol{r}_2$ and $\boldsymbol{e}_{z'}$). But we have not assumed that $|\boldsymbol{r}_1| = 1$.

Let us substitute the unitary description coefficients a_1, b_1, as in

$$| \Psi_1 \rangle = a_1 \, |\uparrow\rangle + b_1 \, |\downarrow\rangle, \tag{4.134}$$

and a_2, b_2, as in

$$| \Psi_2 \rangle = a_2 \, |\uparrow\rangle + b_2 \, |\downarrow\rangle, \tag{4.135}$$

in place of r_1^x, r_1^y, r_1^z, r_2^x, r_2^y, and r_2^z. Using equations (4.109), (4.110), and (4.111) yields

$$\begin{aligned} \boldsymbol{r}_2 \cdot \boldsymbol{r}_1 &= \left(a_2 b_2^* + b_2 a_2^*\right)\left(a_1 b_1^* + b_1 a_1^*\right) \\ &\quad - \left(a_2 b_2^* - b_2 a_2^*\right)\left(a_1 b_1^* - b_1 a_1^*\right) \\ &\quad + \left(a_2 a_2^* - b_2 b_2^*\right)\left(a_1 a_1^* - b_1 b_1^*\right). \end{aligned} \tag{4.136}$$

As we have seen, for every fully polarized state

$$a a^* + b b^* = 1. \tag{4.137}$$

So, we can replace the 1 in $(1 + \boldsymbol{r}_2 \cdot \boldsymbol{r}_1)$ with

$$1 = 1 \cdot 1 = \left(a_2 a_2^* + b_2 b_2^*\right)\left(a_1 a_1^* + b_1 b_1^*\right). \tag{4.138}$$

Now, let us combine (4.136) and (4.138):

$$\begin{aligned} & 1 + \boldsymbol{r}_2 \cdot \boldsymbol{r}_1 \\ &\quad = \left(a_2 a_2^* + b_2 b_2^*\right)\left(a_1 a_1^* + b_1 b_1^*\right) \\ &\quad + \left(a_2 b_2^* + b_2 a_2^*\right)\left(a_1 b_1^* + b_1 a_1^*\right) \\ &\quad - \left(a_2 b_2^* - b_2 a_2^*\right)\left(a_1 b_1^* - b_1 a_1^*\right) \end{aligned}$$

$$+ \left(a_2 a_2^* - b_2 b_2^*\right)\left(a_1 a_1^* - b_1 b_1^*\right). \tag{4.139}$$

The first line is much like the last one, but there are minuses in front of the b-terms in the last one. Also, the third line is much like the second one, but again there are minuses in front of the ba-terms. On the other hand, there is a minus in front of the third line, so we should really flip the minuses and the pluses inside it. All these minuses will result in four merry cancellations, so that the final result will be left with four terms only. It is easy to see what they're going to be: The middle terms, which result from the full expansion of the first and the last line, drop out; and the edge terms, which result from the full expansion of the second and the third line, drop out too, leaving us with

$$1 + \boldsymbol{r}_2 \cdot \boldsymbol{r}_1 = 2a_2 a_2^* a_1 a_1^* + 2b_2 b_2^* b_1 b_1^* + 2a_2 b_2^* b_1 a_1^* + 2b_2 a_2^* a_1 b_1^*. \tag{4.140}$$

The 1/2 in front of $\frac{1}{2}\left(1 + \boldsymbol{r}_2 \cdot \boldsymbol{r}_1\right)$ kills the 2s in (4.140), and so we're left with

$$\begin{aligned} p_{\boldsymbol{r}_2 \leftarrow \boldsymbol{r}_1} &= a_2 a_2^* a_1 a_1^* + b_2 b_2^* b_1 b_1^* + a_2 b_2^* b_1 a_1^* + b_2 a_2^* a_1 b_1^* \\ &= \left(a_2^* a_1 + b_2^* b_1\right)\left(a_2 a_1^* + b_2 b_1^*\right). \end{aligned} \tag{4.141}$$

The second component in this product is a complex conjugate of the first component. At the same time, the first component is

$$a_2^* a_1 + b_2^* b_1 = \Big\langle a_2^* \langle\uparrow| + b_2^* \langle\downarrow| \Big| a_1 |\uparrow\rangle + b_1 |\downarrow\rangle \Big\rangle = \langle \Psi_2 \mid \Psi_1 \rangle, \tag{4.142}$$

whereas the second component is

$$a_2 a_1^* + b_2 b_1^* = \Big\langle a_1^* \langle\uparrow| + b_1^* \langle\downarrow| \Big| a_2 |\uparrow\rangle + b_2 |\downarrow\rangle \Big\rangle = \langle \Psi_1 \mid \Psi_2 \rangle. \tag{4.143}$$

Probability amplitude

And so, we discover the following.

1.
$$\langle \Psi_2 \mid \Psi_1 \rangle = \langle \Psi_1 \mid \Psi_2 \rangle^*. \tag{4.144}$$

This should not come as a surprise because we have already discovered that when converting a vector into its dual form we had to replace the vector's coefficients with their complex conjugates. This result is merely a consequence of that.

2.
$$p_{\boldsymbol{r}_2 \leftarrow \boldsymbol{r}_1} = \langle \tilde{\boldsymbol{p}}_2, \boldsymbol{p}_1 \rangle = \left|\langle \Psi_2 \mid \Psi_1 \rangle\right|^2. \tag{4.145}$$

For this reason the complex number $\langle \Psi_2 \mid \Psi_1 \rangle$ is called a *probability amplitude* for the transition from $\mid \Psi_1 \rangle$ to $\mid \Psi_2 \rangle$. The tradition in quantum mechanics, which goes back all the way to Dirac, is to write and read probability amplitude expressions from right to left.

A special case here is probability amplitudes for transitions to the basis states, $\langle \uparrow \mid \Psi_1 \rangle$ and $\langle \downarrow \mid \Psi_1 \rangle$. These amplitudes are

$$\langle \uparrow \mid \Psi_1 \rangle = \left\langle \langle \uparrow \middle| a_1 \mid \uparrow \rangle + b_1 \mid \downarrow \rangle \right\rangle = a_1, \quad (4.146)$$

$$\langle \downarrow \mid \Psi_1 \rangle = \left\langle \langle \downarrow \middle| a_1 \mid \uparrow \rangle + b_1 \mid \downarrow \rangle \right\rangle = b_1. \quad (4.147)$$

We can therefore write

$$\mid \Psi_1 \rangle = \mid \uparrow \rangle \langle \uparrow \mid \Psi_1 \rangle + \mid \downarrow \rangle \langle \downarrow \mid \Psi_1 \rangle, \quad (4.148)$$

$$\mid \Psi_2 \rangle = \mid \uparrow \rangle \langle \uparrow \mid \Psi_2 \rangle + \mid \downarrow \rangle \langle \downarrow \mid \Psi_2 \rangle, \quad (4.149)$$

$$\langle \Psi_2 \mid \Psi_1 \rangle = \langle \Psi_2 \mid \uparrow \rangle \langle \uparrow \mid \Psi_1 \rangle + \langle \Psi_2 \mid \downarrow \rangle \langle \downarrow \mid \Psi_1 \rangle. \quad (4.150)$$

The last expression has a peculiar interpretation within the lore of quantum mechanics that derives from the "superstition of superposition." It says that on its way from $\mid \Psi_1 \rangle$ to $\mid \Psi_2 \rangle$ a qubit transits both through $\mid \uparrow \rangle$ and $\mid \downarrow \rangle$ at the same time, as if splitting itself and being in these two states simultaneously.

More superstition

Well, as we have seen before, a qubit that is described by $\mid \Psi_1 \rangle = a_1 \mid \uparrow \rangle + b_1 \mid \downarrow \rangle$ can be thought as being *neither* in the $\mid \uparrow \rangle$ state *nor* in the $\mid \downarrow \rangle$ state. It is in a different state altogether that is not clearly expressed by the unitary formalism but, instead, is hidden inside the two complex numbers a_1 and b_1. The state can be always characterized by switching to the density operator formalism and extracting probabilities of all required qubit characteristics from it or by juggling a_1 and b_1 following equations (4.109), (4.110), and (4.111).

The word "transition" hints at some "motion" through intermediate states of the system. But when we think about a state of a qubit as polarization and then the act of the measurement as filtration, there isn't really any motion between the two polarizations involved. Some qubits pass through the filter; some don't. This is the same as filtering photons by a polarizer plate. What is the actual mechanism that makes one specific qubit pass through the filter and another one not is something that quantum mechanics does not really tell us anything about. Being a probability theory, it describes rather than explains anyway.[3]

When contemplating quantum transitions, it is probably best to think of the amplitude $\langle \Psi_2 \mid \Psi_1 \rangle$ as representing just such an act of filtration. The qubit does not really go through $|\uparrow\rangle$ and $|\downarrow\rangle$ any more than it exists in the $|\uparrow\rangle$ and the $|\downarrow\rangle$ states at the same time. These are just mathematical expressions written on paper that derive from the linearity of the theory. It is perhaps better to think of the qubit as ceasing to be in the $\mid \Psi_1\rangle$ state and reappearing in the $\mid \Psi_2\rangle$ state on the other side of the filter, if it makes it to the other side at all, without passing through any intermediate states in between. Its quantum mechanical "transition" is not continuous.

But one should not confuse transition probability amplitudes with a unitary evolution of a qubit. The unitary Hamiltonian evolution *is* continuous. There are no sudden jumps here. The transition amplitudes we talk about refer to probabilities of registering the qubit in such or another state. They refer to the act of measurement, the act that in itself is nonunitary.

4.6 Spinors

How do qubit representations in the Hilbert space transform under the change of the canonical basis in the qubit's physical space?

Transformation of probabilities under rotations

To answer this question, we must first answer another question. How do the probabilities encapsulated in the qubit's fiducial vector change in the same context? The fiducial vector of a qubit is

$$\boldsymbol{p} = \frac{1}{2}\begin{pmatrix} 1 + r^z \\ 1 - r^z \\ 1 + r^x \\ 1 + r^y \end{pmatrix}. \tag{4.151}$$

The components r^x, r^y, and r^z depend on the specific choice of a basis in the qubit's 3D space. The 3D space may be a geometric space, as is the case for neutron spin, or it may be some other parametric space, as is the case for the quantronium circuit. But, mathematically, they're the same: 3D real vector spaces.

[3] This can be said about any other physics theory. Even though some theories may seem to explain things, it is enough that a new, more accurate theory is discovered to relegate the old one to a mere phenomenology. At the end of the day it is safer to leave the difference between "describe" and "explain" to philosophers and focus on "predict" instead.

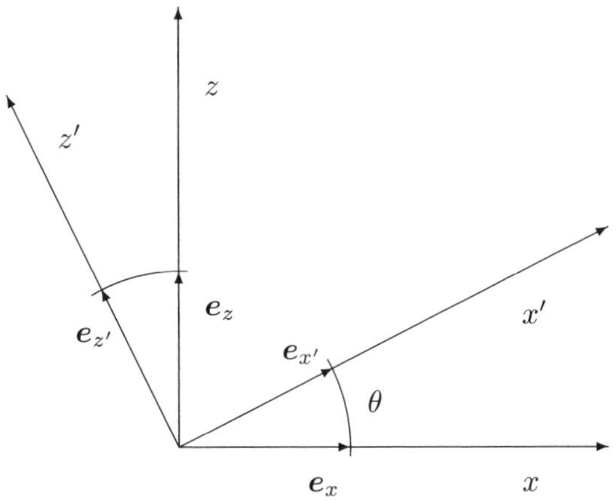

Figure 4.1: A counterclockwise rotation of basis $\boldsymbol{e}_i$ onto $\boldsymbol{e}_{i'}$ by angle θ.

We can rewrite $\boldsymbol{p}$ as follows:

$$\boldsymbol{p} = \frac{1}{2}\begin{pmatrix} 1 + \langle \boldsymbol{\omega}^z, \boldsymbol{r} \rangle \\ 1 - \langle \boldsymbol{\omega}^z, \boldsymbol{r} \rangle \\ 1 + \langle \boldsymbol{\omega}^x, \boldsymbol{r} \rangle \\ 1 + \langle \boldsymbol{\omega}^y, \boldsymbol{r} \rangle \end{pmatrix} = \frac{1}{2}\begin{pmatrix} 1 + \boldsymbol{e}_z \cdot \boldsymbol{r} \\ 1 - \boldsymbol{e}_z \cdot \boldsymbol{r} \\ 1 + \boldsymbol{e}_x \cdot \boldsymbol{r} \\ 1 + \boldsymbol{e}_y \cdot \boldsymbol{r} \end{pmatrix}, \tag{4.152}$$

where $\boldsymbol{\omega}^i$ and $\boldsymbol{e}_i$, for $i \in \{x, y, z\}$, are canonical basis forms and vectors, such that $\boldsymbol{e}_z$ points in the direction of the "magnetic field" used to measure the qubit.

When we rotate the canonical basis, so that $\boldsymbol{e}_i \rightarrow \boldsymbol{e}_{i'}$, then $\boldsymbol{p}$ changes, but $\boldsymbol{r}$ remains the same, because it represents the qubit and its physics, neither of which should depend on our choice of directions. Hence

$$\boldsymbol{p}' = \frac{1}{2}\begin{pmatrix} 1 + \boldsymbol{e}_{z'} \cdot \boldsymbol{r} \\ 1 - \boldsymbol{e}_{z'} \cdot \boldsymbol{r} \\ 1 + \boldsymbol{e}_{x'} \cdot \boldsymbol{r} \\ 1 + \boldsymbol{e}_{y'} \cdot \boldsymbol{r} \end{pmatrix} \tag{4.153}$$

represents the qubit in the same state $\boldsymbol{r}$. The only thing that would have changed is the way we look at it—through a different canonical basis $\boldsymbol{e}_{i'}$ in the qubit's physical space.

Let us rotate the basis by θ in the $\boldsymbol{e}_x \times \boldsymbol{e}_z$ plane as shown in Figure 4.1. The basis vectors $\boldsymbol{e}_x$, $\boldsymbol{e}_z$, $\boldsymbol{e}_{x'}$, and $\boldsymbol{e}_{z'}$ have length 1. In this operation vector $\boldsymbol{e}_y$ remains unchanged; that is, $\boldsymbol{e}_y = \boldsymbol{e}_{y'}$. The new basis vectors $\boldsymbol{e}_{x'}$ and $\boldsymbol{e}_{z'}$ can be expressed in terms of $\boldsymbol{e}_x$ and $\boldsymbol{e}_z$ as follows:

$$\begin{aligned} \boldsymbol{e}_{x'} &= \cos\theta\,\boldsymbol{e}_x + \sin\theta\,\boldsymbol{e}_z, \quad (4.154) \\ \boldsymbol{e}_{z'} &= -\sin\theta\,\boldsymbol{e}_x + \cos\theta\,\boldsymbol{e}_z. \quad (4.155) \end{aligned}$$

The change in the probability vector $\boldsymbol{p}$ is therefore going to be

$$\boldsymbol{p}' = \frac{1}{2}\begin{pmatrix} 1 + \boldsymbol{e}_{z'}\cdot\boldsymbol{r} \\ 1 - \boldsymbol{e}_{z'}\cdot\boldsymbol{r} \\ 1 + \boldsymbol{e}_{x'}\cdot\boldsymbol{r} \\ 1 + \boldsymbol{e}_{y'}\cdot\boldsymbol{r} \end{pmatrix} = \frac{1}{2}\begin{pmatrix} 1 - r^x\sin\theta + r^z\cos\theta \\ 1 + r^x\sin\theta - r^z\cos\theta \\ 1 + r^x\cos\theta + r^z\sin\theta \\ 1 + r^y \end{pmatrix}. \quad (4.156)$$

Let us focus on a simple case. Let $\boldsymbol{r} = \boldsymbol{e}_z$ (meaning that $r^x = r^y = 0$). The unitary representation of this state is $|\uparrow\rangle$. When looked at from the new basis $\boldsymbol{e}_{i'}$ the probability vector evaluates to

$$\boldsymbol{p}' = \frac{1}{2}\begin{pmatrix} 1 + \cos\theta \\ 1 - \cos\theta \\ 1 + \sin\theta \\ 1 \end{pmatrix}. \quad (4.157)$$

In the unitary representation that corresponds to the primed basis—let's call it $|\uparrow\rangle'$ and $|\downarrow\rangle'$—the state is going to be described by

$$|\uparrow\rangle = a\,|\uparrow\rangle' + b\,|\downarrow\rangle', \quad (4.158)$$

where a and b must satisfy equations (4.109), (4.110), and (4.111) from page 126, as well as the normalization condition. In this case

$$\begin{aligned} aa^* + bb^* &= 1, \quad (4.159) \\ aa^* - bb^* &= r^{z'} = \cos\theta, \quad (4.160) \\ ab^* + ba^* &= r^{x'} = \sin\theta, \quad (4.161) \\ \mathrm{i}(ab^* - ba^*) &= r^{y'} = 0. \quad (4.162) \end{aligned}$$

The last condition tells us that ab^* is real. This implies that a and b share the same phase angle. But since Hilbert space vectors are defined up to a constant phase factor anyway, we can just as well ignore this phase factor and assume that both a and b are real. This assumption greatly simplifies our algebra:

$$a^2 + b^2 = 1, \quad (4.163)$$

$$a^2 - b^2 = r^{z'} = \cos\theta, \tag{4.164}$$

$$2ab = r^{x'} = \sin\theta. \tag{4.165}$$

From the normalization condition

$$b^2 = 1 - a^2, \tag{4.166}$$

hence

$$a^2 - b^2 = 2a^2 - 1 = \cos\theta, \tag{4.167}$$

hence

$$a^2 = \frac{\cos\theta + 1}{2} = \cos^2\frac{\theta}{2}, \tag{4.168}$$

and

$$b^2 = 1 - a^2 = 1 - \cos^2\frac{\theta}{2} = \sin^2\frac{\theta}{2}. \tag{4.169}$$

In summary,

$$a = \pm\cos\frac{\theta}{2} \quad \text{and} \quad b = \pm\sin\frac{\theta}{2}. \tag{4.170}$$

How to choose the signs? For $\theta = 0$ we must have

$$|\uparrow\rangle = |\uparrow\rangle', \tag{4.171}$$

which implies that $a = +\cos\frac{\theta}{2}$. Then the $r^{x'}$ equation tells us that, for small positive angles θ, b must have the same sign as a; consequently $b = +\sin\frac{\theta}{2}$.

In summary, *Spinor rotation*

$$|\uparrow\rangle = \cos\frac{\theta}{2}\,|\uparrow\rangle' + \sin\frac{\theta}{2}\,|\downarrow\rangle'. \tag{4.172}$$

If $\boldsymbol{r} = -\boldsymbol{e}_z$, then the reasoning is similar, but we end up with minuses in front of cos and sin in the $r^{z'}$ and $r^{x'}$ equations. The latter tells us that this time a and b must be of the opposite sign, and the former selects sin for a, and so, expecting that for $\theta = 0$ we should have $|\downarrow\rangle = |\downarrow\rangle'$, yields

$$|\downarrow\rangle = -\sin\frac{\theta}{2}\,|\uparrow\rangle' + \cos\frac{\theta}{2}\,|\downarrow\rangle'. \tag{4.173}$$

Because $|\uparrow\rangle$ and $|\downarrow\rangle$ rotate by $\theta/2$ when the physical basis of the qubit rotates by θ, we end up with something strange when a full 360° rotation is performed. The operation maps

$$|\uparrow\rangle \rightarrow -|\uparrow\rangle \quad \text{and} \tag{4.174}$$

$$|\downarrow\rangle \rightarrow -|\downarrow\rangle. \tag{4.175}$$

This mathematical peculiarity is not physically observable, because in the corresponding probability transformation we have a full rotation by 360°.

Matrix

$$\begin{pmatrix} \cos\frac{\theta}{2} & \sin\frac{\theta}{2} \\ -\sin\frac{\theta}{2} & \cos\frac{\theta}{2} \end{pmatrix} \tag{4.176}$$

describes how the unitary qubit representation changes under rotation in the (x, z) plane for $\phi = 0$, as can be seen from Figure 4.1, where the view is from the direction of negative y's. What would the formula be like for other values of ϕ?

Spinor rotation for $\phi \neq 0$

To answer this question, first we must figure out the effect that a frame rotation about the z axis would have on the qubit's unitary components. In this case we have

$$\boldsymbol{e}_{z'} = \boldsymbol{e}_z, \tag{4.177}$$

$$\boldsymbol{e}_{x'} = \cos\phi\boldsymbol{e}_x + \sin\phi\boldsymbol{e}_y, \tag{4.178}$$

$$\boldsymbol{e}_{y'} = -\sin\phi\boldsymbol{e}_x + \cos\phi\boldsymbol{e}_y. \tag{4.179}$$

This produces the following change in the probability vector:

$$\boldsymbol{p}' = \frac{1}{2}\begin{pmatrix} 1 + \boldsymbol{e}_{z'}\cdot\boldsymbol{r} \\ 1 - \boldsymbol{e}_{z'}\cdot\boldsymbol{r} \\ 1 + \boldsymbol{e}_{x'}\cdot\boldsymbol{r} \\ 1 + \boldsymbol{e}_{y'}\cdot\boldsymbol{r} \end{pmatrix} = \frac{1}{2}\begin{pmatrix} 1 + r^z \\ 1 - r^z \\ 1 + r^x\cos\phi + r^y\sin\phi \\ 1 - r^x\sin\phi + r^y\cos\phi \end{pmatrix}. \tag{4.180}$$

Let us assume, for simplicity, that $r^y = 0$, meaning that in the original, unrotated frame $\boldsymbol{r}$ lies in the (x, z) plane (as in the previous example). Then

$$\boldsymbol{p}' = \frac{1}{2}\begin{pmatrix} 1 + r^z \\ 1 - r^z \\ 1 + r^x\cos\phi \\ 1 - r^x\sin\phi \end{pmatrix}. \tag{4.181}$$

Matching this against $|\uparrow\rangle = a\,|\uparrow\rangle' + b\,|\downarrow\rangle'$ and invoking equations (4.109), (4.110), and (4.111), we get

$$aa^* - bb^* = r^{z'} = r^z, \tag{4.182}$$

$$ab^* + ba^* = r^{x'} = r^x\cos\phi, \tag{4.183}$$

$$\mathrm{i}\,(ab^* - ba^*) = r^{y'} = -r^x\sin\phi. \tag{4.184}$$

The first equation can be parameterized by substituting

$$r^z = \cos\theta, \tag{4.185}$$

and

$$|a| = \cos\frac{\theta}{2}, \quad (4.186)$$

$$|b| = \sin\frac{\theta}{2}, \quad (4.187)$$

which results in

$$|a|^2 - |b|^2 = \cos^2\frac{\theta}{2} - \sin^2\frac{\theta}{2} = \cos\theta = r^z. \quad (4.188)$$

Multiplying the third equation by $-\mathrm{i}$ and adding it to the second, we have

$$2ab^* = r^x(\cos\phi + \mathrm{i}\sin\phi) = r^x e^{\mathrm{i}\phi}. \quad (4.189)$$

We can satisfy this equation by making

$$r^x = \sin\theta. \quad (4.190)$$

Then

$$2ab^* = 2\cos\frac{\theta}{2}e^{\mathrm{i}\delta_a}\sin\frac{\theta}{2}e^{-\mathrm{i}\delta_b} = \sin\theta e^{\mathrm{i}(\delta_a-\delta_b)} = \sin\theta e^{\mathrm{i}\phi}, \quad (4.191)$$

where δ_a and δ_b are phases of a and b, respectively. The equation is satisfied when $\delta_a - \delta_b = \phi$, and we can make it so by setting $\delta_a = -\delta_b = \phi/2$. In the end, we find the following transformation of the qubit's unitary coefficients:

$$\begin{pmatrix} \cos\frac{\theta}{2}e^{\mathrm{i}\phi/2} \\ \sin\frac{\theta}{2}e^{-\mathrm{i}\phi/2} \end{pmatrix} = \begin{pmatrix} e^{\mathrm{i}\phi/2} & 0 \\ 0 & e^{-\mathrm{i}\phi/2} \end{pmatrix} \begin{pmatrix} \cos\frac{\theta}{2} \\ \sin\frac{\theta}{2} \end{pmatrix}. \quad (4.192)$$

Now we are ready to generalize matrix (4.176) to a θ rotation in an arbitrary plane ϕ. We do this by rotating by $-\phi$ from the qubit's original plane to the (x, z) plane, then performing the rotation in the (x, z) plane, then rotating the result back by ϕ about the z axis. The resulting matrix is

$$\begin{pmatrix} e^{\mathrm{i}\phi/2} & 0 \\ 0 & e^{-\mathrm{i}\phi/2} \end{pmatrix} \begin{pmatrix} \cos\frac{\theta}{2} & \sin\frac{\theta}{2} \\ -\sin\frac{\theta}{2} & \cos\frac{\theta}{2} \end{pmatrix} \begin{pmatrix} e^{-\mathrm{i}\phi/2} & 0 \\ 0 & e^{\mathrm{i}\phi/2} \end{pmatrix}$$

$$= \begin{pmatrix} \cos\frac{\theta}{2} & e^{\mathrm{i}\phi}\sin\frac{\theta}{2} \\ -e^{-\mathrm{i}\phi}\sin\frac{\theta}{2} & \cos\frac{\theta}{2} \end{pmatrix}. \quad (4.193)$$

The formula can be rewritten with an i put in front of $e^{\mathrm{i}\phi}$:

$$\begin{pmatrix} \cos\frac{\theta}{2} & \mathrm{i}e^{-\mathrm{i}\phi}\sin\frac{\theta}{2} \\ \mathrm{i}e^{\mathrm{i}\phi}\sin\frac{\theta}{2} & \cos\frac{\theta}{2} \end{pmatrix}, \quad (4.194)$$

which is the same as (4.193) on replacing

$$\phi \to \left(\frac{\pi}{2} - \phi\right). \tag{4.195}$$

Whether ϕ is measured in this or in the opposite direction and where it is measured from depends on convention, as well as on whether active or passive transformations are considered. Depending on context and ease of computations, one or the other may be preferable. We will make use of (4.194) in Section 6.4, where we will discuss superconducting controlled-NOT gates.

Spinors and half-vectors

The term *spinors* is used to describe geometric objects that tranform according to equations (4.172), (4.173), (4.192), or (4.193) when the physical basis of the qubit is transformed according to equations (4.154), (4.155), or a combination thereof. In their early history they were called *half-vectors*, because of the half-angles used in their transformation laws.

In any case, we have discovered in this section that the unitary representation of qubits, the qubit Hilbert space, is made of spinors.

Transformation of varsigmas under rotations

Let us now go back to equation (4.156) and work on it some more. In general, a rotation of the canonical basis is described by an orthogonal transformation $\mathbf{\Lambda}$ such that

$$\boldsymbol{e}_{i'} = \sum_j \Lambda_{i'}{}^j \boldsymbol{e}_j \quad \text{and} \tag{4.196}$$

$$\boldsymbol{\omega}^{i'} = \sum_j \boldsymbol{\omega}^j \Lambda_j{}^{i'}, \tag{4.197}$$

where

$$\sum_{k'} \Lambda_i{}^{k'} \Lambda_{k'}{}^j = \delta_i{}^j \quad \text{and} \tag{4.198}$$

$$\sum_{k} \Lambda_{i'}{}^{k} \Lambda_{k}{}^{j'} = \delta_{i'}{}^{j'}. \tag{4.199}$$

We have mentioned this already in Section 1.7, page 28. In this more general case equation (4.156) assumes the following form:

$$\boldsymbol{p}' = \frac{1}{2}\begin{pmatrix} 1 + \langle \boldsymbol{\omega}^{z'}, \boldsymbol{r} \rangle \\ 1 - \langle \boldsymbol{\omega}^{z'}, \boldsymbol{r} \rangle \\ 1 + \langle \boldsymbol{\omega}^{x'}, \boldsymbol{r} \rangle \\ 1 + \langle \boldsymbol{\omega}^{y'}, \boldsymbol{r} \rangle \end{pmatrix}$$

$$= \frac{1}{2}\begin{pmatrix} 1+\sum_i \Lambda_i{}^{z'}\langle\boldsymbol{\omega}^i, \boldsymbol{r}\rangle \\ 1-\sum_i \Lambda_i{}^{z'}\langle\boldsymbol{\omega}^i, \boldsymbol{r}\rangle \\ 1+\sum_i \Lambda_i{}^{x'}\langle\boldsymbol{\omega}^i, \boldsymbol{r}\rangle \\ 1+\sum_i \Lambda_i{}^{y'}\langle\boldsymbol{\omega}^i, \boldsymbol{r}\rangle \end{pmatrix} = \frac{1}{2}\begin{pmatrix} 1+\sum_i \Lambda_i{}^{z'} r^i \\ 1-\sum_i \Lambda_i{}^{z'} r^i \\ 1+\sum_i \Lambda_i{}^{x'} r^i \\ 1+\sum_i \Lambda_i{}^{y'} r^i \end{pmatrix}. \tag{4.200}$$

Now we have to do something a little confusing. Vectors $\boldsymbol{e}_i$ for $i \in \{x, y, z\}$, and $\boldsymbol{e}_{i'}$ for $i' \in \{x', y', z'\}$, and their dual forms $\boldsymbol{\omega}^i$ and $\boldsymbol{\omega}^{i'}$ operate in the physical three-dimensional space. But in the fiducial space we also have canonical vectors and forms we used to decompose the fiducial vector $\boldsymbol{p}$ into its components. We called them $\boldsymbol{e}_i$ and $\boldsymbol{\omega}^i$ for $i \in \{0, 1, 2, 3\}$. To avoid a clash of symbols, we are going to mark them with a hat:

$$\hat{\boldsymbol{e}}_i \quad \text{and} \quad \hat{\boldsymbol{\omega}}^i, \quad \text{where} \quad i = 0, 1, 2, 3.$$

Moreover, we're going to use primed indexes for them in the expression below, because they refer here to measurements made with respect to the primed directions. Using these, we can rewrite our expression for $\boldsymbol{p}'$ as follows:

$$\begin{aligned} \boldsymbol{p}' = \frac{1}{2}\Bigg(& \left(1 + \sum_{i=x,y,z} \Lambda_i{}^{z'} r^i\right)\hat{\boldsymbol{e}}_{0'} + \left(1 - \sum_{i=x,y,z} \Lambda_i{}^{z'} r^i\right)\hat{\boldsymbol{e}}_{1'} \\ & + \left(1 + \sum_{i=x,y,z} \Lambda_i{}^{x'} r^i\right)\hat{\boldsymbol{e}}_{2'} + \left(1 + \sum_{i=x,y,z} \Lambda_i{}^{y'} r^i\right)\hat{\boldsymbol{e}}_{3'}\Bigg). \end{aligned} \tag{4.201}$$

Now let us recall equations (2.59), page 65, namely,

$$\hat{\boldsymbol{e}}_{0'} = \frac{1}{2}\left(\varsigma_{1'} + \varsigma_{z'} - \varsigma_{x'} - \varsigma_{y'}\right), \tag{4.202}$$

$$\hat{\boldsymbol{e}}_{1'} = \frac{1}{2}\left(\varsigma_{1'} - \varsigma_{z'} - \varsigma_{x'} - \varsigma_{y'}\right), \tag{4.203}$$

$$\hat{\boldsymbol{e}}_{2'} = \varsigma_{x'}, \tag{4.204}$$

$$\hat{\boldsymbol{e}}_{3'} = \varsigma_{y'}. \tag{4.205}$$

Using these, we can replace $\hat{\boldsymbol{e}}_i$ with varsigmas, which yields

$$\boldsymbol{p}' = \frac{1}{2}\left(\varsigma_{1'} + \sum_{i=x,y,z} \Lambda_i{}^{x'} r^i \varsigma_{x'} + \sum_{i=x,y,z} \Lambda_i{}^{y'} r^i \varsigma_{y'} + \sum_{i=x,y,z} \Lambda_i{}^{z'} r^i \varsigma_{z'}\right). \tag{4.206}$$

This tells us that

$$r^{x'} = \sum_{i=x,y,z} \Lambda_i{}^{x'} r^i, \tag{4.207}$$

$$r^{y'} = \sum_{i=x,y,z} \Lambda_i{}^{y'} r^i, \tag{4.208}$$

$$r^{z'} = \sum_{i=x,y,z} \Lambda_i{}^{z'} r^i, \tag{4.209}$$

which is what we knew all along, but we can rewrite this expression differently, grouping terms around r^x, r^y, and r^z instead of $\varsigma_{x'}$, $\varsigma_{y'}$, and $\varsigma_{z'}$. This time we get

$$\boldsymbol{p}' = \frac{1}{2}\left(\varsigma_{1'} + r^x \sum_{i'=x',y',z'} \Lambda_x{}^{i'} \varsigma_{i'} + r^y \sum_{i'=x',y',z'} \Lambda_y{}^{i'} \varsigma_{i'} + r^z \sum_{i'=x',y',z'} \Lambda_z{}^{i'} \varsigma_{i'}\right), \tag{4.210}$$

which tells us that

$$\varsigma_x = \sum_{i'=x',y',z'} \Lambda_x{}^{i'} \varsigma_{i'}, \tag{4.211}$$

$$\varsigma_y = \sum_{i'=x',y',z'} \Lambda_y{}^{i'} \varsigma_{i'}, \tag{4.212}$$

$$\varsigma_z = \sum_{i'=x',y',z'} \Lambda_z{}^{i'} \varsigma_{i'}. \tag{4.213}$$

The three varsigmas, indexed with x, y, and z, transform like three-dimensional space vectors under the rotation of the canonical basis in the qubit's physical space, even though they have four components. The varsigmas are therefore peculiar objects. They stand with one leg in the qubit's fiducial space and with the other one in the qubit's physical space. They are subject to transformations in both spaces.

To know that they transform like three-dimensional space vectors in response to rotations will turn out to be useful, especially in more complicated situations where we will have probability matrices that cannot be decomposed into individual probability vectors. Such situations will arise in multiqubit systems.

4.7 Operators and Operands

In the preceding sections of the chapter we have managed to recover an image of pure states (and pure states only) within the formalism of the qubit's Hilbert space that resulted from the unpacking of quaternion symbols into 2×2 complex matrices, which, nota bene, all turned out to be Hermitian.

Eigenvalues and eigenvectors

But so far it has been only the density quaternion, or the quaternion equivalent of the probability vector in the fiducial space of a qubit, that we have played with. What about the other quaternion, the Hamiltonian? Let us recall that

$$\boldsymbol{H} = -\mu \left(B_x \boldsymbol{\sigma}_x + B_y \boldsymbol{\sigma}_y + B_z \boldsymbol{\sigma}_z \right). \tag{4.214}$$

Unpacking the sigmas translates this into

$$\boldsymbol{H} = -\mu \begin{pmatrix} B_z & B_x - iB_y \\ B_x + iB_y & -B_z \end{pmatrix}. \tag{4.215}$$

Since we can think of $\boldsymbol{H}$ as an *operator*, the obvious question to ask is what it does to the vectors of Hilbert space we identified in the previous sections.

Let us consider $\boldsymbol{B} = B_z \boldsymbol{e}_z$ first. For the Hamiltonian constructed from this field

$$\boldsymbol{H}_{\uparrow} \, |\uparrow\rangle = -\mu \begin{pmatrix} B_z & 0 \\ 0 & -B_z \end{pmatrix} \begin{pmatrix} 1 \\ 0 \end{pmatrix} = -\mu B_z \begin{pmatrix} 1 \\ 0 \end{pmatrix} = -\mu B_z \, |\uparrow\rangle. \tag{4.216}$$

Similarly,

$$\boldsymbol{H}_{\uparrow} \, |\downarrow\rangle = -\mu \begin{pmatrix} B_z & 0 \\ 0 & -B_z \end{pmatrix} \begin{pmatrix} 0 \\ 1 \end{pmatrix} = \mu B_z \begin{pmatrix} 0 \\ 1 \end{pmatrix} = \mu B_z \, |\downarrow\rangle. \tag{4.217}$$

We discover that $|\uparrow\rangle$ and $|\downarrow\rangle$ are *eigenvectors* of $\boldsymbol{H}_{\uparrow}$ and the corresponding *eigenvalues* are expectation energies that correspond to these states. In this case the fiducial formalism would have returned

$$\langle E_{\uparrow} \rangle = \langle \boldsymbol{\eta}_{\uparrow}, \boldsymbol{p}_{\uparrow} \rangle = \langle -\mu B_z \boldsymbol{\varsigma}^z, \frac{1}{2} \left(\boldsymbol{\varsigma}_1 + \boldsymbol{\varsigma}_z \right) \rangle = -\mu B_z, \tag{4.218}$$

because $\langle \boldsymbol{\varsigma}^z, \boldsymbol{\varsigma}_z \rangle = 2$ and $\langle \boldsymbol{\varsigma}^z, \boldsymbol{\varsigma}_1 \rangle = 0$.

Similarly,

$$\langle E_{\downarrow} \rangle = \langle \boldsymbol{\eta}_{\uparrow}, \boldsymbol{p}_{\downarrow} \rangle = \langle -\mu B_z \boldsymbol{\varsigma}^z, \frac{1}{2} \left(\boldsymbol{\varsigma}_1 - \boldsymbol{\varsigma}_z \right) \rangle = \mu B_z. \tag{4.219}$$

We have already introduced the notation $|\uparrow\rangle\langle\uparrow|$ and $|\downarrow\rangle\langle\downarrow|$ to describe $\boldsymbol{\rho}_{\uparrow}$ and $\boldsymbol{\rho}_{\downarrow}$ as tensor products of unitary vectors and forms; cf. equation (4.72), page 119, and equation (4.81), page 120. In matrix notation

$$|\uparrow\rangle\langle\uparrow| = \begin{pmatrix} 1 & 0 \\ 0 & 0 \end{pmatrix} \quad \text{and} \quad |\downarrow\rangle\langle\downarrow| = \begin{pmatrix} 0 & 0 \\ 0 & 1 \end{pmatrix}. \tag{4.220}$$

Hamiltonian $\boldsymbol{H}_{\uparrow}$ can therefore be described as

$$\boldsymbol{H}_{\uparrow} = -\mu B_z \, |\uparrow\rangle\langle\uparrow| + \mu B_z \, |\downarrow\rangle\langle\downarrow| \, . \tag{4.221}$$

Measuring the energy of an individual qubit against the Hamiltonian returns either $-\mu B_z$, and in this case the qubit emerges in the $|\uparrow\rangle$ state from the $\boldsymbol{H}_\uparrow$ measuring apparatus, or μB_z, in which case the qubit emerges in the $|\downarrow\rangle$ from the $\boldsymbol{H}_\downarrow$ measuring apparatus. This we know from the experiment.

Measurements as projections

From the unitary point of view, then, the measurement represented by $\boldsymbol{H}_\uparrow$ performed on an individual qubit performs an act of projection. Whatever the state of the qubit was originally, the measurement projects the qubit either on the $|\uparrow\rangle$ or on the $|\downarrow\rangle$ state.

Hamiltonian eigenstates as projection operators

Operators $|\uparrow\rangle\langle\uparrow|$ and $|\downarrow\rangle\langle\downarrow|$ may be thought of as projectors: $\boldsymbol{P}_\uparrow$ and $\boldsymbol{P}_\downarrow$. Hence the Hamiltonian itself becomes a linear combination of projections:

$$\boldsymbol{H}_\uparrow = -\mu B_z \boldsymbol{P}_\uparrow + \mu B_z \boldsymbol{P}_\downarrow. \tag{4.222}$$

The projectors have the following obvious properties:

$$\begin{aligned}
\boldsymbol{P}_\uparrow\boldsymbol{P}_\downarrow &= |\uparrow\rangle\langle\uparrow|\downarrow\rangle\langle\downarrow| = 0, && (4.223)\\
\boldsymbol{P}_\downarrow\boldsymbol{P}_\uparrow &= |\downarrow\rangle\langle\downarrow|\uparrow\rangle\langle\uparrow| = 0, && (4.224)\\
\boldsymbol{P}_\uparrow\boldsymbol{P}_\uparrow &= |\uparrow\rangle\langle\uparrow|\uparrow\rangle\langle\uparrow| = |\uparrow\rangle\langle\uparrow| = \boldsymbol{P}_\uparrow, && (4.225)\\
\boldsymbol{P}_\downarrow\boldsymbol{P}_\downarrow &= |\downarrow\rangle\langle\downarrow|\downarrow\rangle\langle\downarrow| = |\downarrow\rangle\langle\downarrow| = \boldsymbol{P}_\downarrow. && (4.226)
\end{aligned}$$

Also

$$\boldsymbol{P}_\uparrow + \boldsymbol{P}_\downarrow = \begin{pmatrix} 1 & 0 \\ 0 & 0 \end{pmatrix} + \begin{pmatrix} 0 & 0 \\ 0 & 1 \end{pmatrix} = \begin{pmatrix} 1 & 0 \\ 0 & 1 \end{pmatrix} = \mathbf{1}. \tag{4.227}$$

We say that the projectors are orthogonal (because $\boldsymbol{P}_\uparrow\boldsymbol{P}_\downarrow = \boldsymbol{P}_\downarrow\boldsymbol{P}_\uparrow = 0$, this is what *orthogonal* means) and *idempotent* (because $\boldsymbol{P}_\uparrow\boldsymbol{P}_\uparrow = \boldsymbol{P}_\uparrow$. and $\boldsymbol{P}_\downarrow\boldsymbol{P}_\downarrow = \boldsymbol{P}_\downarrow$, this is what *idempotent* means). We can combine the two properties into a single equation, namely,

$$\boldsymbol{P}_i\boldsymbol{P}_j = \delta_{ij}\boldsymbol{P}_i, \tag{4.228}$$

where $i \in \{\uparrow, \downarrow\} \ni j$. Also we say that the set of projectors is *complete* (because $\sum_{i\in\{\uparrow,\downarrow\}} \boldsymbol{P}_i = \mathbf{1}$, this is what *complete* means).

The measurement represented by $\boldsymbol{H}_\uparrow$ defines a complete set of orthogonal projectors. This is true of *any* quantum measurement *within* the unitary formalism. For this reason such measurements are called *orthogonal measurements.*[4]

[4]Modern quantum mechanics also talks about nonorthogonal measurements, which arise when a measured system is coupled to the environment, which means pretty much always. We will not talk about these here, preferring to describe measurements in terms of fiducial vectors instead. A measurement is what fills a fiducial vector with probabilities.

For an arbitrary superposition $a\,|\uparrow\rangle + b\,|\downarrow\rangle$

$$\boldsymbol{H}_\uparrow\,(a\,|\uparrow\rangle + b\,|\downarrow\rangle) = (-\mu B_z \boldsymbol{P}_\uparrow + \mu B_z \boldsymbol{P}_\downarrow)\,(a\,|\uparrow\rangle + b\,|\downarrow\rangle)\,. \tag{4.229}$$

The first projector $\boldsymbol{P}_\uparrow$ kills $|\downarrow\rangle$ but leaves $|\uparrow\rangle$ intact, because it is a projector, and the second projector $\boldsymbol{P}_\downarrow$ does the opposite. So the result is

$$-\mu B_z a\,|\uparrow\rangle + \mu B_z b\,|\downarrow\rangle. \tag{4.230}$$

Expectation values

Let us zap this result from the left-hand side with $\langle\Psi\,|$:

$$\begin{aligned}&(a^*\,\langle\uparrow|\;+b^*\,\langle\uparrow|)\,(a(-\mu B_z)\;|\uparrow\rangle\;+b(\mu B_z)\;|\downarrow\rangle) &(4.231)\\ &= aa^*(-\mu B_z) + bb^*(\mu B_z)\\ &= \frac{1}{2}\,(1+r^z)\,(-\mu B_z) + \frac{1}{2}\,(1-r^z)\,(\mu B_z)\\ &= -\mu B_z r_z\\ &= \langle E\rangle. &(4.232)\end{aligned}$$

In summary,

$$\langle\Psi\mid\boldsymbol{H}_\uparrow\mid\Psi\rangle = \langle E\rangle. \tag{4.233}$$

Transition amplitude on the left-hand side of this equation can be understood as either $\langle\Psi\mid$ acting on $\boldsymbol{H}\mid\Psi\rangle$ or as $\langle\Psi\mid\boldsymbol{H}^\dagger$ acting on $\mid\Psi\rangle$. Both yield the same result, because $\boldsymbol{H}^\dagger = \boldsymbol{H}$.

Is equation (4.233) a happy coincidence, or is there more to it? To answer this question, we should evaluate $\langle\Psi\mid\boldsymbol{H}\mid\Psi\rangle$ for an arbitrary Hamiltonian $\boldsymbol{H}$ and an arbirary vector $\mid\Psi\rangle$.

But rather than jump into the computation head first, we begin by figuring out how individual Pauli matrices affect the basis vectors of the Hilbert space. It is easy to check that

$$\begin{aligned}\boldsymbol{\sigma}_x\;|\uparrow\rangle &= |\downarrow\rangle, &(4.234)\\ \boldsymbol{\sigma}_x\;|\downarrow\rangle &= |\uparrow\rangle, &(4.235)\\ \boldsymbol{\sigma}_y\;|\uparrow\rangle &= \mathrm{i}\;|\downarrow\rangle, &(4.236)\\ \boldsymbol{\sigma}_y\;|\downarrow\rangle &= -\mathrm{i}\;|\uparrow\rangle, &(4.237)\\ \boldsymbol{\sigma}_z\;|\uparrow\rangle &= |\uparrow\rangle, &(4.238)\\ \boldsymbol{\sigma}_z\;|\downarrow\rangle &= -\;|\downarrow\rangle. &(4.239)\end{aligned}$$

Now we are ready to jump:

$$-\mu\,(B_x\boldsymbol{\sigma}_x + B_y\boldsymbol{\sigma}_y + B_z\boldsymbol{\sigma}_z)\,(a\;|\uparrow\rangle\;+\;b\;|\downarrow\rangle)$$

$$= -\mu\Big(B_x\left(a\left|\downarrow\right\rangle + b\left|\uparrow\right\rangle\right) + B_y\left(ia\left|\downarrow\right\rangle - ib\left|\uparrow\right\rangle\right) + B_z\left(a\left|\uparrow\right\rangle - b\left|\downarrow\right\rangle\right)\Big). \tag{4.240}$$

Zapping this from the left-hand side with $a^*\left\langle\uparrow\right| + b^*\left\langle\downarrow\right|$ and making use of $\langle\uparrow|\downarrow\rangle = \langle\downarrow|\uparrow\rangle = 0$, we have

$$-\mu\left(B_x\left(a^*b + b^*a\right) + B_y i\left(b^*a - a^*b\right) + B_z\left(a^*a - b^*b\right)\right). \tag{4.241}$$

Let us recall equations (4.110), (4.111), and (4.109) on page 126:

$$ab^* + ba^* = r^x, \tag{4.242}$$
$$i\left(ab^* - ba^*\right) = r^y, \tag{4.243}$$
$$aa^* - bb^* = r^z. \tag{4.244}$$

Making use of these yields

$$\langle\Psi \mid \boldsymbol{H} \mid \Psi\rangle = -\mu\left(B_x r^x + B_y r^y + B_z r^z\right) = \langle E\rangle. \tag{4.245}$$

Thus the formula holds for any Hermitian operator that represents some measurable quantity.

The following chain of equalities sums up what we have learned about fiducial, quaternion, and unitary pictures of quantum systems:

$$\langle\boldsymbol{\eta}_A, \boldsymbol{p}\rangle = 2\Re\left(\boldsymbol{A}\boldsymbol{\rho}\right) = \mathrm{Tr}\left(\boldsymbol{A}\boldsymbol{\rho}\right) = \langle\Psi \mid \boldsymbol{A} \mid \Psi\rangle = \langle A\rangle. \tag{4.246}$$

How the measurement affects the original pure *state*

Let us still go back to the projection aspect of a measurement on a unitary state. Let us consider an experiment in which an energy measurement is performed on the $\mid\Psi\rangle = a\left|\uparrow\right\rangle + b\left|\downarrow\right\rangle$ state. The measurement may (and usually does) have a side effect of splitting the incident beam of qubits so that all qubits in one beam emerging from the measuring apparatus are in the $\left|\uparrow\right\rangle$ state and all qubits in the other beam emerging from the measuring apparatus are in the $\left|\downarrow\right\rangle$ state. The intensities of both beams are equal to $I_0 a^* a$ and $I_0 b^* b$, respectively, where I_0 is the intensity of the incident beam.

If we were to mix the two beams back together, we would end up with a *mixture* of qubits in both states in the proportions corresponding to the beams intensities. Once the measurement has been performed the original pure state, the superposition, is normally destroyed. But if we were to look at each of the two beams separately, we'd find that the qubits in them are all in pure states, namely, $\left|\uparrow\right\rangle$ in one and $\left|\downarrow\right\rangle$ in the other, new states induced by the measurement, states that incidentally are the eigenstates of the Hamiltonian.

The states are not $\boldsymbol{P}_\uparrow \mid \Psi\rangle$ or $\boldsymbol{P}_\downarrow \mid \Psi\rangle$, because the projections of $\mid \Psi\rangle$,

$$\begin{aligned} \boldsymbol{P}_\uparrow \mid \Psi\rangle &= a\,|\uparrow\rangle \quad \text{and} & (4.247)\\ \boldsymbol{P}_\downarrow \mid \Psi\rangle &= b\,|\downarrow\rangle, & (4.248)\end{aligned}$$

are not normalized. How can we express the states that emerge from the measuring apparatus in terms of the original $\mid \Psi\rangle$ and the projectors?

The expectation value of $\boldsymbol{P}_\uparrow$ is

$$\langle \Psi \mid \boldsymbol{P}_\uparrow \mid \Psi\rangle = \langle \Psi \mid a\,|\uparrow\rangle = a^* a\langle\uparrow|\uparrow\rangle = a^* a = |a|^2 = p_\uparrow, \tag{4.249}$$

and similarly for $\boldsymbol{P}_\downarrow$:

$$\langle \Psi \mid \boldsymbol{P}_\downarrow \mid \Psi\rangle = |b|^2 = p_\downarrow. \tag{4.250}$$

In other words, the expectation value of a projection operator $\boldsymbol{P}_i$ is the probability p_i. Therefore,

$$\frac{\boldsymbol{P}_\uparrow \mid \Psi\rangle}{\sqrt{\langle \Psi \mid \boldsymbol{P}_\uparrow \mid \Psi\rangle}} = \frac{|a|e^{i\phi_a}}{|a|}\,|\uparrow\rangle = e^{i\phi_a}\,|\uparrow\rangle, \tag{4.251}$$

which is $|\uparrow\rangle$ up to the phase factor $e^{i\phi_a}$. Also

$$\frac{\boldsymbol{P}_\downarrow \mid \Psi\rangle}{\sqrt{\langle \Psi \mid \boldsymbol{P}_\downarrow \mid \Psi\rangle}} = \frac{|b|e^{i\phi_b}}{|b|}\,|\downarrow\rangle = e^{i\phi_b}\,|\downarrow\rangle. \tag{4.252}$$

The effect of the measurement performed by $\boldsymbol{H} = -\mu B_z \boldsymbol{P}_\uparrow + \mu B_z \boldsymbol{P}_\downarrow$ on $\mid \Psi\rangle$ then is to produce two beams in states

$$\frac{\boldsymbol{P}_\uparrow \mid \Psi\rangle}{\sqrt{\langle \Psi \mid \boldsymbol{P}_\uparrow \mid \Psi\rangle}} \quad \text{and} \quad \frac{\boldsymbol{P}_\downarrow \mid \Psi\rangle}{\sqrt{\langle \Psi \mid \boldsymbol{P}_\downarrow \mid \Psi\rangle}}, \tag{4.253}$$

with beam intensities equal to $I_0\langle \Psi \mid \boldsymbol{P}_\uparrow \mid \Psi\rangle$ and $I_0\langle \Psi \mid \boldsymbol{P}_\downarrow \mid \Psi\rangle$, respectively.

We can rewrite the above formula in terms of density operators. Let us consider the outcome of the $\boldsymbol{P}_\uparrow$ operation first. Here we find that the state of qubits in the beam produced by the projector is

$$\frac{\boldsymbol{P}_\uparrow \mid \Psi\rangle}{\sqrt{\langle \Psi \mid \boldsymbol{P}_\uparrow \mid \Psi\rangle}} \otimes \frac{\langle \Psi \mid \boldsymbol{P}_\uparrow^\dagger}{\sqrt{\langle \Psi \mid \boldsymbol{P}_\uparrow \mid \Psi\rangle}} = \frac{\boldsymbol{P}_\uparrow \mid \Psi\rangle\langle \Psi \mid \boldsymbol{P}_\uparrow^\dagger}{\langle \Psi \mid \boldsymbol{P}_\uparrow \mid \Psi\rangle} = \frac{\boldsymbol{P}_\uparrow \boldsymbol{\rho} \boldsymbol{P}_\uparrow^\dagger}{\mathrm{Tr}\,(\boldsymbol{P}_\uparrow \boldsymbol{\rho})}. \tag{4.254}$$

Because $\langle \Psi \mid \boldsymbol{P}_\uparrow \mid \Psi\rangle$ and $\mathrm{Tr}\,(\boldsymbol{P}_\uparrow \boldsymbol{\rho})$ are both the same thing, namely, the expectation value of $\boldsymbol{P}_\uparrow$ on $\boldsymbol{\rho}$.

As the same holds for $\boldsymbol{P}_\downarrow$ we can restate equation (4.253) by saying that the effect of the measurement performed by $\boldsymbol{H} = -\mu B_z \boldsymbol{P}_\uparrow + \mu B_z \boldsymbol{P}_\downarrow$ on $\boldsymbol{\rho}$ is to produce two beams in states

$$\frac{\boldsymbol{P}_\uparrow \boldsymbol{\rho} \boldsymbol{P}_\uparrow^\dagger}{\mathrm{Tr}\,(\boldsymbol{P}_\uparrow \boldsymbol{\rho})} \quad \text{and} \quad \frac{\boldsymbol{P}_\downarrow \boldsymbol{\rho} \boldsymbol{P}_\downarrow^\dagger}{\mathrm{Tr}\,(\boldsymbol{P}_\downarrow \boldsymbol{\rho})}. \tag{4.255}$$

Equations (4.255) are more general than equations (4.253), because they can be applied to mixed states as well.

4.8 Properties of the Density Operator

When can an arbitrary complex matrix be a density operator?

Let us suppose we have an arbitrary 2×2 complex matrix $\boldsymbol{\rho}$. What would the matrix have to look like to make it a plausible candidate for a physically meaningful density matrix? Being just 2×2 and complex, clearly, is not enough.

We have answered this question implicitly by developing the whole formalism of quantum mechanics from the probability side rather than from the unitary side, arriving at the expression

$$\boldsymbol{\rho} = \frac{1}{2}\left(\mathbf{1} + r^x \boldsymbol{\sigma}_x + r^y \boldsymbol{\sigma}_y + r^z \boldsymbol{\sigma}_z\right), \tag{4.256}$$

where r^x, r^y, and r^z are parameters that define the probability vector $\boldsymbol{p}$ in such a way that all its entries are guaranteed to be confined to $[0, 1]$ and the first two entries add up to 1.

In other words, we can say that if $\boldsymbol{\rho}$ can be written in the form (4.256) with r^x, r^y and r^z forming a three-dimensional real vector of length less than or equal to 1, then $\boldsymbol{\rho}$ is a plausible density matrix.

So, let us rephrase the original question as follows: What general conditions does matrix $\boldsymbol{\rho}$ have to satisfy to be rewritable in form (4.256)?

Must be Hermitian

The first condition is obvious. Matrix $\boldsymbol{\rho}$ must be Hermitian. This is because all Pauli matrices are Hermitian. If they weren't Hermitian, they couldn't represent Pauli quaternions, and we wouldn't have the mapping from probabilities to 2×2 matrices via quaternions.

Since there are three linearly independent Pauli matrices plus the identity matrix, together they constitute a basis in the space of 2×2 complex Hermitian matrices. Every 2×2 Hermitian matrix must be of the form

$$\boldsymbol{\rho} = a\mathbf{1} + b\boldsymbol{\sigma}_x + c\boldsymbol{\sigma}_y + d\boldsymbol{\sigma}_z, \tag{4.257}$$

where all coefficients, a, b, c, and d are real.

To pin a to 1/2, we request that $\mathrm{Tr}(\boldsymbol{\rho}) = 1$, because all Pauli matrices are traceless and $\mathrm{Tr}(\mathbf{1}) = 2$.

Must have trace equal 1

But this still leaves too much freedom to possible values that b, c, and d may assume.

If $\boldsymbol{\rho}$ corresponds to a valid fiducial state $\boldsymbol{p}$, then for every other valid *pure* qubit state $\boldsymbol{p_n}$ defined by some direction $\boldsymbol{n}$, where $\boldsymbol{n} \cdot \boldsymbol{n} = 1$, we have that (cf. equation (4.133), page 131)

$$p_{\boldsymbol{p_n} \leftarrow \boldsymbol{p}} = \langle \tilde{\boldsymbol{p}}_{\boldsymbol{n}}, \boldsymbol{p} \rangle = \langle \tilde{\boldsymbol{p}}_{\boldsymbol{n}} \rangle_{\boldsymbol{p}} \in [0, 1]. \tag{4.258}$$

It is convenient here to switch $\boldsymbol{p_n}$ and $\boldsymbol{p}$ around, so that we consider

$$\langle \tilde{\boldsymbol{p}}, \boldsymbol{p_n} \rangle = \langle \tilde{\boldsymbol{p}} \rangle_{\boldsymbol{p_n}} \tag{4.259}$$

instead. Since $\boldsymbol{p_n}$ is a pure state we can always find such $| \Psi_{\boldsymbol{n}} \rangle$ that the density operator that corresponds to $\boldsymbol{p_n}$ is $| \Psi_{\boldsymbol{n}} \rangle \langle \Psi_{\boldsymbol{n}} |$. But this is our good old friend, the projector, discussed at some length in Section 4.7. Let us call this projector $\boldsymbol{P_n}$, rather than $\boldsymbol{\rho_n}$. Using this notation, we rewrite the above by

$$\langle \tilde{\boldsymbol{p}} \rangle_{\boldsymbol{p_n}} = \mathrm{Tr}(\boldsymbol{P_n} \boldsymbol{\rho}) = \langle \Psi_{\boldsymbol{n}} | \boldsymbol{\rho} | \Psi_{\boldsymbol{n}} \rangle. \tag{4.260}$$

Because this must be a probability, confined to $[0, 1]$ for any direction $\boldsymbol{n}$, we arrive at the following condition:

$$\forall_{\boldsymbol{n}} \langle \Psi_{\boldsymbol{n}} | \boldsymbol{\rho} | \Psi_{\boldsymbol{n}} \rangle \in [0, 1]. \tag{4.261}$$

The condition in a somewhat relaxed form of

$$\forall_{\boldsymbol{n}} \langle \Psi_{\boldsymbol{n}} | \boldsymbol{\rho} | \Psi_{\boldsymbol{n}} \rangle \geq 0 \quad \text{or} \quad \forall_{\boldsymbol{n}} \mathrm{Tr}(\boldsymbol{P_n} \boldsymbol{\rho}) \geq 0 \tag{4.262}$$

is referred to as *positivity* of the density matrix $\boldsymbol{\rho}$.

It is easy to see that this condition, in combination with the trace condition, is sufficient to enforce $\boldsymbol{r} \cdot \boldsymbol{r} \leq 1$.

Must be positive

Because $\boldsymbol{\rho}$ is Hermitian, it can be diagonalized by "rotating" the basis of the Hilbert space, so that it aligns with the eigenvectors $| \eta_i \rangle$ of $\boldsymbol{\rho}$. Strictly speaking, "rotations" in the Hilbert space are unitary operations. Now we can choose the $\boldsymbol{\rho}$ eigenvectors as some of the $| \Psi_{\boldsymbol{n}} \rangle$ vectors, and for each of them we still expect that

$$\langle \eta_i | \boldsymbol{\rho} | \eta_i \rangle = \rho_i \langle \eta_i | \eta_i \rangle = \rho_i \geq 0, \tag{4.263}$$

where ρ_i are the eigenvalues of $\boldsymbol{\rho}$. The trace condition implies that $\sum_i \rho_i = 1$, which means that all $\rho_i \in [0, 1]$. Hence, the determinant of $\boldsymbol{\rho}$, which is a unitary invariant equal to $\prod_i \rho_i$ must be positive as well.

Let us recall equation (4.64), page 118. If we allow the r^i coefficients to be anything, it corresponds to our general matrix $\boldsymbol{\rho}$ with $a = 1$, which, as we saw above, derived from the trace condition.

We can now use equation (4.64) to evaluate the determinant:

$$\begin{aligned}\det \boldsymbol{\rho} &= \frac{1}{4}\left(\left(1+r^z\right)\left(1-r^z\right) - \left(r^x + i r^y\right)\left(r^x - i r^y\right)\right) \\ &= \frac{1}{4}\left(1 - \left(r^z\right)^2 - \left(r^x\right)^2 - \left(r^y\right)^2\right) \\ &= \frac{1}{4}\left(1 - \boldsymbol{r}\cdot\boldsymbol{r}\right) \geq 0, \end{aligned} \tag{4.264}$$

which implies that

$$\boldsymbol{r}\cdot\boldsymbol{r} \leq 1. \tag{4.265}$$

In summary, the following three conditions characterize a plausible density matrix $\boldsymbol{\rho}$:

1. Matrix $\boldsymbol{\rho}$ must be Hermitian.
2. Trace of $\boldsymbol{\rho}$ must be 1.
3. Matrix $\boldsymbol{\rho}$ must be positive, meaning

$$\forall_{\boldsymbol{n}} \langle \Psi_{\boldsymbol{n}} \mid \boldsymbol{\rho} \mid \Psi_{\boldsymbol{n}} \rangle \geq 0 \quad \text{or} \quad \forall_{\boldsymbol{n}} \operatorname{Tr}\left(\boldsymbol{P}_{\boldsymbol{n}}\boldsymbol{\rho}\right) \geq 0. \tag{4.266}$$

The positivity condition is not an easy condition to use in general, but if we can show that at least one of the eigenvalues of $\boldsymbol{\rho}$ is negative, this disqualifies $\boldsymbol{\rho}$ from being a plausible density matrix right away.

Must be idempotent for pure states

One more property pertains to density operators of pure states only. For such operators we find that

$$\boldsymbol{\rho}\boldsymbol{\rho} = \boldsymbol{\rho} \tag{4.267}$$

because these states are *projectors*.

To prove that $\boldsymbol{\rho}$ of a pure state is idempotent is trivial on the unitary level. Since for a pure state we have that

$$\boldsymbol{\rho} = \mid \Psi\rangle\langle\Psi \mid, \tag{4.268}$$

we can easily see that

$$\boldsymbol{\rho}\boldsymbol{\rho} = \mid \Psi\rangle\langle\Psi \mid \cdot \mid \Psi\rangle\langle\Psi \mid = \mid \Psi\rangle\left(\langle\Psi \mid \Psi\rangle\right)\langle\Psi \mid = \mid \Psi\rangle\left(1\right)\langle\Psi \mid = \mid \Psi\rangle\langle\Psi \mid = \boldsymbol{\rho}. \tag{4.269}$$

It is instructive to repeat this computation on the quaternion level, because this is going to tell us something about the geometric significance of the density operator's idempotence. If $\boldsymbol{\rho} = \left(\mathbf{1} + \sum_i r^i \boldsymbol{\sigma}_i\right)/2$, then

$$\begin{aligned}
\boldsymbol{\rho}\boldsymbol{\rho} & \\
&= \frac{1}{2}\left(\mathbf{1} + \sum_i r^i \boldsymbol{\sigma}_i\right) \cdot \frac{1}{2}\left(\mathbf{1} + \sum_j r^j \boldsymbol{\sigma}_j\right) \\
&= \frac{1}{4}\left(\mathbf{1}\cdot\mathbf{1} + \mathbf{1}\cdot\sum_j r^j \boldsymbol{\sigma}_j + \left(\sum_i r^i \boldsymbol{\sigma}_i\right)\cdot\mathbf{1} + \sum_i\sum_j r^i r^j \boldsymbol{\sigma}_i \cdot \boldsymbol{\sigma}_j\right) \\
&= \frac{1}{4}\left(\mathbf{1} + 2\sum_j r^i \boldsymbol{\sigma}_i + \sum_i\sum_j r^i r^j \left(\delta_{ij}\mathbf{1} + i\sum_k \epsilon_{ijk}\boldsymbol{\sigma}_k\right)\right) \\
&= \frac{1}{4}\left((1 + \boldsymbol{r}\cdot\boldsymbol{r})\,\mathbf{1} + 2\sum_i r^i \boldsymbol{\sigma}_i\right),
\end{aligned} \tag{4.270, 4.271}$$

because $\sum_i \sum_j r^i r^j \epsilon_{ijk} = 0$, on account of $r^i r^j$ being symmetric and ϵ_{ijk} antisymmetric.

Only when $\boldsymbol{r}\cdot\boldsymbol{r} = 1$, that is, only for pure states, do we end up with

$$\boldsymbol{\rho}\boldsymbol{\rho} = \frac{1}{2}\left(\mathbf{1} + \sum_i r^i \boldsymbol{\sigma}_i\right) = \boldsymbol{\rho}; \tag{4.272}$$

otherwise $\boldsymbol{r}\cdot\boldsymbol{r}$ falls short: it does not reach the 1, and we end up with a deformed quaternion that does not represent any state, because its real component is less than 1/2.

Conditions 1 through 3 as well as the observation that a density operator of a pure state is idempotent extend to quantum systems of dimensionality higher than those of single qubits, for example to multiqubit systems and even to infinitely dimensional systems.

4.9 Schrödinger Equation

Like the fiducial vector of a qubit, $\boldsymbol{p}$, and like its corresponding density quaternion $\boldsymbol{\rho}$, a Hilbert space vector $|\,\Psi\rangle$, which describes a pure qubit state, evolves in the presence of a "magnetic field" $\boldsymbol{B}$ as well. The evolution equation can be derived by taking apart the von Neumann equation (3.73) discussed in Section 3.7 on page 104:

$$\frac{\mathrm{d}}{\mathrm{d}t}\boldsymbol{\rho} = -\frac{\mathrm{i}}{\hbar}\left[\boldsymbol{H}, \boldsymbol{\rho}\right]. \tag{4.273}$$

Unpacking the von Neumann equation

The trick is to substitute

$$\boldsymbol{\rho} = \mid \Psi\rangle\langle\Psi \mid . \tag{4.274}$$

We do not even have to split $\mid \Psi\rangle$ into a superposition. Let us work on the left-hand side of the resulting equation first.

$$\frac{\mathrm{d}}{\mathrm{d}t}\boldsymbol{\rho} = \frac{\mathrm{d}}{\mathrm{d}t}\left(\mid \Psi\rangle\langle\Psi \mid\right) = \left(\frac{\mathrm{d}}{\mathrm{d}t} \mid \Psi\rangle\right)\langle\Psi \mid + \mid \Psi\rangle\left(\frac{\mathrm{d}}{\mathrm{d}t}\langle\Psi \mid\right). \tag{4.275}$$

On the right-hand side we have

$$-\frac{\mathrm{i}}{\hbar}\left(\boldsymbol{H} \mid \Psi\rangle\langle\Psi \mid - \mid \Psi\rangle\langle\Psi \mid \boldsymbol{H}\right). \tag{4.276}$$

Combining the two we discover a sum of two equations:

$$\left(\frac{\mathrm{d}}{\mathrm{d}t} \mid \Psi\rangle\right)\langle\Psi \mid \; = \; -\frac{\mathrm{i}}{\hbar}\left(\boldsymbol{H} \mid \Psi\rangle\right)\langle\Psi \mid \quad \text{and} \tag{4.277}$$

$$\mid \Psi\rangle\left(\frac{\mathrm{d}}{\mathrm{d}t}\langle\Psi \mid\right) \; = \; \mid \Psi\rangle\frac{\mathrm{i}}{\hbar}\left(\langle\Psi \mid \boldsymbol{H}\right). \tag{4.278}$$

The equations are duals of each other and they reduce to

$$\frac{\mathrm{d}}{\mathrm{d}t} \mid \Psi\rangle = -\frac{\mathrm{i}}{\hbar}\boldsymbol{H} \mid \Psi\rangle. \tag{4.279}$$

Schrödinger equation

This is the celebrated Schrödinger equation for a qubit. Like the von Neumann equation and its fiducial space equivalent, the equation preserves the purity of the state. We call the evolution *unitary* because it does not affect the length of the state vector $\mid \Psi\rangle$, which remains $\langle\Psi \mid \Psi\rangle = 1$, or... *unity*. The name "unitary" is also used to describe the whole formalism that results from unpacking quaternions into Pauli matrices and that represents quantum states by vectors in the Hilbert space on which the matrices operate, rather than by density quaternions (or operators).

Universality of 2×2 Hamiltonian

Matrix $\boldsymbol{H}$, as parameterized by equation (4.214), is almost the most general 2×2 *Hermitian* matrix. For such a matrix the diagonal elements must be real, because in this case $H_{ij} = H^*_{ji}$ implies $H_{ii} = H^*_{ii}$, and off-diagonal elements must be complex conjugates of their mirror images across the diagonal. This requirement leaves us with four independent parameters, and we find them all here in equation (4.215).

Well, we actually find only three, but let us remember that $\boldsymbol{H}$ represents energy, and energy is defined only up to an additive constant. We can therefore always choose the constant so that $H_{11} = -H_{22}$.

A Hamiltonian is responsible for a unitary evolution of a quantum system. Because $\boldsymbol{H}$ is the most general 2×2 Hamiltonian possible, any other quantum system characterized by two basis states only must be described by a matrix that looks the same. The interpretation of vector $\boldsymbol{B}$, of course, differs from a system to a system, as does the intepretation of $\boldsymbol{r}$, but the equations and their solutions are identical.

This is why whether we talk about a quantronium or about a neutron spin—two systems that couldn't be more different—we end up with exactly the same mathematics, the same dynamics, and the same properties. Whatever can be said about a neutron beam translates immediately into pronouncements that can be made about a statistical ensemble of quantroniums, or two-level molecules, or two-level quantum dots, or any other two-level quantum system.

4.9.1 General Solution of the Schrödinger Equation

Equation (4.279) can be solved for a general case of $\boldsymbol{H}$. We are going to solve it for $\boldsymbol{H}$ = constant first.

We start from a simple discrete approximation of the equation:

$$\frac{\mid \Psi(t+\Delta t)\rangle - \mid \Psi(t)\rangle}{\Delta t} \approx \frac{\mathrm{d}\boldsymbol{H}}{\mathrm{d}t} = -\frac{\mathrm{i}}{\hbar}\boldsymbol{H} \mid \Psi(t)\rangle. \tag{4.280}$$

We can extract $\mid \Psi(t+\Delta t)\rangle$ from it, and this leads to the familiar Euler time step,

Advancing the Schrödinger equation by a single Euler time step

$$\mid \Psi(t+\Delta t)\rangle \approx \mid \Psi(t)\rangle - \frac{\mathrm{i}}{\hbar}\boldsymbol{H} \mid \Psi(t)\rangle \Delta t. \tag{4.281}$$

The solution tells us something quite insightful about the Schrödinger equation to begin with. The equation represents the simplest evolution possible. It says that, evolved over a short time span Δt, vector $\mid \Psi(t+\Delta t)\rangle$ is going to differ from the original vector $\mid \Psi(t)\rangle$ by a small linear correction $-\mathrm{i}\boldsymbol{H} \mid \Psi(t)\rangle \Delta t/\hbar$. There are no fancy second derivatives here as we have in the Newton's equations, no third derivatives as we have in the Lorentz-Abraham equations, no complicated curvature terms and connection symbols as we have in the Einstein's equations of General Relativity. It is amazingly simple.

Let us use this insight to figure out how a quantum system is going to evolve over a longer time stretch.

We start from $t = 0$. After a sufficiently short Δt an initial state $|\,\Psi(0)\rangle$ evolves into

$$|\,\Psi(\Delta t)\rangle \approx \left(\mathbf{1} + \frac{1}{\mathrm{i}\hbar}\boldsymbol{H}\Delta t\right) |\,\Psi(0)\rangle. \tag{4.282}$$

Having made this one time step, we make another one, also of length Δt,

$$\begin{aligned} |\,\Psi(2\Delta t)\rangle &\approx \left(\mathbf{1} + \frac{1}{\mathrm{i}\hbar}\boldsymbol{H}\Delta t\right) |\,\Psi(\Delta t)\rangle \\ &\approx \left(\mathbf{1} + \frac{1}{\mathrm{i}\hbar}\boldsymbol{H}\Delta t\right)\left(\mathbf{1} + \frac{1}{\mathrm{i}\hbar}\boldsymbol{H}\Delta t\right) |\,\Psi(0)\rangle. \end{aligned} \tag{4.283}$$

It is now clear that for every additional time step of length Δt, we're going to act on the initial state $|\,\Psi(0)\rangle$ with a yet another instance of $\left(\mathbf{1} + \frac{1}{\mathrm{i}\hbar}\boldsymbol{H}\Delta t\right)$. In summary,

$$|\,\Psi(n\Delta t)\rangle \approx \left(\mathbf{1} + \frac{1}{\mathrm{i}\hbar}\boldsymbol{H}\Delta t\right)^n |\,\Psi(0)\rangle. \tag{4.284}$$

Advancing the Schrödinger equation by multiple Euler time steps

But $n\Delta t = t$, so

$$|\,\Psi(t)\rangle \approx \left(\mathbf{1} + \frac{1}{\mathrm{i}\hbar}\boldsymbol{H}\frac{t}{n}\right)^n |\,\Psi(0)\rangle. \tag{4.285}$$

The expression is approximate, because it results from taking n finite, though small, time steps of length $\Delta t = t/n$. Clearly, we can only get more accurate by making $\Delta t = t/n$ smaller, which is the same as taking n larger, converging on the exact solution for $n \to \infty$:

$$|\,\Psi(t)\rangle = \lim_{n\to\infty} \left(\mathbf{1} + \frac{1}{\mathrm{i}\hbar}\boldsymbol{H}\frac{t}{n}\right)^n |\,\Psi(0)\rangle. \tag{4.286}$$

The limit can be evaluated as follows. We make use of the familiar Newton formula for $(a+b)^n$:

$$(a+b)^n = \sum_{k=0}^{n} \binom{n}{k} a^{n-k} b^k. \tag{4.287}$$

We can use the formula here because $\mathbf{1}\boldsymbol{H} = \boldsymbol{H}\mathbf{1}$ (the derivation of the Newton formula makes use of $ab = ba$), and this yields

$$\left(\mathbf{1} + \frac{\boldsymbol{H}t}{\mathrm{i}\hbar n}\right)^n = \sum_{k=0}^{n} \binom{n}{k} \mathbf{1}^{n-k} \left(\frac{\boldsymbol{H}t}{\mathrm{i}\hbar n}\right)^k. \tag{4.288}$$

Of course, $\mathbf{1}$ applied any number of times is still $\mathbf{1}$, so we can drop it. We can also unpack the $\binom{n}{k}$ symbol. We end up with

$$\left(\mathbf{1} + \frac{\boldsymbol{H}t}{\mathrm{i}\hbar n}\right)^n = \sum_{k=0}^{n} \frac{n!}{k!(n-k)!} \left(\frac{\boldsymbol{H}t}{\mathrm{i}\hbar n}\right)^k. \tag{4.289}$$

Let us have a closer look at $n!/(k!(n-k)!)$. This can be rewritten as

$$\frac{(n-k+1)(n-k+2)\dots(n-k+k)}{k!}. \tag{4.290}$$

We have k terms in the numerator and they are all of the form $(n - \text{something})$ with the exception of the last one, which is just n. If we were to evaluate this, we'd get

$$n^k + n^{k-1} \times \text{something} + n^{k-2} \times \text{something else} + \dots. \tag{4.291}$$

But there is also n^k in the denominator of $(\boldsymbol{H}t/(\mathrm{i}\hbar n))^k$. For $n \to \infty$ $n^k/n^k = 1$, but all the other terms like $n^{k-1} \times \text{something}/n^k$ become zero. And so, we end up with

$$\lim_{n\to\infty}\left(\mathbf{1} + \frac{\boldsymbol{H}t}{\mathrm{i}\hbar n}\right)^n = \sum_{k=0}^{\infty}\frac{1}{k!}\left(\frac{\boldsymbol{H}t}{i\hbar}\right)^k. \tag{4.292}$$

Ah, but this looks so much like e^x,

$$e^x = \sum_{k=0}^{\infty}\frac{x^k}{k!}. \tag{4.293}$$

Sure, we have just a plain number x in (4.293), but an operator $\boldsymbol{H}t/(\mathrm{i}\hbar)$ in (4.292). We know that $e^n = e \times e \times \dots \times e$ n-times. But what does e^{operator} mean? Well, it means

$$e^{-\mathrm{i}\boldsymbol{H}t/\hbar} \doteq \sum_{k=0}^{\infty}\frac{1}{k!}\left(\frac{\boldsymbol{H}t}{\mathrm{i}\hbar}\right)^k. \tag{4.294}$$

General solution for a constant Hamiltonian

This is how we *define* it; and having done so, we can write the general solution to the Schrödinger equation for $\boldsymbol{H} = \text{constant}$ as

$$\mid \Psi(t)\rangle = e^{-\mathrm{i}\boldsymbol{H}t/\hbar} \mid \Psi(0)\rangle. \tag{4.295}$$

Matrix exponential

The matrix exponential $e^{\boldsymbol{A}}$ does not always have the same properties as an ordinary number exponential e^x. The reason is that matrices do not commute in general, whereas numbers (with the notable exception of quaternions) do. In particular we cannot always write

$$e^{\boldsymbol{A}}e^{\boldsymbol{B}} = e^{\boldsymbol{A}+\boldsymbol{B}}. \tag{4.296}$$

This equation applies only when $\boldsymbol{A}$ and $\boldsymbol{B}$ commute, that is, when $\boldsymbol{AB} = \boldsymbol{BA}$. Of course, since $\boldsymbol{A}$ commutes with itself, we can always write

$$e^{a\boldsymbol{A}}e^{b\boldsymbol{A}} = e^{(a+b)\boldsymbol{A}}. \tag{4.297}$$

But other properties of the exponential, such as $e^0 = 1$ and $e^{-x} = 1/e^x$, still hold with appropriate matrix substitutions for the inverse and for the 1:

$$e^{\mathbf{0}} = \mathbf{1}, \tag{4.298}$$

$$\left(e^{\boldsymbol{A}}\right)^{-1} = e^{-\boldsymbol{A}}. \tag{4.299}$$

Another useful property is that when $\boldsymbol{A}$ and $\boldsymbol{B}$ are *similar*, that is, such that

$$\boldsymbol{B} = \boldsymbol{M}^{-1}\boldsymbol{A}\boldsymbol{M}, \tag{4.300}$$

where $\boldsymbol{M}$ is an arbitrary invertible matrix (sized so that the equation above makes sense), then

$$e^{\boldsymbol{B}} = \boldsymbol{M}^{-1}e^{\boldsymbol{A}}\boldsymbol{M}. \tag{4.301}$$

This is easy to see, because it follows from

$$\boldsymbol{B}\boldsymbol{B}\ldots\boldsymbol{B} = \boldsymbol{M}^{-1}\boldsymbol{A}\boldsymbol{M}\boldsymbol{M}^{-1}\boldsymbol{A}\boldsymbol{M}\ldots\boldsymbol{M}^{-1}\boldsymbol{A}\boldsymbol{M} = \boldsymbol{M}^{-1}\boldsymbol{A}\boldsymbol{A}\ldots\boldsymbol{A}\boldsymbol{M}. \tag{4.302}$$

Applying the above to every term of the exponential expansion yields (4.301). The reason (4.301) is such a useful property is that it lets us evaluate $\exp \boldsymbol{B}$ easily if $\boldsymbol{B}$ is diagonalizable by a similarity transformation to some diagonal matrix $\boldsymbol{A}$, because then $\exp \boldsymbol{A}$ is a diagonal matrix filled with exponentials of the diagonal terms of $\boldsymbol{A}$.

Let us return to the Schrödinger equation and its solution.

What if $\boldsymbol{H}$ is not constant, that is, $\boldsymbol{H} = \boldsymbol{H}(t)$?

General solution for time-dependent Hamiltonian

The problem is still tractable. Much depends on how $\boldsymbol{H}$ varies with t. Let us suppose that $\boldsymbol{H}$ is constant and equal to, say, $\boldsymbol{H}_1$ for $t \in [0, t_1]$. Then $\boldsymbol{H}$ changes rapidly to $\boldsymbol{H}_2$ for $t \in]t_1, t_2]$, and so on. The solution in this case will be

$$\begin{aligned}
\mid \Psi(t)\rangle &= e^{-\mathrm{i}\boldsymbol{H}_1 t/\hbar} \mid \Psi(0)\rangle \quad \text{for} \quad t \in [0, t_1], \\
\mid \Psi(t)\rangle &= e^{-\mathrm{i}\boldsymbol{H}_2(t-t_1)/\hbar} e^{-\mathrm{i}\boldsymbol{H}_1 t_1/\hbar} \mid \Psi(0)\rangle \quad \text{for} \quad t \in]t_1, t_2], \\
\mid \Psi(t)\rangle &= e^{-\mathrm{i}\boldsymbol{H}_3(t-t_2)/\hbar} e^{-\mathrm{i}\boldsymbol{H}_2(t_2-t_1)/\hbar} e^{-\mathrm{i}\boldsymbol{H}_1 t_1/\hbar} \mid \Psi(0)\rangle \quad \text{for} \quad t \in]t_2, t_3], \\
&\ldots
\end{aligned}$$

and so on.

To tackle a more general case, let us chop time into small segments Δt, and let us assume that the Hamiltonian changes sufficiently slowly, so that we can consider it constant within each segment. Following the reasoning presented above, we can write the solution of this equation in the following form:

$$\mid \Psi(t)\rangle = e^{-\mathrm{i}\boldsymbol{H}(t)\Delta t/\hbar} e^{-\mathrm{i}\boldsymbol{H}(t-\Delta t)\Delta t/\hbar} e^{-\mathrm{i}\boldsymbol{H}(t-2\Delta t)\Delta t/\hbar}$$

$$\ldots e^{-\mathrm{i}\boldsymbol{H}(\Delta t)\Delta t/\hbar} e^{-\mathrm{i}\boldsymbol{H}(0)\Delta t/\hbar} \mid \Psi(0)\rangle. \tag{4.303}$$

Of course, $e^{\boldsymbol{H}_1} e^{\boldsymbol{H}_2} \neq e^{\boldsymbol{H}_1+\boldsymbol{H}_2}$, if $\boldsymbol{H}_1$ and $\boldsymbol{H}_2$ do not commute, but if the evolution of $\boldsymbol{H}(t)$ is such that $[\boldsymbol{H}(t_1), \boldsymbol{H}(t_2)] = \boldsymbol{0}$ for each t_1 and t_2, then, in this happy circumstance, we are allowed to gather all the exponents into a sum:

$$\mid \Psi(t)\rangle = e^{-\mathrm{i}(\boldsymbol{H}(t)+\boldsymbol{H}(t-\Delta t)+\boldsymbol{H}(t-2\Delta t)+\ldots+\boldsymbol{H}(\Delta t)+\boldsymbol{H}(0))\Delta t/\hbar} \mid \Psi(0)\rangle. \tag{4.304}$$

The shorter the Δt, the more accurate the expression, so in the limit of $\Delta t \to 0$ we get

$$\mid \Psi(t)\rangle = e^{-\mathrm{i}\left(\int_0^t \boldsymbol{H}(t)\,\mathrm{d}t\right)/\hbar} \mid \Psi(0)\rangle. \tag{4.305}$$

Unitarity of the Schrödinger evolution

Factor $\exp\left(-\mathrm{i}\boldsymbol{H}\Delta t/\hbar\right)$ represents a finite transformation of a quantum system enacted by $\boldsymbol{H}$ that was applied to the system for the duration of Δt. Its characteristic feature is that it does not affect the length of vector $\Psi(t)$:

$$\langle \Psi(\Delta t) \mid \Psi(\Delta t)\rangle = \langle \Psi(0) \mid e^{\mathrm{i}\boldsymbol{H}^\dagger \Delta t/\hbar} e^{-\mathrm{i}\boldsymbol{H}\Delta t/\hbar} \mid \Psi(0)\rangle. \tag{4.306}$$

Because $\boldsymbol{H}$ is Hermitian, $\boldsymbol{H}^\dagger = \boldsymbol{H}$, we find that

$$e^{\mathrm{i}\boldsymbol{H}^\dagger \Delta t/\hbar} e^{-\mathrm{i}\boldsymbol{H}\Delta t/\hbar} = e^{\mathrm{i}\boldsymbol{H}\Delta t/\hbar} e^{-\mathrm{i}\boldsymbol{H}\Delta t/\hbar} = e^{\mathrm{i}(\boldsymbol{H}-\boldsymbol{H})\Delta t/\hbar} = e^{\boldsymbol{0}} = \boldsymbol{1}. \tag{4.307}$$

We could gather the exponents into a sum, because $\boldsymbol{H}$ commutes with itself. And so in the end we get

$$\langle \Psi(\Delta t) \mid \Psi(\Delta t)\rangle = \langle \Psi(0) \mid \Psi(0)\rangle. \tag{4.308}$$

A most general such operation, given by equation (4.303), is a superposition of unitary operations and thus a unitary operation itself. Let us call it $\boldsymbol{U}(t)$, so that

$$\mid \Psi(t)\rangle = \boldsymbol{U}(t) \mid \Psi(0)\rangle. \tag{4.309}$$

Its dual equivalent, $\boldsymbol{U}^\dagger(t)$, given by

$$\begin{aligned} \boldsymbol{U}^\dagger(t) \quad = \quad & e^{\mathrm{i}\boldsymbol{H}(0)\Delta t/\hbar} e^{\mathrm{i}\boldsymbol{H}(\Delta t)\Delta t/\hbar} \\ & \ldots e^{\mathrm{i}\boldsymbol{H}(t-2\Delta t)\Delta t/\hbar} e^{\mathrm{i}\boldsymbol{H}(t-\Delta t)\Delta t/\hbar} e^{\mathrm{i}\boldsymbol{H}(t)\Delta t/\hbar}, \end{aligned} \tag{4.310}$$

evolves a form that is dual to vector $\mid \Psi\rangle$,

$$\langle \Psi(t) \mid = \langle \Psi(0) \mid \boldsymbol{U}^\dagger(t). \tag{4.311}$$

When put together, they annihilate each other from the middle onwards:

$$\langle \Psi(t) \mid \Psi(t)\rangle = \langle \Psi(0) \mid \boldsymbol{U}^\dagger(t)\boldsymbol{U}(t) \mid \Psi(0)\rangle$$

$$\begin{aligned}
&= \langle \Psi(0) \mid e^{\mathrm{i}\boldsymbol{H}(0)\Delta t/\hbar} e^{\mathrm{i}\boldsymbol{H}(\Delta t)\Delta t/\hbar} \\
&\quad \ldots \left(e^{\mathrm{i}\boldsymbol{H}(t-2\Delta t)\Delta t/\hbar} \left(e^{\mathrm{i}\boldsymbol{H}(t-\Delta t)\Delta t/\hbar} \right.\right. \\
&\quad \left(e^{\mathrm{i}\boldsymbol{H}(t)\Delta t/\hbar} e^{-\mathrm{i}\boldsymbol{H}(t)\Delta t/\hbar} \right) \\
&\quad \left.\left. e^{-\mathrm{i}\boldsymbol{H}(t-\Delta t)\Delta t/\hbar} \right) e^{-\mathrm{i}\boldsymbol{H}(t-2\Delta t)\Delta t/\hbar} \right) \\
&\quad \ldots e^{-\mathrm{i}\boldsymbol{H}(\Delta t)\Delta t/\hbar} e^{-\mathrm{i}\boldsymbol{H}(0)\Delta t/\hbar} \mid \Psi(0)\rangle \\
&= \langle \Psi(0) \mid \Psi(0)\rangle. \qquad (4.312)
\end{aligned}$$

A qubit evolution operator $\boldsymbol{U}(t)$ can be represented by a 2×2 complex matrix. The matrices have the following property:

$$\boldsymbol{U}(t)\boldsymbol{U}^{\dagger}(t) = \mathbf{1}, \qquad (4.313)$$

which we have just demonstrated. Matrices that satisfy this property are called *unitary*, and the corresponding operators are called *unitary operators*.

Another way to look at unitary operators is to observe that

$$\boldsymbol{U}^{\dagger} = \boldsymbol{U}^{-1}, \qquad (4.314)$$

that is, the Hermitian conjugate of $\boldsymbol{U}$ is its inverse.

Unitary operators are closely related to orthogonal operators, that is, rotations and reflections. Indeed, if $\boldsymbol{U}$ is real, then $\boldsymbol{U}^{\dagger} = \boldsymbol{U}^{T}$, and equation (4.313) becomes

$$\boldsymbol{U}(t)\boldsymbol{U}^{T}(t) = \mathbf{1}, \qquad (4.315)$$

which defines orthogonal operators.

A combination of $\boldsymbol{U}(t)$ and $\boldsymbol{U}^{\dagger}(t)$ is needed to evolve a density operator made of $\mid \Psi\rangle$ and $\langle \Psi \mid$:

$$\begin{aligned}
\boldsymbol{\rho}(t) &= \mid \Psi(t)\rangle\langle\Psi(t) \mid \\
&= \boldsymbol{U}(t) \mid \Psi(0)\rangle\langle\Psi(0) \mid \boldsymbol{U}^{\dagger}(t) \\
&= \boldsymbol{U}(t)\boldsymbol{\rho}(0)\boldsymbol{U}^{\dagger}(t). \qquad (4.316)
\end{aligned}$$

It is easy to get from here back to the von Neumann equation. Let us focus on a short-time-increment version of $\boldsymbol{U}$, that is,

$$\boldsymbol{U}(\Delta t) = \mathbf{1} + \frac{1}{\mathrm{i}\hbar}\boldsymbol{H}\Delta t \quad \text{and} \qquad (4.317)$$

$$U^\dagger(\Delta t) = \mathbf{1} - \frac{1}{\mathrm{i}\hbar} \boldsymbol{H} \Delta t, \tag{4.318}$$

where, again, we have made use of $\boldsymbol{H} = \boldsymbol{H}^\dagger$. Applying these to $\boldsymbol{\rho}(0)$ yields

$$\begin{aligned}
\boldsymbol{\rho}(\Delta t) &= \boldsymbol{U}(\Delta t)\boldsymbol{\rho}(0)\boldsymbol{U}^\dagger(\Delta t) \\
&= \left(\mathbf{1} + \frac{1}{\mathrm{i}\hbar}\boldsymbol{H}\Delta t\right)\boldsymbol{\rho}(0)\left(\mathbf{1} - \frac{1}{\mathrm{i}\hbar}\boldsymbol{H}\Delta t\right) \\
&= \left(\boldsymbol{\rho}(0) + \frac{\Delta t}{\mathrm{i}\hbar}\boldsymbol{H}\boldsymbol{\rho}(0)\right)\left(\mathbf{1} - \frac{1}{\mathrm{i}\hbar}\boldsymbol{H}\Delta t\right) \\
&= \boldsymbol{\rho}(0) + \frac{\Delta t}{\mathrm{i}\hbar}\boldsymbol{H}\boldsymbol{\rho}(0) - \frac{\Delta t}{\mathrm{i}\hbar}\boldsymbol{\rho}(0)\boldsymbol{H} + \mathcal{O}(\Delta t)^2 \\
&\approx \boldsymbol{\rho}(0) + \frac{1}{\mathrm{i}\hbar}\left[\boldsymbol{H}, \boldsymbol{\rho}(0)\right]\Delta t.
\end{aligned} \tag{4.319}$$

And this implies

$$\frac{\boldsymbol{\rho}(\Delta t) - \boldsymbol{\rho}(0)}{\Delta t} = \frac{1}{\mathrm{i}\hbar}\left[\boldsymbol{H}, \boldsymbol{\rho}(0)\right], \tag{4.320}$$

which is a finite difference approximation of the von Neumann equation.

The 2×2 unitary operators, $\boldsymbol{U}$ and $\boldsymbol{U}^\dagger$, are not Hermitian. Nevertheless, they can be represented in terms of Pauli matrices and mapped onto quaternions assuming that some of the coefficients are complex. For example, the small Δt form of $\boldsymbol{U}$ for a magnetized qubit is given by

$$\begin{aligned}
\boldsymbol{U} &= \mathbf{1} + \frac{1}{\mathrm{i}\hbar}\boldsymbol{H}\Delta t \\
&= \mathbf{1} - \frac{\mu\Delta t}{\mathrm{i}\hbar}\left(B_x\boldsymbol{\sigma}_x + B_y\boldsymbol{\sigma}_y + B_z\boldsymbol{\sigma}_z\right).
\end{aligned} \tag{4.321}$$

Together with $\boldsymbol{U}^\dagger$ they produce a small rotation of the polarization vector $\boldsymbol{r}$ by $\omega_L\Delta t$ about the direction of vector $\boldsymbol{B}$, where $\omega_L = 2\mu B/\hbar$ and B is the length of $\boldsymbol{B}$. Each by itself produces a corresponding small "rotation" of the unitary equivalent of $\boldsymbol{r}$, either $\mid \Psi\rangle$ or $\langle\Psi \mid$, in the spinor space. Their compositions, $\left(\boldsymbol{U}(\Delta t)\right)^n$, are equivalent to multiple small rotations of $\boldsymbol{r}$ and add up to large finite rotations of $\boldsymbol{r}$.

In summary, all that the unitary machinery of quantum mechanics can do to a single qubit is to rotate its polarization vector $\boldsymbol{r}$ in the same way that $2\mu\boldsymbol{r} \times \boldsymbol{B}/\hbar$ does it.

4.9.2 Larmor Precession Revisited

Let us assume that at $t = 0$ the qubit is in an eigenstate of $\boldsymbol{H}$, described by $\mid \Psi_n(0)\rangle$, where index $n = 1, 2$ numbers the two possible eigenstates. Being an eigenstate means that

$$\boldsymbol{H} \mid \Psi_n\rangle = E_n \mid \Psi_n\rangle, \tag{4.322}$$

that is, $\boldsymbol{H}$ applied to its own eigenstate stretches it or shrinks it but does not change its direction. Applying the Schrödinger equation to any of the eigenstates of $\boldsymbol{H}$ results in

$$\mathrm{i}\hbar\frac{\mathrm{d}}{\mathrm{d}t} \mid \Psi_n(t)\rangle = E_n \mid \Psi_n(t)\rangle, \tag{4.323}$$

Hamiltonian eigenstates "vibrate" with eigenfrequency.

which has a simple solution

$$\mid \Psi_n(t)\rangle = e^{-\mathrm{i}E_n t/\hbar} \mid \Psi_n(0)\rangle. \tag{4.324}$$

This is a special case of the general solution given by equation (4.295) on page 155.

Each of the eigenstates appears to "vibrate" with its own eigenfrequency

$$\omega_n = E_n/\hbar. \tag{4.325}$$

The corresponding forms vibrate with the same frequencies but in the opposite direction.

$$\langle\Psi_n(t) \mid = \langle\Psi_n(0) \mid e^{\mathrm{i}E_n t/\hbar} \tag{4.326}$$

Undetectability of eigenfrequencies

In all expressions of the kind $\langle\Psi \mid \boldsymbol{H} \mid \Psi\rangle$ the two vibrations, that of the vector and that of the form, cancel. Similarly, for the density operator $\boldsymbol{\rho} = \mid \Psi\rangle\langle\Psi \mid$ the vibration terms cancel again, so the "vibrations" of the eigenvectors are physically unobservable. The eigenstates just stay put and don't change. To force some visible change, for example, to make the qubit flip, we have to use a different Hamiltonian $\boldsymbol{H}_1$, such that the original eigenstates of $\boldsymbol{H}$ are no longer the eigenstates of $\boldsymbol{H}_1$.

The superposition beats with Larmor frequency.

But now let us take a superposition of $|\uparrow\rangle$ and $|\downarrow\rangle$, for example,

$$|\rightarrow\rangle = \frac{1}{\sqrt{2}}\left(|\uparrow\rangle + |\downarrow\rangle\right), \tag{4.327}$$

and assume that $\boldsymbol{H} = -\mu B_z \boldsymbol{\sigma}_z$. Both $|\uparrow\rangle$ and $|\downarrow\rangle$ are eigenstates of this Hamiltonian, as we saw in Section 4.7, page 142.

Each of the two eigenstates evolves in its own way, independent of the other one, because the Schrödinger equation is linear. If $\mid \Psi(0)\rangle = |\rightarrow\rangle$, then, at some later time t,

$$\mid \Psi(t)\rangle = \frac{1}{\sqrt{2}}\left(e^{\mathrm{i}\mu B_z t/\hbar} |\uparrow\rangle + e^{-\mathrm{i}\mu B_z t/\hbar} |\downarrow\rangle\right). \tag{4.328}$$

Let us invoke again equations (4.109), (4.110), and (4.111) from page 126:

$$aa^* - bb^* = r^z, \tag{4.329}$$
$$ab^* + ba^* = r^x, \tag{4.330}$$
$$\mathrm{i}\,(ab^* - ba^*) = r^y. \tag{4.331}$$

Here $a = \frac{1}{\sqrt{2}} \exp\left(\mathrm{i}\mu B_z t/\hbar\right)$ and $b = \frac{1}{\sqrt{2}} \exp\left(-\mathrm{i}\mu B_z t/\hbar\right)$. We notice that $r^z = 0$, so $\boldsymbol{r}$ remains in the plane perpendicular to $\boldsymbol{e}_z$. But

$$r^x = \frac{1}{2}\left(e^{\mathrm{i}2\mu B_z t/\hbar} + e^{-\mathrm{i}2\mu B_z t/\hbar}\right) = \cos\frac{2\mu B_z t}{\hbar} \quad \text{and} \tag{4.332}$$
$$r^y = \frac{\mathrm{i}}{2}\left(e^{\mathrm{i}2\mu B_z t/\hbar} - e^{-\mathrm{i}2\mu B_z t/\hbar}\right) = -\sin\frac{2\mu B_z t}{\hbar}. \tag{4.333}$$

We see that state $|\rightarrow\rangle$ precesses about the z-axis with Larmor frequency

$$\omega_L = \frac{2\mu B_z}{\hbar}. \tag{4.334}$$

In summary, even though the *phase factors*, with which the eigenstates "vibrate," namely, $\exp\left(-\mathrm{i}E_n t/\hbar\right)$, are invisible in isolation, they become detectable in superpositions, where they manifest as Larmor precession.

Larrmor precession as spinor rotation

Another way of looking at this is to recall equation (4.192) from Section 4.6 that talked about spinors and spinor transformations. There we saw that spinor $|\rightarrow\rangle$ would change its representation on the rotation of the coordinate system about the z axis in exactly the same way as shown by equation (4.328) with $\phi(t) = 2\mu B_z t/\hbar$. We can therefore reinterpret this equation yet again by saying that it describes active rotation of spinor $|\ \Psi(t)\rangle$ as induced by field B_z. And so, yet again do we find that there is more than one way to understand quantum superposition of states.

Because our dynamic equation that described the evolution of qubit probabilities, and then its quaternion equivalent, the von Neumann equation that described the evolution of the density quaternion, were restricted to fully polarized states and did not describe depolarization, they were, in fact, fully equivalent to the Schrödinger equation. The unitary formalism can reproduce all that we have covered in our discussion of qubit dynamics. The difference is that the unitary formalism *hides* vector $\boldsymbol{r}$ inside the two *complex* coefficients a and b that multiply the two basis vectors of the Hilbert space. People who look at superpositions such as $a\,|\uparrow\rangle + b\,|\downarrow\rangle$ often think of a and b as two *real* numbers and apply intuitions that pertain to *real* vector spaces. But there are four real numbers in the two complex coefficients a

and b, which the normalization condition $aa^* + bb^* = 1$ reduces to three, and the phase invariance condition reduces further to just two, the angles ϕ and θ. Qubit dynamics so encoded is equivalent to that described by $\mathrm{d}\boldsymbol{r}/\mathrm{d}t = 2\mu\boldsymbol{r} \times \boldsymbol{B}/\hbar$.

4.10 Single Qubit Gates

The highly suggestive qubit notation employed by the unitary formalism lets us identify qubit states $|\uparrow\rangle$ and $|\downarrow\rangle$ with 0 and 1 of Boolean logic. But, of course, $|\uparrow\rangle$ and $|\downarrow\rangle$ are not 0 and 1. Generally, a qubit state $|\,\Psi\rangle$ corresponds to a three-dimensional vector of length 1 (or of length that is no greater than 1 if we allow for mixtures) that can point in any direction. This, as we have emphasized above, may not be clear within the confines of the unitary formalism, because $\boldsymbol{r}$ is hidden inside a, a^*, b and b^*, but it is there.

Driving qubit evolution

Rabi oscillations, discussed in Chapter 2, Section 2.11 (page 78), and then illustrated with the example of a quantum circuit, the quantronium, discussed in Section 2.12 (page 86), provided us with an example of controllable driven evolution of a qubit that did not result in depolarization—at least on paper. We had actually seen depolarization in the quantronium example. A tip of the qubit's vector $\boldsymbol{r}$ drew a continuous line on the Bloch sphere in the course of the evolution. Rabi oscillations were slow compared to Larmor precession. This feature was what made them controllable and precise. For this reason Rabi oscillations are a preferred method for executing various computational operations on qubits. But it is not impossible to use controlled Larmor precession for this purpose, at least in principle. By lowering the value of the magnetic field $\boldsymbol{B}_{\|}$ we can slow the pace of Larmor precession to the point where it can be controllable.

The NOT *gate*

The simplest logical operation is the NOT gate. We have analyzed it already in Section 2.11. But let us rehash the general idea here. Let us assume that the qubit is in the $|\uparrow\rangle \equiv |\,0\rangle$ state originally and sits in the strong background field $\boldsymbol{B}_{\|}$, storing its computational value. To flip the qubit from $|\,0\rangle$ to $|\,1\rangle$, we can buzz it with $\boldsymbol{B}_{\perp} = B_{\perp}\,(\boldsymbol{e}_x \sin\omega_L t - \boldsymbol{e}_y \cos\omega_L t)$, where $\omega_L = 2\mu B_{\|}/\hbar$, for $\pi\hbar/(2\mu B_{\perp})$ seconds exactly. If the initial state of the qubit was $|\downarrow\rangle \equiv |\,1\rangle$, the same operation would flip it to $|\,0\rangle$. This is what makes this operation a proper NOT gate. It has to do the right thing for both $|\,0\rangle$ and $|\,1\rangle$.

Diagrammatic representation of the NOT *gate*

We can draw a quantum circuit representations for the NOT gate as shown in Figure 4.2.

The lines with arrows symbolize, say, a polarized neutron beam. The box labeled with the logical NOT symbol, $\neg$, stands for, say, a chamber filled with the combination of $\boldsymbol{B}_{\perp}$ and $\boldsymbol{B}_{\|}$ needed to perform the operation. The dimensions of

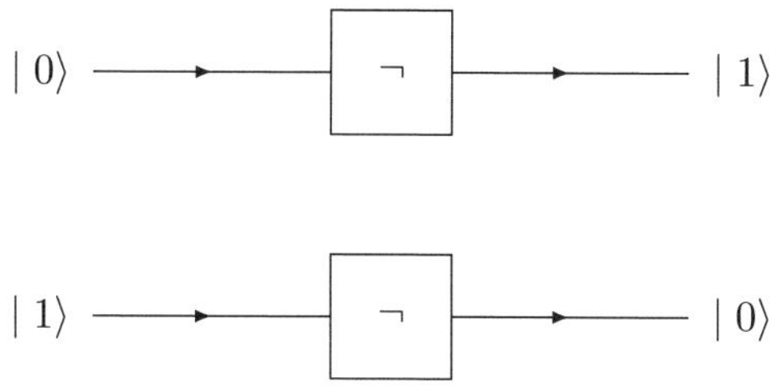

Figure 4.2: Diagrammatic representation of the NOT gate.

the chamber can be set so that, given the beam's velocity v, the neutrons of the beam would spend exactly the right amount of time in the chamber. We have used the word *say* in order to emphasize that what the symbols drawn in Figure 4.2 represent depends on the particular implementation of the qubit.

Arithmetic definition of NOT

The quantum NOT operation can be defined arithmetically as well:

$$\neg \,|\,0\rangle = |\,1\rangle, \tag{4.335}$$

$$\neg \,|\,1\rangle = |\,0\rangle. \tag{4.336}$$

In the world of classical physics this wouldn't be enough to specify the operation uniquely. Every rotation by 180° about an axis perpendicular to $\boldsymbol{e}_z$ could be used to implement the operation and there is an infinite number of such axes bisecting the great circle of the Bloch sphere. But in the peculiar world of quantum physics, $|\,0\rangle$ and $|\,1\rangle$ are the physical basis states and basis states in the Hilbert space of a qubit as well. This means that equations (4.335) and (4.336) should also apply to superpositions of $|\,0\rangle$ and $|\,1\rangle$. Since, as we have seen in Section 4.4 (page 121),

$$|\rightarrow\rangle = \frac{1}{\sqrt{2}}\left(|\,0\rangle + |\,1\rangle\right), \tag{4.337}$$

$$|\leftarrow\rangle = \frac{1}{\sqrt{2}}\left(|\,0\rangle - |\,1\rangle\right), \tag{4.338}$$

$|\rightarrow\rangle$ and $|\leftarrow\rangle$ are invariants of quantum NOT.

this implies that both $|\rightarrow\rangle$ and $|\leftarrow\rangle$ are invariants of quantum NOT:

$$\neg\,|\rightarrow\rangle = \neg\left(\frac{1}{\sqrt{2}}\left(|\,0\rangle + |\,1\rangle\right)\right) = \frac{1}{\sqrt{2}}\left(\neg\,|\,0\rangle + \neg\,|\,1\rangle\right)$$
$$= \frac{1}{\sqrt{2}}\left(|\,1\rangle + |\,0\rangle\right) = |\rightarrow\rangle, \tag{4.339}$$

and

$$\neg\,|{\leftarrow}\rangle = \neg\left(\frac{1}{\sqrt{2}}\left(|\,0\rangle - |\,1\rangle\right)\right) = \frac{1}{\sqrt{2}}\left(\neg\,|\,0\rangle - \neg\,|\,1\rangle\right)$$
$$= \frac{1}{\sqrt{2}}\left(|\,1\rangle - |\,0\rangle\right) = -\;|{\leftarrow}\rangle. \tag{4.340}$$

The minus sign in front of $|{\leftarrow}\rangle$ is not physically detectable. It vanishes when we switch from the unitary description of the qubit to fiducial vectors, so we can ignore it here. In other words, we can say that quantum NOT leaves $\boldsymbol{e}_x$ and $-\boldsymbol{e}_x$ unchanged.

Quantum NOT *swaps* $\boldsymbol{e}_y$ *and* $-\boldsymbol{e}_y$.

Similarly, it is easy to see that

$$|\,\otimes\rangle = \frac{1}{\sqrt{2}}\left(|\,0\rangle + \mathrm{i}\,|\,1\rangle\right) \quad \text{and} \tag{4.341}$$
$$|\,\odot\rangle = \frac{1}{\sqrt{2}}\left(|\,0\rangle - \mathrm{i}\,|\,1\rangle\right), \tag{4.342}$$

imply that

$$\neg\,|\,\otimes\rangle = \mathrm{i}\,|\,\odot\rangle \quad \text{and} \tag{4.343}$$
$$\neg\,|\,\odot\rangle = -\mathrm{i}\,|\,\otimes\rangle. \tag{4.344}$$

We can again ignore the factors "i" and "-i" that appear in front of $|\,\odot\rangle$ and $|\,\otimes\rangle$ on the right-hand side of the equations above, because they vanish when we switch to the fiducial formalism. What the above says, in effect, is that quantum NOT switches $\boldsymbol{e}_y$ to $-\boldsymbol{e}_y$ and vice versa.

Quantum NOT *is a rotation by* 180° *about* $\boldsymbol{e}_x$.

These two additional observations define quantum NOT uniquely as a rotation of the Bloch ball by 180° about $\boldsymbol{e}_x$ and not some other axis.

Square root of NOT

Having narrowed the definition and implementation of quantum NOT so, we can define another quantum gate, which is called *the square root of* NOT. If instead of rotating the Bloch ball by 180° about $\boldsymbol{e}_x$ we were to rotate it by 90° only, we would have to follow this operation with another rotation by 90° in order to complete the NOT. This is shown in Figure 4.3.

The corresponding arithmetic definition of the square root of NOT is

$$\sqrt{\neg}\sqrt{\neg}\,|\,0\rangle = |\,1\rangle, \tag{4.345}$$
$$\sqrt{\neg}\sqrt{\neg}\,|\,1\rangle = |\,0\rangle. \tag{4.346}$$

Hadamard rotation

If we can have the square root of NOT, could we have a nontrivial square root of

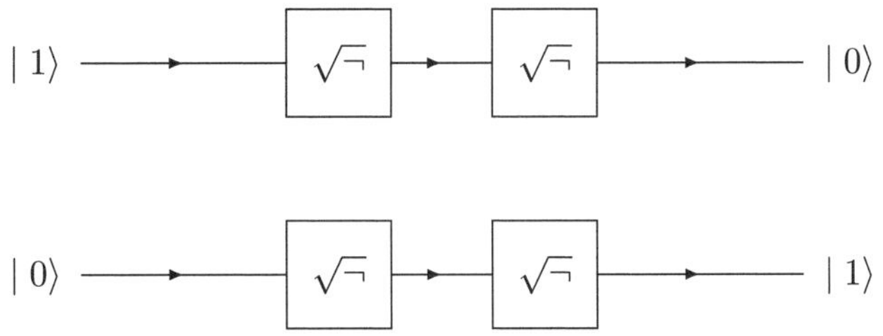

Figure 4.3: The square root of NOT.

one? Such an operation is shown in Figure 4.4. It is called a Hadamard rotation after a French mathematician Jacques-Salomon Hadamard (1865–1963).

The Hadamard rotation $\boldsymbol{H}$[5] rotates the Bloch ball about the direction that bisects the right angle between $-\boldsymbol{e}_x$ and $\boldsymbol{e}_z$, as shown in Figure 4.4, by 180°. Because the direction of the axis of the Hadamard rotation is inclined by 45° with respect to $-\boldsymbol{e}_x$ and $\boldsymbol{e}_z$, the rotation swaps $-\boldsymbol{e}_x$ and $\boldsymbol{e}_z$. The Hadamard rotation also swaps $-\boldsymbol{e}_z$ and $\boldsymbol{e}_x$. Repeating it twice restores the Bloch ball to its original orientation, hence $\boldsymbol{HH} = \boldsymbol{1}$.

Swaps $\boldsymbol{e}_z$ *and* $-\boldsymbol{e}_x$.

Swaps $-\boldsymbol{e}_z$ *and* $\boldsymbol{e}_x$.

We can define the Hadamard rotation diagrammatically as shown in Figure 4.5. The arithmetic definition of the Hadamard rotation is as follows:

$$\boldsymbol{H} \mid 0\rangle = \frac{1}{\sqrt{2}} (\mid 0\rangle + \mid 1\rangle), \tag{4.347}$$

$$\boldsymbol{H} \mid 1\rangle = \frac{1}{\sqrt{2}} (\mid 0\rangle - \mid 1\rangle). \tag{4.348}$$

It is easy to see both from Figure 4.4 and from the following calculation that the Hadamard rotation swaps $\boldsymbol{e}_y$ and $-\boldsymbol{e}_y$:

Swaps $\boldsymbol{e}_y$ *and* $-\boldsymbol{e}_y$.

$$\begin{aligned}\boldsymbol{H} \mid \otimes\rangle &= \boldsymbol{H}\frac{1}{\sqrt{2}}(\mid 0\rangle + \mathrm{i} \mid 1\rangle) = \frac{1}{\sqrt{2}}\left(\frac{1}{\sqrt{2}}(\mid 0\rangle + \mid 1\rangle) + \frac{\mathrm{i}}{\sqrt{2}}(\mid 0\rangle - \mid 1\rangle)\right) \\ &= \frac{1}{\sqrt{2}}\left(\frac{1+\mathrm{i}}{\sqrt{2}} \mid 0\rangle + \frac{1-\mathrm{i}}{\sqrt{2}} \mid 1\rangle\right) = \frac{1+\mathrm{i}}{\sqrt{2}}\frac{1}{\sqrt{2}}(\mid 0\rangle - \mathrm{i} \mid 1\rangle) \\ &= \frac{1+\mathrm{i}}{\sqrt{2}} \mid \odot\rangle. \end{aligned} \tag{4.349}$$

[5] The choice of the letter is unfortunate because it can be confused with Hamiltonian. On the other hand, we do not normally draw a Hamiltonian in quantum computing diagrams, so it is easy to remember that in this context $\boldsymbol{H}$ is the Hadamard rotation.

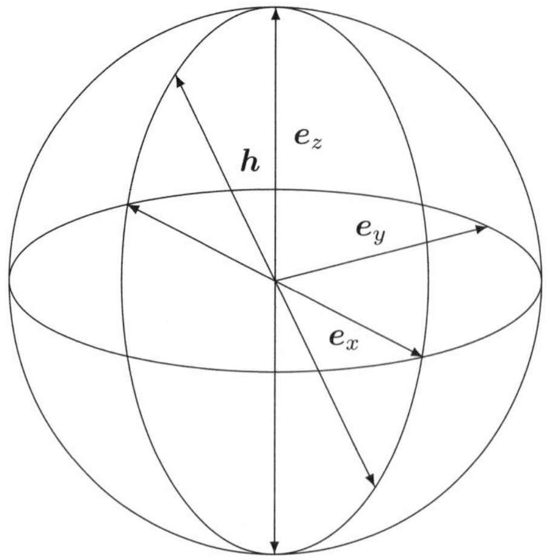

Figure 4.4: The Hadamard rotation $\boldsymbol{H}$ rotates the Bloch ball about vector $\boldsymbol{h}$, which bisects the right angle between $-\boldsymbol{e}_x$ and $\boldsymbol{e}_z$, by 180°. This swaps $-\boldsymbol{e}_z$ and $\boldsymbol{e}_x$. At the same time, $\boldsymbol{e}_z$ and $-\boldsymbol{e}_x$ are swapped, too. Applying $\boldsymbol{H}$ twice results in the rotation by 360° about $\boldsymbol{h}$, which brings the Bloch ball to its original orientation.

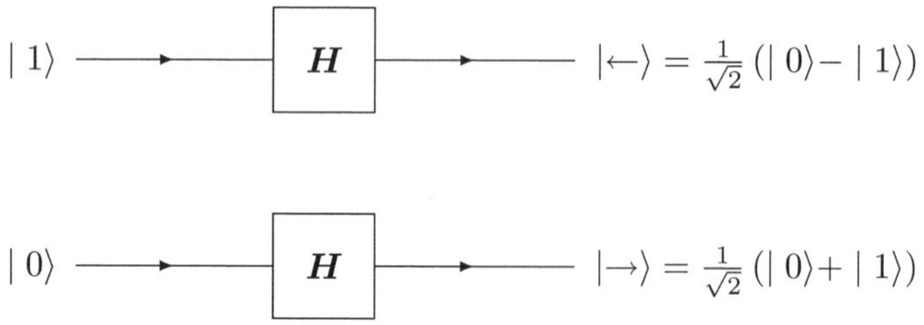

Figure 4.5: Diagrammatic representation of the Hadamard gate.

As before, we can ignore the factor $(1+\mathrm{i})/\sqrt{2}$, because it vanishes in the translation from the unitary to the fiducial description of the qubit.

Similarity to analog computing

Operations such as the Hadamard rotation and the square root of NOT are a reflection of the fact that by using the controlled Larmor precessions about various directions in space, or a combination of the Larmor precession and the Rabi oscillations, we can do with a qubit all that we can do with a ping-pong ball. We can rotate the qubit's Bloch ball in any way we wish, thus implementing an arbitrary continuous mapping between any two points on the Bloch sphere. By adding various dissipative operations such as the measurement, we can dig *into* the interior of the Bloch ball, too.

This makes the qubit a rich object, markedly richer than its classical cousin, the bit, which can assume only one of two discrete values. The riches resemble analog, or fuzzy logic, computing, where a computational element can assume any real value within a certain range. The ping-pong ball, mentioned above, provides us with a full computational equivalent of a single qubit. We will see in Section 7.5 how this and other classical analogies let us implement quantum computations on classical analog computers.

Qubit states can be characterized only by exploring their statistical ensembles.

The only snag is our inability to extract the full information about $\boldsymbol{r}$ from a single measurement. To ascertain the state of the qubit, we have to perform a large number of measurements on it, so as to build up and then explore its statistical ensemble.

There are two fundamental ways to do this.

The first way is to work with a single qubit *sequentially*, as we have seen in the quantronium example. We would prepare the qubit in some well-defined initial state. Then we would perform some operations on the qubit and finally we would send the qubit through the beam-splitter, or an equivalent measuring device, in order to see whether it emerges in the $|\uparrow\rangle$ or in the $|\downarrow\rangle$ state. We would have to perform this procedure, say, 50,000 times, as was the case in the quantronium example, in order to gather sufficient statistics that would give us r^z with reasonable accuracy. We would then have to repeat all these operations without change, but on the output we would modify the beam-splitting chamber so that the qubit would emerge from it in the $|\rightarrow\rangle$ or the $|\leftarrow\rangle$ state. After some 50,000 of such measurements we would get a fair idea about r^x. Then we would have to repeat this whole procedure once more, but this time we would rotate the beam-splitting apparatus so that the qubit would emerge from it in the $|\otimes\rangle$ or in the $|\odot\rangle$ state, which, after sufficient statistics had been collected, would yield r^y.

The other way to approach the measurement of $\boldsymbol{r}$ is to work with millions, perhaps even billions, of identically prepared qubits and send them in the form of a particle beam through various gates all at the same time. It will be then enough to rotate the beam-splitting apparatus in various ways and to measure output beam intensities in order to reconstruct $\boldsymbol{r}$. This is a quick and accurate way of doing things, because the large number of qubits ensures that we can average random errors away and obtain precise distributions. The underlying assumption here is that the qubits do not interact with one another, or, if they do, that such interaction can be averaged away.

4.11 Taking Qubits for a Ride

So far our qubits were stationary. A qubit would hang somewhere suspended in a solution (this is how nuclear magnetic resonance experiments are carried out) or drifting in vacuum (this would be a qubit trapped in a potential well, or a qubit in a particle beam) or printed on a circuit board, and we would subject it to either static or buzzing "magnetic field" in order to manipulate its quantum state.

Manipulating qubits with rapid magnetic field sweeps

But there are also other ways to manipulate qubits. For example, one may subject a qubit to a rapidly sweeping magnetic field. Such manipulations can be faster and more precise than buzzing qubits gently with weak magnetic oscillations [48]. And speed matters for many reasons. Faster gates mean faster computers for starters. But faster gates also mean that quantum information can be processed before the qubits states, the information is encoded on, decohere. Because of qubits' great sensitivity to the environment, quantum computing is always a race against time.

In this section we are going to look at a yet another way of manipulating qubits that is in itself interesting and that may play a role in the future of quantum computing, although there hasn't been much activity in this field yet. We are going to take qubits for a ride—a slow ride in a parameter space.

In the process we are also going to exercise all that we have learned in this chapter: various formulas of the unitary formalism, whatever we learned about the Hamiltonian and its eigenstates, the Schrödinger equation, and so forth.

How would a qubit's state change if we were to drag it from A to B along a certain trajectory through space filled with a "magnetic" field $\boldsymbol{B}$ that varies from location to location?

This question proved remarkably fruitful, and it was not answered until 1984, when Sir Michael V. Berry of Bristol University published a paper in the Proceedings of the Royal Society of London [12]. For this achievement Berry was awarded a prestigious Wolf Prize in 1998 together with Yakir Aharonov, who discovered the related Aharonov-Bohm effect in 1959[6] But the Aharonov-Bohm effect was specific to charged particles in the presence of the magnetic potential, whereas the Berry effect applies to every quantum system, electromagnetic or not. It applies to systems that are not necessarily qubits, too.

Moving the environment around a qubit is equivalent to moving the qubit itself.

In practice, moving a qubit physically from A to B is a difficult, if not impossible, endeavor. Qubits are extremely delicate, so there is no way to move them, even touch them, without destroying their quantum state at the same time. But since

[6] David Bohm was passed over by the awards committee on account of being dead. He died of heart attack in London in 1992.

motion is relative, we can always move the environment around the qubit instead, and this is how the related experiments are carried out.

4.11.1 Dragging a Qubit along an Arbitrary Trajectory

Our starting point is the Schrödinger equation (4.279):

$$\mathrm{i}\hbar\frac{\mathrm{d}}{\mathrm{d}t} \mid \Psi(t)\rangle = \boldsymbol{H}(t) \mid \Psi(t)\rangle. \tag{4.350}$$

The environment the qubit is immersed in is represented by the Hamiltonian $\boldsymbol{H}$. As we move the qubit around a trajectory given by $\boldsymbol{x}(t)$, where $\boldsymbol{x} = (x, y, z)$ is the *guiding vector*, its environment changes, and this can be described by making the Hamiltonian an explicit function of position $\boldsymbol{x}(t)$. Once the Hamiltonian is an explicit function of position, so must be the state vector, and we end up with

$$\mathrm{i}\hbar\frac{\mathrm{d}}{\mathrm{d}t} \mid \Psi(\boldsymbol{x}(t))\rangle = \boldsymbol{H}\left(\boldsymbol{x}(t)\right) \mid \Psi(\boldsymbol{x}(t))\rangle. \tag{4.351}$$

At every point along the trajectory the Hamiltonian matrix has some eigenvectors, two eigenvectors in the case of a qubit. They vary from one point to another changing their direction, though not length, because this is all done within the unitary formalism. Let us call these position-dependent eigenvectors $\mid n(\boldsymbol{x}(t))\rangle$ so that

$$\boldsymbol{H}(\boldsymbol{x}(t)) \mid n(\boldsymbol{x}(t))\rangle = E_n(\boldsymbol{x}(t)) \mid n(\boldsymbol{x}(t))\rangle, \tag{4.352}$$

where $E_n(\boldsymbol{x}(t))$ is the eigenvalue of $\boldsymbol{H}(\boldsymbol{x}(t))$ that corresponds to $\mid n(\boldsymbol{x}(t))\rangle$ at $\boldsymbol{x}(t)$.

Adiabatic evolution

We are going to introduce an important concept of an *adiabatic* motion. Let us suppose that a qubit is in an eigenstate $\mid n(\boldsymbol{x}(0))\rangle$ at the beginning. As we move the qubit ever so gently and *slowly* around $\boldsymbol{x}(t)$, the qubit has enough time to "thermalize" at every $\boldsymbol{x}(t)$, that is, to adjust itself to the local Hamiltonian at this position, so that it remains in an eigenstate, even though the eigenstate itself changes.

Solution to the adiabatic motion problem postulated

Inspired by equation (4.305), we will seek a solution to our adiabatic qubit transfer problem in the form

$$\mid \Psi(t)\rangle = e^{-\mathrm{i}\left(\int_0^t E_n(\boldsymbol{x}(t'))\,\mathrm{d}t'\right)/\hbar} e^{\mathrm{i}\gamma_n(t)} \mid n(\boldsymbol{x}(t))\rangle. \tag{4.353}$$

We have the exponential with the time integral of the eigenvalue here, but we also allow for an additional time-dependent phase factor $e^{\mathrm{i}\gamma_n(t)}$.

Let us plug this solution into the Schrödinger equation. The equation should tell us something about the way the gamma factor, $\gamma_n(t)$, relates to the eigenstates and, possibly, to the eigenenergies as well.

The right-hand side of the Schrödinger equation, $\boldsymbol{H} \mid \Psi\rangle$, is easy. Here we get

$$E_n(\boldsymbol{x}(t)) \mid \Psi(t)\rangle. \tag{4.354}$$

The left-hand side, that is, the time-derivative side of the equation, is somewhat more problematic, because the proposed solution (4.353) depends on time in a rather complicated way. But, at the end of the day there are just three factors here:

$$\mid \Psi(t)\rangle = e^{\text{integral}} e^{\text{gamma}} \mid \text{eigenvector}\rangle, \tag{4.355}$$

so

$$\begin{aligned} \frac{\mathrm{d}}{\mathrm{d}t} \mid \Psi(t)\rangle &= \left(\frac{\mathrm{d}}{\mathrm{d}t} e^{\text{integral}}\right) e^{\text{gamma}} \mid \text{eigenvector}\rangle \\ &\quad + e^{\text{integral}} \left(\frac{\mathrm{d}}{\mathrm{d}t} e^{\text{gamma}}\right) \mid \text{eigenvector}\rangle \\ &\quad + e^{\text{integral}} e^{\text{gamma}} \left(\frac{\mathrm{d}}{\mathrm{d}t} \mid \text{eigenvector}\rangle\right). \end{aligned} \tag{4.356}$$

Of the three time derivatives, the easiest to evaluate is the derivative of the gamma factor. Here it is just

$$\frac{\mathrm{d}}{\mathrm{d}t} e^{\mathrm{i}\gamma_n(t)} = \mathrm{i} e^{\mathrm{i}\gamma_n(t)} \frac{\mathrm{d}}{\mathrm{d}t} \gamma_n(t). \tag{4.357}$$

The other exponential, the one with the integral in it, throws out $-\mathrm{i}/\hbar$, and the exponential itself, and then we have to find a time derivative of the integral. But the time derivative of the integral is the integrated function $\boldsymbol{E_n}$, and so

$$\frac{\mathrm{d}}{\mathrm{d}t} e^{-\mathrm{i}\left(\int_0^t E_n(\boldsymbol{x}(t'))\,\mathrm{d}t'\right)/\hbar} = -\frac{\mathrm{i}}{\hbar} e^{-\mathrm{i}\left(\int_0^t E_n(\boldsymbol{x}(t'))\,\mathrm{d}t'\right)/\hbar} E_n(\boldsymbol{x}(t)). \tag{4.358}$$

Finally, let us have a look at the time derivative of the eigenstate, $\mid n(\boldsymbol{x}(t))\rangle$. The eigenstate is a Hilbert-vector-valued function of position, which then itself is a function of time. The time derivative of $\mid n\rangle$ is therefore

$$\frac{\partial n(\boldsymbol{x}(t))}{\partial \boldsymbol{x}} \cdot \frac{\mathrm{d}\boldsymbol{x}(t)}{\mathrm{d}t}. \tag{4.359}$$

The expression $\partial/\partial\boldsymbol{x}$ is an exotic way of writing a gradient $\boldsymbol{\nabla}$, so we can rewrite the above as

$$\frac{\mathrm{d}}{\mathrm{d}t} \mid n(\boldsymbol{x}(t))\rangle = \mid \boldsymbol{\nabla} n(\boldsymbol{x}(t))\rangle \cdot \frac{\mathrm{d}\boldsymbol{x}(t)}{\mathrm{d}t}. \tag{4.360}$$

Before we go any further, let us consider this expression, $|\boldsymbol{\nabla} n(\boldsymbol{x}(t))\rangle \cdot \mathrm{d}\boldsymbol{x}(t)/\mathrm{d}t$, some more. The reason it requires explaining is that we are mixing here vector objects that belong to different spaces. There are normal three-dimensional geometric space vectors and vector operators in it, namely, $\boldsymbol{\nabla}$ and $\boldsymbol{x}$, and then we have a Hilbert space vector $|\, n\rangle$ in it, too.

Any spinor, including $|\, n\rangle$, can be decomposed into basis spinors, for example, $|\uparrow\rangle$ and $|\downarrow\rangle$:

$$\begin{aligned} |\, n\rangle &= n_\uparrow\, |\uparrow\rangle + n_\downarrow\, |\downarrow\rangle \\ &= \sum_{m=\uparrow,\downarrow} n^m\, |\, m\rangle \\ &= \sum_{m=\uparrow,\downarrow} |\, m\rangle\langle m \,|\, n\rangle. \end{aligned} \tag{4.361}$$

Here n^m does not mean n to the power of m. It means the mth component of n, and, as we have seen in Section 4.5, equations (4.146) and (4.147), it evaluates to $\langle m \,|\, n(\boldsymbol{x})\rangle$, which is a normal complex-valued function of position $\boldsymbol{x}$, a function that can be differentiated.

Taking a three-dimensional space gradient of $|\, n\rangle$ means the following:

$$\begin{aligned} |\, \boldsymbol{\nabla} n(\boldsymbol{x})\rangle &= \sum_{m=\uparrow,\downarrow} \boldsymbol{\nabla} n^m(\boldsymbol{x})\, |\, m\rangle \\ &= \sum_{m=\uparrow,\downarrow} \sum_{i=x,y,z} \frac{\partial n^m(\boldsymbol{x})}{\partial x^i} \boldsymbol{e}_i \otimes |\, m\rangle \\ &= \sum_{m=\uparrow,\downarrow} \sum_{i=x,y,z} \boldsymbol{e}_i \otimes |\, m\rangle \frac{\partial \langle m \,|\, n(\boldsymbol{x})\rangle}{\partial x^i}. \end{aligned} \tag{4.362}$$

This is a tensor product with one leg, $\boldsymbol{e}_i$, standing in the three-dimensional space and the other leg, $|\, m\rangle$, standing in the spinor space. Now we are going to take a three-dimensional space scalar (dot) product of this with

$$\frac{\mathrm{d}\boldsymbol{x}(t)}{\mathrm{d}t} = \sum_{j=x,y,z} \frac{\mathrm{d}x^j(t)}{\mathrm{d}t} \boldsymbol{e}_j. \tag{4.363}$$

Of course, we cannot contract $\boldsymbol{e}_j$ with $|\, m\rangle$. We can contract it only with $\boldsymbol{e}_i$, and so we end up with

$$|\, \boldsymbol{\nabla} n(\boldsymbol{x})\rangle \cdot \frac{\mathrm{d}\boldsymbol{x}(t)}{\mathrm{d}t} = \sum_{m=\uparrow,\downarrow} |\, m\rangle \left(\sum_{i=x,y,z} \frac{\partial \langle m \,|\, n(\boldsymbol{x})\rangle}{\partial x^i} \frac{\mathrm{d}x^i(t)}{\mathrm{d}t} \right). \tag{4.364}$$

The three-dimensional vectors of this expression devour each other in the frenzy of a dot product feeding, and leave a three-dimensional scalar behind. The scalar is

$$\sum_{i=x,y,z} \frac{\partial \langle m \mid n(\boldsymbol{x})\rangle}{\partial x^i} \frac{\mathrm{d}x^i(t)}{\mathrm{d}t}, \tag{4.365}$$

and it is this scalar that is used as the coefficient in the spinor's expansion into the basis spinors of the Hilbert space.

Now we have to put it all together into $\mathrm{i}\hbar \,\mathrm{d} \mid \Psi\rangle/\mathrm{d}t = \boldsymbol{H} \mid \Psi\rangle$:

$$\begin{aligned}
\frac{\mathrm{d}}{\mathrm{d}t} \mid \Psi(t)\rangle &= \frac{\mathrm{d}}{\mathrm{d}t}\left(e^{-\mathrm{i}\left(\int_0^t E_n(\boldsymbol{x}(t'))\,\mathrm{d}t'\right)/\hbar} e^{\mathrm{i}\gamma_n(t)} \mid n(\boldsymbol{x}(t))\rangle\right) \\
&= -\frac{\mathrm{i}}{\hbar} e^{-\mathrm{i}\left(\int_0^t E_n(\boldsymbol{x}(t'))\,\mathrm{d}t'\right)/\hbar} E_n(\boldsymbol{x}(t)) e^{\mathrm{i}\gamma_n(t)} \mid n(\boldsymbol{x}(t))\rangle \\
&\quad + e^{-\mathrm{i}\left(\int_0^t E_n(\boldsymbol{x}(t'))\,\mathrm{d}t'\right)/\hbar} \mathrm{i} e^{\mathrm{i}\gamma_n(t)} \frac{\mathrm{d}}{\mathrm{d}t}\gamma_n(t) \mid n(\boldsymbol{x}(t))\rangle \\
&\quad + e^{-\mathrm{i}\left(\int_0^t E_n(\boldsymbol{x}(t'))\,\mathrm{d}t'\right)/\hbar} e^{\mathrm{i}\gamma_n(t)} \mid \boldsymbol{\nabla} n(\boldsymbol{x}(t))\rangle \cdot \frac{\mathrm{d}\boldsymbol{x}(t)}{\mathrm{d}t} \\
&= -\frac{\mathrm{i}}{\hbar} E_n(\boldsymbol{x}(t)) \mid \Psi(t)\rangle + \mathrm{i}\frac{\mathrm{d}\gamma_n(t)}{\mathrm{d}t} \mid \Psi(t)\rangle \\
&\quad + e^{-\mathrm{i}\left(\int_0^t E_n(\boldsymbol{x}(t'))\,\mathrm{d}t'\right)/\hbar} e^{\mathrm{i}\gamma_n(t)} \mid \boldsymbol{\nabla} n(\boldsymbol{x}(t))\rangle \cdot \frac{\mathrm{d}\boldsymbol{x}(t)}{\mathrm{d}t}.
\end{aligned} \tag{4.366}$$

We are nearly there. Let us multiply this by $\mathrm{i}\hbar$, and then let us match it against $E_n \mid \Psi\rangle$ on the right-hand side:

$$\begin{aligned}
& E_n(\boldsymbol{x}(t)) \mid \Psi(t)\rangle - \hbar\frac{\mathrm{d}\gamma_n(t)}{\mathrm{d}t} \mid \Psi(t)\rangle \\
& + \mathrm{i}\hbar e^{-\mathrm{i}\left(\int_0^t E_n(\boldsymbol{x}(t'))\,\mathrm{d}t'\right)/\hbar} e^{\mathrm{i}\gamma_n(t)} \mid \boldsymbol{\nabla} n(\boldsymbol{x}(t))\rangle \cdot \frac{\mathrm{d}\boldsymbol{x}(t)}{\mathrm{d}t} \\
& = E_n(\boldsymbol{x}(t)) \mid \Psi(t)\rangle.
\end{aligned} \tag{4.367}$$

We can immediately see $E_n \mid \Psi\rangle$ on both sides of the equation. We can also see a lot of other stuff on the left-hand side, which we don't want. With this other stuff out of the way we would get $E_n \mid \Psi\rangle = E_n \mid \Psi\rangle$, which is a perfectly fine way of making the Schrödinger equation happy.

So we arrive at the condition "This other stuff ought to vanish." We translate this into the following equation:

$$-\hbar\frac{\mathrm{d}\gamma_n(t)}{\mathrm{d}t} \mid \Psi(t)\rangle + \mathrm{i}\hbar e^{-\mathrm{i}\left(\int_0^t E_n(\boldsymbol{x}(t'))\,\mathrm{d}t'\right)/\hbar} e^{\mathrm{i}\gamma_n(t)} \mid \boldsymbol{\nabla} n(\boldsymbol{x}(t))\rangle \cdot \frac{\mathrm{d}\boldsymbol{x}(t)}{\mathrm{d}t} = 0. \tag{4.368}$$

But let us recall that $\mid \Psi\rangle$ itself contains three terms. It helps to write them explicitly here:

$$
\begin{aligned}
& -\hbar \frac{\mathrm{d}\gamma_n(t)}{\mathrm{d}t} e^{-\mathrm{i}\left(\int_0^t E_n(\boldsymbol{x}(t'))\,\mathrm{d}t'\right)/\hbar} e^{\mathrm{i}\gamma_n(t)} \mid n(\boldsymbol{x}(t))\rangle \\
& + \mathrm{i}\hbar e^{-\mathrm{i}\left(\int_0^t E_n(\boldsymbol{x}(t'))\,\mathrm{d}t'\right)/\hbar} e^{\mathrm{i}\gamma_n(t)} \mid \boldsymbol{\nabla} n(\boldsymbol{x}(t))\rangle \cdot \frac{\mathrm{d}\boldsymbol{x}(t)}{\mathrm{d}t} \\
& = 0.
\end{aligned} \tag{4.369}
$$

Now we can throw a lot of stuff away. First we divide both sides by the exponentials; then we divide both sides by $\hbar$. We are left with

$$
\frac{\mathrm{d}\gamma_n(t)}{\mathrm{d}t} \mid n(\boldsymbol{x}(t))\rangle = \mathrm{i} \mid \boldsymbol{\nabla} n(\boldsymbol{x}(t))\rangle \cdot \frac{\mathrm{d}\boldsymbol{x}(t)}{\mathrm{d}t}. \tag{4.370}
$$

Finally, we multiply both sides by $\langle n(\boldsymbol{x}(t))|$ from the left. This step devours $|n(\boldsymbol{x}(t))\rangle$ on the left-hand side, leaving a pure time derivative of γ_n,

$$
\frac{\mathrm{d}\gamma_n(t)}{\mathrm{d}t} = \mathrm{i}\langle n(\boldsymbol{x}(t)) \mid \boldsymbol{\nabla} n(\boldsymbol{x}(t))\rangle \cdot \frac{\mathrm{d}\boldsymbol{x}(t)}{\mathrm{d}t}. \tag{4.371}
$$

The differential equation for the geometric phase in adiabatic motion

The line integral solution for the geometric phase

The equation has an obvious solution in the form of a line integral

$$
\gamma_n(C) = \mathrm{i} \int_C \langle n(\boldsymbol{x}) \mid \boldsymbol{\nabla} n(\boldsymbol{x})\rangle \cdot \mathrm{d}\boldsymbol{x}, \tag{4.372}
$$

which says that as we move a qubit along trajectory C, its phase γ_n accumulates contributions of $\langle n(\boldsymbol{x}) \mid \boldsymbol{\nabla} n(\boldsymbol{x})\rangle \cdot \Delta\boldsymbol{x}$ along the line. We note that this accumulation does not depend on how fast or how slow we move the qubit, as long as the changes are adiabatic. In this case $\gamma_n(C)$ depends only on the variation of $|n(\boldsymbol{x})\rangle$ along the trajectory C.

The integral is purely imaginary; therefore $\gamma_n(C)$ is purely real. This fact is important because, if $\gamma_n(C)$ had an imaginary component, the exponential $e^{\mathrm{i}\gamma_n}$ would change the length of vector $\mid \Psi\rangle$, and this situation cannot happen within the confines of the unitary formalism.

That the integral is purely imaginary can be seen as follows:

$$
\begin{aligned}
0 &= \boldsymbol{\nabla} 1 = \boldsymbol{\nabla}\langle n \mid n\rangle \\
&= \langle \boldsymbol{\nabla} n \mid n\rangle + \langle n \mid \boldsymbol{\nabla} n\rangle \\
&= \langle n \mid \boldsymbol{\nabla} n\rangle^* + \langle n \mid \boldsymbol{\nabla} n\rangle \\
&= 2\Re\langle n \mid \boldsymbol{\nabla} n\rangle.
\end{aligned} \tag{4.373}
$$

4.11.2 Closed Trajectory Case

Equation (4.372) can be transformed further if the trajectory C encloses a surface S so that

$$C = \partial S, \tag{4.374}$$

where ∂S means "the edge of S." In this case $\int_C$ becomes $\oint_{\partial S}$, and we can invoke the Stokes theorem that converts a line integral over the edge of a surface into a *curl* integral over the surface itself:

The Stokes theorem can be invoked if the trajectory is closed.

$$\oint_{\partial S} \langle n(\boldsymbol{x}) \mid \boldsymbol{\nabla} n(\boldsymbol{x})\rangle \cdot \mathrm{d}\boldsymbol{x} = \int_S \left(\boldsymbol{\nabla} \times \langle n(\boldsymbol{x}) \mid \boldsymbol{\nabla} n(\boldsymbol{x})\rangle\right) \cdot \mathrm{d}^2\boldsymbol{S}. \tag{4.375}$$

We should again stop here and explain this expression in terms of actual functions, vector and spinor components. What is being differentiated here, and what is being "cross-producted"? $\langle n(\boldsymbol{x}) \mid \boldsymbol{\nabla} n(\boldsymbol{x})\rangle$ is a three-dimensional vector field on the normal three-dimensional space. The field arises in the following way:

$$\langle n(\boldsymbol{x}) \mid \boldsymbol{\nabla} n(\boldsymbol{x})\rangle = \sum_{m=\uparrow,\downarrow} \sum_{i=x,y,z} n^*_m(\boldsymbol{x}) \left(\frac{\partial}{\partial x^i} n^m(\boldsymbol{x})\right) \boldsymbol{e}_i, \tag{4.376}$$

where $n^*_m = \langle n \mid m\rangle$ and $n^m = \langle m \mid n\rangle$. The spinor index m is summed away (we call such an index *saturated*), and we are left with just a three-dimensional space index i. Every i term is then multiplied by $\boldsymbol{e}_i$ so that a vector field is produced. Now we act on the field with the *curl* operator $\boldsymbol{\nabla}\times$. The result is

$$\boldsymbol{\nabla} \times \langle n(\boldsymbol{x}) \mid \boldsymbol{\nabla} n(\boldsymbol{x})\rangle = \sum_{\substack{i = x,y,z \\ j = x,y,z \\ k = x,y,z}} \epsilon_{ijk} \boldsymbol{e}_i \frac{\partial}{\partial x^j} \left(\sum_{m=\uparrow,\downarrow} n^*_m(\boldsymbol{x}) \frac{\partial}{\partial x^k} n^m(\boldsymbol{x}) \right), \tag{4.377}$$

where ϵ_{ijk} is the fully antisymmetric three-dimensional symbol. The spinor index m in the expression is saturated as before. So are the j and k three-dimensional space indexes, but this time they are saturated in the cross-product way so that, for example, the x-component of $\boldsymbol{\nabla} \times \langle n(\boldsymbol{x}) \mid \boldsymbol{\nabla} n(\boldsymbol{x})\rangle$ is

$$\frac{\partial}{\partial y} \left(\sum_{m=\uparrow,\downarrow} n^*_m(\boldsymbol{x}) \frac{\partial}{\partial z} n^m(\boldsymbol{x}) \right) - \frac{\partial}{\partial z} \left(\sum_{m=\uparrow,\downarrow} n^*_m(\boldsymbol{x}) \frac{\partial}{\partial y} n^m(\boldsymbol{x}) \right). \tag{4.378}$$

These expressions may look somewhat tedious, but they should not look scary. They are easy to understand in terms of what is what. And we are going to go some way still toward making them more usable.

The first thing to observe about

$$\sum_{\substack{i=x,y,z\\ j=x,y,z\\ k=x,y,z}} \epsilon_{ijk}\boldsymbol{e}_i \frac{\partial}{\partial x^j}\left(\sum_{m=\uparrow,\downarrow} n_m^*(\boldsymbol{x})\frac{\partial}{\partial x^k}n^m(\boldsymbol{x})\right) \tag{4.379}$$

is that the derivative $\partial/\partial x^j$ is going to hit $n_m^*(\boldsymbol{x})$ first, but when it gets to $\partial n^m(\boldsymbol{x})/\partial x^k$, it'll give us zero, because ϵ_{ijk} is anti-symmetric in j and k, but $\frac{\partial}{\partial x^j}\frac{\partial}{\partial x^k}$ is symmetric in j and k. This is a yet another formulation of the rule that a *curl* of a gradient is zero, $\boldsymbol{\nabla}\times\boldsymbol{\nabla}n^m(\boldsymbol{x}) = 0$.

Therefore we find that

$$\begin{aligned}&\boldsymbol{\nabla}\times\langle n(\boldsymbol{x}) \mid \boldsymbol{\nabla}n(\boldsymbol{x})\rangle\\ &\quad= \sum_{m=\uparrow,\downarrow}\sum_{\substack{i=x,y,z\\ j=x,y,z\\ k=x,y,z}} \epsilon_{ijk}\left(\frac{\partial n_m^*(\boldsymbol{x})}{\partial x^j}\right)\left(\frac{\partial n^m(\boldsymbol{x})}{\partial x^k}\right)\boldsymbol{e}_i.\end{aligned} \tag{4.380}$$

$\langle\boldsymbol{\nabla}n \mid$ is not the same as $\mid\boldsymbol{\nabla}n\rangle$. On the index level one is n_m^*, and the other one is n^m. This is just as well because if they were the same, we would have something like $(\boldsymbol{\nabla}n)\times(\boldsymbol{\nabla}n) = 0$, and the whole computation would be over.

But $\boldsymbol{\nabla}n^m$ is simply $\langle m \mid \boldsymbol{\nabla}n\rangle$ and $\boldsymbol{\nabla}n_m^*$ is $\langle\boldsymbol{\nabla}n \mid m\rangle$. Therefore, we can rewrite the expression for γ_n as follows:

$$\gamma_n(\partial S) = \mathrm{i}\int_S \sum_{m=\uparrow,\downarrow}\langle\boldsymbol{\nabla}n(\boldsymbol{x}) \mid m\rangle\times\langle m \mid \boldsymbol{\nabla}n(\boldsymbol{x})\rangle\cdot \mathrm{d}^2\boldsymbol{S}, \tag{4.381}$$

because $\sum_{ijk}\epsilon_{ijk}\boldsymbol{e}_i u^j v^k = \boldsymbol{u}\times\boldsymbol{v}$.

This does not look simpler or more useful than equation (4.372), but we are now going to invoke two helpful observations.

The first one is that $\mid n\rangle$ is an eigenvector of a local $\boldsymbol{H}$ at every point on surface S, and therefore it is one of $|\uparrow\rangle$ or $|\downarrow\rangle$ at that point. Since $\boldsymbol{\nabla}\langle n \mid n\rangle = 0$, we have that $\langle\boldsymbol{\nabla}n \mid n\rangle = -\langle n \mid \boldsymbol{\nabla}n\rangle$. Therefore

$$\langle\boldsymbol{\nabla}n \mid n\rangle\times\langle n \mid \boldsymbol{\nabla}n\rangle = 0. \tag{4.382}$$

because for any vector $\boldsymbol{v}$ we have that $\boldsymbol{v} \times (-\boldsymbol{v}) = 0$. For this reason we can rewrite our expression for γ_n yet again, as follows:

$$\gamma_n(\partial S) = \mathrm{i} \int_S \sum_{m \neq n} \langle \boldsymbol{\nabla} n(\boldsymbol{x}) \mid m \rangle \times \langle m \mid \boldsymbol{\nabla} n(\boldsymbol{x}) \rangle \cdot \mathrm{d}^2 \boldsymbol{S}, \tag{4.383}$$

where $\mid m\rangle$ is *the other* eigenvector. Even though for two-dimensional systems such as qubits, the sum in this equation reduces to just one component, we are going to keep it because this way the expression is going to be valid for systems with larger number of dimensions, too. In this case $\mid m\rangle$ stands for *all the other* eigenvectors of the local Hamiltonian.

The second observation is that $\mid n\rangle$ must satisfy

$$\boldsymbol{H} \mid n\rangle = E_n \mid n\rangle. \tag{4.384}$$

Let us apply the Nabla operator to both sides:

$$\boldsymbol{\nabla} \left(\boldsymbol{H} \mid n\rangle \right) = \boldsymbol{\nabla} \left(E_n \mid n\rangle \right). \tag{4.385}$$

This yields

$$(\boldsymbol{\nabla} \boldsymbol{H}) \mid n\rangle + \boldsymbol{H} \mid \boldsymbol{\nabla} n\rangle = (\boldsymbol{\nabla} E_n) \mid n\rangle + E_n \mid \boldsymbol{\nabla} n\rangle. \tag{4.386}$$

Let us multiply this equation by $\langle m \mid$ from the left.

$$\langle m \mid \boldsymbol{\nabla} \boldsymbol{H} \mid n\rangle + \langle m \mid \boldsymbol{H} \mid \boldsymbol{\nabla} n\rangle = \boldsymbol{\nabla} E_n \langle m \mid n\rangle + E_n \langle m \mid \boldsymbol{\nabla} n\rangle \tag{4.387}$$

Because $\langle m \mid \boldsymbol{H} = \langle m \mid E_m$ and $\mid m\rangle \neq \mid n\rangle$, we have that $\langle m \mid n\rangle = 0$. In summary,

$$\langle m \mid \boldsymbol{\nabla} \boldsymbol{H} \mid n\rangle + E_m \langle m \mid \boldsymbol{\nabla} n\rangle = E_n \langle m \mid \boldsymbol{\nabla} n\rangle, \tag{4.388}$$

or

$$\langle m \mid \boldsymbol{\nabla} n\rangle = \frac{\langle m \mid \boldsymbol{\nabla} \boldsymbol{H} \mid n\rangle}{E_n - E_m}. \tag{4.389}$$

Surface integral solution for the geometric phase arising from adiabatic motion around a closed loop

A similar expression holds for $\langle \boldsymbol{\nabla} n \mid m\rangle$. This lets us rewrite our equation for $\gamma_n(\partial S)$ yet again:

$$\gamma_n(\partial S) = \mathrm{i} \int_S \sum_{m \neq n} \frac{\langle n \mid \boldsymbol{\nabla} \boldsymbol{H} \mid m\rangle \times \langle m \mid \boldsymbol{\nabla} \boldsymbol{H} \mid n\rangle}{(E_n - E_m)^2} \cdot \mathrm{d}^2 \boldsymbol{S}. \tag{4.390}$$

The equation should be read as follows. At every point of surface S we are going to differentiate the Hamiltonian. This will produce *three* new operators in place of just one; they'll correspond to $\partial \boldsymbol{H}/\partial x$, $\partial \boldsymbol{H}/\partial y$ and $\partial \boldsymbol{H}/\partial z$. For a given state $| n\rangle$ and for each state $| m\rangle \neq | n\rangle$ (in case of a qubit there will be only one such state) we are going to have *three* numbers per point obtained by evaluating transition amplitudes $\langle n | \partial \boldsymbol{H}/\partial x | m\rangle$, $\langle n | \partial \boldsymbol{H}/\partial y | m\rangle$, and $\langle n | \partial \boldsymbol{H}/\partial z | m\rangle$—thus forming a three-dimensional vector at this point. We are also going to have another three numbers obtained by evaluating $\langle m | \partial \boldsymbol{H}/\partial x | n\rangle$, $\langle m | \partial \boldsymbol{H}/\partial y | n\rangle$, and $\langle m | \partial \boldsymbol{H}/\partial z | n\rangle$ forming another three-dimensional vector at this point. We will have to take a cross-product of these two vectors and divide it by $(E_n - E_m)^2$. Then we'll have to take a scalar product of the resulting vector with the surface element $\mathrm{d}^2\boldsymbol{S}$. This will produce a number—just a normal complex number. For each point of surface S this operation has to be repeated for the remaining m-s, if there are such—this is going to be the case only in quantum systems with a larger number of dimensions than single qubits—and the results added. Finally, the operation has to be repeated for every other point of the surface; then the resulting numbers are all summed up, the result is multiplied by "i," and this is our γ_n.

A numerical procedure doing all this can be implemented fairly easily, which implies that we understand the formula in depth. This is not always the case in quantum physics.

Equations (4.372) and (4.390) provide us with a method to evaluate the change of phase that accompanies the movement of a qubit (or any other quantum system, because we were sufficiently general here) along any trajectory and along a trajectory that encloses a surface, respectively. The first equation (4.372) that looks quite simple is also rather general. The second equation (4.390) looks more complicated, but this is because it is more specific.

4.11.3 A Qubit in the Rotating Magnetic Field

Although the title of this section, "Taking Qubits for a Ride," as well as the wording that accompanies the derivation, are highly suggestive of moving the qubits in the physical geometric three-dimensional space, the formulas derived and the reasoning itself are more general. They describe the movement of the qubit or any other quantum system in any parametric space. For example, if a qubit is subjected to slowly varying magnetic field, sufficiently slowly for the qubit's eigenstates to thermalize, here meaning to align with the direction of the magnetic field, at all stages of the evolution, then this can be also thought of as a "movement" of the qubit through ... the magnetic space. The only assumption we have made in deriving equation (4.390) was that the parametric space was three-dimensional, so that both the Stokes formula and cross-product manipulations could be applied.

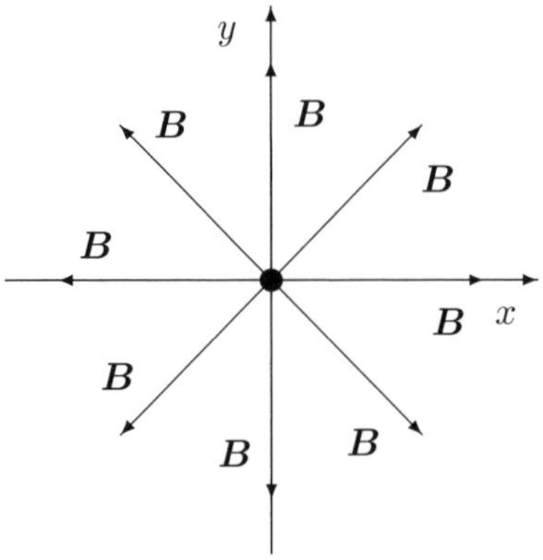

Figure 4.6: Magnetic field $\boldsymbol{B}$ rotates adiabatically around a qubit placed in the center of the figure.

But no such assumption was made in deriving equation (4.372), which is therefore applicable to higher-dimensional parametric spaces.

Rotating magnetic field adiabatically around a qubit

Let us consider a situation in which a magnetic field vector $\boldsymbol{B}$ is rotated adiabatically around a qubit in a plane as shown in Figure 4.6. The field's value does not change throughout the rotation. The operation is equivalent to taking the qubit for a ride along a circle of radius B in the $\boldsymbol{B}$ space as shown in Figure 4.7.

In order to evaluate a contribution that this operation makes to the phase $\gamma_n(\partial S)$—we ought to remember that there is going to be a dynamic phase factor $\exp\left(-\mathrm{i}\int_0^t E_n\left(\boldsymbol{x}(t')\right)\mathrm{d}t'\right)$ in the complete solution for $|\,\Psi(t)\rangle$, too—we need to use equation (4.390). But right here we have a conundrum, because the surface S in the $\boldsymbol{B}$ space, namely the circle of radius B, passes through $\boldsymbol{B}=0$, where $E_\uparrow = E_\downarrow$, and equation (4.390) tells us that we should divide some such rather complicated expression made of a cross-product and transition amplitudes by $(E_\uparrow - E_\downarrow)^2$. Alas, the expression is valid and the same for *any* surface S as long as its edge is the contour along which the qubit moves. So here we can use a different surface, for example, a hemisphere of radius B that stands on the great circle of the sphere, with the qubit moving along the great circle. The hemisphere stays away from $B=0$, and so we don't have the problem. Furthermore, B is the same at every point of the hemisphere, so this makes our calculations easier.

S may be deformed.

B is constant on the hemisphere, and we know its surface area.

We are going to evaluate $\gamma_n(\partial S)$ for $n = \uparrow, \downarrow$. The sum in equation (4.390) reduces to a single component only, because our qubit system is two dimensional. Hence,

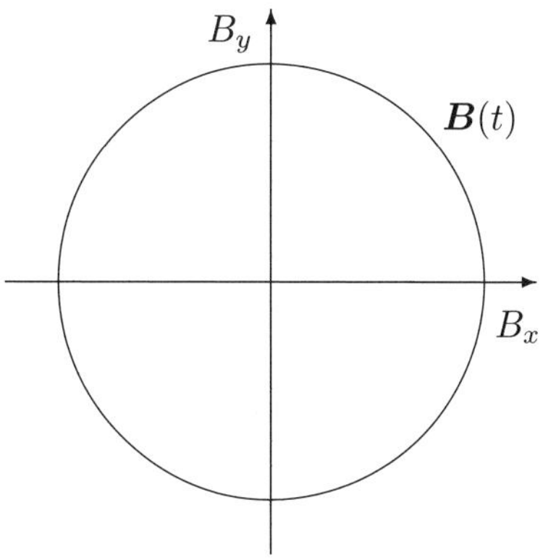

Figure 4.7: A qubit can be thought of as moving along the circle of radius B in the $\boldsymbol{B}$ space.

we have

$$\gamma_\uparrow(\partial S) = \mathrm{i}\int_S \frac{\langle\uparrow|\,\boldsymbol{\nabla H}\,|\downarrow\rangle \times \langle\downarrow|\,\boldsymbol{\nabla H}\,|\uparrow\rangle}{(E_\uparrow - E_\downarrow)^2} \cdot \mathrm{d}^2\boldsymbol{S} \tag{4.391}$$

$$\gamma_\downarrow(\partial S) = \mathrm{i}\int_S \frac{\langle\downarrow|\,\boldsymbol{\nabla H}\,|\uparrow\rangle \times \langle\uparrow|\,\boldsymbol{\nabla H}\,|\downarrow\rangle}{(E_\downarrow - E_\uparrow)^2} \cdot \mathrm{d}^2\boldsymbol{S}. \tag{4.392}$$

Now, let us have a look at $\boldsymbol{\nabla H}$. This is easy to evaluate because

$$\boldsymbol{H} = -\mu\left(B_x\boldsymbol{\sigma}_x + B_y\boldsymbol{\sigma}_y + B_z\boldsymbol{\sigma}_z\right) \quad \text{and} \tag{4.393}$$

$$\boldsymbol{\nabla} = \left(\partial/\partial B_x, \partial/\partial B_y, \partial/\partial B_z\right), \tag{4.394}$$

and so

$$\boldsymbol{\nabla H} = -\mu\begin{pmatrix}\boldsymbol{\sigma}_x\\ \boldsymbol{\sigma}_y\\ \boldsymbol{\sigma}_z\end{pmatrix}. \tag{4.395}$$

Consequently,

$$\gamma_\uparrow(\partial S) = \mathrm{i}\int_S \frac{\mu^2}{(E_\uparrow - E_\downarrow)^2}\begin{pmatrix}\langle\uparrow|\,\boldsymbol{\sigma}_y\,|\downarrow\rangle\langle\downarrow|\,\boldsymbol{\sigma}_z\,|\uparrow\rangle - \langle\uparrow|\,\boldsymbol{\sigma}_z\,|\downarrow\rangle\langle\downarrow|\,\boldsymbol{\sigma}_y\,|\uparrow\rangle\\ \langle\uparrow|\,\boldsymbol{\sigma}_z\,|\downarrow\rangle\langle\downarrow|\,\boldsymbol{\sigma}_x\,|\uparrow\rangle - \langle\uparrow|\,\boldsymbol{\sigma}_x\,|\downarrow\rangle\langle\downarrow|\,\boldsymbol{\sigma}_z\,|\uparrow\rangle\\ \langle\uparrow|\,\boldsymbol{\sigma}_x\,|\downarrow\rangle\langle\downarrow|\,\boldsymbol{\sigma}_y\,|\uparrow\rangle - \langle\uparrow|\,\boldsymbol{\sigma}_y\,|\downarrow\rangle\langle\downarrow|\,\boldsymbol{\sigma}_x\,|\uparrow\rangle\end{pmatrix} \cdot \mathrm{d}^2\boldsymbol{S}, \tag{4.396}$$

and

$$\gamma_\downarrow(\partial S) = \mathrm{i}\int_S \frac{\mu^2}{(E_\uparrow - E_\downarrow)^2} \begin{pmatrix} \langle\downarrow|\,\boldsymbol{\sigma}_y\,|\uparrow\rangle\langle\uparrow|\,\boldsymbol{\sigma}_z\,|\downarrow\rangle - \langle\downarrow|\,\boldsymbol{\sigma}_z\,|\uparrow\rangle\langle\uparrow|\,\boldsymbol{\sigma}_y\,|\downarrow\rangle \\ \langle\downarrow|\,\boldsymbol{\sigma}_z\,|\uparrow\rangle\langle\uparrow|\,\boldsymbol{\sigma}_x\,|\downarrow\rangle - \langle\downarrow|\,\boldsymbol{\sigma}_x\,|\uparrow\rangle\langle\uparrow|\,\boldsymbol{\sigma}_z\,|\downarrow\rangle \\ \langle\downarrow|\,\boldsymbol{\sigma}_x\,|\uparrow\rangle\langle\uparrow|\,\boldsymbol{\sigma}_y\,|\downarrow\rangle - \langle\downarrow|\,\boldsymbol{\sigma}_y\,|\uparrow\rangle\langle\uparrow|\,\boldsymbol{\sigma}_x\,|\downarrow\rangle \end{pmatrix} \cdot \mathrm{d}^2\boldsymbol{S}. \tag{4.397}$$

It is easy to see that $\gamma_\uparrow(\partial S) = -\gamma_\downarrow(\partial S)$. This become evident once the integrals have been written out in detail as above.

Here, $|\uparrow\rangle$ and $|\downarrow\rangle$ are eigenvectors of the local Hamiltonian, which is represented by vector $\boldsymbol{B}$ and may point in any direction, not necessarily in the $\boldsymbol{e}_z$ direction. Hence $|\uparrow\rangle$ and $|\downarrow\rangle$ may not be equivalent to $\binom{1}{0}$ and $\binom{0}{1}$, which are the eigenvectors of $\boldsymbol{\sigma}_z$. Nevertheless $(E_\uparrow - E_\downarrow)^2$ is still $(2\mu B)^2$ and

$$\frac{\mu^2}{(E_\uparrow - E_\downarrow)^2} = \frac{\mu^2}{4\mu^2 B^2} = \frac{1}{4B^2}. \tag{4.398}$$

We may represent $|\uparrow\rangle$ and $|\downarrow\rangle$ by $\binom{1}{0}$ and $\binom{0}{1}$.

At any given point of surface S we *can* rotate our system of coordinates so that $\boldsymbol{H} = -\mu B\boldsymbol{\sigma}_z$ there. This action does not affect $\boldsymbol{\nabla H}$, which still has three nonvanishing components, but it lets us represent $|\uparrow\rangle$ and $|\downarrow\rangle$ at this point and at this point only as $\binom{1}{0}$ and $\binom{0}{1}$. At this point we can evaluate

$$\begin{pmatrix} (1,0)\boldsymbol{\sigma}_y\begin{pmatrix}0\\1\end{pmatrix}(0,1)\boldsymbol{\sigma}_z\begin{pmatrix}1\\0\end{pmatrix} - (1,0)\boldsymbol{\sigma}_z\begin{pmatrix}0\\1\end{pmatrix}(0,1)\boldsymbol{\sigma}_y\begin{pmatrix}1\\0\end{pmatrix} \\ (1,0)\boldsymbol{\sigma}_z\begin{pmatrix}0\\1\end{pmatrix}(0,1)\boldsymbol{\sigma}_x\begin{pmatrix}1\\0\end{pmatrix} - (1,0)\boldsymbol{\sigma}_x\begin{pmatrix}0\\1\end{pmatrix}(0,1)\boldsymbol{\sigma}_z\begin{pmatrix}1\\0\end{pmatrix} \\ (1,0)\boldsymbol{\sigma}_x\begin{pmatrix}0\\1\end{pmatrix}(0,1)\boldsymbol{\sigma}_y\begin{pmatrix}1\\0\end{pmatrix} - (1,0)\boldsymbol{\sigma}_y\begin{pmatrix}0\\1\end{pmatrix}(0,1)\boldsymbol{\sigma}_x\begin{pmatrix}1\\0\end{pmatrix} \end{pmatrix}. \tag{4.399}$$

Here $\binom{1}{0}$ and $\binom{0}{1}$ are eigenvectors of $\boldsymbol{\sigma}_z$. For this reason all occurrences of $\binom{1}{0}\boldsymbol{\sigma}_z\binom{0}{1}$ and $\binom{0}{1}\boldsymbol{\sigma}_z\binom{1}{0}$ must vanish, and by doing so they take down the first and the second row of the large, fat vector above with them. The only row that survives is the third:

$$\begin{aligned} &(1,0)\boldsymbol{\sigma}_x\begin{pmatrix}0\\1\end{pmatrix}(0,1)\boldsymbol{\sigma}_y\begin{pmatrix}1\\0\end{pmatrix} - (1,0)\boldsymbol{\sigma}_y\begin{pmatrix}0\\1\end{pmatrix}(0,1)\boldsymbol{\sigma}_x\begin{pmatrix}1\\0\end{pmatrix} \\ &= (1,0)\begin{pmatrix}0 & 1\\ 1 & 0\end{pmatrix}\begin{pmatrix}0\\1\end{pmatrix}(0,1)\begin{pmatrix}0 & -\mathrm{i}\\ \mathrm{i} & 0\end{pmatrix}\begin{pmatrix}1\\0\end{pmatrix} \\ &\quad - (1,0)\begin{pmatrix}0 & -\mathrm{i}\\ \mathrm{i} & 0\end{pmatrix}\begin{pmatrix}0\\1\end{pmatrix}(0,1)\begin{pmatrix}0 & 1\\ 1 & 0\end{pmatrix}\begin{pmatrix}1\\0\end{pmatrix} \\ &= \mathrm{i} - (-\mathrm{i}) = 2\mathrm{i}. \end{aligned} \tag{4.400}$$

At this point the vector is

$$\begin{pmatrix} 0 \\ 0 \\ 2\mathrm{i} \end{pmatrix} = 2\mathrm{i}\frac{\boldsymbol{B}}{B}. \tag{4.401}$$

By writing it in this form $\boldsymbol{B}/B$ we make it independent of the system of coordinates we have chosen to evaluate this expression, so that we can use it elsewhere, too.

Finally, we wrap it all together and get for $\gamma_\uparrow(\partial S)$ (and for $\gamma_\downarrow = -\gamma_\uparrow$ as well):

$$\gamma_\uparrow(\partial S) = \mathrm{i}\int_S \frac{1}{4B^2}\frac{2\mathrm{i}\boldsymbol{B}}{B}\cdot \mathrm{d}^2\boldsymbol{S} = -\frac{1}{2}\int_S \frac{\boldsymbol{B}}{B^3}\cdot \mathrm{d}^2\boldsymbol{S}. \tag{4.402}$$

The integral $\int_S \left(\boldsymbol{B}/B^3\right)\cdot \mathrm{d}^2\boldsymbol{S}$ represents the *solid angle* that surface S subtends with respect to point $\boldsymbol{B} = 0$. Calling the angle Ω, we get this amazingly beautiful result:

$$\gamma_\uparrow(\partial S) = -\frac{1}{2}\Omega.$$

The phase shift experienced by a qubit in an eigenstate that is moved adiabatically along a closed trajectory in the $\boldsymbol{B}$ space is equal to $\pm 1/2$ (the sign depends on whether it is $|\uparrow\rangle$ or $|\downarrow\rangle$) times the solid angle subtended by the surface enclosed by the trajectory with respect to the $\boldsymbol{B} = 0$ point.

Seldom do we arrive at so startlingly elegant a result in physics or in mathematics. When we do, it is a cause for celebration.

But what about the specific problem of a qubit in a rotating magnetic field? What is the actual number?

The number is easy to evaluate. Our surface in this case is the hemisphere centered on $\boldsymbol{B} = 0$ of a constant radius B. Vector $\boldsymbol{B}$ in this case is parallel to the normal vector at each point of the surface, that is, $\boldsymbol{B} = B\boldsymbol{n}$, and so is parallel to $\mathrm{d}^2\boldsymbol{S} = \boldsymbol{n}\mathrm{d}^2 S$, too. The integral becomes

$$\gamma_\uparrow(\partial S) = -\frac{1}{2}\frac{1}{B^2}\int_S \boldsymbol{n}\cdot\boldsymbol{n}\mathrm{d}^2 S = -\frac{1}{2B^2}2\pi B^2 = -\pi, \tag{4.403}$$

and the phase factor γ is

$$e^{\mathrm{i}\gamma_\uparrow} = e^{-\mathrm{i}\pi} = -1 \tag{4.404}$$

and

$$e^{\mathrm{i}\gamma_\downarrow} = e^{\mathrm{i}\pi} = -1 \tag{4.405}$$

For an arbitrary spinor $|\,\Psi\rangle = a\,|\uparrow\rangle + b\,|\downarrow\rangle$ the final result of the excursion is

$$|\,\Psi\rangle \rightarrow a e^{-\mathrm{i}E_\uparrow \Delta t/\hbar}(-1)\,|\uparrow\rangle + b e^{-\mathrm{i}E_\downarrow \Delta t/\hbar}(-1)\,|\downarrow\rangle, \tag{4.406}$$

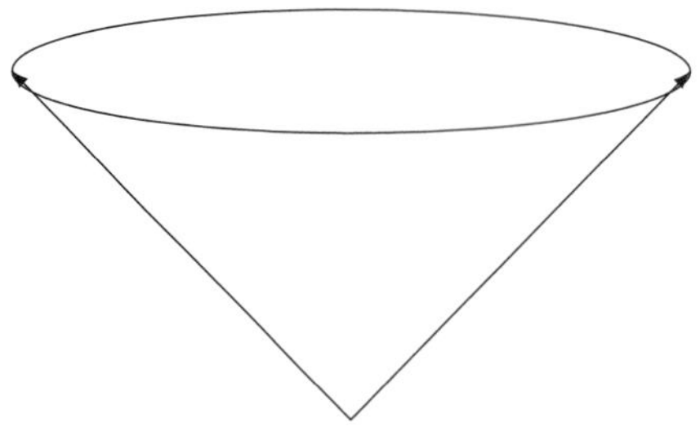

Figure 4.8: An excursion in the $\boldsymbol{B}$ space that results in a smaller solid angle $\Omega < 2\pi$.

where Δt is the time it takes to rotate the field. The time integrals in the dynamic phase part of the expression reduce to just $\exp\left(-\mathrm{i}E_{\uparrow,\downarrow}\Delta t/\hbar\right)$, because B is constant (though $\boldsymbol{B}$ rotates slowly). This corresponds to Larmor precession around each direction of $\boldsymbol{B}$ the qubit goes through. The dynamic factor does depend on time it takes to complete the excursion. On the other hand, the γ factor depends on the solid angle only. This is why it is called the *geometric* phase factor. It is also called the Berry phase factor, or just the Berry phase for short.

In this case, though, the geometric phase factor clearly has no physical effect, because it vanishes in all expressions such as $\boldsymbol{\rho} = | \Psi\rangle\langle\Psi |$ or $|\langle\uparrow, \downarrow| \Psi\rangle|^2$.

But what if an excursion in the $\boldsymbol{B}$ space is around a surface that subtends a solid angle Ω that is less than 2π? In the physical and in the $\boldsymbol{B}$ space this means that we rotate $\boldsymbol{B}$ conically rather than in a plane. This is shown in Figure 4.8.

In this case $e^{\mathrm{i}\gamma} = e^{-\mathrm{i}\Omega/2}$ and $e^{-\mathrm{i}\gamma} = e^{\mathrm{i}\Omega/2}$ no longer overlap, and we get a physically observable effect.

Let us assume we begin the excursion with a qubit in the $|\rightarrow\rangle$ state

$$| \Psi(t = 0)\rangle = \frac{1}{\sqrt{2}} \left(|\uparrow\rangle + |\downarrow\rangle\right). \tag{4.407}$$

At the end of the excursion the qubit ends up in the state given by

$$\frac{1}{\sqrt{2}} \left(e^{\mathrm{i}\mu B\Delta t/\hbar} e^{-\mathrm{i}\Omega/2} |\uparrow\rangle + e^{-\mathrm{i}\mu B\Delta t/\hbar} e^{\mathrm{i}\Omega/2} |\downarrow\rangle\right). \tag{4.408}$$

What is the effect of this change on the $\boldsymbol{r}$ vector that describes the qubit? The z component of $\boldsymbol{r}$ does not change, because

$$r^z = aa^* - bb^* = 0. \tag{4.409}$$

But both r^x and r^y acquire an additional rotation as a result of γ:

$$\begin{aligned} r^x &= ab^* + ba^* \\ &= \frac{1}{2}\left(e^{\mathrm{i}(\mu B\Delta t/\hbar - \Omega/2)} e^{\mathrm{i}(\mu B\Delta t/\hbar - \Omega/2)} + e^{-\mathrm{i}(\mu B\Delta t/\hbar - \Omega/2)} e^{-\mathrm{i}(\mu B\Delta t/\hbar - \Omega/2)}\right) \\ &= \frac{1}{2}\left(e^{\mathrm{i}(2\mu B\Delta t/\hbar - \Omega)} + e^{-\mathrm{i}(2\mu B\Delta t/\hbar - \Omega)}\right) \\ &= \cos(\omega_L \Delta t - \Omega), \end{aligned} \tag{4.410}$$

where $\omega_L = 2\mu B/\hbar$ is the Larmor frequency, and

$$r^y = \mathrm{i}(ab^* - ba^*) = -\sin(\omega_L \Delta t - \Omega). \tag{4.411}$$

We find that the qubit has accumulated an angular lag of Ω in the real physical space. The lag cancels out when $\Omega = 2\pi$.

4.11.4 Observing the Berry Phase Experimentally

Cooper pair box demonstration

In the past 20 years or so, numerous papers have been published about the Berry phase. A search on the automated e-print archives retrieves some 340 papers on the Berry phase or on geometric phase. But few of these are experimental papers that would focus on the demonstration of the Berry phase in quantum systems. The most recent such demonstration, as of the writing of this book, was published toward the end of December 2007 by a joint collaboration of scientists from the famous Eidgenössische Technische Hochschule in Zurich (Einstein's alma mater), Université de Sherbrooke in Québec, University of Waterloo in Ontario, and Yale University [86]. The group demonstrated Berry phase in a superconducting Cooper pair box qubit, similar to the one used in the quantronium device.

Ultracold neutrons demonstration

But here we will discuss an older demonstration of the phenomenon by Richardson, Kilvington, Green, and Lamoreaux [119] that was published in Physical Review Letters in 1988. Richardson and his colleagues from the Institut Laue-Langevin in France and from the Rutherford-Appleton Laboratory in the UK used ultracold neutrons to demonstrate the Berry shift.

Their method was very clean, and the results obtained by the authors agreed with theoretical predictions exceptionally well. This is important because there are certain assumptions in the derivation of the Berry phase that may be questioned. The first one is the notion of an adiabatic evolution of a qubit. The concept itself is a little unclear, and some authors proposed revisions. Alas, a surprising effect of the revisions was that the Berry phase disappeared, so that the revisions themselves had to be revised in order to bring it back [107]. This development points to the crucial importance of experiment in physics. However convincing and mathematically elegant a derivation, there may be always something subtle in it that's missed or the underlying assumptions may not be correct. Experimental verification of a prediction provides us with certainty as to the proposed effect.

The experiment carried out by Richardson, Kilvington, Green, and Lamoraux is additionally of great value and interest to us because it demonstrates what is involved in precise manipulation and control of quantum systems to the level that would make quantum information processing feasible.

Storing neutrons

Ultracold neutrons used in the experiment were neutrons that had been slowed down to less than 5 m/s. The neutrons were stored in a vacuum chamber lined with beryllium and beryllium oxide walls, since ultracold neutrons reflect from such walls with negligible loss for all incident angles. Nevertheless, surface contamination and leaks in the valve that was used to fill and empty the chamber reduced the lifetime of neutrons dramatically, from nearly 10 minutes to only about 80 seconds. But 80 seconds is time aplenty in the quantum domain. We had seen in the quantronium example that the quantum state there decayed within about a *micro*second.

Generating and cooling neutrons

The neutrons were generated by irradiating a 25-liter container filled with liquid deuterium held at 25 K with fast neutrons emitted by a high-flux nuclear reactor. The irradiation procedure resulted in the generation of ultracold and *very* cold neutrons, which were then transported 10 m up through a nickel coated evacuated pipe. Nickel coating for the pipes was chosen again to limit neutron loss. The neutrons lost some of their energy in the process to gravity and were then pushed through a turbine that further slowed the very cold neutrons, thus doubling the number of ultracold neutrons. At the output the ultracold neutron density was about $90/\text{cm}^3$.

Polarizing neutrons

The ultra cold neutrons were then passed through an 800 nm thick magnetically saturated polarizing foil made of cobalt-nickel alloy. Afterwards they were transported to a 5-liter beryllium and beryllium oxide lined chamber along a copper-nickel alloy coated silica guide and pushed into the chamber through holes in the five-layer Mumetal magnetic shield. Mumetal is an 80% nickel-iron alloy that is specially designed for magnetic shielding applications.

Thermalizing and rotating neutrons

It took about 10 seconds to fill the chamber to the density of 10 neutrons per cubic centimeter, whereupon the neutron valve was closed. All neutrons in the chamber were initially polarized in the z direction by a 5-milligauss magnetic field, $\boldsymbol{B} = B_z^0 \boldsymbol{e}_z$. They were allowed to thermalize for 2 seconds and were then subjected to a temporally varying magnetic field for about 7.4 seconds, which was followed by another 2-second wait.

How the field is rotated

The varying magnetic field applied during the middle 7.4-second period was given by

$$\boldsymbol{B} = aB_0\boldsymbol{e}_x \pm (1+\epsilon)B_0 \cos\frac{2\pi t}{T} + B_0 \sin\frac{2\pi t}{T}, \tag{4.412}$$

where a, ϵ, B_0 and T are constants and T is the time of one rotation.

The field rotated in the $\boldsymbol{e}_y \times \boldsymbol{e}_z$ plane and had a nonvanishing x component that allowed the experimenters to control the solid angle Ω. Additionally ϵ was the ellipticity parameter that allowed for variation of B over the path. The solid angle Ω that corresponds to the trajectory was given by

$$\Omega = 2\pi\left(1 \pm \frac{a}{\sqrt{1+a^2}}\right), \tag{4.413}$$

where the sign $\pm$ depends on the sign of the magnetic moment and direction of the rotation.

Extracting the neutrons from the chamber and measuring them

After the excursion the neutron valve was opened, and the same foil that had been used to polarize them in the first place was now used to filter them on the way out, thus performing the measurement. An adiabatic spin flipper was used to filter both $|\uparrow\rangle$ and $|\downarrow\rangle$ states, which were then counted in a helium proportional counter neutron detector for 10 seconds per state.

Why the rotation is done in the $\boldsymbol{e}_y \times \boldsymbol{e}_z$ plane

The magnetic field configuration used in the experiment was rotated by 90° compared to our examples discussed in the previous section. But this was done for a good reason. What we normally measure is the z component of the spin, which we called p^0 and p^1 in Chapter 2. If we were to perform the rotation in the $\boldsymbol{e}_x \times \boldsymbol{e}_y$ plane we would have to have a separate, differently oriented device for measuring spins in the x and y directions. Furthermore, we would have problems with preparation of the initial conditions, too. In the experiment, when the neutrons entered the chamber, they were polarized in the z direction. After thermalization, their $p^0(t=0)$ was 1, and their $r^z(t=0)$ was 1, too. They were then rotated in the $\boldsymbol{e}_y \times \boldsymbol{e}_z$ plane. Their final angle of rotation, θ, could be read from $p^0 = (1+r^z)/2$ and $p^1 = (1-r^z)/2$, namely, $r^z(t=T) = \cos\theta = p^0(T) - p^1(T)$, as measured by the same polarizing foil when the neutrons were extracted from the chamber.

The final angle of rotation contained the accumulated dynamic phase and the geometric phase, namely,

$$\theta_\uparrow = 2\mu \int_0^T B(t)\mathrm{d}t - \Omega, \tag{4.414}$$

and $\theta_\downarrow = -\theta_\uparrow$.

For a multiple number of revolutions of the field, N, both the dynamic and the geometric phases accumulated, so in this case the experimenters got $\theta_\uparrow = N\left(2\mu\int_0^T B(t)\mathrm{d}t - \Omega\right)$.

How the magnetic field was generated and controlled

The magnetic field was generated by running current through three sets of coils, which were placed within the magnetic shield but outside the vacuum chamber containing the neutrons. The coils were perpendicular to each other with accuracy of better than 2° and were calibrated to within 0.1% accuracy. An analog computer was used to control the currents. In turn a timer and a zero-crossing switch were used to control the computer, so that exactly one full rotation, or a multiple thereof, could be generated.

Sources of depolarization

In spite of all the precautions and the use of special materials there was a residual magnetic field of about 10 μG and a nonvanishing gradient in the neutron chamber. The residual field was strong enough to depolarise the neutrons for small values of B_0. The presence of the residual field limited the duration of the rotation T and set a lower limit to B_0.

Another problem was caused by an aluminum can in which the neutron chamber was enclosed. Varying $\boldsymbol{B}$ too quickly would generate eddy currents in the can, which would rapidly depolarize the neutrons, too. So T could not have been made too short. The choice of $T \approx 7.4$ s resulted from this restriction.

Collecting neutron counts

The measurements proceeded as follows. A given set of parameters N, a, and ϵ was fixed. For this set the spin-up and spin-down counts were collected as a function of B_0. Each chamber fill and store cycle yielded a certain number of counts. This process was repeated until between 60,000 and 70,000 counts were collected for *each* (N, a, ϵ, B_0) tuple.

Calibrating the instrument and weighing the counts

The collected counts had to be weighed to correct for the fact that spin-up neutrons were counted first and while they were being counted, the spin-down neutrons were stored in the guide and reflected off the polarizer, thus suffering additional depolarization. Other sources of depolarization had to be included in final data analysis, too. This is normally done by *calibrating* the system against known neutron configurations obtained, for example, by filling the chamber with polarized neutrons, thermalizing them, doing nothing to them for 7.4 seconds, then emptying the chamber and counting the neutrons.

The Berry phase is clearly seen.

In the end, after many days of collecting and processing data, the experimenters arrived at the numbers shown here in Table 4.1.

Very good agreement was obtained between Berry's predictions and observed values of the additional angle due to the geometric phase shift. Ellipticity of the orbit had no effect as long as the solid angle in the $\boldsymbol{B}$ space remained unchanged. The Berry angle was clearly cumulative, as could be seen by comparing data for various values of N.

Table 4.1: Results of the experiment. Here a defines the solid angle Ω, ϵ is the ellipticity; N is the number of rotations; Ω is the theoretical value of the Berry angle (see equation (4.414)), which should be equal to the solid angle and to the solid angle alone; Ω_u is the measured value of the Berry angle obtained by counting spin-up neutrons; and Ω_d is the measured value of the Berry angle obtained by counting spin-down neutrons. We observe increasing depolarization for $N = 3$ and the lack of effect for $\epsilon \neq 0$. Table reprinted with permission from [119].

			Calculated	*Observed*	
a	ϵ	N	$\Omega/2\pi$	$\Omega_u/2\pi$	$\Omega_d/2\pi$
0.000	0.00	1	1.000	1.00 ± 0.01	1.00 ± 0.02
0.000	0.25	1	1.000	1.00 ± 0.03	0.99 ± 0.05
0.000	0.62	1	1.000	1.01 ± 0.03	1.00 ± 0.05
0.000	0.00	2	2.000	2.00 ± 0.03	1.97 ± 0.06
0.000	0.00	3	3.000	2.87 ± 0.15	2.89 ± 0.15
0.268	0.00	1	1.259	1.28 ± 0.01	1.26 ± 0.03
0.577	0.00	1	1.500	1.52 ± 0.02	1.51 ± 0.03
0.577	0.00	2	3.000	3.00 ± 0.03	2.99 ± 0.05
1.000	0.00	1	1.707	1.68 ± 0.01	1.69 ± 0.02
1.732	0.00	1	1.866	1.74 ± 0.15	1.72 ± 0.40
3.732	0.00	1	1.966	1.97 ± 0.01	1.98 ± 0.02

4.11.5 Berry Phase Gates

Examples of Berry phase gates

The Berry phase has been proposed as an additional mechanism for processing quantum information. The device discussed in the previous section constitutes an example of a Berry phase gate. Jones, Vedral, Ekert, and Castagnoli even demonstrated a *conditional* Berry phase gate using nuclear magnetic resonance [71]. Yau, De Poortere, and Shayegan detected signs of the Berry phase in oscillations of the resistance of a mesoscopic gallium arsenide ring embedded in a magnetic field, although this could be demonstrated only after the measured spectra were compared with simulation results [151]. And, as we have remarked at the beginning of Section 4.11.4, most recently[7] Berry phase was demonstrated in a superconducting qubit by a group of ten scientists from ETH, two Canadian universities, Sherbrooke and Waterloo, and Yale [86]. This last demonstration was exceptionally accurate. The collected statistics were 200,000 measured events per graph point, and the accumu-

[7]Too recently, unfortunately, to fully discuss in this text.

lated Berry phase agreed with theory within the experimental error of ± 0.14 radians with angles stretching up to $3\pi/4$, which sets a new impressive benchmark in the art of superconducting qubit control.

But the idea is not without problems.

Berry phase and Larmor precession

The first difficulty is that the Berry phase is always mixed with Larmor precession in a way that Rabi oscillations aren't. Let us recall that Rabi oscillations affect p^0/p^1, whereas Larmor precession leaves p^0 and p^1 intact, unless the background field is rotated by 90° as has been done in the Richardson's experiment. This makes it possible to design a computational system based on Rabi oscillations that uses p^0 and p^1 but ignores p^2 and p^3. Indeed, most quantum algorithms developed so far and their experimental demonstrations do just this.

Addressing individual qubits

Then we have the problem of addressing individual qubits in a quantum register. The register may be a molecule or a collection of atoms trapped in an optical lattice or something else. Rotating magnetic field around the register would perform the operation on all its qubits, not just on a selected qubit. The usefulness of such an operation is likely to be limited. On the other hand, the Rabi oscillations mechanism lets us talk to individual qubits on private channels. If a register is a specially constructed molecule, atoms in various locations within the molecule are sensitive to different Rabi frequencies because of the so-called chemical shifts. By sending signals on these frequencies we can address individual qubits and not have other qubits eavesdrop on the communication. Similarly, we can *read* individual qubits by tuning the receivers to chemically shifted Rabi frequencies. This difficulty could be overcome in quantum electronic circuits if every qubit could be equipped in its own local "magnetic field" circuitry, as has been done, for example, in the quantronium or in the ETH device mentioned above.

Adiabatic transport is slow.

Another problem is adiabatic transport. Berry phase equation does not work for nonadiabatic transport. If a qubit manipulation is too fast, we have to solve the Schrödinger equation exactly without making the adiabatic assumption that the eigenvector remains an eigenvector throughout the whole excursion. For a transport to be adiabatic it must be slow. But a slow qubit manipulation means a slow gate and inevitable high depolarization rate while the gate is being traversed. This problem is shared with Rabi oscillations, though, which are another example of adiabatic qubit manipulation.

Refocusing

It may be that rather than thinking of the Berry phase as a possible gate mechanism, we may have to think about it as a yet another parasitic effect, alongside with Larmor precession, that has to be kept under control while quantum information is being processed. We have already mentioned *refocusing* (discussed more in Section 7.2), which has been invented to control similar problems that affect molecular registers.

Topological quantum computer

But these obstacles are no worse than general difficulties that characterize quantum computing. One *can* think of using Berry phase as a primary computational mechanism. This is how a topological quantum computer, proposed originally by Alexei Kitaev in 1997 [80], and then picked up by Michael Freedman and his colleagues [124], is supposed to work.

At present, a topological quantum computer is a purely theoretical concept, which is not based on the idea of a qubit. But this is fine. The choice of a qubit as a basic computational device is ours to make or to discard. A topological quantum computer, instead, builds on a concept of non-Abelian anyons, which are quasiparticles associated with certain two-dimensional device structures. We do not know (as of February 2008) if non-Abelian anyons can form in such structures, but we do know that Abelian ones can. These are associated with the fractional quantum Hall effect, which was first observed in 1982 by Tsui and Störmer [137] and explained in 1983 by Laughlin [85], for which all three were awarded the 1998 Nobel prize.

Fractional quantum Hall effect

Anyons and braids

Anyons are neither fermions nor bosons.[8] Such objects cannot exist in a three-dimensional space, but they can exist in a two-dimensional one, and, as the previous paragraph asserts, they do. Like fermions, anyons cannot occupy the same state. If we were to think of them as having a continuous pointlike existence—this, however, is classical, not quantum, thinking because trajectories cannot be associated with quantum particles[9]—then their world lines could never cross or merge. But they could braid about each other. And as they braid, they would accumulate Berry phase, thus implementing a computation.

Freedman, Larsen, and Wang have showed that a topological quantum computer so conceived—but it must utilize non-Abelian anyons for this, the Abelian ones wouldn't serve—can perform any computation that a more conventional quantum computer can do [46].

At this stage it is uncertain whether these ideas are at all physical, not only because non-Abelian anyons have not been seen yet, but also because the idea of

[8]We are going to say more about fermions and bosons in Section 5.2.

[9]An alert reader may well ask here, "How so? Haven't we just dragged a qubit along some $\boldsymbol{x}(t)$ to see what this would do to its phase?" We have, but at no stage did we really identify $\boldsymbol{x}(t)$ as being a point in the physical three-dimensional space or made any use of it. When it came to experimental demonstrations of the phenomenon, we rotated magnetic fields around a qubit or, as it was done in the ETH experiment, we varied some circuit parameters. An attempt to grab a qubit, for example, a neutron, in some tweezers and then to drag it around in space would almost certainly dephase the qubit. A particle subjected to such exertions becomes classicalized. It is no longer a quantum object.

braiding anyon trajectories is profoundly classical. Can this at all be done (and how) without destroying a quantum state of the anyons? How could we "move" one anyon and not move all others at the same time? But where Nature lacks, human ingenuity may provide. Various suggestions on how such a system could be made of more conventional components have been put forward and continue to be actively pursued.

Some very elegant mathematics is involved in all this, which is among the reasons that many beautiful papers had been written on the subject. But Nature does not always care about mathematical beauty. The gritty reality of dephasing and depolarization, as well as the horrid messiness of entanglement, especially when mixtures are concerned—about which more in the next chapter—attests to Nature's total disregard for good taste.

5 The Biqubit

5.1 Entangled States

Whereas a single, isolated qubit is mathematically equivalent to a classical magnetic dipole, or, to be even more laconic, a ping-pong ball—be it with some read-out and statistical complications—a system of two qubits, a *biqubit*, is equivalent to two classical magnetic dipoles, or ping-pong balls, occasionally only. At other times it displays behavior that, although possible to simulate and understand classically in some respects [133, 103, 104, 105, 24], is otherwise rather puzzling and stirred a great deal of theoretical and experimental investigations toward the end of the twentieth century, from which the ideas of quantum computing eventually arose.

Two separate qubits

But let us begin by considering two *separate* qubits, that is, qubits that *are* equivalent to two classical magnetic dipoles.

Using the elementary laws of probability calculus, as we have discussed in Section 1.9, page 37, we would describe the qubits in terms of a tensor product of their fiducial vectors. We would say that the state of the system comprising qubits A and B is

$$\boldsymbol{p}_A \otimes \boldsymbol{p}_B. \tag{5.1}$$

Measurements on this system would then be expressed by a tensor product of two forms: the first one, $\boldsymbol{\omega}_A$, describing a measurement on $\boldsymbol{p}_A$ and the second one, $\boldsymbol{\omega}_B$, describing a measurement on $\boldsymbol{p}_B$. So, for example, the probability that qubit A is detected with its spin pointing up is

$$p_A^{\uparrow} = p_A^0 = \langle \boldsymbol{\omega}_A^0, \boldsymbol{p}_A \rangle, \tag{5.2}$$

where $\boldsymbol{\omega}_A^0$ is a canonical form in the space of qubit A.

The probability that qubit B is detected with its spin pointing to the right is

$$p_B^{\rightarrow} = p_B^2 = \langle \boldsymbol{\omega}_B^2, \boldsymbol{p}_B \rangle, \tag{5.3}$$

and the probability that qubit A is detected with its spin pointing up while qubit B is detected with its spin pointing to the right—of all other possible two-qubit combinations—is

$$p_A^{\uparrow} p_B^{\rightarrow} = p_A^0 p_B^2 = \langle \boldsymbol{\omega}_A^0, \boldsymbol{p}_A \rangle \langle \boldsymbol{\omega}_B^2, \boldsymbol{p}_B \rangle = \langle \boldsymbol{\omega}_A^0 \otimes \boldsymbol{\omega}_B^2, \boldsymbol{p}_A \otimes \boldsymbol{p}_B \rangle. \tag{5.4}$$

Biqubit energy form

A two-qubit energy form is more complicated, because energy is an additive quantity. Energy of a two-qubit system is a sum of energies of the two qubits, as long as they don't interact with each other.

This can be captured in the following way. Let us recall that

$$\langle \varsigma^i, \varsigma_j \rangle = 2\delta^i{}_j, \quad i, j = 1, x, y, z. \tag{5.5}$$

Therefore

$$\langle \varsigma^1, \boldsymbol{p} \rangle = \langle \varsigma^1, \frac{1}{2} \left(\varsigma_1 + r^x \varsigma_x + r^y \varsigma_y + r^z \varsigma_z \right) \rangle = 1. \tag{5.6}$$

If $\boldsymbol{\eta}_A$ and $\boldsymbol{\eta}_B$ are energy forms acting on the fiducial vector of qubits A and B, respectively, the energy form for a system of two separate noninteracting qubits is

$$\boldsymbol{\eta}_A \otimes \varsigma_B^1 + \varsigma_A^1 \otimes \boldsymbol{\eta}_B. \tag{5.7}$$

What if the qubits do interact with each other? Then the energy form may have an additional term that couples to both qubits simultaneously,

$$\boldsymbol{\eta}_A \otimes \varsigma_B^1 + \varsigma_A^1 \otimes \boldsymbol{\eta}_B + \boldsymbol{\eta}_{AB}, \tag{5.8}$$

where $\boldsymbol{\eta}_{AB}$ is some nontrivial form of rank two, the latter meaning a form that acts on objects such as $\boldsymbol{p}_A \otimes \boldsymbol{p}_B$. Physicists often tend toward sloppy notation that does not emphasize these things, and, in effect, the equations produced may be hard to understand.

When evaluating biqubit expressions, we must always remember that only forms operating on qubit A can be applied to this qubit. Expressions such as

$$\langle \boldsymbol{\eta}_A, \boldsymbol{p}_B \rangle \tag{5.9}$$

make no sense, because form $\boldsymbol{\eta}_A$ does *not* operate in the space of states of qubit B.

Extending ⊗ to quaternions and beyond

We can switch between fiducial and quaternion formalisms by converting qubit probability vectors $\boldsymbol{p}$ to the corresponding quaternions $\boldsymbol{\rho}$, keeping at the same time the tensor product symbol in place:

$$\boldsymbol{p}_A \otimes \boldsymbol{p}_B \rightarrow \boldsymbol{\rho}_A \otimes \boldsymbol{\rho}_B. \tag{5.10}$$

Then we can substitute Pauli matrices in place of quaternion units $\boldsymbol{\sigma}_x$, $\boldsymbol{\sigma}_y$, and $\boldsymbol{\sigma}_z$, so that the tensor product of quaternions becomes the tensor product of Pauli matrices.

We *must not* yield to the temptation of just multiplying $\boldsymbol{\rho}_A$ by $\boldsymbol{\rho}_B$, regardless of whether the sigmas are thought of as quaternions or Pauli matrices. The *product in waiting* must wait. The two separate vector spaces are mapped on two separate quaternion spaces, and these in turn are mapped on two separate spaces of 2×2 complex matrices.

To emphasize this point, we're going to add subscripts A and B to the sigmas as well, namely,

$$\begin{aligned} \boldsymbol{p}_A &\rightarrow \frac{1}{2}\left(\mathbf{1}_A + r_A^x \boldsymbol{\sigma}_{xA} + r_A^y \boldsymbol{\sigma}_{y\,A} + r_A^z \boldsymbol{\sigma}_{zA}\right), \quad (5.11) \\ \boldsymbol{p}_B &\rightarrow \frac{1}{2}\left(\mathbf{1}_B + r_B^x \boldsymbol{\sigma}_{xB} + r_B^y \boldsymbol{\sigma}_{y\,B} + r_B^z \boldsymbol{\sigma}_{zB}\right). \quad (5.12) \end{aligned}$$

Matrices $\boldsymbol{\sigma}_{iA}$ and $\boldsymbol{\sigma}_{iB}$ look the same as normal Pauli matrices $\boldsymbol{\sigma}_i$ but operate in different spaces. This must be emphasized *ad nauseam*.

Having unpacked the sigmas onto 2×2 matrices, we can go further and express our qubit states in terms of Hilbert space vectors, $|\,\Phi\rangle_A$ and $|\,\Psi\rangle_B$, such that

$$\begin{aligned} |\,\Phi\rangle_A \otimes \langle\Phi\,|_A &= \boldsymbol{\rho}_A \quad \text{and} \\ |\,\Psi\rangle_B \otimes \langle\Psi\,|_B &= \boldsymbol{\rho}_B. \end{aligned}$$

This will work only when both constitutent states $\boldsymbol{p}_A$ and $\boldsymbol{p}_B$ are pure. And so, for pure states we can write

$$\begin{aligned} \boldsymbol{\rho}_A \otimes \boldsymbol{\rho}_B &= (|\,\Phi\rangle_A \otimes \langle\Phi\,|_A) \otimes (|\,\Psi\rangle_B \otimes \langle\Psi\,|_B) \\ &= (|\,\Phi\rangle_A \otimes |\,\Psi\rangle_B) \otimes (\langle\Phi\,|_A \otimes \langle\Psi\,|_B). \quad (5.13) \end{aligned}$$

In effect, we end up with tensor products of two Hilbert space vectors or forms representing a biqubit system.

In summary, our chain of mappings from measurable probabilities for a system of two qubits in pure states to highly abstract (though convenient) states in the Hilbert space looks as follows:

$$\boldsymbol{p}_A \otimes \boldsymbol{p}_B \rightarrow \boldsymbol{\rho}_A \otimes \boldsymbol{\rho}_B \rightarrow |\,\Phi\rangle_A \otimes |\,\Psi\rangle_B. \quad (5.14)$$

Let

$$\begin{aligned} |\,\Phi\rangle_A &= a\,|\!\uparrow\rangle_A + b\,|\!\downarrow\rangle_A \quad \text{and} \quad (5.15) \\ |\,\Psi\rangle_B &= c\,|\!\uparrow\rangle_B + d\,|\!\downarrow\rangle_B, \quad (5.16) \end{aligned}$$

where a, b, c, and d are four complex numbers such that $aa^* + bb^* = 1$ and $cc^* + dd^* = 1$. Then

$$\begin{aligned} |\,\Phi\rangle_A \otimes |\,\Psi\rangle_B &= ac\,|\!\uparrow\rangle_A \otimes |\!\uparrow\rangle_B + ad\,|\!\uparrow\rangle_A \otimes |\!\downarrow\rangle_B \\ &\quad + bc\,|\!\downarrow\rangle_A \otimes |\!\uparrow\rangle_B + bd\,|\!\downarrow\rangle_A \otimes |\!\downarrow\rangle_B. \quad (5.17) \end{aligned}$$

Normalization conditions imposed on the Hilbert space states of individual qubits result in the following normalization condition of the biqubit state:

$$\begin{aligned} & ac(ac)^* + ad(ad)^* + bc(bc)^* + bd(bd)^* \\ &= aa^*(cc^* + dd^*) + bb^*(cc^* + dd^*) \\ &= aa^* + bb^* = 1. \end{aligned} \tag{5.18}$$

When is a pure biqubit state separable?

In summary, we find that a pure biqubit state

$$\alpha \,|\uparrow\rangle_A \otimes |\uparrow\rangle_B + \beta\,|\uparrow\rangle_A \otimes |\downarrow\rangle_B + \gamma\,|\downarrow\rangle_A \otimes |\uparrow\rangle_B + \delta\,|\downarrow\rangle_A \otimes |\downarrow\rangle_B \tag{5.19}$$

such that

$$\alpha = ac, \tag{5.20}$$
$$\beta = ad, \tag{5.21}$$
$$\gamma = bc, \tag{5.22}$$
$$\delta = bd, \tag{5.23}$$

where $aa^* + bb^* = cc^* + dd^* = 1$ represents two *separate* qubits.

Dividing the first equation by the second one and then the third equation by the fourth one yields

$$\alpha/\beta = c/d = \gamma/\delta, \tag{5.24}$$

which is equivalent to

$$\alpha\delta - \beta\gamma = 0, \tag{5.25}$$

or

$$\det \begin{pmatrix} \alpha & \beta \\ \gamma & \delta \end{pmatrix} = 0. \tag{5.26}$$

This is a simple criterion that we can use to check whether a given pure (because here we're within the unitary formalism) biqubit state can be separated into two independent qubits at all. To be more precise, this is a *necessary* though not a *sufficient* condition. But *necessary* is good enough if we want to prove that a given pure biqubit state is not separable.

Example of an inseparable biqubit state

To see how the criterion works, let us consider the following biqubit state:

$$|\,\Psi^-\rangle_{AB} = \frac{1}{\sqrt{2}}\left(|\uparrow\rangle_A \otimes |\downarrow\rangle_B - |\downarrow\rangle_A \otimes |\uparrow\rangle_B\right). \tag{5.27}$$

For this state $\alpha = \delta = 0$ but $\beta = 1/\sqrt{2} = -\gamma$. The state is normalized because $\beta^2 + \gamma^2 = 1$ but

$$\det \begin{pmatrix} \alpha & \beta \\ \gamma & \delta \end{pmatrix} = -\beta\gamma = \frac{1}{2}. \tag{5.28}$$

This, then, is an example of a pure biqubit state that *cannot* be split into two separate qubits.

Can such states exist? If so, what do they mean, and what is their fiducial, or observable, representation?

To arrive at the fiducial representation of the biqubit state, we have to convert it to a density operator and then to a quaternion representation. We begin with

$$\begin{aligned}
\boldsymbol{\rho}_{AB} &= |\,\Psi^-\rangle_{AB} \otimes {}_{AB}\langle\Psi^-\,| \\
&= \frac{1}{\sqrt{2}}\left(|\uparrow\rangle_A\otimes|\downarrow\rangle_B - |\downarrow\rangle_A\otimes|\uparrow\rangle_B\right)\otimes\frac{1}{\sqrt{2}}\left({}_A\langle\uparrow|\otimes{}_B\langle\downarrow| - {}_A\langle\downarrow|\otimes{}_B\langle\uparrow|\right) \\
&= \frac{1}{2}\Big(\left(|\uparrow\rangle_A\otimes{}_A\langle\uparrow|\right)\otimes\left(|\downarrow\rangle_B\otimes{}_B\langle\downarrow|\right) - \left(|\uparrow\rangle_A\otimes{}_A\langle\downarrow|\right)\otimes\left(|\downarrow\rangle_B\otimes{}_B\langle\uparrow|\right) \\
&\quad - \left(|\downarrow\rangle_A\otimes{}_A\langle\uparrow|\right)\otimes\left(|\uparrow\rangle_B\otimes{}_B\langle\downarrow|\right) + \left(|\downarrow\rangle_A\otimes{}_A\langle\downarrow|\right)\otimes\left(|\uparrow\rangle_B\otimes{}_B\langle\uparrow|\right)\Big).
\end{aligned} \tag{5.29}$$

Invoking expressions we have arrived at in Section 4.3, page 117, we can convert this readily to 2×2 matrices.

$$\begin{aligned}
\boldsymbol{\rho}_{AB} &= \frac{1}{2}\left(\begin{pmatrix}1&0\\0&0\end{pmatrix}_A\otimes\begin{pmatrix}0&0\\0&1\end{pmatrix}_B - \begin{pmatrix}0&1\\0&0\end{pmatrix}_A\otimes\begin{pmatrix}0&0\\1&0\end{pmatrix}_B\right. \\
&\quad\left. - \begin{pmatrix}0&0\\1&0\end{pmatrix}_A\otimes\begin{pmatrix}0&1\\0&0\end{pmatrix}_B + \begin{pmatrix}0&0\\0&1\end{pmatrix}_A\otimes\begin{pmatrix}1&0\\0&0\end{pmatrix}_B\right).
\end{aligned} \tag{5.30}$$

Now we use equations 4.53 on page 116 to express the matrices above in terms of Pauli matrices:

$$\begin{aligned}
\boldsymbol{\rho}_{AB} = \frac{1}{2}\Bigg(&\frac{1}{2}\left(\mathbf{1}_A + \boldsymbol{\sigma}_{zA}\right)\otimes\frac{1}{2}\left(\mathbf{1}_B - \boldsymbol{\sigma}_{zB}\right) \\
&-\frac{1}{2}\left(\boldsymbol{\sigma}_{xA} + \mathrm{i}\boldsymbol{\sigma}_{yA}\right)\otimes\frac{1}{2}\left(\boldsymbol{\sigma}_{xB} - \mathrm{i}\boldsymbol{\sigma}_{yB}\right) \\
&-\frac{1}{2}\left(\boldsymbol{\sigma}_{xA} - \mathrm{i}\boldsymbol{\sigma}_{yA}\right)\otimes\frac{1}{2}\left(\boldsymbol{\sigma}_{xB} + \mathrm{i}\boldsymbol{\sigma}_{yB}\right) \\
&+\frac{1}{2}\left(\mathbf{1}_A - \boldsymbol{\sigma}_{zA}\right)\otimes\frac{1}{2}\left(\mathbf{1}_B + \boldsymbol{\sigma}_{zB}\right)\Bigg),
\end{aligned} \tag{5.31}$$

which simplifies eventually to

$$\boldsymbol{\rho}_{AB} = \frac{1}{4}\left(\mathbf{1}_A\otimes\mathbf{1}_B - \boldsymbol{\sigma}_{xA}\otimes\boldsymbol{\sigma}_{xB} - \boldsymbol{\sigma}_{yA}\otimes\boldsymbol{\sigma}_{yB} - \boldsymbol{\sigma}_{zA}\otimes\boldsymbol{\sigma}_{zB}\right). \tag{5.32}$$

At this stage we can convert this to measurable probabilities by replacing Pauli sigmas with Pauli varsigmas:

$$\boldsymbol{p}_{AB} = \frac{1}{4}\left(\varsigma_{1A} \otimes \varsigma_{1B} - \varsigma_{xA} \otimes \varsigma_{xB} - \varsigma_{yA} \otimes \varsigma_{yB} - \varsigma_{zA} \otimes \varsigma_{zB}\right). \tag{5.33}$$

Biqubit probability matrix

A general fiducial matrix for a biqubit has the following interpretation:

$$\boldsymbol{p}_{AB} = \begin{pmatrix} p^{00} & p^{01} & p^{02} & p^{03} \\ p^{10} & p^{11} & p^{12} & p^{13} \\ p^{20} & p^{21} & p^{22} & p^{23} \\ p^{30} & p^{31} & p^{32} & p^{33} \end{pmatrix} = \begin{pmatrix} p^{\uparrow\uparrow} & p^{\uparrow\downarrow} & p^{\uparrow\rightarrow} & p^{\uparrow\otimes} \\ p^{\downarrow\uparrow} & p^{\downarrow\downarrow} & p^{\downarrow\rightarrow} & p^{\downarrow\otimes} \\ p^{\rightarrow\uparrow} & p^{\rightarrow\downarrow} & p^{\rightarrow\rightarrow} & p^{\rightarrow\otimes} \\ p^{\otimes\uparrow} & p^{\otimes\downarrow} & p^{\otimes\rightarrow} & p^{\otimes\otimes} \end{pmatrix}. \tag{5.34}$$

It has 16 entries and specifies the biqubit system *entirely* in terms of probabilities of detecting qubit A's "spin" against one "direction" and qubit B's "spin" against some other "direction" at the same time—with "directions" for both qubits being $\pm\boldsymbol{e}_z$ ($\uparrow$, $\downarrow$), $\boldsymbol{e}_x$ ($\rightarrow$) and $\boldsymbol{e}_y$ ($\otimes$). The measurements on each qubit are carried out the same way as before, but this time it is *not* enough to measure each qubit separately. To fully characterize the state, we need 16 probabilities of two specific events happening simultaneously.

How simultaneous do they have to be? Simultaneous enough so that, say, qubit B does not have a chance to interact with the environment (including qubit A) after qubit A has been measured.

How to get the actual numbers, p^{ij}, from the varsigma expression? This is easier than it may seem at first glance.

Let us recall that

$$p^{ij} = \langle \boldsymbol{\omega}^i \otimes \boldsymbol{\omega}^j, \boldsymbol{p}_{AB} \rangle, \tag{5.35}$$

where $\boldsymbol{\omega}^i$ are the canonical forms. The canonical forms are not dual to Pauli vectors, but they are dual to canonical vectors $\boldsymbol{e}_i$ and we know how to express Pauli vectors in terms of canonical vectors, because equations (2.55) on page 65 specify the procedure, namely,

$$\varsigma_1 = \boldsymbol{e}_0 + \boldsymbol{e}_1 + \boldsymbol{e}_2 + \boldsymbol{e}_3, \tag{5.36}$$

$$\varsigma_x = \boldsymbol{e}_2, \tag{5.37}$$

$$\varsigma_y = \boldsymbol{e}_3, \tag{5.38}$$

$$\varsigma_z = \boldsymbol{e}_0 - \boldsymbol{e}_1. \tag{5.39}$$

The trick here is to replace the varsigmas in equation (5.33) with canonical vectors. The replacement should flush the actual probabilities right away.

Having made the corresponding substitutions, we obtain

$$\begin{aligned} \boldsymbol{p}_{AB} &= \frac{1}{4}\Big((\boldsymbol{e}_{0A} + \boldsymbol{e}_{1A} + \boldsymbol{e}_{2A} + \boldsymbol{e}_{3A}) \otimes (\boldsymbol{e}_{0B} + \boldsymbol{e}_{1B} + \boldsymbol{e}_{2B} + \boldsymbol{e}_{3B}) \\ &\quad - \boldsymbol{e}_{2A} \otimes \boldsymbol{e}_{2B} - \boldsymbol{e}_{3A} \otimes \boldsymbol{e}_{3B} \\ &\quad - (\boldsymbol{e}_{0A} - \boldsymbol{e}_{1A}) \otimes (\boldsymbol{e}_{0B} - \boldsymbol{e}_{1B}) \Big). \end{aligned} \tag{5.40}$$

The first term corresponds to a 4×4 matrix of ones:

$$\begin{aligned} & (\boldsymbol{e}_{0A} + \boldsymbol{e}_{1A} + \boldsymbol{e}_{2A} + \boldsymbol{e}_{3A}) \otimes (\boldsymbol{e}_{0B} + \boldsymbol{e}_{1B} + \boldsymbol{e}_{2B} + \boldsymbol{e}_{3B}) \\ &= \begin{pmatrix} 1 & 1 & 1 & 1 \\ 1 & 1 & 1 & 1 \\ 1 & 1 & 1 & 1 \\ 1 & 1 & 1 & 1 \end{pmatrix}. \end{aligned} \tag{5.41}$$

The second and third terms correspond to matrices of zeros with 1 in the $(2,2)$ and $(3,3)$ positions, respectively:

$$\boldsymbol{e}_{2A} \otimes \boldsymbol{e}_{2B} = \begin{pmatrix} 0 & 0 & 0 & 0 \\ 0 & 0 & 0 & 0 \\ 0 & 0 & 1 & 0 \\ 0 & 0 & 0 & 0 \end{pmatrix}, \quad \boldsymbol{e}_{3A} \otimes \boldsymbol{e}_{3B} = \begin{pmatrix} 0 & 0 & 0 & 0 \\ 0 & 0 & 0 & 0 \\ 0 & 0 & 0 & 0 \\ 0 & 0 & 0 & 1 \end{pmatrix}. \tag{5.42}$$

The last term corresponds to a matrix that looks as follows.

$$\begin{aligned} & (\boldsymbol{e}_{0A} - \boldsymbol{e}_{1A}) \otimes (\boldsymbol{e}_{0B} - \boldsymbol{e}_{1B}) \\ &= \boldsymbol{e}_{0A} \otimes \boldsymbol{e}_{0B} - \boldsymbol{e}_{0A} \otimes \boldsymbol{e}_{1B} - \boldsymbol{e}_{1A} \otimes \boldsymbol{e}_{0B} + \boldsymbol{e}_{1A} \otimes \boldsymbol{e}_{1B} \\ &= \begin{pmatrix} 1 & -1 & 0 & 0 \\ -1 & 1 & 0 & 0 \\ 0 & 0 & 0 & 0 \\ 0 & 0 & 0 & 0 \end{pmatrix}. \end{aligned} \tag{5.43}$$

Combining the matrices yields

$$\begin{aligned} \boldsymbol{p}_{AB} &= \frac{1}{4}\left(\begin{pmatrix} 1 & 1 & 1 & 1 \\ 1 & 1 & 1 & 1 \\ 1 & 1 & 1 & 1 \\ 1 & 1 & 1 & 1 \end{pmatrix} - \begin{pmatrix} 0 & 0 & 0 & 0 \\ 0 & 0 & 0 & 0 \\ 0 & 0 & 1 & 0 \\ 0 & 0 & 0 & 0 \end{pmatrix} \right. \\ &\quad \left. - \begin{pmatrix} 0 & 0 & 0 & 0 \\ 0 & 0 & 0 & 0 \\ 0 & 0 & 0 & 0 \\ 0 & 0 & 0 & 1 \end{pmatrix} - \begin{pmatrix} 1 & -1 & 0 & 0 \\ -1 & 1 & 0 & 0 \\ 0 & 0 & 0 & 0 \\ 0 & 0 & 0 & 1 \end{pmatrix} \right) \end{aligned}$$

$$= \frac{1}{4}\begin{pmatrix} 0 & 2 & 1 & 1 \\ 2 & 0 & 1 & 1 \\ 1 & 1 & 0 & 1 \\ 1 & 1 & 1 & 0 \end{pmatrix} = \begin{pmatrix} p^{\uparrow\uparrow} & p^{\uparrow\downarrow} & p^{\uparrow\rightarrow} & p^{\uparrow\otimes} \\ p^{\downarrow\uparrow} & p^{\downarrow\downarrow} & p^{\downarrow\rightarrow} & p^{\downarrow\otimes} \\ p^{\rightarrow\uparrow} & p^{\rightarrow\downarrow} & p^{\rightarrow\rightarrow} & p^{\rightarrow\otimes} \\ p^{\otimes\uparrow} & p^{\otimes\downarrow} & p^{\otimes\rightarrow} & p^{\otimes\otimes} \end{pmatrix}. \tag{5.44}$$

Quantum correlations

Now we can finally get down to physics. And the physics that emerges from this matrix is most peculiar. First, we note that

$$p^{\uparrow\uparrow} = p^{\downarrow\downarrow} = p^{\rightarrow\rightarrow} = p^{\otimes\otimes} = 0. \tag{5.45}$$

If the biqubit is made of two neutrons and has been prepared in this special state, the probability of finding the neutrons aligned, that is, with their spins both pointing up or both pointing down, or both pointing right, or both pointing "into the page," is zero. Regardless of how we orient the measuring apparatuses, the qubits (neutrons in this case) always come out pointing in the opposite directions—even if the measuring polarizers are far away from each other, but this only as long as the neutrons are still described by $(|\uparrow\rangle_A\otimes|\downarrow\rangle_B - |\downarrow\rangle_A\otimes|\uparrow\rangle_B)/\sqrt{2}$, which is not going to be forever, because this peculiar biqubit state is going to depolarize faster even than single-qubit polarized states.

The upper left corner of matrix $\boldsymbol{p}_{AB}$ is normalized. States

$$|\uparrow\rangle_A\otimes|\uparrow\rangle_B, \qquad |\uparrow\rangle_A\otimes|\downarrow\rangle_B,$$
$$|\downarrow\rangle_A\otimes|\uparrow\rangle_B, \quad \text{and} \quad |\downarrow\rangle_A\otimes|\downarrow\rangle_B$$

constitute the physical basis of the system, and

$$p^{\uparrow\uparrow} + p^{\uparrow\downarrow} + p^{\downarrow\uparrow} + p^{\downarrow\downarrow} = 1. \tag{5.46}$$

The remaining entries in the matrix, outside the diagonal, describe measurements against directions that are perpendicular to each other. For example, if qubit A is measured against $\boldsymbol{e}_x$ and qubit B is measured against $\boldsymbol{e}_y$, then the probability associated with the outcome is $p^{\rightarrow\otimes} = 1/4$. This result is consistent with the assumption that both qubits in this state must always be found pointing in the opposite directions. For example, if qubit A is found pointing in the $\boldsymbol{e}_x$ direction, which on the whole is going to be half of all cases associated with this measurement, then qubit B should point in the $-\boldsymbol{e}_x$ direction, but when measured against the $\boldsymbol{e}_y$ direction, half of all qubits B will emerge pointing in the $\boldsymbol{e}_y$ direction and the other half pointing in the $-\boldsymbol{e}_y$ direction. And so the probabilty of finding that qubit A points in the $\boldsymbol{e}_x$ direction and qubit B points in the $\boldsymbol{e}_y$ direction is $1/2 \times 1/2 = 1/4$.

Bell states

There are three other similar states, whose probability matrix can be computed the same way. The first is

$$
\begin{aligned}
\mid \Psi^{+}\rangle_{AB} &= \frac{1}{\sqrt{2}}\left(|\uparrow\rangle_A \otimes |\downarrow\rangle_B + |\downarrow\rangle_A \otimes |\uparrow\rangle_A\right), \qquad (5.47)\\
\boldsymbol{\rho}_{AB} &= \frac{1}{4}\left(\mathbf{1}_A \otimes \mathbf{1}_B + \boldsymbol{\sigma}_{xA} \otimes \boldsymbol{\sigma}_{xB} + \boldsymbol{\sigma}_{y_A} \otimes \boldsymbol{\sigma}_{y_B} - \boldsymbol{\sigma}_{zA} \otimes \boldsymbol{\sigma}_{zB}\right), \qquad (5.48)\\
\boldsymbol{p}_{AB} &= \begin{pmatrix} p^{\uparrow\uparrow} & p^{\uparrow\downarrow} & p^{\uparrow\rightarrow} & p^{\uparrow\otimes} \\ p^{\downarrow\uparrow} & p^{\downarrow\downarrow} & p^{\downarrow\rightarrow} & p^{\downarrow\otimes} \\ p^{\rightarrow\uparrow} & p^{\rightarrow\downarrow} & p^{\rightarrow\rightarrow} & p^{\rightarrow\otimes} \\ p^{\otimes\uparrow} & p^{\otimes\downarrow} & p^{\otimes\rightarrow} & p^{\otimes\otimes} \end{pmatrix} = \frac{1}{4}\begin{pmatrix} 0 & 2 & 1 & 1 \\ 2 & 0 & 1 & 1 \\ 1 & 1 & 2 & 1 \\ 1 & 1 & 1 & 2 \end{pmatrix}. \qquad (5.49)
\end{aligned}
$$

This state describes a biqubit system of "total spin 1" but with a zero component in the $\boldsymbol{e}_z$ direction. When projected on either $\boldsymbol{e}_x$ or $\boldsymbol{e}_y$ both 1/2-spins of the biqubit align and add up. But when projected on $\boldsymbol{e}_z$ the spins counteralign, and so the projection of the total spin on this direction is zero.

Coming back to $\left(|\uparrow\rangle_A \otimes |\downarrow\rangle_B - |\downarrow\rangle_A \otimes |\uparrow\rangle_B\right)/\sqrt{2}$, which was characterized by 1/2-spins counteraligning in any direction, we would label that biqubit state with "total spin zero."

The next state is

$$
\begin{aligned}
\mid \Phi^{-}\rangle_{AB} &= \frac{1}{\sqrt{2}}\left(|\uparrow\rangle_A \otimes |\uparrow\rangle_B - |\downarrow\rangle_A \otimes |\downarrow\rangle_A\right), \qquad (5.50)\\
\boldsymbol{\rho}_{AB} &= \frac{1}{4}\left(\mathbf{1}_A \otimes \mathbf{1}_B - \boldsymbol{\sigma}_{xA} \otimes \boldsymbol{\sigma}_{xB} + \boldsymbol{\sigma}_{y_A} \otimes \boldsymbol{\sigma}_{y_B} + \boldsymbol{\sigma}_{zA} \otimes \boldsymbol{\sigma}_{zB}\right), \qquad (5.51)\\
\boldsymbol{p}_{AB} &= \begin{pmatrix} p^{\uparrow\uparrow} & p^{\uparrow\downarrow} & p^{\uparrow\rightarrow} & p^{\uparrow\otimes} \\ p^{\downarrow\uparrow} & p^{\downarrow\downarrow} & p^{\downarrow\rightarrow} & p^{\downarrow\otimes} \\ p^{\rightarrow\uparrow} & p^{\rightarrow\downarrow} & p^{\rightarrow\rightarrow} & p^{\rightarrow\otimes} \\ p^{\otimes\uparrow} & p^{\otimes\downarrow} & p^{\otimes\rightarrow} & p^{\otimes\otimes} \end{pmatrix} = \frac{1}{4}\begin{pmatrix} 2 & 0 & 1 & 1 \\ 0 & 2 & 1 & 1 \\ 1 & 1 & 0 & 1 \\ 1 & 1 & 1 & 2 \end{pmatrix}. \qquad (5.52)
\end{aligned}
$$

This state is similar to $\left(|\uparrow\rangle_A \otimes |\downarrow\rangle_B + |\downarrow\rangle_A \otimes |\uparrow\rangle_B\right)/\sqrt{2}$, meaning that it is also a "total spin 1" state, but this time it is the $\boldsymbol{e}_x$ component of the spin that is missing. When projected on this direction, both spins counteralign, but they align when projected on $\boldsymbol{e}_z$ or $\boldsymbol{e}_x$.

The final state is

$$| \Phi^+\rangle_{AB} = \frac{1}{\sqrt{2}} (|\uparrow\rangle_A \otimes |\uparrow\rangle_B + |\downarrow\rangle_A \otimes |\downarrow\rangle_A), \quad (5.53)$$

$$\boldsymbol{\rho}_{AB} = \frac{1}{4}\left(\mathbf{1}_A \otimes \mathbf{1}_B + \boldsymbol{\sigma}_{xA} \otimes \boldsymbol{\sigma}_{xB} - \boldsymbol{\sigma}_{yA} \otimes \boldsymbol{\sigma}_{yB} + \boldsymbol{\sigma}_{zA} \otimes \boldsymbol{\sigma}_{zB}\right), \quad (5.54)$$

$$\boldsymbol{p}_{AB} = \begin{pmatrix} p^{\uparrow\uparrow} & p^{\uparrow\downarrow} & p^{\uparrow\rightarrow} & p^{\uparrow\otimes} \\ p^{\downarrow\uparrow} & p^{\downarrow\downarrow} & p^{\downarrow\rightarrow} & p^{\downarrow\otimes} \\ p^{\rightarrow\uparrow} & p^{\rightarrow\downarrow} & p^{\rightarrow\rightarrow} & p^{\rightarrow\otimes} \\ p^{\otimes\uparrow} & p^{\otimes\downarrow} & p^{\otimes\rightarrow} & p^{\otimes\otimes} \end{pmatrix} = \frac{1}{4}\begin{pmatrix} 2 & 0 & 1 & 1 \\ 0 & 2 & 1 & 1 \\ 1 & 1 & 2 & 1 \\ 1 & 1 & 1 & 0 \end{pmatrix}, \quad (5.55)$$

which is like the other two states above, meaning, a "total spin 1" state, with the $\boldsymbol{e}_y$ component missing.

The unitary formalism does a rather good job of hiding the full physical characterizations of these states. We have to go through a number of quite complicated transformations to arrive at the probability matrices, which, after all, are what is measured. To facilitate similar computations in future, we have listed the auxiliary matrices used in computing biqubit probabilities in Appendix B.

The density matrix (or quaternion) formalism is somewhat better. With a little practice one can see the actual probability matrices hiding inside $\boldsymbol{\rho}_{AB}$.

The fiducial formalism gives us the probability matrices explicitly. Alas, for three qubits, $\boldsymbol{p}_{ABC}$ is going to be a cube, and a hypercube for four qubits and at this stage the fiducial representation quickly becomes too complex, whereas the unitary representation continues to be manageable. This is precisely because it hides information.

Entanglement

States $| \Psi^-\rangle_{AB}$ (equation (5.27)), $| \Psi^+\rangle_{AB}$ (equation (5.47)), $| \Phi^-\rangle_{AB}$ (equation (5.50)) and $| \Phi^+\rangle_{AB}$ (equation (5.53)) are said to be *entangled*. The verb *entangle* has two meanings:

1. *Wrap or twist together*—there is a degree of togetherness in these states that, as we'll see later, cannot be explained by naive classical physics reasoning based on the concept of *local realism*[1] and leads to amusing paradoxes.

2. *Involve in a perplexing or troublesome situation*—and this meaning is right on the spot, too.

[1]It can, however, be explained by less naive classical physics reasoning that abandons the locality assumption [8] [14] [36].

The paradoxes mentioned above have been perplexing physics community ever since John Stewart Bell (1928–1990) came up with their concise mathematical characterization in 1964[2] [7], [8]. To make matters worse, they are not just theoretical paradoxes to be contemplated by arm-chair philosophers. They have all been confirmed by elaborate experiments [4] and have become fundamental to quantum computing and quantum communications.

In memory of John Bell, states $|\,\Psi^-\rangle_{AB}$, $|\,\Psi^+\rangle_{AB}$, $|\,\Phi^-\rangle_{AB}$, and $|\,\Phi^+\rangle_{AB}$ are called *Bell states*.

How do Bell states differ from separable biqubit states? Well, we already know that they can't be separated from the unitary formalism, but how does this manifest on the fiducial formalism level?

Fiducial representation of a separable biqubit

A biqubit state that is made of two separate qubits, each in its own well-defined state that may be a mixture, has the following fiducial representation:

$$\begin{aligned}
\boldsymbol{p}_{AB} &= \boldsymbol{p}_A \otimes \boldsymbol{p}_B \\
&= \frac{1}{4}\left(\varsigma_{1A} + \sum_{i=x,y,z} r_A^i \varsigma_{iA}\right) \otimes \left(\varsigma_{1B} + \sum_{j=x,y,z} r_B^j \varsigma_{jB}\right) \\
&= \frac{1}{4}\left(\varsigma_{1A} \otimes \varsigma_{1B} + \sum_{j=x,y,z} r_B^j \varsigma_{1A} \otimes \varsigma_{jB} + \sum_{i=x,y,z} r_A^i \varsigma_{iA} \otimes \varsigma_{1B}\right. \\
&\quad \left. + \sum_{i=x,y,z} \sum_{j=x,y,z} r_A^i r_B^j \varsigma_{iA} \otimes \varsigma_{jB}\right), \qquad (5.56)
\end{aligned}$$

where $\boldsymbol{r}_A^2 \leq 1$ and $\boldsymbol{r}_B^2 \leq 1$, too, and where the equality would hold for pure single qubit constituents. Analogous expressions can be constructed for the quaternion and density matrix formalisms by merely replacing varsigmas with sigmas.

Simple separability

We call such state a *simple separable* state. Its individual constituents may not be pure, and the resulting biqubit may not be pure either, but it is made of just one pair of well-defined separate qubits, pure or not.

Comparing the above expression with Bell states and remembering that varsigmas constitute bases in the fiducial spaces of both qubits, we find the first important difference: all Bell states are of the form

$$\frac{1}{4}\left(\varsigma_{1A} \otimes \varsigma_{1B} \pm \varsigma_{xA} \otimes \varsigma_{xB} \pm \varsigma_{yA} \otimes \varsigma_{yB} \pm \varsigma_{zA} \otimes \varsigma_{zB}\right). \qquad (5.57)$$

[2] The paradoxes themselves go back to Einstein, Podolsky, and Rosen, who discussed one such paradox in their paper in 1935 [38]. But EPR, as the trio is affectionately called, did not produce an experimentally verifiable formula that could be used to check which way the chips fall.

There are no $\varsigma_{1A} \otimes \varsigma_{jB}$ and no $\varsigma_{iA} \otimes \varsigma_{1B}$ terms here. This immediately suggests that $\boldsymbol{r}_A = \boldsymbol{r}_B = 0$. And yet $r_A^i r_B^i = \pm 1$. How can this be?

Mixing simple separable states

Let us consider a mixture of two simple separable biqubit states:

$$P_\alpha \boldsymbol{p}_{A\alpha} \otimes \boldsymbol{p}_{B\alpha} + P_\beta \boldsymbol{p}_{A\beta} \otimes \boldsymbol{p}_{B\beta}, \tag{5.58}$$

where $P_\alpha + P_\beta = 1$ and both are positive.

The fiducial representation of the mixture is

$$\begin{aligned}\frac{1}{4}\Bigg(&\varsigma_{1A} \otimes \varsigma_{1B} \\ &+ \sum_{i=x,y,z} \left(P_\alpha r_{A\alpha}^i + P_\beta r_{A\beta}^i\right) \varsigma_{iA} \otimes \varsigma_{1B} \\ &+ \sum_{i=x,y,z} \left(P_\alpha r_{B\alpha}^i + P_\beta r_{B\beta}^i\right) \varsigma_{1A} \otimes \varsigma_{iB} \\ &+ \sum_{i,j=x,y,z} \left(P_\alpha r_{A\alpha}^i r_{B\alpha}^j + P_\beta r_{A\beta}^i r_{B\beta}^j\right) \varsigma_{iA} \otimes \varsigma_{jB}\Bigg).\end{aligned} \tag{5.59}$$

In general

$$P_\alpha r_{A\alpha}^i r_{B\alpha}^j + P_\beta r_{A\beta}^i r_{B\beta}^j \neq \left(P_\alpha r_{A\alpha}^i + P_\beta r_{A\beta}^i\right) \cdot \left(P_\alpha r_{B\alpha}^j + P_\beta r_{B\beta}^j\right). \tag{5.60}$$

Separability

Therefore, if we want to admit a description of mixtures of simple separable biqubit states, each of which also may be made of two qubits in some mixed states, and such a *finite* mixture is called a *separable state*, we must allow the following, more general fiducial representation:

$$\begin{aligned}\boldsymbol{p}_{AB} = \frac{1}{4}\Bigg(&\alpha\, \varsigma_{1A} \otimes \varsigma_{1B} + \sum_{j=x,y,z} r_B^j \varsigma_{1A} \otimes \varsigma_{jB} + \sum_{i=x,y,z} r_A^i \varsigma_{iA} \otimes \varsigma_{1B} \\ &+ \sum_{i,j=x,y,z} x_{AB}^{ij} \varsigma_{iA} \otimes \varsigma_{jB}\Bigg),\end{aligned} \tag{5.61}$$

where α may not necessarily be 1 and x_{AB}^{ij} may be independent of r_A^i and r_B^j. There would be nine such x_{AB}^{ij} coefficients, plus three r_A^i coefficients and three r_B^j coefficients, plus the α—altogether sixteen *real* numbers are therefore needed to fully characterize a biqubit state. The number may be reduced to fifteen by imposing a normalization condition such as

$$p^{\uparrow\uparrow} + p^{\uparrow\downarrow} + p^{\downarrow\uparrow} + p^{\downarrow\downarrow} = 1, \tag{5.62}$$

which implies that $\alpha = 1$.

Can entangled states be faked?

It is possible to construct complicated biqubit mixtures made of many components, more than just two, that can get pretty close to an entangled state such as the Bell state $| \Psi^- \rangle_{AB}$. Hence, it may be sometimes difficult to distinguish between entangled and mixed states experimentally, especially if neither is pure.

Let us take a 50/50 mixture of two simple separable biqubit states described by the following two pairs of vectors,

$$(\boldsymbol{r}_{A\alpha}, \boldsymbol{r}_{B\alpha}) \quad \text{and} \quad (\boldsymbol{r}_{A\beta}, \boldsymbol{r}_{B\beta}), \tag{5.63}$$

such that

$$\boldsymbol{r}_{A\alpha} = -\boldsymbol{r}_{A\beta} \quad \text{and} \quad \boldsymbol{r}_{B\alpha} = -\boldsymbol{r}_{B\beta}. \tag{5.64}$$

For this mixture we find that

$$P_\alpha r^i_{A\alpha} + P_\beta r^i_{A\beta} = 0, \tag{5.65}$$

$$P_\alpha r^i_{B\alpha} + P_\beta r^i_{B\beta} = 0, \tag{5.66}$$

but

$$\begin{aligned} x^{ij}_{AB} &= P_\alpha r^i_{A\alpha} r^j_{B\alpha} + P_\beta r^i_{A\beta} r^j_{B\beta} \\ &= 0.5 \left(r^i_{A\alpha} r^j_{B\alpha} + (-r^i_{A\alpha})(-r^j_{B\alpha}) \right) \\ &= r^i_{A\alpha} r^j_{B\alpha}. \end{aligned} \tag{5.67}$$

So, here we end up with a state that looks similar to $| \Psi^- \rangle_{AB}$. Its $\boldsymbol{r}_A$ and $\boldsymbol{r}_B$ vanish, but its $\boldsymbol{x}_{AB}$ does not. At the same time, though, its $\boldsymbol{x}_{AB}$ is not the same as the one we have for $| \Psi^- \rangle_{AB}$. We cannot make a diagonal matrix with none of the diagonal elements vanishing out of $r^i_{A\alpha} r^j_{B\alpha}$.

It turns out that $\boldsymbol{p}_{AB}$ of an entangled state cannot be reproduced by any finite mixture of simple separable biqubit states [67]. But this is not a criterion that is easy to use, especially when we deal with experimental data that is always contaminated with some error.

Fiducial criteria for simply separable biqubits

For a simple separable biqubit state the following trivial observations hold.

If r^i_A and r^j_B do not vanish and $x^{ij}_{AB} = r^i_A r^j_B$, then the biqubit is clearly separable into two individual qubits. If one or both r^i_A and r^j_B vanish, then one or both constituent qubits are fully depolarized. In this case x^{ij}_{AB} must vanish, too, if the biqubit is to be separated into two individual qubits.

The separability of x^{ij}_{AB} into r^i_A and r^j_B can be tested easily. The following must hold:

$$\boldsymbol{r}_A \cdot \boldsymbol{x}_{AB} = r^2_A \boldsymbol{r}_B \quad \text{and} \tag{5.68}$$

$$\boldsymbol{x}_{AB} \cdot \boldsymbol{r}_B = \boldsymbol{r}_A r_B^2. \tag{5.69}$$

Another obvious feature of a separable $\boldsymbol{x}_{AB}$ in this context is that

$$\det \boldsymbol{x}_{AB} = \sum_{i,j,k\in\{x,y,z\}} \epsilon_{ijk} x^{xi} x^{yj} x^{zk} = \sum_{i,j,k\in\{x,y,z\}} \epsilon_{ijk} r_A^x r_B^i r_A^y r_B^j r_A^z r_B^k = 0, \tag{5.70}$$

because ϵ_{ijk} is fully antisymmetric whereas $r_B^i r_B^j r_B^k$ is fully symmetric. The vanishing of $\det \boldsymbol{x}_{AB}$ is a *necessary* condition but not a sufficient one. Still, it is good enough if we just want to check whether a given state can at all be separated into two individual qubits.

The above considerations are presented in terms of varsigma coefficients. But when the actual measurements are made, the probabilities that are assembled into a matrix $\boldsymbol{p}^{ij}$ are not varsigma coefficients. They are canonical coefficients instead. How are we to find r_A^i, r_B^j and x_{AB}^{ij}? This can be done easily by contracting $\boldsymbol{p}$ with appropriate combinations of Pauli forms, namely,

$$r_A^i = \langle \varsigma_A^i \otimes \varsigma_B^1, \boldsymbol{p}_{AB} \rangle, \tag{5.71}$$

$$r_B^j = \langle \varsigma_A^1 \otimes \varsigma_B^j, \boldsymbol{p}_{AB} \rangle, \tag{5.72}$$

$$x_{AB}^{ij} = \langle \varsigma_A^i \otimes \varsigma_B^j, \boldsymbol{p}_{AB} \rangle, \tag{5.73}$$

where i and j run through x, y and z.

Another way is to replace canonical vectors in $\boldsymbol{p}_{AB} = \sum_{ij} p_{AB}^{ij} \boldsymbol{e}_{iA} \otimes \boldsymbol{e}_{jB}$ with Pauli vectors using equations (2.59) on page 65, since Pauli vectors are duals of Pauli forms.

5.2 Pauli Exclusion Principle

But how do we know entangled states exist at all? They can't be made by merely placing two qubits next to each other. This yields a nonentangled biqubit only, which is just two separate qubits.

Multielectron atoms posed a conundrum.

The idea that pairs of elementary particles can be entangled goes all the way back to Wolfgang Pauli (1900–1958). Today we know that we can entangle even "macroscopic" objects [72, 11, 134], but back in 1925 quantum mechanics was strictly a science of atoms and elementary particles. After Niels Bohr (1885–1962) presented his model of the hydrogen atom in 1913 that correctly accounted for hydrogen atom's energy levels—well, at least until physicists had a closer look and found that there was more to it—people turned to other atoms, trying to understand their structure. It soon became apparent that even as simple an atom as helium was immensely more complicated than hydrogen. For starters, helium had two separate families of energy levels with seldom observed transitions between one and the other. For a while people even thought that there were two different types of helium configurations. Other atoms' spectra are even more complex.

Eventually Pauli figured out that he could account for some features of alcaline atoms' spectra if he postulated that (1) electrons had an additional as yet unrecognized degree of freedom (later called *spin*) and (2)

Pauli exclusion principle

> there can never be two or more equivalent electrons in an atom for which ... the values of all quantum numbers ... are the same. If an electron is present in the atom for which these quantum numbers ... have definite values, this state is *occupied.* [108].

At the time, quantum mechanics did not exist in the form we know it today. Werner Karl Heisenberg (1901–1976) was yet to publish his 1925 paper on "matrix mechanics," and Erwin Schrödinger was yet to publish his 1926 paper on "wave mechanics." When Paul Dirac (1902–1984) finally merged matrix and wave mechanics into quantum mechanics in 1926 and Kronig, Uhlenbeck, and Goudsmit identified Pauli's additional degree of freedom as spin, it became clear that the way to express Pauli's principle was to antisymmetrize the multielectron wave function. For example, for two electrons A and B whose individual wave functions may be $\mid \Psi\rangle_A$ and $\mid \Phi\rangle_B$, their combined wave function in an atom would have to be

The Pauli exclusion principle implies entanglement.

$$\frac{1}{\sqrt{2}}\left(\mid \Psi\rangle_A \otimes \mid \Phi\rangle_B - \mid \Phi\rangle_A \otimes \mid \Psi\rangle_B\right). \tag{5.74}$$

This way, if both functions are identical, meaning that all quantum numbers inside $\mid \Psi\rangle$ and $\mid \Phi\rangle$ are the same, so that $\mid \Psi\rangle = \mid \Phi\rangle$, the combined wave function is zero. This trick captures the Pauli principle automatically.

In summary, the spectra of multielectron atoms tell us that entangled states such as $\left(|\uparrow\rangle_A \otimes |\downarrow\rangle_B - |\downarrow\rangle_A \otimes |\uparrow\rangle_B\right)/\sqrt{2}$ exist.

The Pauli exclusion principle applies to identical particles only, meaning that they must be both of the same type, for example, two electrons, two neutrons, or two protons. They must be fermions, too: their spin must be an odd multiple of $\hbar/2$. If the particles are bosons, which means that their spin is an even multiple of $\hbar/2$, examples of such particles are photons and alpha particles, then their wave functions must be symmetric. This concept has important macroscopic consequences in terms of statistics and phenomena such as superconductivity and superfluidity.

In layman terms fermions hate being like the other guys. They're individualistic. If there is a fermion nearby that does something and we happen to be an identical

Fermions and bosons

fermion, we'll do our best to dress differently, drive a different car, look the other way, and preferably get out of the neighborhood as soon as an opportunity arises. On the other hand, bosons love to be together and to be alike: "I'm having what she's having." If a boson is driving on a freeway, a whole pack of bosons soon will be driving in the same direction right next to one another—this is how superfluidity works.

Cats are fermions. Dogs are bosons.

Identical particles are not very useful for computing because they cannot be addressed individually. In quantum computing the best results are obtained if qubits are associated with different quantum objects, for example, with different nuclei in a molecule. But identical particles can be combined into collective states. Here a quantum state of interest may be associated not with an individual electron but with a large number of electrons all forced into a single quantum configuration.

Bi-Fermions become bosons.

But how can electrons be forced into a collective if they are fermions? This can happen only if there is an intermediary agent present. The agent acts like a glue. Electrons in a crystal lattice of niobium, for example, are held together by phonons. Pairs of such phonon-glued electrons, called Cooper pairs, behave like bosons, because their combined spin is an even multiple of $\hbar/2$. Since Cooper pairs are bosons, large collective states are possible. These states are responsible for superconductivity.

Entanglement is fragile.

Because all elementary particles can be classified either as fermions or as bosons and these can exist only in entangled states, an obvious question is why we don't normally see entanglement in the macroscopic world around us, apart from the rather special phenomena of superfluidity and superconductivity, both of which require extremely low temperatures.

The answer is that entanglement is technically a superposition. It is a superposition in the biqubit Hilbert space. The biqubit superpositions decay just as quickly as (if not more so than) single-qubit superpositions. The more a biqubit interacts with the environment, the faster its entangled state flips onto an unentangled one. And so, after a short while we end up with a mixture of separable biqubits.

This topic will be investigated in more depth in Section 5.11.

5.3 A Superconducting Biqubit

The two quantum electronic devices discussed in this section were constructed to demonstrate biqubit entanglement.

The first one was made by a group of scientists from the University of Maryland in 2003 [11] and the second one three years later by a group from the University of California, Santa Barbara [134].

Two Josephson junction regimes

A Josephson junction, which is constructed by inserting a very thin insulator between two superconductors, is characterized by the *critical current*. The junction is biased by pushing current through it. Up to the critical current, pairs of electrons, the Cooper pairs, flow through the junction unimpeded and without any voltage drop across the junction. This is called the DC Josephson junction regime. But when the current exceeds the critical current, then the voltage drop across the junction suddenly appears, and the current begins to oscillate rapidly. This is called the AC Josephson junction regime.

Tilted washboard potential

When the junction is biased somewhat below the critical current, that is, still in the DC Josephson regime, the junction's inductance and capacitance form an anharmonic LC resonator with an anharmonic potential U that can be approximated by a cubic function of position within the junction, as shown in Figure 5.1 (A). This is sometimes called a "tilted washboard potential." Strictly speaking, U is a function of the *phase* γ of the Cooper pairs wave function within the circuit (they all have the same wave function), but within the junction the phase changes approximately linearly with position, and outside the junction the phase is approximately constant. The left bend of the washboard curve forms a natural potential well, within which discrete energy levels form; and the right bend forms a natural potential barrier, through which Cooper pairs trapped in the well may tunnel. The height $\Delta U(I)$ of the potential barrier is a function of the junction bias current given by

$$\Delta U(I) = \frac{2\sqrt{2} I_0 \Phi_0}{3\pi} \left(1 - \frac{I}{I_0}\right)^{3/2}, \tag{5.75}$$

where I_0 is the critical current and $\Phi_0 = h/2e$, where h is the Planck constant. The potential barrier vanishes when $I \to I_0$. The energy at the bottom of the well corresponds to the classical plasma oscillation frequency $\omega_p(I)$ given by

$$\omega_p(I) = \sqrt{\frac{2\sqrt{2}\pi I_0}{\Phi_0 C}} \left(1 - \frac{I}{I_0}\right)^{1/4}, \tag{5.76}$$

where C is the junction capacitance. The $\omega_p(I)$ also vanishes when $I \to I_0$, but more slowly than $\Delta U(I)$.

Observing the energy levels

The discrete energy levels within the well can be observed by adding small microwave pulses to the bias DC current. If the frequency of the pulse matches the transition frequency between the two levels drawn in the diagram, Cooper pairs trapped within the lower level absorb the energy and jump to the upper level, which is characterized by a large tunneling rate. And so, they then tunnel through the barrier. This tunneling is illustrated by the right-pointing arrow in Figure 5.1 (A). After the tunneling event the junction becomes momentarily depleted of carriers and behaves like an open switch across which a macroscopically measurable voltage develops.

Phase qubits

Such qubits, which are made of collective excitations of up to 10^9 paired electrons, depending on the size of the junction, are called *phase qubits.* Recent advances in phase qubit technology made it possible to carry out their full characterization. The group from Santa Barbara was able to view the full traversal of the qubit, across the Bloch ball during a Ramsey measurement [135]. Like the quantronium circuit, phase qubits are macroscopic devices that behave quantum-mechanically. Also, like the quantronium circuit, they have to be immersed in a cryobath, close to absolute zero, to work.

Maryland biqubit

But let us return to the Maryland group biqubit shown in Figure 5.1 (B). Here we have two phase qubits characterized by their critical currents I_{c1} and I_{c2} of $14.779\,\mu$A and $15.421\,\mu$A, respectively, both shunted by capacitors C_j of 4.8 pF each, which help stabilize the qubits, and coupled through the capacitor C_c of 0.7 pF.

Qubit #1, on the left, is DC biased, and the bias current that flows through it, I_{b1}, is $14.630\,\mu$A. Qubit #2, on the right, is biased with a linear ramp that, in effect, allows for repetitive scanning of the biqubit parameters.

Figure 5.1 (C) shows a photograph of the device. Two coupling capacitors in the device are visible in the center of the photograph. The role of the lower capacitor is to short out parasitic inductance in the ground line. It forms the effective C_c together with the upper capacitor. The Josephson junctions are inside the two narrow horizontal features on both sides of the photograph, one of which is shown in magnification in Figure 5.1 (D). Here we can see two strips made of niobium that overlap in a square box a little to the left of the center of Figure 5.1 (D). This is the junction itself. The two overlapping niobium strips are separated by a thin layer of aluminum oxide. Each Josephson junction is quite large, the side of the box being $10\,\mu$m long. The distance between the two junctions is 0.7 mm, which is huge by quantum mechanics standards. It is a macroscopic distance.

The biqubit is observed in a way that is similar to a single-phase qubit observation technique. It has a certain characteristic and discrete energy spectrum. A microwave signal I_m (see Figure 5.1 (B)) is applied through the bias lines to the right qubit directly, but through the coupling capacitor to the left qubit, too. So, in effect, the whole biqubit is irradiated. This process induces transitions from

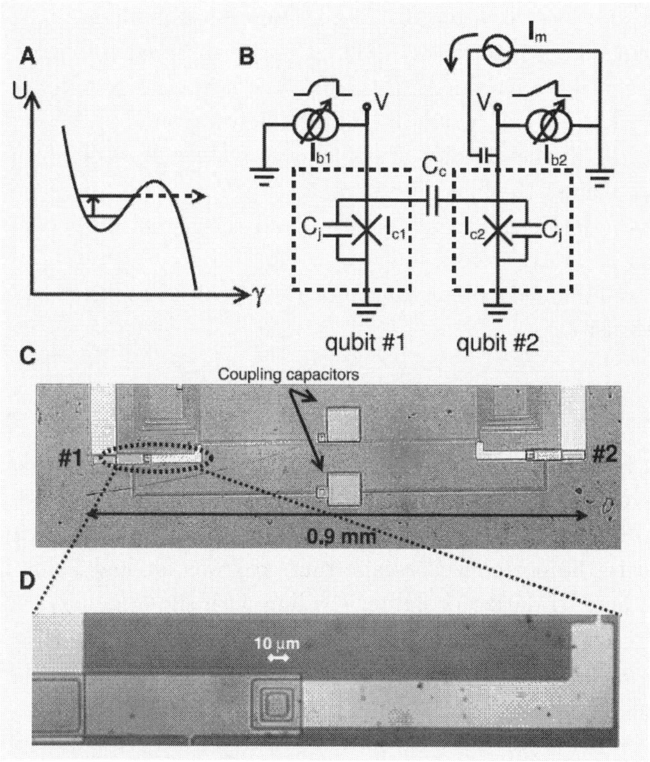

Figure 5.1: A Josephson junction biqubit. (A) illustrates how a two-level quantum system, a qubit, forms within the junction when it is biased in a special way. (B) shows the schematic diagram of the device. (C) and (D) show the photographs of the device. From [11]. Reprinted with permission from AAAS.

the ground state of the biqubit to its higher energy states, but only if the applied microwave signal has its frequency matching the energy gap. Higher energy states are closer to the knees of the washboard potentials for both qubits, and so they have a higher probability of escaping from the potential wells of both junctions. This creates an open circuit condition, because the junctions run out of carriers, and this in turn manifests as surges of DC voltage V in the right-hand side of the circuit and can be detected easily.

Ramping I_{b2} for a microwave signal of fixed frequency has the effect of changing the parameters of the biqubit so that eventually we come across the ones for which the energy absorption takes place. This is illustrated in Figure 5.2 (A). Here the microwave frequency f is set to 4.7 GHz. Δ is the observed escape rate normalized against the escape rate measured in the absence of the microwave agitation and is given by $\Delta = (\Gamma_m - \Gamma)/\Gamma$, where Γ_m is the microwave-induced escape rate and Γ is the escape rate in the absence of the microwave signal. The normalized escape rate is plotted against the bias current I_{b2}.

Let us call the lower energy state $|\,0\rangle_{1,2}$ and the higher energy state $|\,1\rangle_{1,2}$ for qubits #1 and #2, respectively.

There is a well-defined Lorentzian absorption peak in Figure 5.2 (A) that corresponds to a transition

$$|\,0\rangle_1 \otimes |\,0\rangle_2 \rightarrow \frac{1}{\sqrt{2}} \left(|\,0\rangle_1 \otimes |\,1\rangle_2 - |\,1\rangle_1 \otimes |\,0\rangle_2 \right). \tag{5.77}$$

By varying both f and I_{b2} and repeating the measurement for each point in the (f, I_{b2}) plane up to 100,000 times, the researchers arrived at the histogram map reproduced in Figure 5.2 (B). The original histogram was in color, which was made to correspond to the normalized escape rate: red was for high, blue for low. This is reproduced here by darker or lighter shading with the originally red peaks along the solid white lines showing as darker patches. The somewhat darker large patch in the middle of the diagram, between the dotted lines, corresponds to a blue valley in the original.

For a given value of I_{b2} there are two absorption peaks that correspond to transitions:

$$|\,0\rangle_1 \otimes |\,0\rangle_2 \rightarrow \frac{1}{\sqrt{2}} \left(|\,0\rangle_1 \otimes |\,1\rangle_2 \pm |\,1\rangle_1 \otimes |\,0\rangle_2 \right). \tag{5.78}$$

Avoided crossing indicates existence of entangled states.

The tiny white circles mark the exact measured locations of the absorption peaks. The solid white lines mark theoretically predicted[3] locations of the absorption peaks for the transitions to the entangled states investigated here. The black dashed lines correspond to theoretically evaluated transitions between the ground state $|\,0\rangle_1 \otimes |\,0\rangle_2$ and the two *unentangled* states $|\,0\rangle_1 \otimes |\,1\rangle_2$ and $|\,1\rangle_1 \otimes |\,0\rangle_2$. The black lines cross, whereas the white ones do not. It is the experimental observation of the avoided crossing that confirms the existence of the entangled states.

The agreement between the measured data and the theoretical predictions based on the assumption that the transitions are from the ground state to one of the two *entangled* states is exceptional.

Lifetime of the entangled states

The measured dependency of the transition rate on the bias current I_{b2} can be

[3]The theoretical analysis of the biqubit and the resulting computations are nontrivial and beyond the scope of this text. Interested readers will find more details in [11].

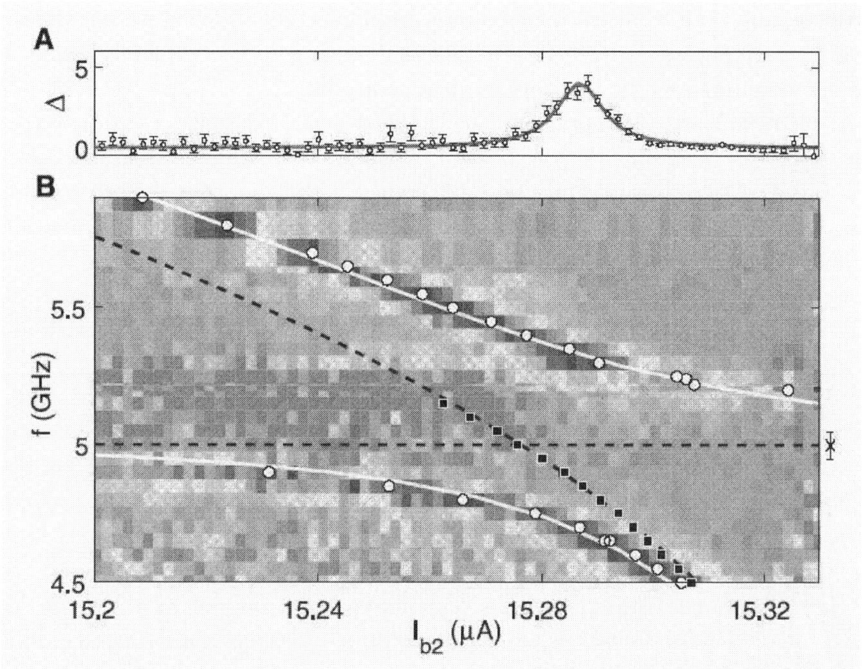

Figure 5.2: (A) shows the absorption spectrum of the biqubit for a fixed microwave frequency $f = 4.7\,\text{GHz}$ in function of the Josephson junction bias current I_{b2}. (B) is a two-dimensional map obtained by varying both the microwave frequency f and the Josephson junction bias current I_{b2}. From [11]. Reprinted with permission from AAAS.

Fourier transformed into the frequency space. The width of the absorption peak in this space yields the lifetime of the entangled state, which turns out to be about 2 ns. After this time, the state decays into a mixture of separable states.

What is so remarkable about this beautiful experiment (and device) is that not only does it demonstrate the existence of entangled states but it also entangles two heavy *macroscopic* objects separated by a large *macroscopic* distance of 0.7 mm.[4] Cooling to near-absolute zero enables the entangled state of the two qubits to stretch this far. Cooling freezes out interactions with the environment that would otherwise destroy both the entanglement and the individual qubit states themselves.

[4] A similar demonstration was carried out by a group from the NEC Tsukuba laboratory [106].

We see here again that entangled states and quantum behavior are not restricted to very small objects such as elementary particles and to very small distances such as are encountered in the interiors of alcaline atoms.

Energy spectra are not enough.

On the other hand, if the only evidence in favor of the existence of entangled states were atomic or electronic device spectra, we might be justified in holding back our enthusiasm. After all, one could perhaps come up with another theory that would reproduce the observed spectra in some other way. To be truly convinced that entanglement is not just an artifact of quantum calculus but a physical phenomenon, we need more evidence. We need to construct an entangled state and then perform a full set of measurements on it so as to reproduce probability matrices such as the ones derived for the Bell states in Section 5.1.

Measurements of quantum correlations

Such measurements have been done for pairs of protons and pairs of photons, the latter separated by a distance as large as 600 m [5]. They have been done even for two macroscopic samples of caesium gas each comprising 10^{12} atoms [72]. And the Santa Barbara group did a measurement for a phase biqubit, but it took three years of technology and methodology improvements after the Maryland biqubit demonstration to get there.

A full set of quantum state measurements that reproduces its whole probability vector or density matrix is called quantum state tomography. It is a difficult measurement. In our quantronium example we saw only the r^z component of the polarization vector $\boldsymbol{r}$. But here we need to measure r^x, r^y, and r^z for both qubits as well as correlation coefficients x^{ij}.

The Santa Barbara researchers demonstrated full quantum state tomography for a single qubit first [135] and then for a biqubit [134]. The biqubit device they used in their experiments is shown in Figure 5.3.

Santa Barbara biqubit

A distinguishing feature of this device is qubit isolation. We no longer apply the bias, the measuring pulse, and the microwave signal directly to the qubit's circuit. Instead, each qubit is manipulated and measured through the loop inductance of $L = 850\,\mathrm{pH}$. The shunting capacitors that stabilize the qubits both have the capacitance $C = 1.3\,\mathrm{pF}$ and the coupling capacitor is $C_x = 3\,\mathrm{fF}$. Amorphous silicon nitride is used as a dielectric in the shunting capacitors, because its loss tangent is very small, on the order of 10^{-4}, which yields a fairly long energy relaxation time of about 170 ns.

When operating normally both qubits are biased as shown in Figure 5.3 (B), but they are conditioned so that only the two lowest energy levels, $|\,0\rangle$ and $|\,1\rangle$, are filled. When the qubits are measured, a strong current pulse I_z is applied that changes the junction bias as shown in Figure 5.3 (C). The $|\,1\rangle$ state is then flushed

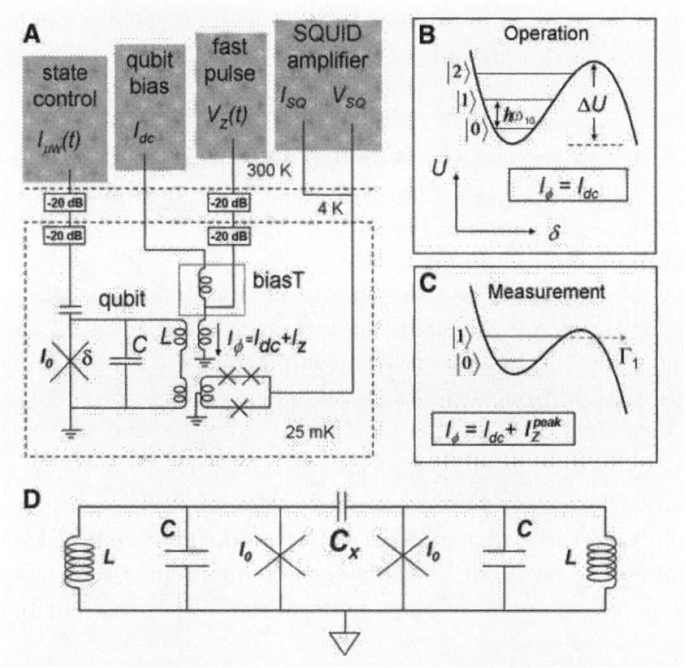

Figure 5.3: Schematic diagram of the Santa Barbara biqubit. (A) is a single qubit diagram that shows qubit manipulation and measurement circuitry, (B) and (C) are the LC resonator potential during qubit manipulation and measurement, respectively, and (D) is a simplified biqubit diagram that shows the coupling capacitor C_x. From [134]. Reprinted with permission from AAAS.

out of the cubic well, and this event is picked up by a SQUID amplifier.[5]

Both qubits are biased so that the transition between each qubit's $\mid 0\rangle$ and $\mid 1\rangle$ states occurs at $\omega_{10} = 2\pi \times 5.1\,\mathrm{GHz}$. An experiment begins by freezing and waiting—in the process both qubits drop naturally to $\mid 0\rangle$. Section 5.11.4, page 273, will explain in more detail how this happens, but at this stage we are content with the intuitive understanding that when things are left on their own in a cool place, they calm down. Now we flip qubit #2 by sending it an appropriate Rabi pulse of 10 ns duration. The biqubit ends up in the $\mid 0\rangle \otimes \mid 1\rangle$ state. Because this

[5] A SQUID (the acronym stands for Superconducting Quantum Interference Device) is another extremely sensitive Josephson junction device that is used to measure changes in the magnetic field flux.

state is not an eigenstate of the biqubit Hamiltonian, the biqubit evolves as follows:

$$|\,\Psi_{12}(t)\rangle = \cos\left(\frac{St}{2\hbar}\right)|\,0\rangle\otimes|\,1\rangle - \mathrm{i}\sin\left(\frac{St}{2\hbar}\right)|\,1\rangle\otimes|\,0\rangle, \tag{5.79}$$

where $S/h = 10\,\mathrm{MHz}$. So we don't have to do anything to rotate the biqubit other than wait a certain t. This is a common practice when manipulating biqubits and, by itself, constitutes a nontrivial biqubit gate. We will dwell on this some more in Section 6.2.2, page 296.

Probability measurements

Having waited a given t, we can measure the qubit; and by repeating the experiment 1,000 times, we can arrive at the probabilities p^{00}, p^{01}, p^{10}, and p^{11}. The observed probabilities are consistent with the idea that the biqubit becomes entangled after about 16 ns, forming a state described by

$$\frac{1}{\sqrt{2}}\left(|\,0\rangle\otimes|\,1\rangle - \mathrm{i}\,|\,1\rangle\otimes|\,0\rangle\right). \tag{5.80}$$

To fully diagnose the state and make sure that the qubit is indeed entangled, we have to measure the probabilities of finding the qubits in states such as $|\rightarrow\rangle$ and $|\otimes\rangle$ as well. In other words, we must perform the full tomography of the biqubit state.

We do this by subjecting the biqubit to a yet another Rabi pulse that rotates its individual qubits by 90° about the $\boldsymbol{e}_x$ or $\boldsymbol{e}_y$ directions prior to the measurement.

Repeating biqubit preparation and measurement procedures 20,000 times for each combination of directions, we obtain probabilities $p^{\uparrow\uparrow}$, $p^{\uparrow\rightarrow}$, and similar, and then assemble them into a density matrix shown in Figure 5.4.

The upper pannel of Figure 5.4 shows the results obtained from the raw probability data, and the lower pannel shows the density matrix "corrected" for known inefficiencies of single qubit measurements.

Fidelity

The expression

$$F = \mathrm{Tr}\sqrt{\boldsymbol{\sigma}^{1/2}\boldsymbol{\rho}_{\mathrm{exp}}\boldsymbol{\sigma}^{1/2}}, \tag{5.81}$$

where $\boldsymbol{\sigma}$ is the theoretically expected density matrix and $\boldsymbol{\rho}_{\mathrm{exp}}$ is the experimentally measured one, provides us with a convenient estimate of the combined accuracy of the state preparation and its tomography in the form of a single number. This number is called the *fidelity* of the reconstructed (from the measurements) quantum state. If $\boldsymbol{\rho}_{\mathrm{exp}} = \boldsymbol{\sigma}$, the fidelity is 100%.

The fidelity of the state reconstructed in Figure 5.4 (B), the one in the upper panel, is 75%. The fidelity of the state corrected for the known measurement inefficiencies of single qubits, shown in the lower panel, is 87%.

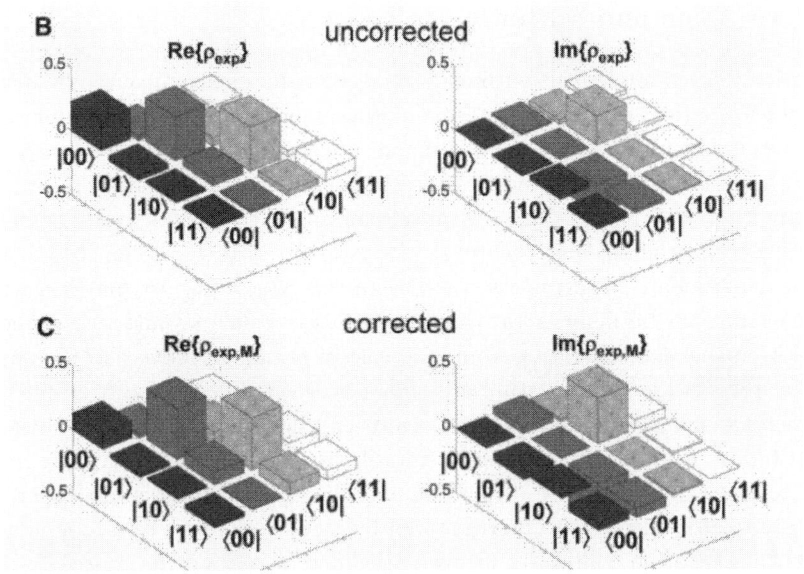

Figure 5.4: State tomography of the Santa Barbara biqubit. From [134]. Reprinted with permission from AAAS.

An in-depth theoretical analysis of the device that takes various environmental effects and known characteristics of the device into account shows that the fidelity should be 89%. This leaves only 3% of the *infidelity* unaccounted for, which is an impressive quantum device modeling result.

Is the observed state indeed entangled? An inspection of the density matrix suggests so. The imaginary components $\mid 01\rangle\langle 10 \mid$ and $\mid 10\rangle\langle 01 \mid$ have almost the same value as real components $\mid 01\rangle\langle 01 \mid$ and $\mid 10\rangle\langle 10 \mid$, which is what we would expect for the state $(\mid 0\rangle\otimes \mid 1\rangle - i \mid 1\rangle\otimes \mid 0\rangle)/\sqrt{2}$.

Are the qubits really entangled?

But isn't it possible that a mixture could be constructed that would get pretty close to the observed density matrix? Is there a way to demonstrate unequivocally that the observed state is indeed entangled by some well-posed criterion other than just looking at and comparing the bars on the graph? This turns out to be a nontrivial question, the answer to which was found only in 1996. In Section 5.10, page 247, we discuss a solution to this problem.

5.4 An Atom and a Photon

Demonstrating entanglement with elaborate quantum electronic circuits, although of obvious practical interest, may leave one pondering whether a short circuit or some specific circuit design feature is not responsible for the observed correlations rather than fundamental physics. After all, these are complicated devices. Their fabrication is difficult and their operation complex. This is, of course, not likely: numerous tests, checks, and theoretical analysis go into the design of the device and the experiment. Nevertheless, yielding to this possibility, we may ask whether entanglement can be demonstrated using just two atoms or just two elementary particles. Such a demonstration would, at least in principle, prove that the observed behavior reflects the law of nature and is not an electronic artifact.

Photons are good for entanglement experiments.

Numerous experiments of this type have been performed and are still being performed today. Photons are especially suitable because they are relatively immune to environmental decoherence—a photon may travel almost undisturbed across the whole observable universe, to be registered by an astronomer's telescope, still with sufficient information content to let us make inferences about its source.

The experiment discussed in this section was carried out by Blinov, Moehring, Duan, and Monroe from the University of Michigan in 2004 [13]. It is perhaps one of the cleanest and most elegant demonstrations of entanglement. In the experiment a single atom and a single photon emitted by the atom are entangled and measured.

The apparatus

Figure 5.5 shows a schematic diagram of an apparatus used in the measurement. A single positively charged ion of cadmium, $^{111}\mathrm{Cd}^+$, is held in an asymmetric-quadrupole radio frequency trap about 0.7 mm across, to which a magnetic field of approximately 0.7 Gauss is applied in order to provide the $\boldsymbol{e}_z$ direction. The ion is manipulated by a combination of optical and microwave pulses. In response to the manipulations the ion emits a single photon, which is collected by an $f/2.1$ imaging lens and directed toward a $\lambda/2$ waveplate. The waveplate is used to rotate the photon polarization, which in this setting is like switching from $p^{\uparrow_{\text{atom}}\uparrow_{\text{photon}}}$ to, say, $p^{\uparrow_{\text{atom}}\rightarrow_{\text{photon}}}$. The photon is then directed toward a polarizing beamsplitter, marked as "PBS" in the diagram, and then sent toward one of the two photon-counting photomultiplier tubes (PMTs) that can detect a single photon with about 20% efficiency. The PMT detector D1 is set up to detect photons polarized in the plane of the figure; these are called $|\,V\rangle$-photons. The other detector, D2, is set up to detect photons polarized in the plane perpendicular to the plane of the figure; these are called $|\,H\rangle$-photons. The $|\,V\rangle$ and $|\,H\rangle$ states of the photon are like qubit states $|\,0\rangle$ and $|\,1\rangle$.

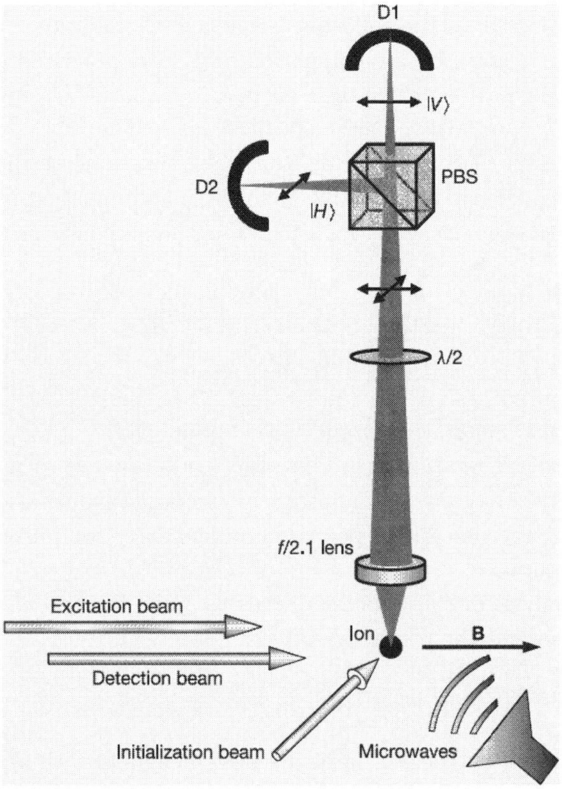

Figure 5.5: Schematic diagram of the apparatus used in the atom-photon entanglement experiment. Reprinted by permission from Macmillan Publishers Ltd: Nature [13], © 2004.

The purpose of the experiment is to demonstrate the quantum entanglement between the ion and the photon emitted by it. Whereas the state of the photon is measured by the PMT detectors shown on top of the diagram, reading the state of the ion is performed with a specially polarized 200 μs optical detection pulse beamed at the ion. The ion responds to the pulse by fluorescing differently depending on its state. Here the ion qubit read-out efficiency is greater than 95%.

The measurement

Prior to the read-out the ion's quantum state can be subjected to a Rabi rotation by irradiating it with a microwave pulse. In this way we can measure, say, $p^{\uparrow \text{atom} \uparrow \text{photon}}$ and $p^{\rightarrow \text{atom} \uparrow \text{photon}}$ as well.

State preparation

The ion is subjected to the following sequence of operations. First, the ion is initialized in the $|\uparrow\rangle$ state by a combination of a $30\,\mu$s polarized optical pulse and a $15\,\mu$s microwave rotation. Then, it is excited to a short-lived higher energy state called $^2\mathrm{P}_{3/2} \mid 2,1\rangle$ by a 50 ns polarized optical pulse. The $^2\mathrm{P}_{3/2} \mid 2,1\rangle$ state decays after about 3 ns either back to $|\uparrow\rangle$, or to a state with a somewhat higher energy, here called $|\downarrow\rangle$. The decay of $^2\mathrm{P}_{3/2} \mid 2,1\rangle$ to $|\uparrow\rangle$ is accompanied by the emission of an $\mid H\rangle$ photon, and the decay of $^2\mathrm{P}_{3/2} \mid 2,1\rangle$ to $|\downarrow\rangle$ is accompanied by the emission of a $\mid V\rangle$ photon:

$$^2\mathrm{P}_{3/2} \mid 2,1\rangle \;\rightarrow\; |\uparrow\rangle + \mid H\rangle, \tag{5.82}$$

$$^2\mathrm{P}_{3/2} \mid 2,1\rangle \;\rightarrow\; |\downarrow\rangle + \mid V\rangle. \tag{5.83}$$

The energy gap that separates $|\uparrow\rangle$ and $|\downarrow\rangle$ is about 1 MHz.

After the initial preparation procedure the ion is allowed to rest for about $1\,\mu$s and then is irradiated again with another $15\,\mu$s microwave rotation pulse. Finally the ion is irradiated with a $200\,\mu$s polarized optical detection pulse that lets us read the ion resident qubit.

Entangled state

Theoretical analysis of this process reveals that the ion and the emitted photon must be entangled and the resulting state is

$$\left(\mid H\rangle\otimes|\uparrow\rangle + 2\mid V\rangle\otimes|\downarrow\rangle\right)/\sqrt{3}. \tag{5.84}$$

Because the photon detector is markedly less efficient than the ion qubit detector, the experimenters measure probabilities of detecting an atomic qubit state $|\uparrow\rangle$ or $|\downarrow\rangle$ *conditioned* upon detecting an emitted photon either in the $\mid H\rangle$ or in the $\mid V\rangle$ states, given 1,000 successful trials.

Probability matrix

The results for the original basis, such that there was no atomic or photonic rotation prior to the measurement, are shown in Figure 5.6. Here we find that

$$\begin{pmatrix} p^{\uparrow,H} & p^{\uparrow,V} \\ p^{\downarrow,H} & p^{\downarrow,V} \end{pmatrix} \equiv \begin{pmatrix} p^{\uparrow\uparrow} & p^{\uparrow\downarrow} \\ p^{\downarrow\uparrow} & p^{\downarrow\downarrow} \end{pmatrix} = \begin{pmatrix} 0.97 \pm 0.01 & 0.06 \pm 0.01 \\ 0.03 \pm 0.01 & 0.94 \pm 0.01 \end{pmatrix}. \tag{5.85}$$

Now the experimenters rotate the $\lambda/2$ waveplate and the atomic qubit (by applying a microwave pulse after the emission) so as to rotate both through a Bloch angle of 90° eventually. The rotation of the atom-resident qubit is not clean, though. It is loaded with an additional phase factor due to the phase of the microwave signal. This is really the angle that the atomic qubit subtends with the photonic qubit in the $\boldsymbol{e}_x \times \boldsymbol{e}_y$ plane. The angle can be adjusted, and varying it results in the correlation fringes shown in Figure 5.7 (a). The fringes correspond to $p^{\rightarrow\rightarrow}$ and $p^{\rightarrow\leftarrow}$.

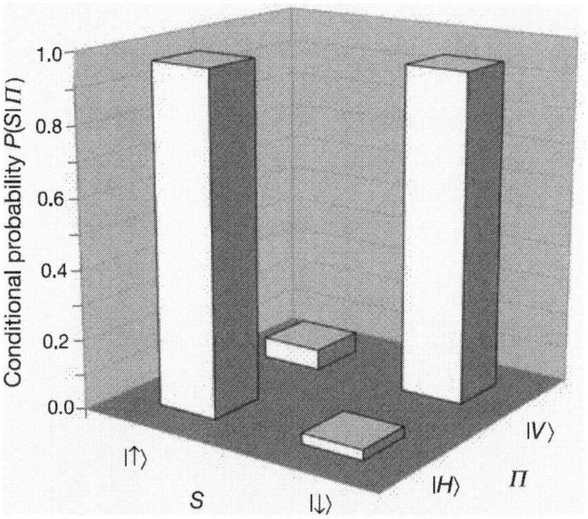

Figure 5.6: Measured conditional probabilities in the original basis—no atomic or photonic qubit rotation before the measurement. Reprinted by permission from Macmillan Publishers Ltd: Nature [13], © 2004.

By locking ourselves on the point of highest correlation in Figure 5.7 (a), we can finally arrive at the results shown in Figure 5.7 (b).

Here the probabilities are

$$\begin{pmatrix} p^{\rightarrow\rightarrow} & p^{\rightarrow\leftarrow} \\ p^{\leftarrow\rightarrow} & p^{\leftarrow\leftarrow} \end{pmatrix} = \begin{pmatrix} 0.89 \pm 0.01 & 0.06 \pm 0.01 \\ 0.11 \pm 0.01 & 0.94 \pm 0.01 \end{pmatrix}. \tag{5.86}$$

This is not full tomography, as we have seen done for the Santa Barbara qubit, but we can clearly observe correlations. Both qubits appear aligned for both measurement angles, whereas the probability of finding the qubits counteraligned is very low in both cases.

5.5 A Biqubit in a Rotated Frame

Rotating frames for particles A and B

Let us consider the Bell state $|\Psi^-\rangle_{AB}$ for which

$$\boldsymbol{p}_{AB} = \frac{1}{4}\left(\varsigma_{1A} \otimes \varsigma_{1B} - \varsigma_{xA} \otimes \varsigma_{xB} - \varsigma_{yA} \otimes \varsigma_{yB} - \varsigma_{zA} \otimes \varsigma_{zB}\right). \tag{5.87}$$

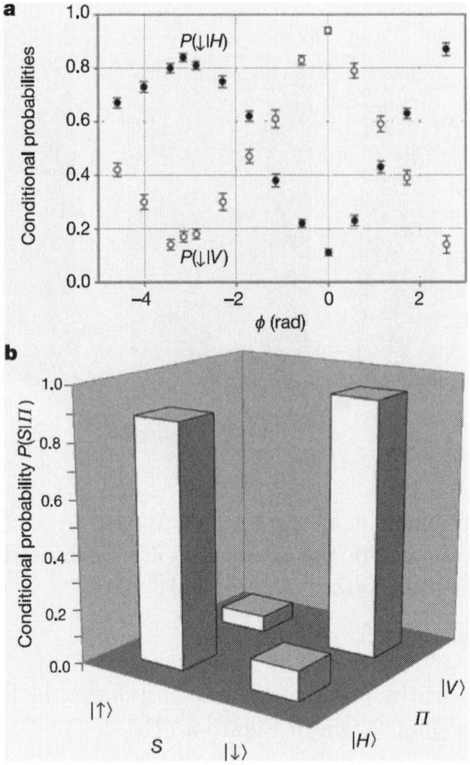

Figure 5.7: Measured conditional probabilities after a Bloch rotation of both qubits by 90°. Reprinted by permission from Macmillan Publishers Ltd: Nature [13], © 2004.

We have seen in Section 4.6, page 134, that Pauli vectors ς_i transform like normal three-dimensional vectors under rotations. So, if we were to rotate a polarization filter of qubit A in the $\boldsymbol{e}_x \times \boldsymbol{e}_z$ plane by angle θ_A, we'd find that

$$\varsigma_{xA} = \cos\theta_A \varsigma_{x'A} - \sin\theta_A \varsigma_{z'A} \quad \text{and} \tag{5.88}$$

$$\varsigma_{zA} = \sin\theta_A \varsigma_{x'A} + \cos\theta_A \varsigma_{z'A}. \tag{5.89}$$

Similarly, if we were to rotate the polarization filter of qubit B in the same plane by angle θ_B, we'd find that

$$\varsigma_{xB} = \cos\theta_B \varsigma_{x'B} - \sin\theta_B \varsigma_{z'B} \quad \text{and} \tag{5.90}$$

$$\varsigma_{zB} = \sin\theta_B \varsigma_{x'B} + \cos\theta_B \varsigma_{z'B}. \tag{5.91}$$

How do the two rotations affect probability readings for the biqubit?

We can find $\boldsymbol{p}'_{AB}$ by rotating its component varsigmas and remembering that

$$\varsigma_{1A} = \varsigma_{1'A}, \tag{5.92}$$
$$\varsigma_{yA} = \varsigma_{y'A}, \tag{5.93}$$
$$\varsigma_{1B} = \varsigma_{1'B}, \tag{5.94}$$
$$\varsigma_{yB} = \varsigma_{y'B}. \tag{5.95}$$

This yields

$$\begin{aligned}\boldsymbol{p}'_{AB} = \frac{1}{4}\Big(&\varsigma_{1'A} \otimes \varsigma_{1'B} - \varsigma_{y'A} \otimes \varsigma_{y'B} \\ &- (\cos\theta_A \varsigma_{x'A} - \sin\theta_A \varsigma_{z'A}) \otimes (\cos\theta_B \varsigma_{x'B} - \sin\theta_B \varsigma_{z'B}) \\ &- (\sin\theta_A \varsigma_{x'A} + \cos\theta_A \varsigma_{z'A}) \otimes (\sin\theta_B \varsigma_{x'B} + \cos\theta_B \varsigma_{z'B})\Big).\end{aligned} \tag{5.96}$$

Let us gather terms that multiply $\varsigma_{x'A} \otimes \varsigma_{x'B}$, $\varsigma_{x'A} \otimes \varsigma_{z'B}$, $\varsigma_{z'A} \otimes \varsigma_{x'B}$, and $\varsigma_{z'A} \otimes \varsigma_{z'B}$:

$$\begin{aligned}\boldsymbol{p}'_{AB} = \frac{1}{4}\Big(&\varsigma_{1'A} \otimes \varsigma_{1'B} - \varsigma_{y'A} \otimes \varsigma_{y'B} \\ &- (\cos\theta_A \cos\theta_B + \sin\theta_A \sin\theta_B)\, \varsigma_{x'A} \otimes \varsigma_{x'B} \\ &- (-\cos\theta_A \sin\theta_B + \sin\theta_A \cos\theta_B)\, \varsigma_{x'A} \otimes \varsigma_{z'B} \\ &- (-\sin\theta_A \cos\theta_B + \cos\theta_A \sin\theta_B)\, \varsigma_{z'A} \otimes \varsigma_{x'B} \\ &- (\sin\theta_A \sin\theta_B + \cos\theta_A \cos\theta_B)\, \varsigma_{z'A} \otimes \varsigma_{z'B}\Big).\end{aligned} \tag{5.97}$$

Because

$$\sin(\theta_A - \theta_B) = \sin\theta_A \cos\theta_B - \cos\theta_A \sin\theta_B \quad \text{and} \tag{5.98}$$
$$\cos(\theta_A - \theta_B) = \cos\theta_A \cos\theta_B + \sin\theta_A \sin\theta_B, \tag{5.99}$$

we find that

$$\begin{aligned}\boldsymbol{p}'_{AB} = \frac{1}{4}\Big(&\varsigma_{1'A} \otimes \varsigma_{1'B} - \varsigma_{y'A} \otimes \varsigma_{y'B} \\ &- \cos(\theta_A - \theta_B)\, \varsigma_{x'A} \otimes \varsigma_{x'B} \\ &- \sin(\theta_A - \theta_B)\, \varsigma_{x'A} \otimes \varsigma_{z'B}\end{aligned}$$

$$
\begin{aligned}
&+\sin(\theta_A-\theta_B)\,\varsigma_{z'A}\otimes\varsigma_{x'B}\\
&-\cos(\theta_A-\theta_B)\,\varsigma_{z'A}\otimes\varsigma_{z'B}\Big).
\end{aligned}
\tag{5.100}
$$

To extract the probabilities, we need to switch from the rotated Pauli vectors $\varsigma_{i'}$ to the rotated canonical basis vectors $\boldsymbol{e}_{\alpha'}, \alpha' = 0, 1, 2, 3$, as we did in Section 5.1. This leads to the following mappings between $\varsigma_{i'A}\otimes\varsigma_{j'B}$ and probability matrices:

$$
\varsigma_{1'A}\otimes\varsigma_{1'B} = \begin{pmatrix} 1 & 1 & 1 & 1\\ 1 & 1 & 1 & 1\\ 1 & 1 & 1 & 1\\ 1 & 1 & 1 & 1 \end{pmatrix}, \tag{5.101}
$$

$$
\varsigma_{x'A}\otimes\varsigma_{x'B} = \begin{pmatrix} 0 & 0 & 0 & 0\\ 0 & 0 & 0 & 0\\ 0 & 0 & 1 & 0\\ 0 & 0 & 0 & 0 \end{pmatrix}, \tag{5.102}
$$

$$
\varsigma_{y'A}\otimes\varsigma_{y'B} = \begin{pmatrix} 0 & 0 & 0 & 0\\ 0 & 0 & 0 & 0\\ 0 & 0 & 0 & 0\\ 0 & 0 & 0 & 1 \end{pmatrix}, \tag{5.103}
$$

$$
\varsigma_{z'A}\otimes\varsigma_{z'B} = \begin{pmatrix} 1 & -1 & 0 & 0\\ -1 & 1 & 0 & 0\\ 0 & 0 & 0 & 0\\ 0 & 0 & 0 & 0 \end{pmatrix}. \tag{5.104}
$$

Additionally we need to find matrices for $\varsigma_{x'A}\otimes\varsigma_{z'B}$ and $\varsigma_{z'A}\otimes\varsigma_{x'B}$. Since $\varsigma_{x'A,B} = \boldsymbol{e}_{2'A,B}$ and $\varsigma_{z'A,B} = \boldsymbol{e}_{0'A,B} - \boldsymbol{e}_{1'A,B}$, it is easy to see that

$$
\varsigma_{x'A}\otimes\varsigma_{z'B} = \begin{pmatrix} 0 & 0 & 0 & 0\\ 0 & 0 & 0 & 0\\ 1 & -1 & 0 & 0\\ 0 & 0 & 0 & 0 \end{pmatrix} \quad \text{and} \tag{5.105}
$$

$$
\varsigma_{z'A}\otimes\varsigma_{x'B} = \begin{pmatrix} 0 & 0 & 1 & 0\\ 0 & 0 & -1 & 0\\ 0 & 0 & 0 & 0\\ 0 & 0 & 0 & 0 \end{pmatrix}. \tag{5.106}
$$

Combining all these results yields

$$
\boldsymbol{p}'_{AB} = \frac{1}{4}\left(\begin{pmatrix} 1 & 1 & 1 & 1\\ 1 & 1 & 1 & 1\\ 1 & 1 & 1 & 1\\ 1 & 1 & 1 & 1 \end{pmatrix} - \begin{pmatrix} 0 & 0 & 0 & 0\\ 0 & 0 & 0 & 0\\ 0 & 0 & 0 & 0\\ 0 & 0 & 0 & 1 \end{pmatrix}\right.
$$

$$
\begin{aligned}
& - \cos(\theta_A - \theta_B) \begin{pmatrix} 0 & 0 & 0 & 0 \\ 0 & 0 & 0 & 0 \\ 0 & 0 & 1 & 0 \\ 0 & 0 & 0 & 0 \end{pmatrix} \\
& - \cos(\theta_A - \theta_B) \begin{pmatrix} 1 & -1 & 0 & 0 \\ -1 & 1 & 0 & 0 \\ 0 & 0 & 0 & 0 \\ 0 & 0 & 0 & 0 \end{pmatrix} \\
& - \sin(\theta_A - \theta_B) \begin{pmatrix} 0 & 0 & 0 & 0 \\ 0 & 0 & 0 & 0 \\ 1 & -1 & 0 & 0 \\ 0 & 0 & 0 & 0 \end{pmatrix} \\
& + \sin(\theta_A - \theta_B) \begin{pmatrix} 0 & 0 & 1 & 0 \\ 0 & 0 & -1 & 0 \\ 0 & 0 & 0 & 0 \\ 0 & 0 & 0 & 0 \end{pmatrix} \Bigg) . \qquad (5.107)
\end{aligned}
$$

We could just add it all up now, but it is convenient at this stage to make use of the following trigonometric identities:

$$
\frac{1 - \cos(\theta_A - \theta_B)}{2} = \sin^2 \frac{\theta_A - \theta_B}{2}, \qquad (5.108)
$$

$$
\frac{1 + \cos(\theta_A - \theta_B)}{2} = \cos^2 \frac{\theta_A - \theta_B}{2}. \qquad (5.109)
$$

Similar formulas for $1 \pm \sin(\theta_A - \theta_B)$ are somewhat clumsier because we end up with a 90° angle thrown in:

$$
\frac{1 - \sin(\theta_A - \theta_B)}{2} = \cos^2 \frac{\theta_A - \theta_B + 90^\circ}{2}, \qquad (5.110)
$$

$$
\frac{1 + \sin(\theta_A - \theta_B)}{2} = \sin^2 \frac{\theta_A - \theta_B + 90^\circ}{2}. \qquad (5.111)
$$

Let $\theta_A - \theta_B = \theta_{AB}$ for short. Then we obtain the following result:

$$
\boldsymbol{p}'_{AB} = \frac{1}{2} \begin{pmatrix} \sin^2 \frac{\theta_{AB}}{2} & \cos^2 \frac{\theta_{AB}}{2} & \sin^2 \frac{\theta_{AB}+90^\circ}{2} & \frac{1}{2} \\ \cos^2 \frac{\theta_{AB}}{2} & \sin^2 \frac{\theta_{AB}}{2} & \cos^2 \frac{\theta_{AB}+90^\circ}{2} & \frac{1}{2} \\ \cos^2 \frac{\theta_{AB}+90^\circ}{2} & \sin^2 \frac{\theta_{AB}+90^\circ}{2} & \sin^2 \frac{\theta_{AB}}{2} & \frac{1}{2} \\ \frac{1}{2} & \frac{1}{2} & \frac{1}{2} & 0 \end{pmatrix}. \qquad (5.112)
$$

For $\theta_A = \theta_B$, so that $\theta_{AB} = 0$, the probability matrix does not change at all, which implies that for the Bell state $|\,\Psi^-\rangle_{AB}$ the probability of finding both qubits aligned in *any* direction—not just the original $\boldsymbol{e}_x$, $\boldsymbol{e}_y$, and $\boldsymbol{e}_z$—is zero.

Quantum correlations for angles other than 0° and 90°

But evaluating probabilities at angles other than $\theta_{AB} = 0$ or $\theta_{AB} = 90°$ (which is what basically sits in terms such as $p^{\uparrow\rightarrow}$) reveals something peculiar. Let us evaluate

$$p^{\uparrow\nearrow} + p^{\nearrow\rightarrow} - p^{\uparrow\rightarrow}, \tag{5.113}$$

where $\nearrow$ stands for a polarization axis that is tilted by 45°. This can be rewritten as

$$p^{\uparrow\uparrow}(0°, 45°) + p^{\uparrow\uparrow}(45°, 90°) - p^{\uparrow\uparrow}(0°, 90°), \tag{5.114}$$

where the first angle in the bracket is θ_A and the second angle is θ_B, or in terms of θ_{AB}:

$$\begin{aligned} & p^{\uparrow\uparrow}(45°) + p^{\uparrow\uparrow}(45°) - p^{\uparrow\uparrow}(90°) \\ & = \frac{1}{2}\sin^2\frac{45°}{2} + \frac{1}{2}\sin^2\frac{45°}{2} - \frac{1}{2}\sin^2\frac{90°}{2} \\ & = \sin^2(\pi/8) - 0.5 \cdot \sin^2(\pi/4) = -0.10355. \end{aligned} \tag{5.115}$$

So what?

The fact that this number is negative has dumbfounded a whole generation of physicists. It tells us something unexpected about the world of quantum physics.

5.6 Bell Inequality

Let us consider

$$p^{\uparrow_A, \nearrow_B} + p^{\nearrow_A, \rightarrow_B} \tag{5.116}$$

for the Bell state $| \Psi^-\rangle_{AB}$. We have marked the arrows clearly with A and B, because soon we are going to replace B with A.

Entanglement lets us measure qubit A twice.

For the Bell state $| \Psi^-\rangle_{AB}$ qubit B is always an inverted image of qubit A. When measured, it always points in the opposite direction, regardless of which measurement direction we choose.

So, we may hypothesize that if we were to measure qubit A first, this would force qubit B automatically into a reverse image of A. Therefore, performing a measurement on A first and then on B is really like performing a measurement on A, then flipping what comes out and measuring it again the way that B would be measured.

But we could just as well reverse the order of measurements on A and B or perform both measurements simultaneously. The result in terms of probabilities still ought to be the same, as long as there isn't enough time between one and the other measurement for the other qubit to interact with the environment, since this may change the state of the other qubit in a way that no longer depends on the first qubit only.

Some quantum mechanics purists object to this reasoning, saying that quantum mechanics does not allow for a single qubit to be measured twice against polarizers at two different angles, and this is indeed true. But here the same quantum mechanics provides us with a mechanism that lets us overcome this restriction—the mechanism is entanglement, entanglement that makes qubit B an inverted copy of qubit A. By performing simultanous measurements on A and B we can, in effect, measure, say, A against two different angles at the same time.

Let us then substitute a counteraligned qubit A in place of qubit B:

$$\nearrow_B \longrightarrow \swarrow_A, \tag{5.117}$$

$$\rightarrow_B \longrightarrow \leftarrow_A . \tag{5.118}$$

But if qubit A passes through the $\swarrow$ polarizer, it is the same as to say that it would *not* pass through the $\nearrow_A$ polarizer, so, in terms of the actual probabilities, we can write

$$p^{\swarrow_A} = p^{\neg\nearrow_A}, \tag{5.119}$$

where $\neg$ is a Boolean *not*[6].

Similarly,

$$p^{\leftarrow_A} = p^{\neg\rightarrow_A}. \tag{5.120}$$

In effect, we can rewrite our original expression (5.116) as

$$p^{\uparrow_A,\neg\nearrow_A} + p^{\nearrow_A,\neg\rightarrow_A}. \tag{5.121}$$

The first term of (5.121) is the probability that qubit A passes through the $\uparrow$ polarizer and at the same time fails to pass through the $\nearrow$ polarizer. The second term is the probability that qubit A passes through the $\nearrow$ polarizer and at the same time fails to pass through the $\rightarrow$ polarizer.

The following—though, as it will turn out, physically incorrect—argument asserts that the sum of the two probabilities should be no less than the probability that qubit A passes through the $\uparrow$ polarizer and at the same time fails to pass through the $\rightarrow$ polarizer, $p^{\uparrow_A,\neg\rightarrow_A}$. Why should it be so?

The assumption behind the Bell inequality

We assume that there is a device inside a qubit that *determines* whether the qubit passes or does not pass through the polarizer, depending on its internal state,

[6] A fiducial vector of probabilities is always normalized so that for the principal direction, such as $\nearrow$, we have that $p^{\nearrow_A} + p^{\swarrow_A} = p^{\nearrow_A} + p^{\neg\nearrow_A} = 1$.

which is classical and well defined. A qubit that passes through the $\uparrow$ polarizer and at the same time fails to pass through the $\rightarrow$ one—let us group such qubits into a $\{\uparrow, \neg\rightarrow\}$ set—*may or may not* pass through the $\nearrow$ polarizer, if we were to subject it to such a measurement, depending on the condition of its internal device—strictly speaking, we would need to have a triqubit system to perform such experiment. We can therefore split the $\{\uparrow, \neg\rightarrow\}$ set into two *disjoint* subsets,

$$\{\uparrow, \neg\rightarrow\} = \{\uparrow, \nearrow, \neg\rightarrow\} \cup \{\uparrow, \neg\nearrow, \neg\rightarrow\}, \tag{5.122}$$

assuming at the same time that

$$\{\uparrow, \nearrow, \neg\rightarrow\} \cap \{\uparrow, \neg\nearrow, \neg\rightarrow\} = \varnothing, \tag{5.123}$$

where $\varnothing$ is the empty set.

The condition of the qubit's internal device splits all qubits, not only the ones accounted for above, into two disjoint sets: Those that pass through the $\nearrow$ polarizer, which form a $\{\nearrow\}$ set, and those that do not, which form a $\{\neg\nearrow\}$ set. And, since both sets complement each other, we have that

$$\{\nearrow\} \cup \{\neg\nearrow\} = \Omega \quad \text{and} \tag{5.124}$$

$$\{\nearrow\} \cap \{\neg\nearrow\} = \varnothing, \tag{5.125}$$

where Ω is the set of all qubits.

Qubits that have passed through the $\uparrow$ polarizer but fail to pass through the $\nearrow$ polarizer form a $\{\uparrow, \neg\nearrow\}$ set, which is contained in the $\{\neg\nearrow\}$ set,

$$\{\uparrow, \neg\nearrow\} \subset \{\neg\nearrow\}. \tag{5.126}$$

Qubits that have failed to pass through the $\rightarrow$ polarizer but pass through the $\nearrow$ one form a $\{\nearrow, \neg\rightarrow\}$ set, which is contained in the $\{\nearrow\}$ set,

$$\{\nearrow, \neg\rightarrow\} \subset \{\nearrow\}. \tag{5.127}$$

Clearly,

$$\{\uparrow, \neg\nearrow, \neg\rightarrow\} \subset \{\uparrow, \neg\nearrow\}, \tag{5.128}$$

and

$$\{\uparrow, \nearrow, \neg\rightarrow\} \subset \{\nearrow, \neg\rightarrow\}. \tag{5.129}$$

This is illustrated in Figure 5.8.

From the figure we can easily see that

$$\{\uparrow, \neg\rightarrow\} = (\{\uparrow, \nearrow, \neg\rightarrow\} \cup \{\uparrow, \neg\nearrow, \neg\rightarrow\}) \subset (\{\uparrow, \neg\nearrow\} \cup \{\nearrow, \neg\rightarrow\}). \tag{5.130}$$

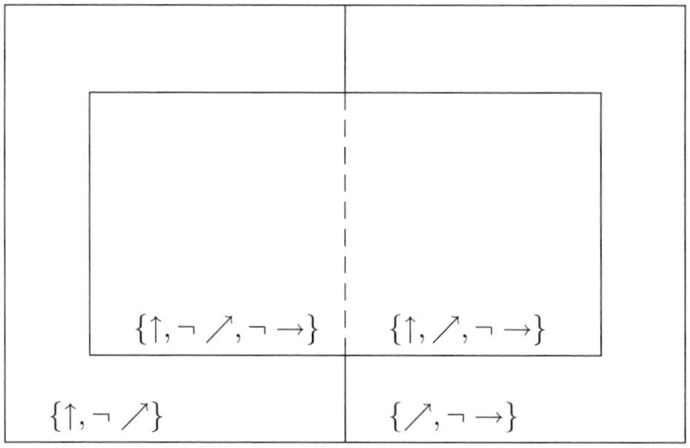

Figure 5.8: A classical look at the Bell inequality. We have four boxes labeled by the sets, with qubits belonging to one or more. Boxes labeled by $\{\uparrow, \neg \nearrow\}$ and $\{\nearrow, \neg \rightarrow\}$ are *disjoint*. The inner box corresponds to qubits in the $\{\uparrow, \neg \rightarrow\}$-category. We subdivide it into two *disjoint* boxes labeled by $\{\uparrow, \neg \nearrow, \neg \rightarrow\}$ and $\{\uparrow, \nearrow, \neg \rightarrow\}$. Qubits in the left box, labeled by $\{\uparrow, \neg \nearrow, \neg \rightarrow\}$ also belong to the $\{\uparrow, \neg \nearrow\}$ box and qubits in the right box, labeled by $\{\uparrow, \nearrow, \neg \rightarrow\}$ also belong to the $\{\nearrow, \neg \rightarrow\}$ box.

We convert this to probabilities, by taking counts, C, of each set and dividing them by the number of all qubits, $C(\Omega)$.

$$\frac{C(\{\uparrow, \neg \rightarrow\})}{C(\Omega)} \leq \frac{C(\{\uparrow, \neg \nearrow\})}{C(\Omega)} + \frac{C(\{\nearrow, \neg \rightarrow\})}{C(\Omega)}, \tag{5.131}$$

where we continue to make use of the assumption that $\{\uparrow, \neg \nearrow\}$ and $\{\nearrow, \neg \rightarrow\}$ are disjoint. This translates into

$$p^{\uparrow_A, \neg \rightarrow_A} \leq p^{\uparrow_A, \neg \nearrow_A} + p^{\nearrow_A, \neg \rightarrow_A}. \tag{5.132}$$

Substituting qubit B back in place of qubit A, and rearranging all probabilities to one side, yields *Bell inequality*

$$p^{\uparrow_A, \nearrow_B} + p^{\nearrow_A, \rightarrow_B} - p^{\uparrow_A, \rightarrow_B} \geq 0. \tag{5.133}$$

Ah, but this contradicts equation (5.115), where we have found that

$$p^{\uparrow_A, \nearrow_B} + p^{\nearrow_A, \rightarrow_B} - p^{\uparrow_A, \rightarrow_B} < 0. \tag{5.134}$$

What is amiss?

Inequality (5.133) is one of the celebrated Bell inequalities. Bell and others—for example, Clauser, Hold, Horne, and Shimony [23]—and more recently Greenberger, Horne, and Zeilinger [54] [53] and Hardy [59]—demonstrated several such inequalities and other algebraic expressions that purport to flesh out the difference between classical and quantum physics.

We cannot split qubits into disjoint sets until after the measurement.

Let us have a closer look at the reasoning that led us to inequality (5.133). The way we transformed biqubit probabilities into single-qubit probabilities was actually OK, even though it looked somewhat unnerving. The resulting probabilities all check out. What does not check out is the division of qubits into *disjoint* sets depending on the hypothetical state of some internal device, and not on the outcome of the actual measurement. In the quantum reality the same qubit in the same quantum state may pass through the $\nearrow$ polarizer and then it may not. Each qubit, therefore, belongs both to the $\{\nearrow\}$ set and to the $\{\neg\nearrow\}$ set, until the actual measurement is made. But we do not make this measurement in this experiment. We use the concepts of $\{\uparrow, \nearrow, \neg\rightarrow\}$ and $\{\uparrow, \neg\nearrow, \neg\rightarrow\}$ only as an intermediate step in our reasoning. Therefore the simple analysis that leads to inequality (5.133) is physically incorrect. We've been trapped into classical thinking about qubits and biqubits in terms of *things* that can be divided into sets based on an internal switch that *is going to* determine their behavior.

There is no such switch.

We cannot divide qubits on grounds other than the counts actually registered in an experiment. If we do not carry out the measurement, the property in question does not exist, and hence it cannot be used to split the set of all qubits.

Explanation of the inequality in terms of the polarization vector

What does exist is vector $\boldsymbol{r}$ that describes a quantum state of each qubit in the $\{\uparrow, \neg\rightarrow\}$ set. If the vector points in a direction other than $\nearrow$ or $\swarrow$, and we may expect this to be the case for almost every qubit in this set, then there is a nonzero probability that the qubit will pass through the $\nearrow$ polarizer, but there is also a non-zero probability that it will not. So, the *same qubit* in the $\{\uparrow, \neg\rightarrow\}$ set may contribute to both the $\{\uparrow, \neg\nearrow\}$ set and to the $\{\nearrow, \neg\rightarrow\}$ set, which are disjoint, because they are based on the actual measurement. This has the effect of *swelling the volume* of the $\{\uparrow, \neg\rightarrow\}$ set with respect to the hypothetical $\nearrow$ measurement, and so in some cases we end up with the violation of the Bell inequality.

The Bell inequality holds for some angles.

The Bell inequality (5.133) holds for some angles; in particular it holds for θ_{AB} of $0°$, $90°$, and $180°$, that is, the angles we have in our basic fiducial matrix of a biqubit.[7] But it does not hold for some other angles, most notably for the

[7] This is why the problem was not noticed in 1935 when Einstein, Podolsky, and Rosen first considered a biqubit measurement of the kind discussed here. It was only in 1964 that Bell noticed the discrepancy between predictions of quantum mechanics and predictions based on "local

combination of θ_{AB} of 45° and 90°.

What is it, then, that the Bell inequality (5.133) and its violation in quantum physics illustrated by inequality (5.115) tell us?

What does Bell inequality violation tell us?

They tell us more than one thing, and this is the problem. The discrepancy is the result of various aspects of quantum physics combining, which leads to confusion.

At first glance we may conclude—and this is the commonly accepted lore—that the physical process underlying the projection of a qubit in some quantum state onto a specific direction of a polarizer is truly random, that is, that there cannot be a hidden device inside the qubit that *determines* (with 100% accuracy) how the qubit is going to align. In quantum mechanics parlance we say that

> no deterministic local hidden parameters theory can explain the quantum mechanical result,

the "local hidden parameter" being the hypothetical device that sits inside the qubit.

But this result does not exclude the possiblity of nondeterministic local hidden parameters, and it does not exclude the possiblity of deterministic nonlocal hidden parameters either.

Polarization vector is a non deterministic, local, hidden parameter.

We do have a "local hidden parameter" inside a qubit. It is vector $\boldsymbol{r}$. It is "hidden": we cannot get at it in a single measurement. To evaluate all three components of $\boldsymbol{r}$, we must explore the whole statistical ensemble of the qubit with instruments designed to measure them, and no instrument can measure all three at once either. The parameter, however, does not *determine* how a qubit is going to behave when measured. It provides us only with probabilities of various outcomes. It is not a deterministic parameter.

Non local, deterministic, hidden parameter theory is another option.

We also have a deterministic nonlocal theory, much favored by Bell [8], that can be used to reproduce violations of Bell inequalities and other similar expressions [36], which is due primarily to Erwin Schrödinger (1887–1961), Louis de Broglie (1982–1987), David Bohm (1917-1992) and Basil Hiley [14]. The theory has some troubling implications and is not well known. Some physicists may have heard of it, but few studied it in depth. This is a great pity because the theory is physically and logically unassailable on account of being derived entirely and solely from the fundamental equations of quantum mechanics, so all that quantum mechanics predicts, the theory predicts too.

The theory provides "classical dynamic explanations" for all quantum phenomena such as spin, probabilities, measurements, collapse of the wave function, interference

---

realism" and encapsulated the discrepancy in the form of experimentally testable inequalities.

fringes, and nonlocality—and, who knows, it may even be true. But true or not, it makes good reading and gives one plenty of food for thought.

The troubling implication of the theory is that on the fundamental level it treats the whole universe like a single indivisible object—nothing can be truly isolated from the rest. Such isolation and subsequent identification of the isolated components as, for example, "individual electrons," becomes possible in thermodynamic limit only, hence the title of Bohm's and Hiley's book *The Undivided Universe.* The other troubling implication is a violation of special relativity on the fundamental level. But special relativity is still recovered on the level of expectation values, which is what we see in the macroscopic domain.

Yet, these are the implications of quantum theory taken to its logical conclusion, even if somewhat beyond the point where most physicists are prepared to go—and, needless to say, we have already arrived at quite similar conclusions when studying the inseparability of a biqubit. These will be explored further in the next section.

Why this particular picture should be more troubling to a majority of physicists than, say, multiple universes, for which there is not a shred of experimental evidence, or geometric dimensions in excess of 3 + 1, for which there is not a shred of experimental evidence either, is hard to tell. If, as Einstein commented, Bohm's theory (not called "Bohm's" back then) is *too cheap*, the other theories seem far *too expensive*. At the same time they all, including Bohm's, suffer from the shared fundamental malady of dragging macroscopic, classical concept of space-time into their framework.

5.7 Nonlocality

When applied to a biqubit system in the Bell state $| \Psi^- \rangle_{AB}$, the violation of Bell inequality (5.133) illustrates one more aspect of quantum physics: the non locality of a biqubit.

What does it mean?

Quantum mechanical description of a biqubit ignores distance.

The quantum mechanical description of a biqubit is extremely primitive. There is nothing here about the actual physical location of the biqubit components. Let us suppose that both qubit components are separated by a large *macroscopic* distance, for example, 600 m, as has been demonstrated in a fairly recent photon experiment by Aspelmeyer and his 12 colleagues from the Institut für Experimentalphysik, Universität Wien in Austria [5], or even a planetary-scale distance, as is planned for a forthcoming satellite-based experiment. Can it really be that qubit B measured characteristics end up being always opposite of qubit A's, if, as the Bell inequality tells us, they are made at random at the point of the measurement rather than due to some "deterministic hidden parameter" inside the qubit? How can qubit B know *instantaneously* what state qubit A has been filtered into? Surely, there ought to be some retardation terms inserted into the probability matrix, to the effect that qubit B would learn about qubit A's encounter with the polarization filter after, say, x_{AB}/c only, where x_{AB} is the distance between the qubits and c is the speed of light.

EPR paradox

This is exactly the objection that Einstein, Podolsky and Rosen brought up in 1935 [38]. Their conclusion was that since no information could travel faster than light, there had to be a "deterministic hidden parameter" inside a qubit that would predetermine the way both qubits would interact with the polarizing filters. Bell inequalities and numerous subsequent experiments [4] demonstrated clearly that there could not be such a "deterministic hidden parameter" inside a qubit. At the most fundamental level quantum mechanics of multiple qubits seems to contradict special theory of relativity. This somewhat superficial observation is further confirmed by a more formal proof provided by Hardy [59] and the realization of Hardy's *gedanken* experiment by Irvine, Hodelin, Simon, and Bouwmeester [69].

It is well worth having a quick look at how the Bohm theory explains what happens.

Bohm's explanation

In the Bohm theory every quantum particle is associated with a field, called *quantum potential*, that stretches all the way to infinity and does not diminish with distance. The field can be derived from the Schrödinger and Dirac equations on fairly standard substitutions similar to what physicists do within the so-called WKB approximation. But in the Bohm theory we don't approximate. We calculate things exactly. Taken to this level, fundamental equations of quantum mechanics can be interpreted as equations that describe congruences of trajectories as determined by the quantum potential and various other externally applied fields, electric, magnetic, gravitational. The interaction of the particle with the quantum potential is instantaneous, like in the Newtonian theory of gravity, meaning that whatever the field "touches" and however far away has an instantaneous effect on the particle.

A measuring apparatus is also a quantum object, and so it has a quantum potential field associated with it, too.

A biqubit confronted by two widely spaced polarizers is a system of two qubits and two polarizers all joined with the fabric of the shared nonlocal quantum potential. Whatever happens to qubit A is immediately and instantaneously transmitted to qubit B and vice versa—but not only this. A configuration of both polarizers is also transmitted to both qubits, even before they arrive at their respective points of measurement, and has an effect on how they align.

A heuristic analysis presented in [14], as well as detailed calculations presented by Durt and Pierseaux [36], shows how this mutual and instantaneous coupling of all four partners results in the violation of Bell inequalities and other similar expressions.

The model is quite telling. Not only does it point to the instantaneous interaction between qubits A and B, something that we are forced to expect as soon as we learn about entanglement, and find about its various experimental demonstrations, but, just as important, it tells us that we must be careful when thinking about the act of measurement itself. It is a physical interaction between two or more physical systems. If all participants to the measurement have nonlocal feelers, we have to consider the possibility of a biqubit adjusting itself to a configuration of the measuring instruments prior to the actual measurement.

The resulting "conspiracy of nature" gives us the quantum reality that puzzles us so at every step.

It is not necessary to believe in the Bohm's theory to appreciate various important and interesting points the model makes, just as it is not necessary to be a Christian to appreciate wisdom of Christ's parables.

How else can we think of an entangled biqubit?

Let's get rid of space-time.

Another approach would be to be more radical about the notion of space-time itself. After all, space-time is a macroscopic construction that requires macroscopic rulers and clocks to define. But we can't take these into the quantum domain, so we should not drag the classical fabric of space-time into the quantum domain either. Yet this is what just about all present-day theories do, perhaps with the notable exception of Smolin's "loop quantum gravity" [130] and "spin networks" of Roger Penrose [110].

How could we replace classical macroscopic space-time in the quantum domain? We could think of a graph of interactions. Quantum systems, qubits, biqubits, and n-qubits would not be embedded in any space-time. There would not be any distance between them. Instead they would exchange various properties with each other. In some cases the exchanges would be intense; in other cases they would be weak. The exchanges might be ordered, too, though not necessarily strongly. In the thermodynamic limit the graph may turn into space and time. The weakly or seldom interacting quantum objects might appear as being far away from one other, the strongly or frequently interacting ones as being close. The ordering of the graph might turn into macroscopic time. Projections of interactions between quantum systems onto the macroscopic time and space so constructed might acquire some randomness, which is what we, the creatures of Macroscopia, see when we look at the quantum world.

Everything in the system would stay together at the most fundamental level, and this would explain the nonlocality of biqubits—their eventual separation into two independent particles occurring only at the macroscopic level as the result of interaction with great many other nodes of the graph.

Can it really be that all the universe, all physical reality, somehow exists in a single distance-less pot, and its macroscopically observable spatial and temporal extent is an illusion built from myriad interactions within the pot?

Photons experience neither time nor space.

Why not? We all know about photons. In a photon's system of reference time stops because of relativistic time dilation. A photon does not experience time. Similarly, in a photon's system of reference the whole world is squeezed into a point because of Lorentz contraction. From the photon's personal viewpoint it is everywhere at the same time. Could it be that the photon is right?[8]

In some sense this picture is not so distant from Bohm's theory. It is like Bohm's theory with the space and time taken out of it, so that its quantum potential *can* be non local.

It is not uncommon in theoretical physics that different conceptual and mathematical frameworks turn out to be equivalent and lead to identical physics.

At the end of the day none of the above may be true. But it is certainly true that Bell inequalities and quantum physics force us to radically revise our often naive notions about the nature of reality.

5.8 Single-Qubit Expectation Values

For a separable biqubit state described by $\boldsymbol{p}_A \otimes \boldsymbol{p}_B$ it is easy to extract probabilities that refer to just one of the qubits. We did something similar when defining energy form for a system of two separate noninteracting qubits in Section 5.1, page 191. The trick is to contract the biqubit with $\varsigma^1_{A,B}$ and make use of $\langle \varsigma^i, \varsigma_j \rangle = 2\delta^i{}_j$, where $i, j = 1, x, y, z$. And so we have

$$\langle \varsigma^1_B, \boldsymbol{p}_A \otimes \boldsymbol{p}_B \rangle = \boldsymbol{p}_A, \quad (5.135)$$
$$\langle \varsigma^1_A, \boldsymbol{p}_A \otimes \boldsymbol{p}_B \rangle = \boldsymbol{p}_B. \quad (5.136)$$

[8]Penrose attempted to capture this very idea in his twistor theory of 1967, which more recently, in 2003, was picked up by Witten, who developed it further into a *twistor string theory* [146]. The twistor-based Theory of Everything is (3+1)-dimensional, which has made some people, including Penrose, extremely happy, and others, who have already made bets on real estate in 10 dimensions, uncomfortable.

We can think of $\langle \varsigma^1_B, \boldsymbol{p}_A \otimes \boldsymbol{p}_B \rangle$ as an expectation value for the measurement ς^1_B on $\boldsymbol{p}_A \otimes \boldsymbol{p}_B$. Translating this into the language of quaternions, we find

$$2\Re\left(\mathbf{1}_B \cdot (\boldsymbol{\rho}_A \otimes \boldsymbol{\rho}_B)\right) = \boldsymbol{\rho}_A 2\Re\left(\mathbf{1}_B \cdot \boldsymbol{\rho}_B\right) = \boldsymbol{\rho}_A 2\Re \boldsymbol{\rho}_B = \boldsymbol{\rho}_A, \tag{5.137}$$

and, switching from quaternions to Pauli matrices,

$$\mathrm{Tr}_B\left(\boldsymbol{\rho}_A \otimes \boldsymbol{\rho}_B\right) = \boldsymbol{\rho}_A \mathrm{Tr}_B \boldsymbol{\rho}_B = \boldsymbol{\rho}_A. \tag{5.138}$$

Partial trace

The last expression can be rewritten as

$$\boldsymbol{\rho}_A = \mathrm{Tr}_B \boldsymbol{\rho}_{AB}, \tag{5.139}$$

where for a separable biqubit $\boldsymbol{\rho}_{AB} = \boldsymbol{\rho}_A \otimes \boldsymbol{\rho}_B$. Symbol Tr_B means "taking trace over the space dimensions that pertain to qubit B" and is referred to as a *partial trace* operation. In the physicists' parlance we also talk about *tracing particle B out* of the biqubit state.

Individual qubits in Bell states are chaotic.

An interesting insight is gained by tracing qubit B out of the Bell state $|\,\Psi^-\rangle_{AB}$. We remind the reader that

$$|\,\Psi^-\rangle_{AB} \equiv \frac{1}{4}\left(\varsigma_{1A} \otimes \varsigma_{1B} - \varsigma_{xA} \otimes \varsigma_{xB} - \varsigma_{yA} \otimes \varsigma_{yB} - \varsigma_{zA} \otimes \varsigma_{zB}\right). \tag{5.140}$$

Contracting it with ς^1_B yields

$$\frac{1}{4}\left(\varsigma_{1A} \cdot 2\right) = \frac{1}{2}\begin{pmatrix} 1 \\ 1 \\ 1 \\ 1 \end{pmatrix}, \tag{5.141}$$

which implies that $\boldsymbol{r}_A = 0$. This is a completely depolarized state, a total mixture. Similarly, we'd find that particle B, when looked at separately from particle A appears completely depolarized. Yet the biqubit itself is thoroughly polarized. It is in a *pure* state.

Several important conclusions can be drawn from the observation. The first conclusion is that this particular biqubit state, the Bell state $|\,\Psi^-\rangle_{AB}$, does not provide us with means of transferring useful information from a point where qubit A is measured to a point where qubit B is measured. At every point of measurement the measured qubit, be it A or B, appears completely chaotic. Only afterwards, when the results of the measurements for qubits A and B are *compared*, do we realize that the two qubits were entangled and in a *pure* biqubit state.

The second conclusion is that if we were to associate specific information with the biqubit, as we have associated 0 with $|\uparrow\rangle$ and 1 with $|\downarrow\rangle$ previously, the information would be contained in the *entanglement*, that is, in the biqubit correlations and not in the individual qubits of the system.

Mixed states in unitary formalism

The third conclusion is that although the unitary formalism does not, at first glance, let us discuss mixtures and therefore the act of measurement either, here we have a unitary model that captures a mixed state, too. It does so by entangling a qubit with another qubit. The result of the entanglement is that the state of each individual qubit becomes mixed. We can therefore generate mixed states of quantum subsystems within the unitary formalism by viewing them as part of larger and *pure* unitary systems. For this reason the procedure is called *purification.*

Purification

Purification lets us model the act of measurement as entanglement of the measured quantum object with the measuring apparatus—the quantum state of the whole remaining pure and unitary, but the quantum state of the measured object decaying into a mixture.

We will discuss this in more detail later, but first let us apply the $\varsigma^1_{A,B}$ measurement to a most general biqubit state, which, as we saw earlier, is

$$\begin{aligned} \boldsymbol{p}_{AB} &= \frac{1}{4}\Bigg(\varsigma_{1A}\otimes\varsigma_{1B} + \sum_{i=x,y,z} r^i_A\,\varsigma_{iA}\otimes\varsigma_{1B} + \sum_{i=x,y,z} r^i_B\,\varsigma_{1A}\otimes\varsigma_{iB} \\ &\quad + \sum_{i,j=x,y,z} x^{ij}_{AB}\,\varsigma_{iA}\otimes\varsigma_{jB}\Bigg). \end{aligned} \tag{5.142}$$

Contracting $\boldsymbol{p}_{AB}$ with ς^1_B yields

$$\boldsymbol{p}_A = \langle\varsigma^1_B, \boldsymbol{p}_{AB}\rangle = \frac{1}{2}\left(\varsigma_{1A} + \sum_{i=x,y,z} r^i_A\varsigma_{iA}\right), \tag{5.143}$$

and contracting $\boldsymbol{p}_{AB}$ with ς^1_A yields

$$\boldsymbol{p}_B = \langle\varsigma^1_A, \boldsymbol{p}_{AB}\rangle = \frac{1}{2}\left(\varsigma_{1B} + \sum_{i=x,y,z} r^i_B\varsigma_{iB}\right). \tag{5.144}$$

The meaning of $r^i_{A,B}$

The two formulas for $\boldsymbol{p}_A$ and $\boldsymbol{p}_B$ provide us with a new interpretation of $r^i_{A,B}$ in a biqubit. The biqubit coefficients $r^i_{A,B}$ encode the results of *separate* measurements on qubits A and B. We can read these directly and easily from equation (5.142). Because they correspond to individual qubits at both ends of a biqubit, they must both lie within the respective single-qubit Bloch balls. We must have that $\boldsymbol{r}_A \cdot \boldsymbol{r}_A \leq 1$ and $\boldsymbol{r}_B \cdot \boldsymbol{r}_B \leq 1$. As we have seen previously, $\boldsymbol{r}_A$ and $\boldsymbol{r}_B$ mix the same way they do in single-qubit systems.

This again confirms that if we focus on one component of a biqubit and ignore the other one, we have no means of telling whether the qubit is entangled with another qubit or just mixed for some other reason.

5.9 Classification of Biqubit States

A biqubit system appears quite simple on first inspection. Yet, when investigated in more depth, it reveals a great deal of complexity. The complexity derives primarily from the many ways in which biqubits can be mixed: both on the level of individual qubits, which is described by $\boldsymbol{r}_A$ and $\boldsymbol{r}_B$, and on the level of biqubits themselves, which is captured by $\boldsymbol{x}_{AB}$. On top of this we have pure states and entangled states, and the latter can be mixed, too.

Biqubits not scrutinized until 1990s

Yet, it is still surprising to learn that most work on biqubit separability and classification began only in the mid-1990s, some 70 years after the birth of quantum mechanics and after Pauli's discovery of his exclusion principle. The reason is that prior to that time, most physicists considered biqubits within the confines of unitary formalism, where matters are greatly simplified. It was the new interest in Bell inequalities and related biqubit probability measurements that made physicists ask whether the correlations they observed were indeed of quantum nature or merely due to classical mixing. This fundamental question proved remarkably difficult to answer.

At the same time, the topic is central to quantum computing, because everything in quantum computing is done with biqubits. One reason is that a biqubit gate, called the controlled-NOT gate, is universal to quantum computing.[9] The other reason is that the moment we enter the domain of experimental quantum physics we have to abandon the comfortable, idealistic world of the unitary formalism and face the reality of fully blown probability theory, or the *density operator* theory as the physicists prefer to call it in this context. Just as working cars cannot be designed without taking friction into account, similarly, working quantum computers cannot be designed without taking into account depolarization, dissipation, and other nonunitary phenomena. We saw it first thing in the quantronium example.

Biqubit normalization

Let us go back to the biqubit representation given by equation (5.61):

[9] We will talk about this in the next chapter.

$$
\begin{aligned}
\boldsymbol{p}_{AB} &= \frac{1}{4}\Bigg(\varsigma_{1A} \otimes \varsigma_{1B} + \sum_{j=x,y,z} r^j_B \varsigma_{1A} \otimes \varsigma_{jB} + \sum_{i=x,y,z} r^i_A \varsigma_{iA} \otimes \varsigma_{1B} \\
&\quad + \sum_{i,j=x,y,z} x^{ij}_{AB} \varsigma_{iA} \otimes \varsigma_{jB} \Bigg),
\end{aligned} \tag{5.145}
$$

where we have normalized the probabilities so that α, the coefficient in front of $\varsigma_{1A} \otimes \varsigma_{1B}$, is 1.

We have seen in Section 5.5 how to change the frame in which a biqubit is measured and what effect this has on equation (5.145). The basic idea there was that the three Pauli vectors labeled by x, y, and z behaved under rotations like normal three-dimensional vectors that pointed in the x, y, and z directions.

Equation (5.145) evaluates probabilities $\boldsymbol{p}_{AB}$ in terms of Pauli vectors attached at two different locations, the location of qubit A and the location of qubit B. The two Pauli frames don't have to be oriented the same way and can be rotated independently of each other—exactly as we did in Section 5.5.

An arbitrary rotation in the 3D space can be characterized by providing three Euler angles. The two independent rotations, one for qubit A and the other one for qubit B, are therefore specified by six Euler angles.[10] We can always choose the six Euler angles so as to kill the six off-diagonal elements of matrix x^{ij}_{AB}.

Matrix x^{ij}_{AB} does not have to be symmetric for this, and we will not end up with complex numbers on the diagonal either, because here we manipulate both frames independently. If we wanted to diagonalize matrix x^{ij}_{AB} by performing an *identical* rotation on both frames, then the matrix would have to be symmetric for this to work. The reason is that we would have only *three* Euler angles to play with, and with these we could kill only *three* off-diagonal elements.

Having diagonalized matrix x^{ij}_{AB}, we end up with only 9 real numbers (that aren't zero) in $\boldsymbol{p}_{AB}$ in place of the original 15. And so, it turns out that of the 15 degrees of freedom that characterize the biqubit in the fiducial formalism, 6 are of purely geometric character and can be eliminated or otherwise modified by rotating frames against which the biqubit components are measured. But the remaining 9 degrees of freedom are physical.

Once we have x^{ij}_{AB} in the diagonal form, we can switch around the labels on the directions x, y, and z so as to rewrite the $\sum_{ij} x^{ij}_{AB} \varsigma_{iA} \otimes \varsigma_{jB}$ term in the following form:

$$
\operatorname{sign}(\det \boldsymbol{x}_{AB}) \left(\kappa^x_{AB} \varsigma_{xA} \otimes \varsigma_{xB} + \kappa^y_{AB} \varsigma_{yA} \otimes \varsigma_{yB} + \kappa^z_{AB} \varsigma_{zA} \otimes \varsigma_{zB} \right), \tag{5.146}
$$

[10] We discuss Euler angles in detail in Section 6.2.1.

where the *kappa* coefficients are ordered as follows:

$$\kappa^x_{AB} \geq \kappa^y_{AB} \geq \kappa^z_{AB}. \tag{5.147}$$

The κ coefficients are the same as the x^{ii}_{AB} coefficients after the diagonalization up to their sign and ordering.

But even now we may have some freedom left. For example, if all κ^i_{AB} are zero, then we can rotate both frames as much as we wish without changing x^{ij}_{AB} or κ^i_{AB} at all. Furthermore, we're still left with the freedom to reflect rather than rotate the varsigmas. For example,

$$\varsigma_{xA} \rightarrow -\varsigma_{xA} \quad \text{and} \tag{5.148}$$

$$\varsigma_{xB} \rightarrow -\varsigma_{xB} \tag{5.149}$$

leave κ^x_{AB} unchanged.

When such additional freedoms are left after diagonalization of x^{ij}_{AB}, we use them to kill as many remaining $r^i_{A,B}$ as possible, usually starting with $r^y_{A,B}$, then proceeding to $r^z_{A,B}$.

The purpose of all these manipulations is to "normalize" the qubit's representation and remove any dependence on geometry and choices of directions.

The resulting probability matrix is

$$\begin{aligned} \boldsymbol{p}_{AB} &= \frac{1}{4}\Bigg(\varsigma_{1A} \otimes \varsigma_{1B} + \sum_{i=x,y,z} r^i_A \varsigma_{iA} \otimes \varsigma_{1B} + \sum_{j=x,y,z} r^j_B \varsigma_{1A} \otimes \varsigma_{jB} \\ &\quad + \operatorname{sign}(\det \boldsymbol{x}_{AB}) \sum_{k=x,y,z} \kappa^k_{AB} \varsigma_{kA} \otimes \varsigma_{kB}\Bigg), \end{aligned} \tag{5.150}$$

where we must remember that (x, y, z) directions at A and B may not necessarily be the same, so that $p^{\uparrow\uparrow}$ does not mean a probability of finding both qubit components pointing in the same direction. It means finding qubit A pointing in the $\boldsymbol{e}_{zA}$ direction and qubit B pointing in the $\boldsymbol{e}_{zB}$ direction. Also, we note that coefficients r^i_A and r^j_B may no longer be the same as they were in the original version of $\boldsymbol{p}_{AB}$. Rotating Pauli vectors $\varsigma_{iA,B}$ changes not only x^{ij}_{AB}, but also r^i_A and r^j_B.

Englert-Metwally classification

Now, following Englert and Metwally [39], we divide all possible biqubit states into several classes.

Class A defined by $\kappa^x_{AB} = \kappa^y_{AB} = \kappa^z_{AB} = 0$. The completely chaotic state with all coefficients equal zero belongs to this class. Since we end up with $x^{ij}_{AB} = 0$

for this class, we can still rotate both frames, at A and B, and kill $r^{y,z}_{A,B}$. The following remains:

$$\boldsymbol{p}_{AB} = \frac{1}{4}\left(\varsigma_{1A}\otimes\varsigma_{1B} + r^x_A\varsigma_{xA}\otimes\varsigma_{1B} + r^x_B\varsigma_{1A}\otimes\varsigma_{xB}\right), \tag{5.151}$$

$$\begin{pmatrix} p^{\uparrow\uparrow} & p^{\uparrow\downarrow} & p^{\uparrow\rightarrow} & p^{\uparrow\otimes} \\ p^{\downarrow\uparrow} & p^{\downarrow\downarrow} & p^{\downarrow\rightarrow} & p^{\downarrow\otimes} \\ p^{\rightarrow\uparrow} & p^{\rightarrow\downarrow} & p^{\rightarrow\rightarrow} & p^{\rightarrow\otimes} \\ p^{\otimes\uparrow} & p^{\otimes\downarrow} & p^{\otimes\rightarrow} & p^{\otimes\otimes} \end{pmatrix}$$
$$= \frac{1}{4}\begin{pmatrix} 1 & 1 & 1+r^x_B & 1 \\ 1 & 1 & 1+r^x_B & 1 \\ 1+r^x_A & 1+r^x_A & 1+r^x_A+r^x_B & 1+r^x_A \\ 1 & 1 & 1+r^x_B & 1 \end{pmatrix}, \tag{5.152}$$

where $r^x_A \geq 0$ and $r^x_B \geq 0$. This is the most general normalized state in this class. It is a strange state, with well-defined local single-qubit states, but with no biqubit correlations at all, not even classical ones of the form $r^x_A r^x_B \varsigma_{xA}\otimes\varsigma_{xB}$.

Class B$^+$ defined by $\kappa^x_{AB} = \kappa^y_{AB} = \kappa^z_{AB} = \kappa > 0$ and $\det \boldsymbol{x}_{AB} > 0$.

Class B$^-$ defined by $\kappa^x_{AB} = \kappa^y_{AB} = \kappa^z_{AB} = \kappa > 0$ and $\det \boldsymbol{x}_{AB} < 0$.

Here we still have the freedom to kill additionally three $r^i_{A,B}$. Following an established convention we choose

$$\begin{aligned}\boldsymbol{p}_{AB} \quad = \quad & \frac{1}{4}\Big(\varsigma_{1A}\otimes\varsigma_{1B} \\ & +r^x_A\varsigma_{xA}\otimes\varsigma_{1B} \\ & +\varsigma_{1A}\otimes\left(r^x_B\varsigma_{xB}+r^z_B\varsigma_{zB}\right) \\ & \pm\kappa\left(\varsigma_{xA}\otimes\varsigma_{xB}+\varsigma_{yA}\otimes\varsigma_{yB}+\varsigma_{zA}\otimes\varsigma_{zB}\right)\Big),\end{aligned} \tag{5.153}$$

$$\begin{pmatrix} p^{\uparrow\uparrow} & p^{\uparrow\downarrow} & p^{\uparrow\rightarrow} & p^{\uparrow\otimes} \\ p^{\downarrow\uparrow} & p^{\downarrow\downarrow} & p^{\downarrow\rightarrow} & p^{\downarrow\otimes} \\ p^{\rightarrow\uparrow} & p^{\rightarrow\downarrow} & p^{\rightarrow\rightarrow} & p^{\rightarrow\otimes} \\ p^{\otimes\uparrow} & p^{\otimes\downarrow} & p^{\otimes\rightarrow} & p^{\otimes\otimes} \end{pmatrix} \tag{5.154}$$
$$= \frac{1}{4}\begin{pmatrix} 1+r^z_B\pm\kappa & 1-r^z_B\mp\kappa & 1+r^x_B & 1 \\ 1+r^z_B\mp\kappa & 1-r^z_B\pm\kappa & 1+r^x_B & 1 \\ 1+r^x_A+r^z_B & 1+r^x_A-r^z_B & 1+r^x_A+r^x_B\pm\kappa & 1+r^x_A \\ 1+r^z_B & 1-r^z_B & 1+r^x_B & 1\pm\kappa \end{pmatrix},$$

where $r_A^x \geq 0$, and when $r_A^x > 0$, then $r_B^z \geq 0$, and when $r_A^x = 0$, then $r_B^x \geq 0$ and $r_B^z = 0$.

Two important families of states,

$$\frac{1}{4}\left(\varsigma_{1A} \otimes \varsigma_{1B} \pm \kappa\left(\varsigma_{xA} \otimes \varsigma_{xB} + \varsigma_{yA} \otimes \varsigma_{yB} + \varsigma_{zA} \otimes \varsigma_{zB}\right)\right), \tag{5.155}$$

Werner states

belong to class $\mathrm{B}^{\pm}$. They are called *Werner states*. A B^{-} Werner state with $\kappa = 1$ is the Bell state $|\Psi^-\rangle_{AB}$.

Class C defined by $\kappa_{AB}^x = \kappa > \kappa_{AB}^y = \kappa_{AB}^z = 0$.

Here $\det \boldsymbol{x}_{AB} = 0$, because $\kappa_{AB}^y = \kappa_{AB}^z = 0$. This leaves us with enough freedom to clean up $r_{A,B}^y$, too—such is the established choice—so that the state looks as follows:

$$\begin{aligned}\boldsymbol{p}_{AB} = \frac{1}{4}\Big(&\varsigma_{1A} \otimes \varsigma_{1B} \\ &+ \left(r_A^x \varsigma_{xA} + r_A^z \varsigma_{zA}\right) \otimes \varsigma_{1B} \\ &+ \varsigma_{1A} \otimes \left(r_B^x \varsigma_{xB} + r_B^z \varsigma_{zB}\right) \\ &\pm \kappa \varsigma_{xA} \otimes \varsigma_{xB}\Big),\end{aligned} \tag{5.156}$$

$$\begin{pmatrix} p^{\uparrow\uparrow} & p^{\uparrow\downarrow} & p^{\uparrow\rightarrow} & p^{\uparrow\otimes} \\ p^{\downarrow\uparrow} & p^{\downarrow\downarrow} & p^{\downarrow\rightarrow} & p^{\downarrow\otimes} \\ p^{\rightarrow\uparrow} & p^{\rightarrow\downarrow} & p^{\rightarrow\rightarrow} & p^{\rightarrow\otimes} \\ p^{\otimes\uparrow} & p^{\otimes\downarrow} & p^{\otimes\rightarrow} & p^{\otimes\otimes} \end{pmatrix} = \frac{1}{4}\begin{pmatrix} 1 + r_A^z + r_B^z & 1 + r_A^z - r_B^z & 1 + r_B^x + r_A^z & 1 + r_A^z \\ 1 - r_A^z + r_B^z & 1 - r_A^z - r_B^z & 1 + r_B^x - r_A^z & 1 - r_A^z \\ 1 + r_A^x + r_B^z & 1 + r_A^x - r_B^z & 1 + r_A^x + r_B^x \pm \kappa & 1 + r_A^x \\ 1 + r_B^z & 1 - r_B^z & 1 + r_B^x & 1 \end{pmatrix}, \tag{5.157}$$

where $r_{A,B}^z \geq 0$ and $r_A^x \geq 0$, and when $r_A^x = 0$, then $r_B^x \geq 0$.

A simple separable biqubit state $\boldsymbol{p}_A \otimes \boldsymbol{p}_B$ defined solely by $\boldsymbol{r}_A$ and $\boldsymbol{r}_B$ belongs to this class. We can rotate both frames so that $\boldsymbol{r}_A = r_A^x \boldsymbol{e}_{xA}$ and $\boldsymbol{r}_B = r_B^x \boldsymbol{e}_{xB}$. Then

$$\boldsymbol{p}_{AB} = \frac{1}{4}\left(\varsigma_{1A} \otimes \varsigma_{1B} + r_A^x \varsigma_{xA} \otimes \varsigma_{1B} + r_B^x \varsigma_{1A} \otimes \varsigma_{xB} + r_A^x r_B^x \varsigma_{xA} \otimes \varsigma_{xB}\right). \tag{5.158}$$

Class D$^+$ defined by $\kappa^x_{AB} > \kappa^y_{AB} = \kappa^z_{AB} = \kappa > 0$ and $\det \boldsymbol{x}_{AB} > 0$.

Class D$^-$ defined by $\kappa^x_{AB} > \kappa^y_{AB} = \kappa^z_{AB} = \kappa > 0$ and $\det \boldsymbol{x}_{AB} < 0$.

Here we have less freedom left after the diagonalization of x^{ij}_{AB}, and the only $r^i_{A,B}$ that we can get rid of is r^y_A. The resulting state looks as follows:

$$\begin{aligned}\boldsymbol{p}_{AB} &= \frac{1}{4}\Big(\varsigma_{1A} \otimes \varsigma_{1B} \\ &\quad + \left(r^x_A \varsigma_{xA} + r^z_A \varsigma_{zA}\right) \otimes \varsigma_{1B} \\ &\quad + \varsigma_{1A} \otimes \left(r^x_B \varsigma_{xB} + r^y_B \varsigma_{yB} + r^z_B \varsigma_{zB}\right) \\ &\quad \pm \left(\kappa^x \varsigma_{xA} \otimes \varsigma_{xB} + \kappa \left(\varsigma_{yA} \otimes \varsigma_{yB} + \varsigma_{zA} \otimes \varsigma_{zB}\right)\right)\Big),\end{aligned} \tag{5.159}$$

$$\begin{pmatrix} p^{\uparrow\uparrow} & p^{\uparrow\downarrow} & p^{\uparrow\rightarrow} & p^{\uparrow\otimes} \\ p^{\downarrow\uparrow} & p^{\downarrow\downarrow} & p^{\downarrow\rightarrow} & p^{\downarrow\otimes} \\ p^{\rightarrow\uparrow} & p^{\rightarrow\downarrow} & p^{\rightarrow\rightarrow} & p^{\rightarrow\otimes} \\ p^{\otimes\uparrow} & p^{\otimes\downarrow} & p^{\otimes\rightarrow} & p^{\otimes\otimes} \end{pmatrix}$$
$$= \frac{1}{4}\begin{pmatrix} 1 + r^z_A + r^z_B \pm \kappa & 1 + r^z_A - r^z_B \mp \kappa & 1 + r^z_A + r^x_B & 1 + r^z_A + r^y_B \\ 1 - r^z_A + r^z_B \mp \kappa & 1 - r^z_A - r^z_B \pm \kappa & 1 - r^z_A + r^x_B & 1 - r^z_A + r^y_B \\ 1 + r^x_A + r^z_B & 1 + r^x_A - r^z_B & 1 + r^x_A + r^x_B \pm \kappa^x & 1 + r^x_A + r^y_B \\ 1 + r^z_B & 1 - r^z_B & 1 + r^x_B & 1 + r^y_B \pm \kappa \end{pmatrix}, \tag{5.160}$$

where $r^x_A \geq 0$, $r^z_A \geq 0$ and $r^y_B \geq 0$. When $r^x_A = 0$, then $r^x_B \geq 0$. When $r^z_A = 0$ and $r^y_B = 0$, then $r^z_B \geq 0$.

All pure states belong to class D$^-$. Their generic form is

All pure biqubits are D$^-$.

$$\begin{aligned}\boldsymbol{p}_{AB} &= \frac{1}{4}\Big(\varsigma_{1A} \otimes \varsigma_{1B} \\ &\quad + r\left(\varsigma_{xA} \otimes \varsigma_{1B} - \varsigma_{1A} \otimes \varsigma_{xB}\right) \\ &\quad - \left(\varsigma_{xA} \otimes \varsigma_{xB} + \sqrt{1 - r^2}\left(\varsigma_{yA} \otimes \varsigma_{yB} + \varsigma_{zA} \otimes \varsigma_{zB}\right)\right)\Big),\end{aligned} \tag{5.161}$$

where $r \in [0, 1]$.

It is easy to see why this must be the generic form of a pure state. Pure states are described by four complex numbers, or eight real numbers, constrained by one normalization condition. This leaves seven real numbers. But six of these can be eliminated by frame rotations, so that we end up with just one generic parameter. The generic parameter is r in the above equation. Thus, the dimensionality is just right.

Next we check whether this is indeed a pure state. We can do so by demonstrating that its corresponding density quaternion is idempotent, meaning that $\boldsymbol{\rho\rho} = \boldsymbol{\rho}$.

We are going to demonstrate this here. Let us begin by replacing varsigmas with sigmas:

$$\begin{aligned}\boldsymbol{\rho}_{AB} &= \frac{1}{4}\Big(\mathbf{1}_A \otimes \mathbf{1}_B \\ &\quad + r\left(\boldsymbol{\sigma}_{xA} \otimes \mathbf{1}_B - \mathbf{1}_A \otimes \boldsymbol{\sigma}_{xB}\right) \\ &\quad - \boldsymbol{\sigma}_{xA} \otimes \boldsymbol{\sigma}_{xB} - \sqrt{1-r^2}\left(\boldsymbol{\sigma}_{yA} \otimes \boldsymbol{\sigma}_{yB} + \boldsymbol{\sigma}_{zA} \otimes \boldsymbol{\sigma}_{zB}\right)\Big). \end{aligned} \tag{5.162}$$

We organize the computation by introducing

$$a = \left(\boldsymbol{\sigma}_{xA} \otimes \mathbf{1}_B - \mathbf{1}_A \otimes \boldsymbol{\sigma}_{xB}\right), \tag{5.163}$$

$$b = \boldsymbol{\sigma}_{xA} \otimes \boldsymbol{\sigma}_{xB}, \tag{5.164}$$

$$c = \left(\boldsymbol{\sigma}_{yA} \otimes \boldsymbol{\sigma}_{yB} + \boldsymbol{\sigma}_{zA} \otimes \boldsymbol{\sigma}_{zB}\right). \tag{5.165}$$

Then

$$\begin{aligned}\boldsymbol{\rho}_{AB} \cdot \boldsymbol{\rho}_{AB} &= \frac{1}{16}\left(\mathbf{1}_A \otimes \mathbf{1}_B + ra - b - \sqrt{1-r^2}c\right) \\ &\quad \times \left(\mathbf{1}_A \otimes \mathbf{1}_B + ra - b - \sqrt{1-r^2}c\right) \\ &= \frac{1}{16}\Big(\mathbf{1}_A \otimes \mathbf{1}_B + 2ra - 2b - 2\sqrt{1-r^2}c \\ &\quad + r^2a^2 + b^2 + (1-r^2)c^2 \\ &\quad - r(ab + ba) \\ &\quad - r\sqrt{1-r^2}(ac + ca) \\ &\quad + \sqrt{1-r^2}(bc + cb)\Big). \end{aligned} \tag{5.166}$$

The trick is now to remember that A sigmas must multiply other A sigmas only and the same holds for B sigmas.

It is easy to see that

$$b^2 = \boldsymbol{\sigma}_{xA} \otimes \boldsymbol{\sigma}_{xB} \cdot \boldsymbol{\sigma}_{xA} \otimes \boldsymbol{\sigma}_{xB} = \mathbf{1}_A \otimes \mathbf{1}_B. \tag{5.167}$$

This is because $\boldsymbol{\sigma}_{i\,A,B}\boldsymbol{\sigma}_{i\,A,B} = \mathbf{1}_{A,B}$, for $i = x, y, z$.

It is almost as easy to see that

$$c^2 = 2\left(\mathbf{1}_A \otimes \mathbf{1}_B - \boldsymbol{\sigma}_{xA} \otimes \boldsymbol{\sigma}_{xB}\right) = 2\left(\mathbf{1}_A \otimes \mathbf{1}_B - b\right). \tag{5.168}$$

This is, first, because of the above, and second, because

$$\boldsymbol{\sigma}_{y\,A} \otimes \boldsymbol{\sigma}_{y\,B} \cdot \boldsymbol{\sigma}_{z\,A} \otimes \boldsymbol{\sigma}_{z\,B} = \boldsymbol{\sigma}_{z\,A} \otimes \boldsymbol{\sigma}_{z\,B} \cdot \boldsymbol{\sigma}_{y\,A} \otimes \boldsymbol{\sigma}_{y\,B} = -\boldsymbol{\sigma}_{x\,A} \otimes \boldsymbol{\sigma}_{x\,B}. \tag{5.169}$$

The minus here comes from i^2.

And it is childishly easy to see that

$$a^2 = c^2 = 2\left(\mathbf{1}_A \otimes \mathbf{1}_B - b\right). \tag{5.170}$$

This is because

$$\mathbf{1}_A \otimes \boldsymbol{\sigma}_{xB} \cdot \mathbf{1}_A \otimes \boldsymbol{\sigma}_{xB} = \mathbf{1}_A \otimes \mathbf{1}_B \tag{5.171}$$

and because

$$\mathbf{1}_A \otimes \boldsymbol{\sigma}_{xB} \cdot \boldsymbol{\sigma}_{xA} \otimes \mathbf{1}_B = \boldsymbol{\sigma}_{xA} \otimes \boldsymbol{\sigma}_{xB} = b. \tag{5.172}$$

Let us then add

$$\begin{aligned}
&r^2a^2 + b^2 + (1 - r^2)c^2 \\
&\quad = r^2 2\left(\mathbf{1}_A \otimes \mathbf{1}_B - b\right) + \mathbf{1}_A \otimes \mathbf{1}_B + (1 - r^2)2\left(\mathbf{1}_A \otimes \mathbf{1}_B - b\right) \\
&\quad = 3\mathbf{1}_A \otimes \mathbf{1}_B - 2b.
\end{aligned} \tag{5.173}$$

Thus,

$$\begin{aligned}
&\mathbf{1}_A \otimes \mathbf{1}_B + 2ra - 2b - 2\sqrt{1 - r^2}c + r^2a^2 + b^2 + (1 - r^2)c^2 \\
&\quad = 4\mathbf{1}_A \otimes \mathbf{1}_B + 2ra - 4b - 2\sqrt{1 - r^2}c.
\end{aligned} \tag{5.174}$$

For perfect happiness we still have to generate additional $2ra$ and additional $-2\sqrt{1 - r^2}c$ using the remaining three anticommutator terms.

Let us observe that

$$bc = cb = -c. \tag{5.175}$$

This is because

$$\boldsymbol{\sigma}_x \boldsymbol{\sigma}_y = -\boldsymbol{\sigma}_y \boldsymbol{\sigma}_x = \mathrm{i}\boldsymbol{\sigma}_z \tag{5.176}$$

and

$$\boldsymbol{\sigma}_x \boldsymbol{\sigma}_z = -\boldsymbol{\sigma}_z \boldsymbol{\sigma}_x = -\mathrm{i}\boldsymbol{\sigma}_y. \tag{5.177}$$

So bc is merely going to swap $\boldsymbol{\sigma}_{y\,A} \otimes \boldsymbol{\sigma}_{y\,B}$ and $\boldsymbol{\sigma}_{z\,A} \otimes \boldsymbol{\sigma}_{z\,B}$ and throw $\mathrm{i}^2 = (-\mathrm{i})^2 = -1$ in front. Consequently

$$bc + cb = -2c. \tag{5.178}$$

Similarly,

$$ab = ba = -a. \tag{5.179}$$

This is because

$$\boldsymbol{\sigma}_{x\,A} \otimes \mathbf{1}_B \cdot \boldsymbol{\sigma}_{x\,A} \otimes \boldsymbol{\sigma}_{x\,B} = \mathbf{1}_A \otimes \boldsymbol{\sigma}_{x\,B}. \tag{5.180}$$

Consequently,

$$ab + ba = -2a \tag{5.181}$$

and

$$-r(ab + ba) + \sqrt{1-r^2}(bc + cb) = 2ra - 2\sqrt{1-r^2}c. \tag{5.182}$$

Adding this to

$$4\mathbf{1}_A \otimes \mathbf{1}_B + 2ra - 4b - 2\sqrt{1-r^2}c \tag{5.183}$$

yields

$$4\mathbf{1}_A \otimes \mathbf{1}_B + 4ra - 4b - 4\sqrt{1-r^2}c, \tag{5.184}$$

which, when divided by 16, returns the original $\boldsymbol{\rho}_{AB}$.

We are left with one more term, namely,

$$r\sqrt{1-r^2}(ac + ca), \tag{5.185}$$

and this term vanishes, because here we have just one $\boldsymbol{\sigma}_{x\,A,B}$ from a multiplying one of the $\boldsymbol{\sigma}_{y,z\,A,B}$ from c, first from the left, in ac, and then from the right, in *ca*. But different sigmas anticommute, so this kills the whole term.

This computation, although tedious, is also instructive. Apart from demonstrating that the quaternion of state (5.162) is idempotent, and so the state itself is pure, the example also shows how to divide a lengthy computation of this nature into smaller, manageable chunks and how to perform the computation itself by using quaternion rules only and not Pauli matrices.

Class E$^+$ defined by $\kappa = \kappa^x_{AB} = \kappa^y_{AB} > \kappa^z_{AB}$ and $\det \boldsymbol{x}_{AB} > 0$.

Class E$^-$ defined by $\kappa = \kappa^x_{AB} = \kappa^y_{AB} > \kappa^z_{AB}$ and $\det \boldsymbol{x}_{AB} < 0$.

These two classes are similar to D$^\pm$. The difference is that whereas previously we had $\kappa = \kappa^y = \kappa^z$, here we have that $\kappa = \kappa^x = \kappa^y$ instead and κ^z is different.

This, as before, lets us kill r^y_A only, and so we end up with

$$\begin{aligned} \boldsymbol{p}_{AB} = \frac{1}{4}\Big(& \varsigma_{1A} \otimes \varsigma_{1B} \\ & + \left(r^x_A \varsigma_{xA} + r^z_A \varsigma_{zA}\right) \otimes \varsigma_{1B} \\ & + \varsigma_{1A} \otimes \left(r^x_B \varsigma_{xB} + r^y_B \varsigma_{yB} + r^z_B \varsigma_{zB}\right) \\ & \pm \left(\kappa\left(\varsigma_{xA} \otimes \varsigma_{xB} + \varsigma_{yA} \otimes \varsigma_{yB}\right) + \kappa^z \varsigma_{zA} \otimes \varsigma_{zB}\right)\Big), \end{aligned} \tag{5.186}$$

$$\begin{pmatrix} p^{\uparrow\uparrow} & p^{\uparrow\downarrow} & p^{\uparrow\rightarrow} & p^{\uparrow\otimes} \\ p^{\downarrow\uparrow} & p^{\downarrow\downarrow} & p^{\downarrow\rightarrow} & p^{\downarrow\otimes} \\ p^{\rightarrow\uparrow} & p^{\rightarrow\downarrow} & p^{\rightarrow\rightarrow} & p^{\rightarrow\otimes} \\ p^{\otimes\uparrow} & p^{\otimes\downarrow} & p^{\otimes\rightarrow} & p^{\otimes\otimes} \end{pmatrix}$$
$$= \frac{1}{4}\begin{pmatrix} 1 + r^z_A + r^z_B \pm \kappa^z & 1 + r^z_A - r^z_B \mp \kappa^z & 1 + r^z_A + r^x_B & 1 + r^z_A + r^y_B \\ 1 - r^z_A + r^z_B \mp \kappa^z & 1 - r^z_A - r^z_B \pm \kappa^z & 1 - r^z_A + r^x_B & 1 - r^z_A + r^y_B \\ 1 + r^x_A + r^z_B & 1 + r^x_A - r^z_B & 1 + r^x_A + r^x_B \pm \kappa & 1 + r^x_A + r^y_B \\ 1 + r^z_B & 1 - r^z_B & 1 + r^x_B & 1 + r^y_B \pm \kappa \end{pmatrix}, \tag{5.187}$$

where $r^x_A \geq 0$ and $r^y_B \geq 0$, and when $r^x_A = 0$, then $r^y_B = 0$, too, and $r^x_B \geq 0$. Additionally $r^z_A \geq 0$; and when $r^z_A = 0$, then $r^z_B \geq 0$.

Class F$^+$ defined by $\kappa^x_{AB} > \kappa^y_{AB} > \kappa^z_{AB}$ and $\det \boldsymbol{x}_{AB} > 0$.

Class F$^-$ defined by $\kappa^x_{AB} > \kappa^y_{AB} > \kappa^z_{AB}$ and $\det \boldsymbol{x}_{AB} < 0$.

With all three kappas different we get no freedom to kill any components of $\boldsymbol{r}$. We can deploy reflections in order to make as many of r^x_A, r^x_B, r^y_A, r^y_B, r^z_A and r^z_B as possible positive—in preference of the order listed.

The resulting probability matrix is

$$\begin{aligned}\boldsymbol{p}_{AB} &= \frac{1}{4}\Big(\varsigma_{1A}\otimes\varsigma_{1B} \\ &\quad + \left(r^x_A\varsigma_{xA} + r^y_A\varsigma_{yA} + r^z_A\varsigma_{zA}\right)\otimes\varsigma_{1B} \\ &\quad + \varsigma_{1A}\otimes\left(r^x_B\varsigma_{xB} + r^y_B\varsigma_{yB} + r^z_B\varsigma_{zB}\right) \\ &\quad \pm\left(\kappa^x\varsigma_{xA}\otimes\varsigma_{xB} + \kappa^y\varsigma_{yA}\otimes\varsigma_{yB} + \kappa^z\varsigma_{zA}\otimes\varsigma_{zB}\right)\Big),\end{aligned} \tag{5.188}$$

$$\begin{pmatrix} p^{\uparrow\uparrow} & p^{\uparrow\downarrow} & p^{\uparrow\rightarrow} & p^{\uparrow\otimes} \\ p^{\downarrow\uparrow} & p^{\downarrow\downarrow} & p^{\downarrow\rightarrow} & p^{\downarrow\otimes} \\ p^{\rightarrow\uparrow} & p^{\rightarrow\downarrow} & p^{\rightarrow\rightarrow} & p^{\rightarrow\otimes} \\ p^{\otimes\uparrow} & p^{\otimes\downarrow} & p^{\otimes\rightarrow} & p^{\otimes\otimes} \end{pmatrix}$$
$$= \frac{1}{4}\begin{pmatrix} 1 + r^z_A + r^z_B \pm \kappa^z & 1 + r^z_A - r^z_B \mp \kappa^z & 1 + r^z_A + r^x_B & 1 + r^z_A + r^y_B \\ 1 - r^z_A + r^z_B \mp \kappa^z & 1 - r^z_A - r^z_B \pm \kappa^z & 1 - r^z_A + r^x_B & 1 - r^z_A + r^y_B \\ 1 + r^x_A + r^z_B & 1 + r^x_A - r^z_B & 1 + r^x_A + r^x_B \pm \kappa^x & 1 + r^x_A + r^y_B \\ 1 + r^y_A + r^z_B & 1 + r^y_A - r^z_B & 1 + r^y_A + r^x_B & 1 + r^y_A + r^y_B \pm \kappa^y \end{pmatrix}. \tag{5.189}$$

Unitary invariance of families

Classes A through F are subdivided into families defined by the values of the parameters that characterize each class. Unitary transformations *do not* take a member of a family outside the family. In other words, the families are unitary invariants. Because of this, the classification is not just a superficial division based on what the probability matrices look like, when expressed in the appropriately rotated varsigma bases.

Still, so far we have not found a way to tell whether a given biqubit state is entangled or just a fanciful mixture that looks similar to an entangled state, but isn't. We're going to find the key to this question in the next section.

5.10 Separability

The Englert and Metwally classification of biqubit states presented in the previous section must be supplemented with two additional conditions. The first one is obvious: every term of matrix $\boldsymbol{p}_{AB}$ must be restricted to $[0 \ldots 1]$, because every term of the matrix is a probability. This imposes restrictions on the κ coefficients, together with more obvious restrictions on the $\boldsymbol{r}_{A,B}$ vectors deriving from their interepretation discovered in Section 5.8, "Single qubit expectation values," page 233.

The second condition is less obvious: it may happen that a state described by $\boldsymbol{p}_{AB}$ looks perfectly OK at first glance, but is, in fact, unphysical.

$\boldsymbol{p}_{AB}$ matrices may be unphysical.

An example of such a state is a Werner state of class B^+ with $\kappa = 1$,

$$\boldsymbol{p}^+_{AB} = \frac{1}{4}\left(\varsigma_{1A} \otimes \varsigma_{1B} + \varsigma_{xA} \otimes \varsigma_{xB} + \varsigma_{yA} \otimes \varsigma_{yB} + \varsigma_{zA} \otimes \varsigma_{zB}\right). \tag{5.190}$$

Its explicit probability matrix is

$$\frac{1}{4}\begin{pmatrix} 2 & 0 & 1 & 1 \\ 0 & 2 & 1 & 1 \\ 1 & 1 & 2 & 1 \\ 1 & 1 & 1 & 2 \end{pmatrix}. \tag{5.191}$$

The matrix looks perfectly acceptable: all terms are within $[0 \ldots 1]$. But its physics is incorrect. What we have here is a system of two 1/2-spins that align in every direction. When both spins are measured against $\boldsymbol{e}_z$, they come out aligned. When they are measured against $-\boldsymbol{e}_z$, they come out aligned, too. This is also the case for every other direction. So this is a system of spin 1 that is spherically symmetric. But a spherically symmetric system cannot have spin 1. Only a system of spin 0 may be spherically symmetric.

A reader who took to heart our admonitions in Section 2.1, "The Evil Quanta," page 41, may object here and say that we should not be hasty in declaring what is and what is not physical *ex cathedra,* that this should be decided by an experiment rather than aesthetic or even mathematical considerations. Even the most beautiful and convincing mathematics is useless if it is derived from incorrect assumptions. Well, we can say with certainty that such a state has never been observed.

If the argument about the symmetry of this state not being physical is not convincing enough, here we have another argument. The probability of a transition between this state and Bell state $|\,\Psi^-\rangle_{AB}$ is negative.

Negative transition probabilities

The Bell state $|\,\Psi^-\rangle_{AB}$ is a perfectly legitimate, pure, and experimentally observed state that defines a certain direction in the biqubit Hilbert space. Its corresponding projection operator is $|\ \Psi^-\rangle_{AB} \otimes \langle \Psi^-\ |_{AB}$, and it corresponds physically to a state with spin 0. It would be perfectly OK for state $\boldsymbol{p}^+_{AB}$ to have zero probability of transition to the Bell state, but a negative probability is clearly unphysical.

Transition probability to the Bell state can be evaluated similarly to the way we did it for single-qubit states in Section 4.5, "Probability Amplitudes," page 130. All we need to do is to evaluate the following bracket:

$$\begin{aligned}
&\langle \tilde{\boldsymbol{p}}^-_{AB}, \boldsymbol{p}^+_{AB}\rangle \\
&\quad = \left\langle \frac{1}{4}\left(\varsigma^1_A \otimes \varsigma^1_B - \varsigma^x_A \otimes \varsigma^x_B - \varsigma^y_A \otimes \varsigma^y_B - \varsigma^z_A \otimes \varsigma^z_B\right), \right. \\
&\qquad \left. \frac{1}{4}\left(\varsigma_{1A} \otimes \varsigma_{1B} + \varsigma_{xA} \otimes \varsigma_{xB} + \varsigma_{yA} \otimes \varsigma_{yB} + \varsigma_{zA} \otimes \varsigma_{zB}\right) \right\rangle \\
&\quad = \frac{1}{16}\Big(\langle \varsigma^1_A, \varsigma_{1A}\rangle\langle \varsigma^1_B, \varsigma_{1B}\rangle - \langle \varsigma^x_A, \varsigma_{xA}\rangle\langle \varsigma^x_B, \varsigma_{xB}\rangle \\
&\qquad\qquad - \langle \varsigma^y_A, \varsigma_{yA}\rangle\langle \varsigma^y_B, \varsigma_{yB}\rangle - \langle \varsigma^z_A, \varsigma_{zA}\rangle\langle \varsigma^z_B, \varsigma_{zB}\rangle \Big) \\
&\quad = \frac{1}{16}\left(2 \times 2 - 2 \times 2 - 2 \times 2 - 2 \times 2\right) \\
&\quad = -\frac{1}{2}.
\end{aligned} \tag{5.192}$$

Let us observe that $\boldsymbol{p}^+_{AB}$ can be obtained from $\boldsymbol{p}^-_{AB}$ by replacing, for example,

$$\begin{aligned}
\varsigma_{xA} &\rightarrow -\varsigma_{xA} \\
\varsigma_{yA} &\rightarrow -\varsigma_{yA} \\
\varsigma_{zA} &\rightarrow -\varsigma_{zA}
\end{aligned} \tag{5.193}$$

while leaving all $\varsigma_{iB}, i = x, y, z$ intact, or the other way round, but not both.

Such an operation would *not* produce a weird unphysical state if performed on a simple separable biqubit state

$$\frac{1}{2}\left(\varsigma_{1A} + \sum_{i=x,y,z} r^i_A \varsigma_{iA}\right) \otimes \frac{1}{2}\left(\varsigma_{1B} + \sum_{i=x,y,z} r^i_B \varsigma_{iB}\right), \tag{5.194}$$

because it would be equivalent to replacing $\boldsymbol{r}_A$ with $-\boldsymbol{r}_A$, and this would still be a perfectly normal physical state made of qubit B in the same state as before and qubit A pointing in the opposite direction to the one that qubit A pointed to originally.

It would not produce a weird unphysical state if performed on a general separable biqubit state, a state that is a finite mixture of simple separable states, for the same reason. It would be equivalent to replacing $\boldsymbol{r}_{Ai}$ with $-\boldsymbol{r}_{Ai}$ for every mixture component labeled by i.

Yet, when applied to an entangled state, it produces an unphysical state.

Separability criterion

It turns out [113] [66] [39] that this is a common feature of all biqubit entangled states. This amazing property was discovered by the family of Horodeckis from the University of Gdańsk in Poland and by Asher Peres from Technion in Haifa.

Why is it so? The reason is that an entangled biqubit can be thought of as a "new compound particle in the making" or an "old compound particle in the breaking." The latter is a more common experimental situation, but they are similar, because the making of a compound particle is the same as the breaking of a compound particle viewed backwards in time.

When a compound particle breaks the spins of the constituents must be aligned just so, in order to conserve various quantum numbers, of which angular momentum is one, and to which spin contributes. If one of the constituents gets switched the other way, artificially, the rules break—producing, as we have seen, a spin-1 system that is spherically symmetric—and we end up with an unphysical configuration.

But for a biqubit that is separable, the same does not hold. Here there is no need to adhere to various conservation principles and such. The two constituents of the biqubit are fully independent and may point whichever way they wish.

A theorem by Durt, about which we will comment later (page 258), puts the above observations in a more formal framework, stating that a nontrivial interaction is always needed to entangle two quantum systems.

How to use the Peres-Horodeckis criterion

Whereas the Peres-Horodeckis criterion of separability is simple and elegant, it is not always easy to find whether a given state produced by operation (5.193) is physical or not. But we can use the bridges that lead from fiducial vectors to Pauli matrices to help ourselves in this task.

The bridges in this case work as follows.

Unitary description of biqubits by 4 × 4 matrices

Let us go back to a simple separable biqubit described by

$$\boldsymbol{p}_{AB} = \boldsymbol{p}_A \otimes \boldsymbol{p}_B. \tag{5.195}$$

Let us consider a simple energy form

$$\boldsymbol{\eta}_{AB} = \boldsymbol{\eta}_A \otimes \boldsymbol{\varsigma}^1_B + \boldsymbol{\varsigma}^1_A \otimes \boldsymbol{\eta}_B. \tag{5.196}$$

The expectation value of energy on $\boldsymbol{p}_{AB}$ is

$$\begin{aligned} &\langle \boldsymbol{\eta}_{AB}, \boldsymbol{p}_{AB} \rangle \qquad (5.197)\\ &= \langle \boldsymbol{\eta}_A, \boldsymbol{p}_A \rangle \langle \boldsymbol{\varsigma}^1_B, \boldsymbol{p}_B \rangle + \langle \boldsymbol{\varsigma}^1_A, \boldsymbol{p}_A \rangle \langle \boldsymbol{\eta}_B, \boldsymbol{p}_B \rangle = -\boldsymbol{\mu}_A \cdot \boldsymbol{B}_A - \boldsymbol{\mu}_B \cdot \boldsymbol{B}_B, \end{aligned}$$

where $\boldsymbol{B}_A$ and $\boldsymbol{B}_B$ correspond to $\boldsymbol{B}$ in locations A and B, respectively. As a reminder, we recall that $\langle \varsigma^1, \boldsymbol{p}\rangle$ is 1, because there is a 1/2 in front of ς_1 inside $\boldsymbol{p}$.

Let us now switch to the quaternion image of the same. Here we have that

$$\boldsymbol{\rho}_{AB} = \boldsymbol{\rho}_A \otimes \boldsymbol{\rho}_B \tag{5.198}$$

and

$$\boldsymbol{H}_{AB} = \boldsymbol{H}_A \otimes \boldsymbol{1}_B + \boldsymbol{1}_A \otimes \boldsymbol{H}_B. \tag{5.199}$$

Multiplying $\boldsymbol{H}_{AB}$ by $\boldsymbol{\rho}_{AB}$ yields

$$\begin{aligned}
&\boldsymbol{H}_{AB}\boldsymbol{\rho}_{AB} \\
&\quad = (\boldsymbol{H}_A \otimes \boldsymbol{1}_B + \boldsymbol{1}_A \otimes \boldsymbol{H}_B)\cdot(\boldsymbol{\rho}_A \otimes \boldsymbol{\rho}_B) \\
&\quad = (\boldsymbol{H}_A\boldsymbol{\rho}_A) \otimes (\boldsymbol{1}_B\boldsymbol{\rho}_B) + (\boldsymbol{1}_A\boldsymbol{\rho}_A) \otimes (\boldsymbol{H}_B\boldsymbol{\rho}_B)/
\end{aligned} \tag{5.200}$$

Since $\boldsymbol{1}\boldsymbol{\rho} = \boldsymbol{\rho}$, we can simplify this to

$$\boldsymbol{H}_{AB}\boldsymbol{\rho}_{AB} = (\boldsymbol{H}_A\boldsymbol{\rho}_A) \otimes \boldsymbol{\rho}_B + \boldsymbol{\rho}_A \otimes (\boldsymbol{H}_B\boldsymbol{\rho}_B). \tag{5.201}$$

Because, in general

$$\begin{aligned}
&\boldsymbol{H}\boldsymbol{\rho} \\
&\quad = -\frac{\mu}{2}(\boldsymbol{B}\cdot\boldsymbol{r})\,\boldsymbol{1} + \text{terms multiplied by sigmas} \\
&\quad = \frac{1}{2}\langle E\rangle\,\boldsymbol{1} + \text{terms multiplied by sigmas},
\end{aligned} \tag{5.202}$$

we get that

$$\begin{aligned}
&\boldsymbol{H}_{AB}\boldsymbol{\rho}_{AB} \\
&\quad = \left(\frac{1}{2}\langle E_A\rangle\,\boldsymbol{1}_A + \ldots\right) \otimes \boldsymbol{\rho}_B + \boldsymbol{\rho}_A \otimes \left(\frac{1}{2}\langle E_B\rangle\,\boldsymbol{1}_B + \ldots\right) \\
&\quad = \left(\frac{1}{2}\langle E_A\rangle\,\boldsymbol{1}_A + \ldots\right) \otimes \left(\frac{1}{2}\boldsymbol{1}_B + \ldots\right) + \left(\frac{1}{2}\boldsymbol{1}_A + \ldots\right) \otimes \left(\frac{1}{2}\langle E_B\rangle\,\boldsymbol{1}_B + \ldots\right) \\
&\quad = \frac{1}{4}\left(\langle E_A\rangle + \langle E_B\rangle\right)\boldsymbol{1}_A \otimes \boldsymbol{1}_B + \ldots,
\end{aligned} \tag{5.203}$$

where "..." is a shortcut for terms that have some sigmas in them, be it one or two. So clearly, extracting what stands in front of $\boldsymbol{1}_A \otimes \boldsymbol{1}_B$ and multiplying it by 4 this time yields the right answer.

This is therefore the operation that we need here:

$$\langle E_{AB}\rangle = 4\Re_A\Re_B\left(\boldsymbol{H}_{AB}\boldsymbol{\rho}_{AB}\right). \tag{5.204}$$

Going one step further and thinking of sigmas as matrices rather than quaternion units, we replace each $2\Re$ with its own trace operation, which yields

$$\langle E_{AB}\rangle = \mathrm{Tr}_A \mathrm{Tr}_B \left(\boldsymbol{H}_{AB}\boldsymbol{\rho}_{AB}\right). \tag{5.205}$$

Various operations on tensor products of two qubits can be rewritten in the form of simple matrix and vector calculus; for example, a tensor product of two 2×2 matrices can be represented by a single 4×4 matrix.

This is how it works.

Let us consider a tensor product of two unitary vectors in a two-dimensional Hilbert space, for example, $\mid u\rangle \otimes \mid v\rangle$. Let

$$\mid u\rangle = u_0 \mid 0\rangle + u_1 \mid 1\rangle \tag{5.206}$$

and

$$\mid v\rangle = v_0 \mid 0\rangle + v_1 \mid 1\rangle. \tag{5.207}$$

Then

$$\begin{aligned} \mid u\rangle \otimes \mid v\rangle &= (u_0 \mid 0\rangle + u_1 \mid 1\rangle) \otimes (v_0 \mid 0\rangle + v_1 \mid 1\rangle) \\ &= u_0 v_0 \mid 0\rangle \otimes \mid 0\rangle + u_0 v_1 \mid 0\rangle \otimes \mid 1\rangle + u_1 v_0 \mid 1\rangle \otimes \mid 0\rangle + u_1 v_1 \mid 1\rangle \otimes \mid 1\rangle \\ &= u_0 v_0 \mid 00\rangle + u_0 v_1 \mid 01\rangle + u_1 v_0 \mid 10\rangle + u_1 v_1 \mid 11\rangle, \end{aligned} \tag{5.208}$$

where we have used a shorthand $\mid 00\rangle$ for $\mid 0\rangle \otimes \mid 0\rangle$ and for other combinations of the basis vectors.

But we can always reinterpret binary sequences as decimal numbers, namely,

$$\begin{aligned} 00 &\equiv 0, \\ 01 &\equiv 1, \\ 10 &\equiv 2, \\ 11 &\equiv 3. \end{aligned}$$

So we can rewrite our tensor product of $\mid u\rangle \otimes \mid v\rangle$ as

$$u_0 v_0 \mid \mathbf{0}\rangle + u_0 v_1 \mid \mathbf{1}\rangle + u_1 v_0 \mid \mathbf{2}\rangle + u_1 v_1 \mid \mathbf{3}\rangle, \tag{5.209}$$

Computational basis

where

$$\begin{aligned} |\,\mathbf{0}\rangle &\equiv |\,0\rangle \otimes |\,0\rangle, \\ |\,\mathbf{1}\rangle &\equiv |\,0\rangle \otimes |\,1\rangle, \\ |\,\mathbf{2}\rangle &\equiv |\,1\rangle \otimes |\,0\rangle, \\ |\,\mathbf{3}\rangle &\equiv |\,1\rangle \otimes |\,1\rangle, \end{aligned}$$

or we can also represent it in terms of a "column vector":

$$|\,u\rangle \otimes |\,v\rangle \equiv \begin{pmatrix} x_{\mathbf{0}} \\ x_{\mathbf{1}} \\ x_{\mathbf{2}} \\ x_{\mathbf{3}} \end{pmatrix} = \begin{pmatrix} u_0 v_0 \\ u_0 v_1 \\ u_1 v_0 \\ u_1 v_1 \end{pmatrix}. \tag{5.210}$$

This can be obtained by the following operation as well:

$$\begin{pmatrix} u_0 \begin{pmatrix} v_0 \\ v_1 \end{pmatrix} \\ u_1 \begin{pmatrix} v_0 \\ v_1 \end{pmatrix} \end{pmatrix}. \tag{5.211}$$

We note the difference between the tensor product and the so-called direct sum of two vectors, which would be

$$|\,u\rangle \oplus |\,v\rangle \equiv \begin{pmatrix} u_0 \\ u_1 \\ v_0 \\ v_1 \end{pmatrix}. \tag{5.212}$$

Not infrequently people ask why a quantum mechanical system of two qubits is described by the tensor (direct) product and not by the direct sum of two qubits. In classical mechanics, for example, a system of two material points is described by the direct sum of the particles' guiding vectors. The reason is that quantum mechanics is a probability theory, and a probability of some combined outcome in a multicomponent system is a *product* of probabilities of outcomes pertaining to each component.

Now let us consider a tensor product of two 2×2 matrices. These are representations of operators acting on unitary vectors in the two-dimensional Hilbert space, namely,

$$\begin{pmatrix} a_{00} & a_{01} \\ a_{10} & a_{11} \end{pmatrix} \equiv |\,0\rangle a_{00} \langle 0\,| + |\,0\rangle a_{01} \langle 1\,| + |\,1\rangle a_{10} \langle 0\,| + |\,1\rangle a_{11} \langle 1\,| = \boldsymbol{a}. \tag{5.213}$$

Let $\boldsymbol{b}$ be a similar matrix/operator. Then

$$\begin{aligned}
\boldsymbol{a}\otimes\boldsymbol{b} &= (\mid 0\rangle a_{00}\langle 0\mid + \mid 0\rangle a_{01}\langle 1\mid + \mid 1\rangle a_{10}\langle 0\mid + \mid 1\rangle a_{11}\langle 1\mid) \\
&\quad \otimes (\mid 0\rangle b_{00}\langle 0\mid + \mid 0\rangle b_{01}\langle 1\mid + \mid 1\rangle b_{10}\langle 0\mid + \mid 1\rangle b_{11}\langle 1\mid) \\
&= \mid 00\rangle a_{00}b_{00}\langle 00\mid + \mid 00\rangle a_{00}b_{01}\langle 01\mid + \mid 01\rangle a_{00}b_{10}\langle 00\mid + \mid 01\rangle a_{00}b_{11}\langle 01\mid \\
&\quad + \mid 00\rangle a_{01}b_{00}\langle 10\mid + \mid 00\rangle a_{01}b_{01}\langle 11\mid + \mid 01\rangle a_{01}b_{10}\langle 10\mid + \mid 01\rangle a_{01}b_{11}\langle 11\mid \\
&\quad + \mid 10\rangle a_{10}b_{00}\langle 00\mid + \mid 10\rangle a_{10}b_{01}\langle 01\mid + \mid 11\rangle a_{10}b_{10}\langle 00\mid + \mid 11\rangle a_{10}b_{11}\langle 01\mid \\
&\quad + \mid 10\rangle a_{11}b_{00}\langle 10\mid + \mid 10\rangle a_{11}b_{01}\langle 11\mid + \mid 11\rangle a_{11}b_{10}\langle 10\mid + \mid 11\rangle a_{11}b_{11}\langle 11\mid \\
&= \mid \mathbf{0}\rangle a_{00}b_{00}\langle \mathbf{0}\mid + \mid \mathbf{0}\rangle a_{00}b_{01}\langle \mathbf{1}\mid + \mid \mathbf{1}\rangle a_{00}b_{10}\langle \mathbf{0}\mid + \mid \mathbf{1}\rangle a_{00}b_{11}\langle \mathbf{1}\mid \\
&\quad + \mid \mathbf{0}\rangle a_{01}b_{00}\langle \mathbf{2}\mid + \mid \mathbf{0}\rangle a_{01}b_{01}\langle \mathbf{3}\mid + \mid \mathbf{1}\rangle a_{01}b_{10}\langle \mathbf{2}\mid + \mid \mathbf{1}\rangle a_{01}b_{11}\langle \mathbf{3}\mid \\
&\quad + \mid \mathbf{2}\rangle a_{10}b_{00}\langle \mathbf{0}\mid + \mid \mathbf{2}\rangle a_{10}b_{01}\langle \mathbf{1}\mid + \mid \mathbf{3}\rangle a_{10}b_{10}\langle \mathbf{0}\mid + \mid \mathbf{3}\rangle a_{10}b_{11}\langle \mathbf{1}\mid \\
&\quad + \mid \mathbf{2}\rangle a_{11}b_{00}\langle \mathbf{2}\mid + \mid \mathbf{2}\rangle a_{11}b_{01}\langle \mathbf{3}\mid + \mid \mathbf{3}\rangle a_{11}b_{10}\langle \mathbf{2}\mid + \mid \mathbf{3}\rangle a_{11}b_{11}\langle \mathbf{3}\mid \\
&\equiv \begin{pmatrix} a_{00}b_{00} & a_{00}b_{01} & a_{01}b_{00} & a_{01}b_{01} \\ a_{00}b_{10} & a_{00}b_{11} & a_{01}b_{10} & a_{01}b_{11} \\ a_{10}b_{00} & a_{10}b_{01} & a_{11}b_{00} & a_{11}b_{01} \\ a_{10}b_{10} & a_{10}b_{11} & a_{11}b_{10} & a_{11}b_{11} \end{pmatrix}.
\end{aligned} \tag{5.214}$$

This matrix can be also obtained by the following operation:

$$\begin{pmatrix} a_{00}\begin{pmatrix} b_{00} & b_{01} \\ b_{10} & b_{11} \end{pmatrix} & a_{01}\begin{pmatrix} b_{00} & b_{01} \\ b_{10} & b_{11} \end{pmatrix} \\ a_{10}\begin{pmatrix} b_{00} & b_{01} \\ b_{10} & b_{11} \end{pmatrix} & a_{11}\begin{pmatrix} b_{00} & b_{01} \\ b_{10} & b_{11} \end{pmatrix} \end{pmatrix}. \tag{5.215}$$

Finally, let us observe that

$$\begin{aligned}
&\mathrm{Tr}_a \mathrm{Tr}_b\, \boldsymbol{a}\otimes\boldsymbol{b} \\
&\quad = \mathrm{Tr}_a\boldsymbol{a}\cdot\mathrm{Tr}_b\boldsymbol{b} = (a_{00}+a_{11})\cdot(b_{00}+b_{11}) \\
&\quad = a_{00}b_{00} + a_{00}b_{11} + a_{11}b_{00} + a_{11}b_{11}
\end{aligned} \tag{5.216}$$

is the same as a single trace of the large 4×4 matrix.

Matrix multiplication of the 4×4 matrix by the 4-slot vector yields another 4-slot vector, which, when contracted with a 4-slot form, produces a number that is the same as would be obtained from, say, operation such as $\langle\Psi_{AB}\mid \boldsymbol{\rho}_{AB}\mid \Psi_{AB}\rangle$. So, instead of working with explicit tensor products of 2×2 Pauli matrices, we can switch to these 4×4 matrices; and instead of calculating double trace operations such as $\mathrm{Tr}_A\mathrm{Tr}_B$, we can calculate single traces of the corresponding 4×4 matrices.

If the 4×4 matrix in question is made of the density quaternion $\boldsymbol{\rho}_{AB}$, then it must satisfy the same requirements that we had discovered for single qubit density operators:

Properties of 4×4 density matrices

1. It must be Hermitian.
2. Its trace must be 1.
3. It must be positive.
4. It must be idempotent if it describes a pure state.

4×4 matrix invariants tell us if $\boldsymbol{\rho}_{AB}$ is physical.

Returning to the Peres-Horodeckis criterion for biqubit separability, we can fairly easily check that a given 4×4 density matrix of some biqubit state is *not* positive by calculating its determinant. If the determinant is negative, it means that either one or three eigenvalues of the matrix are negative, which implies that on the corresponding eigenvectors the expectation values of $\boldsymbol{\rho}_{AB}$ are negative, too, which in turn implies that the state described by $\boldsymbol{\rho}_{AB}$ is unphysical, because these expectation values are supposed to be transition probabilities.

It may happen that the 4×4 density matrix has either two or four negative eigenvalues, in which case its determinant is still positive. But the case with 4 negative eigenvalues being negative can be easily eliminated, because in this case the trace would be negative, too.

The case with two negative eigenvalues is harder. Here we may have to look at other matrix invariants or simply find all eigenvalues explicitly.

Let us see how this works in practice.

Appendix C lists 4×4 matrices that represent some tensor products of Pauli matrices. Let us rewrite the density operator of Bell state $|\Psi^-\rangle$ in the 4×4 form. We start from

$$\boldsymbol{\rho}^- = \frac{1}{4}\left(\mathbf{1} \otimes \mathbf{1} - \boldsymbol{\sigma}_x \otimes \boldsymbol{\sigma}_x - \boldsymbol{\sigma}_y \otimes \boldsymbol{\sigma}_y - \boldsymbol{\sigma}_z \otimes \boldsymbol{\sigma}_z\right). \tag{5.217}$$

Using equations (C.13), (C.16), (C.19), and (C.22) on pages 399 and 401 results in

$$\begin{aligned}\boldsymbol{\rho}^- &= \frac{1}{4}\left(\begin{pmatrix} 1 & 0 & 0 & 0 \\ 0 & 1 & 0 & 0 \\ 0 & 0 & 1 & 0 \\ 0 & 0 & 0 & 1 \end{pmatrix} - \begin{pmatrix} 0 & 0 & 0 & 1 \\ 0 & 0 & 1 & 0 \\ 0 & 1 & 0 & 0 \\ 1 & 0 & 0 & 0 \end{pmatrix}\right. \\ &\qquad \left. - \begin{pmatrix} 0 & 0 & 0 & -1 \\ 0 & 0 & 1 & 0 \\ 0 & 1 & 0 & 0 \\ -1 & 0 & 0 & 0 \end{pmatrix} - \begin{pmatrix} 1 & 0 & 0 & 0 \\ 0 & -1 & 0 & 0 \\ 0 & 0 & -1 & 0 \\ 0 & 0 & 0 & 1 \end{pmatrix}\right)\end{aligned}$$

$$= \frac{1}{4}\begin{pmatrix} 0 & 0 & 0 & 0 \\ 0 & 2 & -2 & 0 \\ 0 & -2 & 2 & 0 \\ 0 & 0 & 0 & 0 \end{pmatrix}. \tag{5.218}$$

The determinant of the matrix is zero, but it is easy to see that three eigenvalues of the matrix are zero and the remaining one is +1. So the matrix makes a viable representation of a density operator. On no biqubit projector state $\boldsymbol{P}$ do we find that $\mathrm{Tr}\,(\boldsymbol{P}\boldsymbol{\rho}^-) < 0$.

But now let us have a look at the Werner "state"

$$\boldsymbol{\rho}^+ = \frac{1}{4}\left(\mathbf{1}\otimes\mathbf{1} + \boldsymbol{\sigma}_x\otimes\boldsymbol{\sigma}_x + \boldsymbol{\sigma}_y\otimes\boldsymbol{\sigma}_y + \boldsymbol{\sigma}_z\otimes\boldsymbol{\sigma}_z\right). \tag{5.219}$$

This time the corresponding 4×4 matrix is

$$\begin{aligned}\boldsymbol{\rho}^+ &= \frac{1}{4}\left(\begin{pmatrix} 1 & 0 & 0 & 0 \\ 0 & 1 & 0 & 0 \\ 0 & 0 & 1 & 0 \\ 0 & 0 & 0 & 1 \end{pmatrix} + \begin{pmatrix} 0 & 0 & 0 & 1 \\ 0 & 0 & 1 & 0 \\ 0 & 1 & 0 & 0 \\ 1 & 0 & 0 & 0 \end{pmatrix}\right.\\ &\qquad \left. + \begin{pmatrix} 0 & 0 & 0 & -1 \\ 0 & 0 & 1 & 0 \\ 0 & 1 & 0 & 0 \\ -1 & 0 & 0 & 0 \end{pmatrix} + \begin{pmatrix} 1 & 0 & 0 & 0 \\ 0 & -1 & 0 & 0 \\ 0 & 0 & -1 & 0 \\ 0 & 0 & 0 & 1 \end{pmatrix}\right)\\ &= \frac{1}{4}\begin{pmatrix} 2 & 0 & 0 & 0 \\ 0 & 0 & 2 & 0 \\ 0 & 2 & 0 & 0 \\ 0 & 0 & 0 & 2 \end{pmatrix}.\end{aligned} \tag{5.220}$$

The determinant of the

$$\frac{1}{4}\begin{pmatrix} 0 & 2 & 0 \\ 2 & 0 & 0 \\ 0 & 0 & 2 \end{pmatrix} \tag{5.221}$$

submatrix is $-1/8$. The algebraic complement of the $1/2$ in the upper left corner of the 4×4 matrix is therefore $(-1)^{1+1}(-1/8) = -1/8$. Hence, using the Laplace expansion formula, we get as the determinant of the 4×4 matrix

$$\frac{1}{2}\cdot\left(-\frac{1}{8}\right) = -\frac{1}{16}. \tag{5.222}$$

It is *negative*, which implies that either one or three eigenvalues of the 4×4 matrix are negative. Therefore, the matrix does not qualify as a plausible density operator representation, because a biqubit projector state $\boldsymbol{P}$ exists such that

$$\mathrm{Tr}\left(\boldsymbol{P}\boldsymbol{\rho}^+\right) < 0. \tag{5.223}$$

Derivation of the Peres-Horodeckis criterion

It is enlightening to review the mathematical reasoning that led the family of Horodeckis to conclude that the operation described in this section—that is, switching the sign in front of sigmas (or varsigmas) that refer to one of the two qubits, but not both, and then checking if the resulting new state is physical—yields a necessary and sufficient condition for separability of the original state [66].

First, let us recapitulate some basic terminology. A state $\boldsymbol{\rho}_{AB}$ is called separable if it can be represented by a finite mixture of *simple* separable states, namely, if

$$\boldsymbol{\rho}_{AB} = \sum_{i=1}^{k} p^i_{AB} \boldsymbol{\rho}_{iA} \otimes \boldsymbol{\rho}_{iB}. \tag{5.224}$$

A state $\boldsymbol{\rho}_{AB}$ is physical if for *any* biqubit projector $\boldsymbol{P}_{AB} = |\,\Psi_{AB}\rangle \otimes \langle \Psi_{AB}\,|$

$$\mathrm{Tr}\left(\boldsymbol{P}_{AB}\boldsymbol{\rho}_{AB}\right) \geq 0. \tag{5.225}$$

The density operator $\boldsymbol{\rho}_{AB}$ that satisfies this condition is also called *positive*. In this case this is a mathematical way of saying that it is physical.

Operators form their own Hilbert space.

Linear operators that act on the qubit Hilbert space themselves form a Hilbert space of their own in which a scalar product

$$\langle \boldsymbol{A}, \boldsymbol{B} \rangle = \mathrm{Tr}\left(\boldsymbol{B}^\dagger \boldsymbol{A}\right) \tag{5.226}$$

can be defined, where the † operation indicates Hermitian adjoint, that is, a matrix transposition combined with complex conjugation of the matrix terms (it does not affect any of the Pauli matrices, which are all Hermitian).

Positive and completely positive maps

Qubit operators may be transformed into one another by the action of linear *maps*. The maps are said to be *positive* if they convert positive operators into some other positive operators. Maps are said to be *completely positive* if their tensor product with the identity is a positive map in the larger biqubit operator space. For example, if $\boldsymbol{\Lambda}_A$ is a map, then it is said to be completely positive if

$$\boldsymbol{\Lambda}_A \otimes \boldsymbol{1}_B \tag{5.227}$$

is positive on the space of biqubit operators.

The reasoning now runs as follows. First we observe that $\boldsymbol{\rho}_{AB}$ must be separable if and only if

$$\mathrm{Tr}\left(\boldsymbol{H}_{AB}\boldsymbol{\rho}_{AB}\right) \geq 0 \tag{5.228}$$

for *any* Hermitian operator $\boldsymbol{H}_{AB}$, such that

$$\mathrm{Tr}\left(\boldsymbol{H}_{AB}\boldsymbol{P}_A \otimes \boldsymbol{P}_B\right) \geq 0, \tag{5.229}$$

where $\boldsymbol{P}_A$ and $\boldsymbol{P}_B$ are arbitrary projectors in the space of qubits A and B, respectively.

This observation seems quite obvious, at least in one direction. However, there is a fairly simple proof that derives from certain general properties of convex spaces equipped in a scalar product. And a space of separable states has these two properties, thereby demonstrating the veracity of the statement in both directions, meaning, *if* and *only if,* too.

A theorem proven by Jamiołkowski in 1972 [82] is now invoked that translates the condition $\mathrm{Tr}\,(\boldsymbol{H}_{AB}\boldsymbol{\rho}_{AB}) \geq 0$ into the language of positive maps with a specific expression in place of $\boldsymbol{H}_{AB}$, namely,

Jamiołkowski theorem

$$\mathrm{Tr}\,(\boldsymbol{P}_{AB}\,(\mathbf{1}_A \otimes \boldsymbol{\Lambda}_B\boldsymbol{\rho}_{AB})) \geq 0, \tag{5.230}$$

where $\boldsymbol{P}_{AB}$ is a one-dimensional Hermitian projector in the biqubit space and $\boldsymbol{\Lambda}_B$ is an arbitrary positive map in the space of qubit B. The projector can be dropped from this condition because it does not affect the positivity of $\mathbf{1}_A \otimes \boldsymbol{\Lambda}_B\boldsymbol{\rho}_{AB}$. Hence, we end up with the following theorem: $\boldsymbol{\rho}_{AB}$ *is separable if and only if* $\mathbf{1}_A \otimes \boldsymbol{\Lambda}_B\boldsymbol{\rho}_{AB}$ *is positive for* any *positive map* $\boldsymbol{\Lambda}_B$.

At this stage a theorem by Strømer and Woronowicz [149] is invoked that says that *any* positive map $\boldsymbol{\Lambda}_B$ in two- and three-dimensional Hilbert spaces is of the form

Strømer-Woronowicz theorem

$$\boldsymbol{\Lambda}_B = \boldsymbol{X}_B + \boldsymbol{Y}_B\boldsymbol{T}_B, \tag{5.231}$$

where $\boldsymbol{X}_B$ and $\boldsymbol{Y}_B$ are *completely positive* maps and $\boldsymbol{T}_B$ is a matrix transposition in the space of qubit B. Thus,

$$\begin{aligned} \mathbf{1}_A \otimes \boldsymbol{\Lambda}_B &= \mathbf{1}_A \otimes (\boldsymbol{X}_B + \boldsymbol{Y}_B\boldsymbol{T}_B) \\ &= \mathbf{1}_A \otimes \boldsymbol{X}_B + \mathbf{1}_A \otimes \boldsymbol{Y}_B\boldsymbol{T}_B. \end{aligned} \tag{5.232}$$

Because $\boldsymbol{X}_B$ and $\boldsymbol{Y}_B$ are *completely positive,* their tensor products with $\mathbf{1}_A$ are positive maps that do not change positivity of $\boldsymbol{\rho}_{AB}$. But $\mathbf{1}_A \otimes \boldsymbol{T}_B$ is the only term above that may change the positivity of $\boldsymbol{\rho}_{AB}$. Hence, we are left with the following criterion that has been literally distilled from the original formulation with $\boldsymbol{H}_{AB}$: $\boldsymbol{\rho}_{AB}$ *is separable if and only if* $\mathbf{1}_A \otimes \boldsymbol{T}_B\boldsymbol{\rho}_{AB}$ *is positive.*

The *partial transposition* operation $\mathbf{1}_A \otimes \boldsymbol{T}_B$ (it is called "partial" because it affects only one of the two qubits) does not affect $\mathbf{1}_B$, $\boldsymbol{\sigma}_{xB}$, and $\boldsymbol{\sigma}_{zB}$, but it affects $\boldsymbol{\sigma}_{yB}$, because this Pauli matrix is antisymmetric:

Partial transposition operation

$$\boldsymbol{\sigma}_y^T = \begin{pmatrix} 0 & -\mathrm{i} \\ \mathrm{i} & 0 \end{pmatrix}^T = \begin{pmatrix} 0 & \mathrm{i} \\ -\mathrm{i} & 0 \end{pmatrix} = -\boldsymbol{\sigma}_y. \tag{5.233}$$

Indeed, instead of replacing all three sigmas (or varsigmas) for particle A (or B, but not both) with minus sigmas (or minus varsigmas), we could have replaced $\boldsymbol{\sigma}_y$ with $-\boldsymbol{\sigma}_y$. This would have converted the Bell state $\boldsymbol{p}_{AB}^{-}$ into an unphysical state. But why should the y direction be the only one blessed so? After all, it is up to us to define *which* direction in space happens to be y.

The physical equivalent of the partial transpose operation is to reverse the polarity of one of the qubits that make a biqubit completely (all three sigmas are reversed) or to reflect its state in a mirror that is placed in the $\boldsymbol{e}_x \times \boldsymbol{e}_z$ plane (only $\boldsymbol{\sigma}_y$ is reversed.)

The partial transpose condition works *only* for two- and three-dimensional Hilbert spaces. For higher-dimensional spaces the condition $\mathrm{Tr}\left(\mathbf{1}_A \otimes \boldsymbol{T}_B \boldsymbol{\rho}_{AB}\right) \geq 0$ is a *necessary* condition for $\boldsymbol{\rho}_{AB}$ to be separable, but not a sufficient one.

In summary, the Horodeckis proof tells us that the partial transpose criterion works because it is a thorough distillation—meaning that all that's left after various irrelevant stuff is removed—of the more obvious but harder to apply criterion (5.228).

Physical meaning of entangled states

The Horodeckis theorem and its physical interpretation give us deep insight into the nature of entangled states. Such states are qualitatively different from separable states and, as the result, are difficult to concoct. It is not enough to just bring two qubits together. All we'll end up with will be a separable state of two qubits or a mixture of such states. To produce an entangled state, we have to make them interact in a nontrivial manner. For example, we may have to split a composite particle, or make an atom emit a photon, or make the two qubits dance together in a special way. Only then, as we have seen in Section 5.4, are we going to produce an entangled state.

Durt theorem

A theorem proven by Thomas Durt of the Free University of Brussels in 2003 [35] illustrates this more formally. The theorem states that for a bipartite evolution given by

$$\mathrm{i}\hbar \frac{\partial}{\partial t} \mid \Psi(t)\rangle_{AB} = \boldsymbol{H}_{AB}(t) \mid \Psi(t)\rangle_{AB} \tag{5.234}$$

all the product states remain product states during the interaction if and only if $\boldsymbol{H}_{AB}(t)$ can be factorized as follows:

$$\boldsymbol{H}_{AB}(t) = \boldsymbol{H}_A(t) \otimes \mathbf{1}_B + \mathbf{1}_A \otimes \boldsymbol{H}_B(t), \tag{5.235}$$

which is equivalent to the energy form (5.7) presented earlier, page 192, that described two non interacting qubits. In other words, to entangle two qubits, we *must* activate a nontrivial interaction $\boldsymbol{\eta}_{AB}$—compare with equation (5.8)—between them.

This observation is then extended to arbitrary mixtures and their von Neumann evolution.

The theorem is valid back to front as well. It tells us that "in quantum mechanics to interact means nearly always to entangle." Entanglement-free interaction between quantum systems is not possible unless at least one of them is classicalized.

5.11 Impure Quantum Mechanics

Traditional unitary quantum mechanics is like Newtonian mechanics without friction. It is a highly idealized picture that, while capturing a great many phenomena quite well, does not correspond entirely to the real world.

In Section 2.12 (pages 86 to 93), Figure 2.9 (page 89), we saw a vivid demonstration of the problem. The amplitude of the observed Rabi oscillations diminished exponentially on the time scale of about a microsecond. The quantronium qubit that was prepared in a fully polarized state gradually lost its polarization. Our model could not account for the phenomenon. The $\boldsymbol{r} \times \boldsymbol{B}$ term ensured that only the direction and not the length of the polarization vector $\boldsymbol{r}$ would change.

Then we revisited the issue in Section 4.9.1, where we discussed a general solution to the Schrödinger equation and discovered that the equation preserved the unitarity of states being evolved; see equation (4.308), page 157.

For many years physicists were baffled by this conundrum. How could it be that quantum physics, which explained so many phenomena so well, could not account for depolarization, dephasing, decoherence, and ultimately the very act of measurement, the act that is rather fundamental to physics. Yet, the Schrödinger equation, with its many variants (Pauli, Dirac), a rich assortment of Hamiltonians, and added complexities of quantum field theories and statistical physics, was the only fundamental quantum equation known to work. It was also, as we observed in Section 4.9, the *simplest* possible equation to evolve a quantum state.

Eventually a solution derived from an observation we made in Section 5.8, "Single Qubit Expectation Values." There we noted that a *pure* biqubit maximally entangled state might look like two qubits, both in completely chaotic mixed states, when the constituent qubits are measured separately, and the purity of the biqubit state asserts itself as *correlations* between the otherwise random outcomes of measurements made on the constituent qubits.

Purification

It turns out that, mathematically, every mixed quantum state can be embedded in a larger system, such that the larger system is unitary and its internal entanglement results in the observed mixed state of the embedded component. The mathematical procedure of finding such an embedding is called *purification*. Thus, we can state that every *impure* quantum state can be *purified*.

Purifying a single qubit

The statement can be demonstrated easily for a single qubit. Let us recall equation (5.161) in Section 5.9 that represented *every* possible *pure* biqubit state in the following form of the D$^-$ class.

$$\begin{aligned} \boldsymbol{p}_{AB} &= \frac{1}{4}\Big(\varsigma_{1A}\otimes\varsigma_{1B} \\ &\quad +r\left(\varsigma_{xA}\otimes\varsigma_{1B}-\varsigma_{1A}\otimes\varsigma_{xB}\right) \\ &\quad -\left(\varsigma_{xA}\otimes\varsigma_{xB}+\sqrt{1-r^2}\left(\varsigma_{yA}\otimes\varsigma_{yB}+\varsigma_{zA}\otimes\varsigma_{zB}\right)\right)\Big) \end{aligned} \tag{5.236}$$

As we have learned in Section 5.8, the way to eliminate the *other* qubit from a biqubit is to contract the whole biqubit with the other qubit's ς^1.

Let us extract qubit A from $\boldsymbol{p}_{AB}$ given by (5.236):

$$\begin{aligned} \boldsymbol{p}_A &= \langle \varsigma^1_B, \boldsymbol{p}_{AB}\rangle \\ &= \frac{1}{4}\left(2\varsigma_{1A}+2r\varsigma_{xA}\right)=\frac{1}{2}\left(\varsigma_{1A}+r\varsigma_{xA}\right). \end{aligned} \tag{5.237}$$

Similarly,

$$\boldsymbol{p}_B=\langle\varsigma^1_A,\boldsymbol{p}_{AB}\rangle=\frac{1}{2}\left(\varsigma_{1B}-r\varsigma_{xB}\right). \tag{5.238}$$

They are both mixed-state qubits, one pointing in the $\boldsymbol{e}_{xA}$ direction and the other one in the $-\boldsymbol{e}_{xB}$ direction. But let us recall that we have obtained equation (5.236) by rotating the varsigmas pertaining to each qubit independently so as to eliminate as many terms from $\boldsymbol{p}_{AB}$ as possible. Hence, $\boldsymbol{e}_{xA}$ and $-\boldsymbol{e}_{xB}$ do not have to point in the same direction. Thus, any two qubits in mixed states and of the same polarization value r, pointing in arbitrary directions can be made to look like they are constituents of a pure biqubit state, if we remember to add the third term in equation (5.236), the one that describes the correlation between the qubits.

This is an algebraic, not a physical, procedure. It does not tell us how, given any two physical qubits in the required mixed states, we can add the third term

$$\frac{1}{4}\left(\varsigma_{xA}\otimes\varsigma_{xB}+\sqrt{1-r^2}\left(\varsigma_{yA}\otimes\varsigma_{yB}+\varsigma_{zA}\otimes\varsigma_{zB}\right)\right), \tag{5.239}$$

that will make the combined state pure. But here the Durt theorem comes to the rescue. It tells us that any nontrivial interaction between qubits is bound to entangle them. To engineer the interaction and to control it with sufficient precision to produce just the state (5.236) may be difficult but does not seem impossible.

Purification is not unique. The same mixed state of a subcomponent can be made a part of various larger systems, in each case contributing to some pure state of the whole. In the case of the biqubit state discussed above, we can purify qubit A by coupling it to any of the possible qubits B pointing in various directions, as long as their polarization value r is the same as that of A. We can also purify qubit A by coupling it to larger systems, for example, to biqubits or triqubits or n-qubits.

So here is an idea. Why not employ this mechanism to include mixed states, at least such as may be generated by quantum processes, into the framework of the unitary formalism? We can combine a quantum subsystem, which is to find itself in a mixed state later, with any number of other quantum systems at the beginning. We can then use the unitary formalism and the Schrödinger equation to describe the evolution of the whole combined system—preserving its purity—only to find that our specific subsystem ends up in a mixed state. If we are not specifically interested in the subsystem's environment, we can kill the purifying components with the ς^1 forms, or, in the parlance of density matrix theory, we can *trace them out*. What's left would be a quantum evolution that leaves our subsystem in a mixed state.

This should work with one snag. The evolution produced will usually differ in details depending on the type of the *bath* the mixed-state subsystem has been embedded in and interactions employed. But this is also the case with classical dissipative systems, which, similarly, evolve differently depending on the specific dissipation mechanism.

Purifying the universe

By enclosing boxes within larger boxes we are led to believe, as we have remarked earlier, that everything can be explained by the unitary formalism, until we try to embed the universe as a whole in order to explain why it is in a mixed state. We arrive at a surprising conclusion that the universe must be in a pure state, because there is nothing out there to entangle it with, unless we invoke a multitude of other universes with which ours may be entangled. This is entertaining stuff, and it's fine as an example of extrapolation and exploration of ideas, but it is not exactly natural science unless some experimental evidence can be delivered in support of such concepts.

Still, for a less ambitious task of modeling a depolarizing qubit, the program discussed here is workable. In the following sections, we'll have a closer look at the most basic results.

5.11.1 Nonunitary Evolution

Let us consider a system of two qubits, of which one is going to stand for an "environment" and the other will be subject to some evolution and observation. Although the model is simple, it is sufficient to demonstrate a number of important nonunitary features that a quantum system entangled with its environment may display.

Let the density operator of the "environment" qubit at $t = 0$ be $\boldsymbol{\rho}_E(0)$ and the density operator of the qubit we want to measure independently of the environment qubit be $\boldsymbol{\rho}_A(0)$ at $t = 0$. Both $\boldsymbol{\rho}_E(0)$ and $\boldsymbol{\rho}_A(0)$ are pure states, and the initial combined state is simple and separable (cf. page 201):

$$\boldsymbol{\rho}_{AE}(0) = \boldsymbol{\rho}_A(0) \otimes \boldsymbol{\rho}_E(0). \tag{5.240}$$

We assume that the biqubit is fully isolated and that its evolution is unitary, as given by equation (4.316), page 158:

$$\boldsymbol{\rho}_{AE}(t) = \boldsymbol{U}_{AE}(t)\boldsymbol{\rho}_{AE}(0)\boldsymbol{U}^{\dagger}_{AE}(t) = \boldsymbol{U}_{AE}(t)\Big(\boldsymbol{\rho}_A(0) \otimes \boldsymbol{\rho}_E(0)\Big)\boldsymbol{U}^{\dagger}_{AE}(t). \tag{5.241}$$

Unless the evolution is trivial, the qubits will become entangled in its course and, when measured individually, will appear in mixed states, but the combined system remains pure. What will the evolution of qubit A, ignoring the environment qubit B, look like?

To answer the question, we rewrite equation (5.241) in more detail. Let Latin indexes i, j, k, and l label basis states in the Hilbert space of qubit A, and let Greek indexes α, β, γ, and δ label basis states in the Hilbert space of qubit E. Then, since $\boldsymbol{U}_{AE}$ is ultimately a linear operation, we can represent it as

$$\boldsymbol{U}_{AE}(t) = \sum_{i,\alpha,\beta,j} \mid i\rangle \mid \alpha\rangle U_{i\alpha\beta j}(t)\langle \beta \mid \langle j \mid, \tag{5.242}$$

where

$$U_{i\alpha\beta j}(t) = \langle i \mid \langle \alpha \mid \boldsymbol{U}_{AE}(t) \mid \beta\rangle \mid j\rangle. \tag{5.243}$$

Furthermore, since $\boldsymbol{\rho}_E(0)$ is pure, we can rewrite it as

$$\boldsymbol{\rho}_E(0) = \mid \Psi_E(0)\rangle\langle\Psi_E(0) \mid. \tag{5.244}$$

Substituting these in equation (5.241) yields

$$\rho_{AE}(t) = \left(\sum_{i,\alpha,\beta,j} \mid i\rangle \mid \alpha\rangle U_{i\alpha\beta j}(t) \langle \beta \mid \langle j \mid \right)$$

$$\rho_A(0) \otimes \Big(\mid \Psi_E(0)\rangle\langle\Psi_E(0) \mid \Big) \left(\sum_{k,\gamma,\delta,l} \mid k\rangle \mid \gamma\rangle U^\dagger_{k\gamma\delta l}(t)\langle\delta \mid \langle l \mid \right). \tag{5.245}$$

Contractions of $\mid \Psi_E(0)\rangle$ and $\langle\Psi_E(0) \mid$ with appropriate bras and kets of the Greek index type produce numbers, which multiply the $U_{i\alpha\beta j}$ and $U^\dagger_{k\gamma\delta l}$ terms resulting in

$$\rho_{AE}(t) = \sum_{i,\alpha,\beta,j,k,\gamma,\delta,l} \Big(\mid i\rangle \mid \alpha\rangle U_{i\alpha\beta j}\langle\beta \mid \Psi_E(0)\rangle\langle j \mid \Big) \rho_A(0) \Big(\mid k\rangle\langle\Psi_E(0) \mid \gamma\rangle U^\dagger_{k\gamma\delta l}\langle\delta \mid \langle l \mid \Big). \tag{5.246}$$

Now we are going to trace out the environment qubit. To do so, we take the unsaturated dangling bras and kets of the Greek kind,

$$\ldots \mid \alpha\rangle \ldots \langle\delta \mid \ldots,$$

and turn them on each other:

$$\mid \alpha\rangle\langle\delta \mid \rightarrow \langle\alpha \mid \delta\rangle = \delta_{\alpha\delta}, \tag{5.247}$$

where $\delta_{\alpha\delta}$ is the Kronecker delta. In expressions involving sums over all indexes, this trick produces a trace. We obtain

$$\begin{aligned}
\rho_A(t) &= \sum_{i,\alpha,\beta,j,k,\gamma,\delta,l} \delta_{\alpha\delta} \Big(\mid i\rangle U_{i\alpha\beta j}\langle\beta \mid \Psi_E(0)\rangle\langle j \mid \Big) \rho_A(0) \Big(\mid k\rangle\langle\Psi_E(0) \mid \gamma\rangle U^\dagger_{k\gamma\delta l}\langle l \mid \Big) \\
&= \sum_{i,\alpha,\beta,j,k,\gamma,l} \Big(\mid i\rangle U_{i\alpha\beta j}\langle\beta \mid \Psi_E(0)\rangle\langle j \mid \Big) \rho_A(0) \Big(\mid k\rangle\langle\Psi_E(0) \mid \gamma\rangle U^\dagger_{k\gamma\alpha l}\langle l \mid \Big) \\
&= \sum_\alpha \sum_{i,j,k,l} \left(\mid i\rangle \left(\sum_\beta U_{i\alpha\beta j}\langle\beta \mid \Psi_E(0)\rangle \right) \langle j \mid \right) \rho_A(0) \left(\mid k\rangle \left(\sum_\gamma \langle\Psi_E(0) \mid \gamma\rangle U^\dagger_{k\gamma\alpha l} \right) \langle l \mid \right).
\end{aligned} \tag{5.248}$$

The term

$$\mid i\rangle \left(\sum_\beta U_{i\alpha\beta j}\langle\beta \mid \Psi_E(0)\rangle \right) \langle j \mid \tag{5.249}$$

represents a matrix element of an operator. The only unsaturated index here is α. So, we call the operator itself $\boldsymbol{M}_\alpha$, and what's in the brackets is $\langle i \mid \boldsymbol{M}_\alpha \mid j\rangle$. The operator acts on $\boldsymbol{\rho}_A(0)$. Similarly,

$$\mid k\rangle \left(\sum_\gamma \langle \Psi_E(0) \mid \gamma\rangle U^\dagger_{k\gamma\alpha l}\right) \langle l \mid \tag{5.250}$$

represents a matrix element of $\boldsymbol{M}^\dagger_\alpha$, and what's in the brackets is $\langle k \mid \boldsymbol{M}^\dagger_\alpha \mid l\rangle$. This operator acts on $\boldsymbol{\rho}_A(0)$ from the right. In summary, we find that

$$\boldsymbol{\rho}_A(t) = \sum_\alpha \boldsymbol{M}_\alpha(t)\boldsymbol{\rho}_A(0)\boldsymbol{M}^\dagger_\alpha(t). \tag{5.251}$$

Quantum operations

Operations of the form given by equation (5.251) are called by some people *quantum operations*, by others *super operations*, and yet by others just *linear maps*. They map $\boldsymbol{\rho}_A(0)$ onto $\boldsymbol{\rho}_A(t)$. They are linear, obviously, but not, in general, unitary.

Let us call a map given by (5.251) $\mathfrak{A}$ and reserve square brackets for its argument $\boldsymbol{\rho}(0)$.

The fancy symbol $\mathfrak{A}$ is a Gothic "A." Mathematicians and physicists resort to Gothic letters only when they have run out of other options and are getting desperate. On the other hand, some mathematicians prefer to use normal letters and brackets for everything, because, after all, just about everything in mathematics is a map of some sort. But this produces formulas that can be difficult to read, because it is hard to see at first what's what. Our preference is for a moderately baroque notation that emphasizes geometric and transformation properties of various objects.

But let us get back to $\mathfrak{A}$.

Operator sum representation of a quantum operation

Its formal definition is

$$\boldsymbol{\rho}(t) = \mathfrak{A}\left[\boldsymbol{\rho}(0)\right] = \sum_\alpha \boldsymbol{M}_\alpha(t)\boldsymbol{\rho}_A(0)\boldsymbol{M}^\dagger_\alpha(t). \tag{5.252}$$

Map $\mathfrak{A}$ must satisfy certain conditions if it is to be physical; that is, we must ensure that $\mathfrak{A}\left[\boldsymbol{\rho}(0)\right]$ produced by it is still a valid density operator. The Peres-Horodeckis map would not qualify here—we cannot produce an unphysical state by *evolving*, nonunitarily but still physically, an entangled biqubit.

Properties of $\boldsymbol{M}_\alpha$

Let us recall equation (5.243). It implies that

$$\langle i \mid \boldsymbol{M}_\alpha \mid j\rangle = \sum_\beta U_{i\alpha\beta j}\langle \beta \mid \Psi_E(0)\rangle = \langle i \mid \langle \alpha \mid \boldsymbol{U}_{AE} \mid \Psi_E(0)\rangle \mid j\rangle, \tag{5.253}$$

which yields

$$M_\alpha = \langle \alpha \mid U_{AE} \mid \Psi_E(0) \rangle. \tag{5.254}$$

This expression is a symbolic abbreviation often used in place of the more detailed, but less readable, (5.249) and (5.250).

Now it is easy to demonstrate that

$$\sum_\alpha M_\alpha^\dagger M_\alpha = \mathbf{1}_A. \tag{5.255}$$

This property will come handy in showing that map $\mathfrak{A}$ is physical.

We begin by expanding

$$\begin{aligned} \sum_\alpha M_\alpha^\dagger M_\alpha &= \sum_\alpha \langle \Psi_E(0) \mid U_{AE}^\dagger \mid \alpha \rangle \langle \alpha \mid U_{AE} \mid \Psi_E(0) \rangle \\ &= \langle \Psi_E(0) \mid U_{AE}^\dagger U_{AE} \mid \Psi_E(0) \rangle. \end{aligned} \tag{5.256}$$

The term $M_\alpha^\dagger M_\alpha$ implies multiplication of the M_α matrices in the (i, j) space, that is, in the qubit A's space. Additionally, the sum over $\mid \alpha \rangle \langle \alpha \mid$ produces multiplication in the (α, β) space, that is, in the qubit E's space. In effect, $U_{AE}^\dagger U_{AE}$ is indeed a full multiplication in both spaces, and therefore it must yield $\mathbf{1}_A \otimes \mathbf{1}_E$, because U_{AE} is unitary. So, we obtain

$$\begin{aligned} \sum_\alpha M_\alpha^\dagger M_\alpha &= \langle \Psi_E(0) \mid \mathbf{1}_A \otimes \mathbf{1}_E \mid \Psi_E(0) \rangle \\ &= \mathbf{1}_A \langle \Psi_E(0) \mid \Psi_E(0) \rangle = \mathbf{1}_A. \end{aligned} \tag{5.257}$$

Armed with this fact we can immediately show that map $\mathfrak{A}$ preserves trace of $\rho_A(0)$. Let us recall that $\mathrm{Tr}(AB) = \mathrm{Tr}(BA)$ and that $\mathrm{Tr}(A+B) = \mathrm{Tr}(A) + \mathrm{Tr}(B)$. Hence,

$$\begin{aligned} &\mathrm{Tr}_A \mathfrak{A}[\rho_A(0)] = \mathrm{Tr} \sum_\alpha M_\alpha \rho_A(0) M_\alpha^\dagger \\ &= \sum_\alpha \mathrm{Tr}\left(M_\alpha \rho_A(0) M_\alpha^\dagger\right) = \sum_\alpha \mathrm{Tr}\left(\rho_A(0) M_\alpha M_\alpha^\dagger\right) \\ &= \sum_\alpha \mathrm{Tr}\left(\rho_A(0) M_\alpha^\dagger M_\alpha\right) = \mathrm{Tr}\left(\sum_\alpha \rho_A(0) M_\alpha^\dagger M_\alpha\right) \\ &= \mathrm{Tr}\left(\rho_A(0) \sum_\alpha M_\alpha^\dagger M_\alpha\right) = \mathrm{Tr}\left(\rho_A(0) \mathbf{1}_A\right) \\ &= \mathrm{Tr}\left(\rho_A(0)\right). \end{aligned} \tag{5.258}$$

Quantum operations preserve physicality of $\boldsymbol{\rho}$.

It is easy to see that $\mathfrak{A}$ preserves positivity of $\boldsymbol{\rho}_A(0)$. Let us consider

$$\langle \Psi_A \mid \boldsymbol{\rho}_A(t) \mid \Psi_A \rangle \tag{5.259}$$

for an arbitrary vector $\mid \Psi_A \rangle$ in the A space. On inserting the $\boldsymbol{\rho}_A(t)$ evolution in terms of the $\boldsymbol{M}_\alpha$s operators, we obtain

$$\sum_\alpha \langle \Psi_A \mid \boldsymbol{M}_\alpha \boldsymbol{\rho}_A(0) \boldsymbol{M}_\alpha^\dagger \mid \Psi_A \rangle. \tag{5.260}$$

But $\boldsymbol{M}_\alpha^\dagger \mid \Psi_A \rangle$ is some other vector $\mid \Phi_\alpha \rangle$ in the A space; and, since $\boldsymbol{\rho}_A(0)$ *is* positive, we find that each $\langle \Phi_\alpha \mid \boldsymbol{\rho}_A(0) \mid \Phi_\alpha \rangle$ term is positive and so their sum is positive, too, which implies that $\langle \Psi_A \mid \boldsymbol{\rho}_A(t) \mid \Psi_A \rangle$ is positive.

Finally, we can demonstrate that $\mathfrak{A}$ preserves hermiticity of $\boldsymbol{\rho}$, meaning that if $\boldsymbol{\rho}(0) = \boldsymbol{\rho}^\dagger(0)$, then $\boldsymbol{\rho}(t) = \boldsymbol{\rho}^\dagger(t)$:

$$\begin{aligned} \boldsymbol{\rho}^\dagger(t) &= \left(\sum_\alpha \boldsymbol{M}_\alpha \boldsymbol{\rho}(0) \boldsymbol{M}_\alpha^\dagger \right)^\dagger = \sum_\alpha \left(\boldsymbol{M}_\alpha \boldsymbol{\rho}(0) \boldsymbol{M}_\alpha^\dagger \right)^\dagger \\ &= \sum_\alpha \left(\boldsymbol{M}_\alpha^\dagger \right)^\dagger \boldsymbol{\rho}^\dagger(0) \boldsymbol{M}_\alpha^\dagger = \sum_\alpha \boldsymbol{M}_\alpha \boldsymbol{\rho}(0) \boldsymbol{M}_\alpha^\dagger \\ &= \boldsymbol{\rho}(t), \end{aligned} \tag{5.261}$$

where we have used $(\boldsymbol{AB})^\dagger = \boldsymbol{B}^\dagger \boldsymbol{A}^\dagger$ and $\boldsymbol{\rho}^\dagger(0) = \boldsymbol{\rho}(0)$ in the third line.

In summary, whatever $\mathfrak{A}$ does to $\boldsymbol{\rho}_A(0)$, the resulting new operator can be still interpreted as a density operator.

The *operator sum* representation of $\mathfrak{A}$—this is how it is called—in terms of operators $\boldsymbol{M}_\alpha$, as given by equation (5.251), is not unique, because the same nonunitary evolution may result from various models in which a qubit entangles with the environment.

5.11.2 Depolarization

Let us consider the phenomenon of depolarization that we saw in Section 2.12, page 86. The unitary description of a single-qubit, alone, cannot describe the gradual depolarization we observed. The single-qubit Schrödinger equation predicts undamped Rabi oscillations and undamped Larmor precession for such a system.

But when the single qubit is a part of a larger system, then its behavior may change dramatically.

Physical meaning of $\boldsymbol{\sigma}_x$

The unitary formalism describes polarization changes in terms of rotations and flips. A typical example of a flip operation is the $\boldsymbol{\sigma}_x$ Pauli matrix, the NOT gate,

$$\boldsymbol{\sigma}_x \begin{pmatrix} a \\ b \end{pmatrix} = \begin{pmatrix} 0 & 1 \\ 1 & 0 \end{pmatrix} \begin{pmatrix} a \\ b \end{pmatrix} = \begin{pmatrix} b \\ a \end{pmatrix}. \tag{5.262}$$

What does this operation do to vector $\boldsymbol{r}$? It swaps a and b, and therefore it changes the sign of r^z and r^y:

$$r^z = aa^* - bb^* \quad \rightarrow \quad bb^* - aa^* = -r^z, \tag{5.263}$$

$$r^x = ab^* + ba^* \quad \rightarrow \quad ba^* + ab^* = r^x, \tag{5.264}$$

$$r^y = \mathrm{i}(ab^* - ba^*) \quad \rightarrow \quad \mathrm{i}(ba^* - ab^*) = -r^y. \tag{5.265}$$

But other flips are possible, too. Let us have a look at what $\boldsymbol{\sigma}_z$ does to a qubit. *Physical meaning of $\boldsymbol{\sigma}_z$*

$$\boldsymbol{\sigma}_z \begin{pmatrix} a \\ b \end{pmatrix} = \begin{pmatrix} 1 & 0 \\ 0 & -1 \end{pmatrix} \begin{pmatrix} a \\ b \end{pmatrix} = \begin{pmatrix} a \\ -b \end{pmatrix} \tag{5.266}$$

What is the effect of this operation on $\boldsymbol{r}$?

$$r^z = aa^* - bb^* \quad \rightarrow \quad aa^* - bb^* = r^z, \tag{5.267}$$

$$r^x = ab^* + ba^* \quad \rightarrow \quad -ab^* - ba^* = -r^x, \tag{5.268}$$

$$r^y = \mathrm{i}(ab^* - ba^*) \quad \rightarrow \quad \mathrm{i}(-ab^* + ba^*) = -r^y. \tag{5.269}$$

The operation changes the sign of r^x and r^y while leaving r^z intact.

Finally, $\boldsymbol{\sigma}_y$ does the following *Physical meaning of $\boldsymbol{\sigma}_y$*

$$\boldsymbol{\sigma}_y \begin{pmatrix} a \\ b \end{pmatrix} = \begin{pmatrix} 0 & -\mathrm{i} \\ \mathrm{i} & 0 \end{pmatrix} \begin{pmatrix} a \\ b \end{pmatrix} = \begin{pmatrix} -\mathrm{i}b \\ \mathrm{i}a \end{pmatrix}. \tag{5.270}$$

This translates into

$$r^z = aa^* - bb^* \quad \rightarrow \quad bb^* - aa^* = -r^z, \tag{5.271}$$

$$r^x = ab^* + ba^* \quad \rightarrow \quad -ba^* - ab^* = -r^x, \tag{5.272}$$

$$r^y = \mathrm{i}(ab^* - ba^*) \quad \rightarrow \quad -\mathrm{i}(ba^* - ab^*) = r^y. \tag{5.273}$$

In summary, $\boldsymbol{\sigma}_x$ rotates $\boldsymbol{r}$ by 180° around the x axis, $\boldsymbol{\sigma}_y$ rotates $\boldsymbol{r}$ by 180° around the y axis, and $\boldsymbol{\sigma}_z$ rotates $\boldsymbol{r}$ by 180° around the z axis.

Every one of these three transformations is a unitary transformation, because

1. every Pauli matrix is Hermitian, and
2. the square of every Pauli matrix is 1.

Therefore for every Pauli matrix

$$\boldsymbol{\sigma}_i \boldsymbol{\sigma}_i^\dagger = \boldsymbol{\sigma}_i \boldsymbol{\sigma}_i = \mathbf{1}, \tag{5.274}$$

which is a sufficient condition for $\boldsymbol{\sigma}_i$ to be unitary.

Let us suppose the qubit is a part of a larger system of three qubits, with the other two qubits providing a simplistic model of an "environment." The basis states of the environment are

$$\begin{aligned} \mid 0_E\rangle \otimes \mid 0_E\rangle &\equiv \mid \mathbf{0}_E\rangle, \\ \mid 0_E\rangle \otimes \mid 1_E\rangle &\equiv \mid \mathbf{1}_E\rangle, \\ \mid 1_E\rangle \otimes \mid 0_E\rangle &\equiv \mid \mathbf{2}_E\rangle, \\ \mid 1_E\rangle \otimes \mid 1_E\rangle &\equiv \mid \mathbf{3}_E\rangle, \end{aligned}$$

where the notation on the right-hand side is a simplified way to denote the environment biqubit basis states.

Let the initial state of the whole system be

$$\mid \Psi_A\rangle \otimes \mid \mathbf{0}_E\rangle. \tag{5.275}$$

Unitary model We consider a transformation $\boldsymbol{U}_{AE}$ of this system into

$$\boldsymbol{U}_{AE} \mid \Psi_A\rangle \mid \mathbf{0}_E\rangle = \sqrt{1-\kappa} \mid \Psi_A\rangle \otimes \mid \mathbf{0}_E\rangle \\ + \sqrt{\frac{\kappa}{3}} \boldsymbol{\sigma}_x \mid \Psi_A\rangle \otimes \mid \mathbf{1}_E\rangle + \sqrt{\frac{\kappa}{3}} \boldsymbol{\sigma}_y \mid \Psi_A\rangle \otimes \mid \mathbf{2}_E\rangle + \sqrt{\frac{\kappa}{3}} \boldsymbol{\sigma}_z \mid \Psi_A\rangle \otimes \mid \mathbf{3}_E\rangle, \tag{5.276}$$

where κ is the probability that the qubit is going to flip, where we allow it to flip about the x, y or z axis with equal probability of $\kappa/3$. The probability that the qubit is not going to flip is $1-\kappa$. If the qubit does not flip, the environment stays in the $\mid \mathbf{0}_E\rangle$ state. If the qubit flips about the x axis, the environment switches to the $\mid \mathbf{1}_E\rangle$ state. If the qubit flips about the y axis, the environment switches to the $\mid \mathbf{2}_E\rangle$ state. And if the qubit flips about the z axis, the environment switches to the $\mid \mathbf{3}_E\rangle$ state. Thus, the environment responds differently to every possible flip, recording, as it were, what has happened.

The initial and final state of this operation are pure triqubit states. Therefore, the operation itself is unitary in the triqubit space, but it is not going to be unitary in the single qubit space. The corresponding map $\mathfrak{A}$ will have the following $\boldsymbol{M}_\alpha$ operator representation:

$$\boldsymbol{M}_{\mathbf{0}} = \langle \mathbf{0}_E \mid \boldsymbol{U}_{AE} \mid \mathbf{0}_E\rangle = \sqrt{1-\kappa}\,\mathbf{1}_A, \tag{5.277}$$

$$M_1 = \langle \mathbf{1}_E \mid U_{AE} \mid \mathbf{0}_E \rangle = \sqrt{\frac{\kappa}{3}}\sigma_x, \quad (5.278)$$

$$M_2 = \langle \mathbf{2}_E \mid U_{AE} \mid \mathbf{0}_E \rangle = \sqrt{\frac{\kappa}{3}}\sigma_y, \quad (5.279)$$

$$M_3 = \langle \mathbf{3}_E \mid U_{AE} \mid \mathbf{0}_E \rangle = \sqrt{\frac{\kappa}{3}}\sigma_z. \quad (5.280)$$

The resulting transformation of ρ_A is

Quantum operation

$$\begin{aligned} \rho_A(t) &= \mathfrak{A}[\rho_A(0)] \\ &= M_0\rho_A(0)M_0^\dagger + M_1\rho_A(0)M_1^\dagger + M_2\rho_A(0)M_2^\dagger + M_3\rho_A(0)M_3^\dagger \\ &= (1-\kappa)\rho_A(0) + \frac{\kappa}{3}\sigma_x\rho_A(0)\sigma_x + \frac{\kappa}{3}\sigma_y\rho_A(0)\sigma_y + \frac{\kappa}{3}\sigma_z\rho_A(0)\sigma_z. \end{aligned} \quad (5.281)$$

Let us apply this formula first to a general case of $\rho_A = \frac{1}{2}(\mathbf{1} + \boldsymbol{r}\cdot\boldsymbol{\sigma})$. We can always rotate our system of coordinates so that $\boldsymbol{e}_z$ is aligned with $\boldsymbol{r}$. Without a loss of generality we can simplify it to $\rho_A = \frac{1}{2}(\mathbf{1} + r\sigma_z)$. Because Pauli matrices anti-commute and square to $\mathbf{1}$, we find that

$$\sigma_x\sigma_z\sigma_x = -\sigma_z\sigma_x\sigma_x = -\sigma_z, \quad (5.282)$$

$$\sigma_y\sigma_z\sigma_y = -\sigma_z\sigma_y\sigma_y = -\sigma_z, \quad (5.283)$$

$$\sigma_z\sigma_z\sigma_z = \sigma_z. \quad (5.284)$$

Hence,

Depolarization shrinks $\boldsymbol{r}$.

$$\begin{aligned} \mathfrak{A}[\rho_A] &= (1-\kappa)\rho_A + \frac{\kappa}{3}\sigma_x\rho_A\sigma_x + \frac{\kappa}{3}\sigma_y\rho_A\sigma_y + \frac{\kappa}{3}\sigma_z\rho_A\sigma_z \\ &= \frac{1}{2}\left(\mathbf{1} + r\left(1 - \frac{4\kappa}{3}\right)\sigma_z\right). \end{aligned} \quad (5.285)$$

We find that, although vector $\boldsymbol{r}$ does not change its direction, it shrinks.

If $\mathfrak{A}$ is a continuous process, we can think of it in terms of a certain probability Γ of $\mathfrak{A}$ happening to the qubit per unit time. In this case $\kappa = \Gamma\Delta t$. As time goes by, the process repeats every Δt beginning with $t = 0$, at which time $r = r(0)$. After the first Δt, the original r shrinks to $r(\Delta t) = (1 - 4\Gamma\Delta t/3)r(0)$. After the second Δt, this new r shrinks to $r(2\Delta t) = (1-4\Gamma\Delta t/3)r(\Delta t) = (1-4\Gamma\Delta t/3)(1-4\Gamma\Delta t/3)r(0)$.

After n such applications of $\mathfrak{A}$, the length of the polarization vector, r, will have shrunk to $r(n\Delta t) = (1 - 4\Gamma\Delta t/3)^n r(0)$.

Let $\Delta t = t/n$. Then

$$r(t) = \left(1 - \frac{4}{3}\frac{\Gamma t}{n}\right)^n r(0). \quad (5.286)$$

$\boldsymbol{r}$ shrinks exponentially.

The expression gets more accurate with the shrinking of Δt and with $n \to \infty$. In the limit we get

$$r(t) = \lim_{n\to\infty}\left(1 - \frac{4}{3}\frac{\Gamma t}{n}\right)^n r(0) = \mathrm{e}^{-4\Gamma t/3} r(0), \quad (5.287)$$

where we have explored the same trick that gave us equation (4.292) on page 155. We see that the qubit depolarizes exponentially. This is indeed what we saw in Section 2.12.

5.11.3 Dephasing

Dephasing results from entanglement with environment

Our next model is quite different. This time we are going to investigate a possible effect that entanglement itself with a biqubit environment E has on the third qubit A, the state of which does not change nominally. Growing entanglement, which manifests in the environment flipping its state, is the only thing that happens. There are no random spin-flips of qubit A here.

Unitary model

Assuming that the initial state of the qubit-biqubit system is $|\, 0_A\rangle \,|\, \mathbf{0}_E\rangle$ or $|\, 1_A\rangle \,|\, \mathbf{0}_E\rangle$, the final state is going to be

$$\boldsymbol{U}_{AE} :|\, 0_A\rangle \,|\, \mathbf{0}_E\rangle \;\to\; \sqrt{1-\kappa}\,|\, 0_A\rangle \,|\, \mathbf{0}_E\rangle + \sqrt{\kappa}\,|\, 0_A\rangle \,|\, \mathbf{1}_E\rangle, \quad (5.288)$$

$$\boldsymbol{U}_{AE} :|\, 1_A\rangle \,|\, \mathbf{0}_E\rangle \;\to\; \sqrt{1-\kappa}\,|\, 1_A\rangle \,|\, \mathbf{0}_E\rangle + \sqrt{\kappa}\,|\, 1_A\rangle \,|\, \mathbf{2}_E\rangle. \quad (5.289)$$

In other words, there is a probability κ that the environment is going to flip upon having been entangled with qubit A from $|\, \mathbf{0}_E\rangle$ to $|\, \mathbf{1}_E\rangle$, if the state of qubit A is $|\, 0_A\rangle$. There is a similar probability κ that the environment is going to flip from $|\, \mathbf{0}_E\rangle$ to $|\, \mathbf{2}_E\rangle$, if the state of qubit A is $|\, 1_A\rangle$. But then, there is also some probability $1-\kappa$, in both cases, that the environment is going to stay as it is.

The operator-sum representation of this interaction will have three terms, with $\boldsymbol{M}_i$ given by

$$\boldsymbol{M}_0 = \langle \mathbf{0}_E \,|\, \boldsymbol{U}_{AE} \,|\, \mathbf{0}_E\rangle, \quad (5.290)$$

$$\boldsymbol{M}_1 = \langle \mathbf{1}_E \,|\, \boldsymbol{U}_{AE} \,|\, \mathbf{0}_E\rangle, \quad (5.291)$$

$$\boldsymbol{M}_2 = \langle \mathbf{2}_E \,|\, \boldsymbol{U}_{AE} \,|\, \mathbf{0}_E\rangle, \quad (5.292)$$

where $\boldsymbol{M}_i$ are operators that act in the space of qubit A. There is no $\boldsymbol{M}_3$ term here because we don't make any use of $|\, \mathbf{3}_E\rangle$ in our definition of $\boldsymbol{U}_{AE}$. Qubit A entangles with a three-dimensional subspace of biqubit E.

It is easy to see that

$$\langle \mathbf{0}_E \mid \boldsymbol{U}_{AE} \mid \mathbf{0}_E \rangle = \sqrt{1-\kappa}\, \mathbf{1}_A. \tag{5.293}$$

For $\boldsymbol{M}_1$, we find that it acts on $\mid 0_A \rangle$ as follows:

$$\begin{aligned} &\langle \mathbf{1}_E \mid \boldsymbol{U}_{AE} \mid 0_A \rangle \mid \mathbf{0}_E \rangle \\ &= \langle \mathbf{1}_E \mid \left(\sqrt{1-\kappa} \mid 0_A \rangle \mid \mathbf{0}_E \rangle + \sqrt{\kappa} \mid 0_A \rangle \mid \mathbf{1}_E \rangle \right) \\ &= \sqrt{\kappa} \mid 0_A \rangle \end{aligned} \tag{5.294}$$

but produces zero when acting on $\mid 1_A \rangle$. The matrix representation of $\boldsymbol{M}_1$ is therefore

$$\boldsymbol{M}_1 = \sqrt{\kappa} \begin{pmatrix} 1 & 0 \\ 0 & 0 \end{pmatrix}. \tag{5.295}$$

On the other hand, $\boldsymbol{M}_2$ produces zero when acting on $\mid 0_A \rangle$; but when it acts on $\mid 1_A \rangle$, it produces

$$\begin{aligned} &\langle \mathbf{2}_E \mid \boldsymbol{U}_{AE} \mid 1_A \rangle \mid \mathbf{0}_E \rangle \\ &= \langle \mathbf{2}_E \mid \left(\sqrt{1-\kappa} \mid 1_A \rangle \mid \mathbf{0}_E \rangle + \sqrt{\kappa} \mid 1_A \rangle \mid \mathbf{2}_E \rangle \right) \\ &= \sqrt{\kappa} \mid 1_A \rangle. \end{aligned} \tag{5.296}$$

Hence, its matrix representation is

$$\boldsymbol{M}_2 = \sqrt{\kappa} \begin{pmatrix} 0 & 0 \\ 0 & 1 \end{pmatrix}. \tag{5.297}$$

The resulting map $\mathfrak{A}$ is

$$\begin{aligned} \mathfrak{A}[\boldsymbol{\rho}_A] &= \boldsymbol{M}_0 \boldsymbol{\rho}_A \boldsymbol{M}_0^\dagger + \boldsymbol{M}_1 \boldsymbol{\rho}_A \boldsymbol{M}_1^\dagger + \boldsymbol{M}_2 \boldsymbol{\rho}_A \boldsymbol{M}_2^\dagger \\ &= (1-\kappa)\, \boldsymbol{\rho}_A + \kappa \begin{pmatrix} 1 & 0 \\ 0 & 0 \end{pmatrix} \boldsymbol{\rho}_A \begin{pmatrix} 1 & 0 \\ 0 & 0 \end{pmatrix} + \kappa \begin{pmatrix} 0 & 0 \\ 0 & 1 \end{pmatrix} \boldsymbol{\rho}_A \begin{pmatrix} 0 & 0 \\ 0 & 1 \end{pmatrix}. \end{aligned} \tag{5.298}$$

Because

$$\begin{pmatrix} 1 & 0 \\ 0 & 0 \end{pmatrix} \begin{pmatrix} \rho_{00} & \rho_{01} \\ \rho_{10} & \rho_{11} \end{pmatrix} \begin{pmatrix} 1 & 0 \\ 0 & 0 \end{pmatrix} = \begin{pmatrix} \rho_{00} & 0 \\ 0 & 0 \end{pmatrix} \tag{5.299}$$

and

$$\begin{pmatrix} 0 & 0 \\ 0 & 1 \end{pmatrix} \begin{pmatrix} \rho_{00} & \rho_{01} \\ \rho_{10} & \rho_{11} \end{pmatrix} \begin{pmatrix} 0 & 0 \\ 0 & 1 \end{pmatrix} = \begin{pmatrix} 0 & 0 \\ 0 & \rho_{11} \end{pmatrix}, \tag{5.300}$$

we find the following expression for $\mathfrak{A}$:

Quantum operation

$$\mathfrak{A}\left[\boldsymbol{\rho}_A\right] = (1-\kappa)\,\boldsymbol{\rho}_A + \kappa \begin{pmatrix} \rho_{00} & 0 \\ 0 & \rho_{11} \end{pmatrix} = \begin{pmatrix} \rho_{00} & (1-\kappa)\rho_{01} \\ (1-\kappa)\rho_{10} & \rho_{11} \end{pmatrix}. \tag{5.301}$$

As before, we are going to assume that the probability of such an entanglement happening to the qubit-biqubit system per unit time is Γ, so that $\kappa = \Gamma\Delta t$. We are going to reason here the same way we reasoned in the previous section about depolarization. After a short time Δt, the density matrix of qubit A becomes

$$\boldsymbol{\rho}_A(\Delta t) = \begin{pmatrix} \rho_{00}(0) & (1-\Gamma\Delta t)\rho_{01}(0) \\ (1-\Gamma\Delta t)\rho_{10}(0) & \rho_{11}(0) \end{pmatrix}. \tag{5.302}$$

Then after two such time intervals,

$$\boldsymbol{\rho}_A(2\Delta t) = \begin{pmatrix} \rho_{00}(0) & (1-\Gamma\Delta t)^2\rho_{01}(0) \\ (1-\Gamma\Delta t)^2\rho_{10}(0) & \rho_{11}(0) \end{pmatrix}. \tag{5.303}$$

And after n intervals such that $t = n\Delta t$, we find that

$$\boldsymbol{\rho}_A(t) = \begin{pmatrix} \rho_{00}(0) & (1-\Gamma\Delta t)^{t/\Delta t}\rho_{01}(0) \\ (1-\Gamma\Delta t)^{t/\Delta t}\rho_{10}(0) & \rho_{11}(0) \end{pmatrix}. \tag{5.304}$$

In the limit $\Delta t \rightarrow 0$, we obtain

$$\lim_{\Delta t\rightarrow 0} (1-\Gamma\Delta t)^{t/\Delta t} = \mathrm{e}^{-\Gamma t}. \tag{5.305}$$

Thus

$$\boldsymbol{\rho}_A(t) = \begin{pmatrix} \rho_{00}(0) & \mathrm{e}^{-\Gamma t}\rho_{01}(0) \\ \mathrm{e}^{-\Gamma t}\rho_{10}(0) & \rho_{11}(0) \end{pmatrix}. \tag{5.306}$$

Dephasing kills r^x and r^y.

The exponential vanishing of the off-diagonal terms implies the exponential vanishing of x and y components of the polarization vector. Here we observe not so much depolarization as exponentially rapid projection of the qubit's polarization onto the $\boldsymbol{e}_z$ direction. This happens *not* because a force or a torque has been applied to the qubit but because the qubit has become entangled with the biqubit that, in this simplified model, represents the environment.

Dephasing as a model of measurement

The phenomenon discussed here may be thought of as simplistic unitary model of the measurement process. When a qubit that may be polarized in any direction encounters the measuring apparatus, it entangles with that apparatus. The effect of the entanglement is an almost instantaneous projection of the qubit onto the measurement direction of the apparatus, represented here by $\boldsymbol{e}_z$. Any information contained in r^x and r^y becomes lost in the process. Only information contained in r^z survives. In order to recover all information that characterizes the qubit, we have to repeat the measurements for the other two directions on the statistical ensemble of identically prepared qubits.

The meaning of the term "dephasing"

As we have seen in Section 4.9.2, page 160, the unitary formalism encodes information about r^x and r^y as phase difference between $|\uparrow\rangle$ and $|\downarrow\rangle$ within the *superposition* that represents the qubit state. From the unitary formalism's point of view, the loss of r^x and r^y means the loss of knowledge about the phase difference, hence the term *dephasing* or *phase damping* that physicists use sometimes when discussing this process.

Measuring τ

Figure 2.9(B) in Chapter 2, Section 2.12, page 89, illustrates a method of measuring the decoherence time $\tau = 1/\Gamma$. In the Ramsey experiment we flip the qubit from its $|\uparrow\rangle$ state to its $|\rightarrow\rangle$ state first—in the unitary formalism; this is $\left(|\uparrow\rangle + |\downarrow\rangle\right)/\sqrt{2}$. Then we let it precess about $\boldsymbol{B}_{\|}$ while the transverse buzzing field $\boldsymbol{B}_{\perp}$ is switched off. After some time, we turn the buzzing field $\boldsymbol{B}_{\perp}$ back on. If the qubit's polarization has rotated around the equator of the Bloch sphere by a multiple of 2π, we are back to the starting point, and the polarization continues on its march toward the south pole. If the qubit's polarization has rotated around the equator of the Bloch sphere by an odd multiple of π, the qubit's polarization will move back toward the north pole. If the measurements are repeated for increasing time between the two buzzing signals, we end up with "Ramsey fringes." The probability curve looks like a sinusoid.

The curve in Figure 2.9(B) looks somewhat like a sinusoid, but it is damped. The damping here is exponential and derives from the dephasing of the qubit. The superposition state $\left(|\uparrow\rangle + |\downarrow\rangle\right)/\sqrt{2}$ does not last, as it rotates around the equator of the Bloch sphere. Because of the qubit's entanglement with the environment, it decays exponentially toward either $|\uparrow\rangle$ or $|\downarrow\rangle$, so that when the second buzzing signal is switched on, the polarization may no longer be on the equator of the Bloch sphere. The result is that as the time between the two Rabi signals is extended, the qubit's final state becomes increasingly chaotic. The qubit's decoherence time τ can be read from the envelope of the curve, and it can be easily seen to be on the order of a few microseconds.

5.11.4 Spontaneous Emission

Spontaneous emission occurs when a quantum system that is in a higher energy state initially decays all of a sudden and for no apparent reason to a lower energy state, emitting some energy quanta in the process. The quanta may be photons, but they may be other particles, too.

A reverse process to spontaneous emission is spontaneous absorption. Here a quantum system absorbs energy quanta from its environment and upgrades itself to a higher energy state.

Both processes are different from unitary absorption or emission of energy as described by the Schrödinger equation. The difference is that here we don't have an obvious driver. We don't know the reasons for the decay or for absorption. There may not be any reasons, or, as we will see shortly, the reason may be entanglement with the environment. In both cases, the resulting process is nonunitary and is described in terms of map $\mathfrak{A}$.

Unitary model

The simplest possible model of spontaneous emission is given by the following two equations that define $\boldsymbol{U}_{AE}$:

$$\begin{aligned} \boldsymbol{U}_{AE} \mid 0_A\rangle \mid \mathbf{0}_E\rangle &= \mid 0_A\rangle \mid \mathbf{0}_E\rangle, & (5.307)\\ \boldsymbol{U}_{AE} \mid 1_A\rangle \mid \mathbf{0}_E\rangle &= \sqrt{1-\kappa} \mid 1_A\rangle \mid \mathbf{0}_E\rangle + \sqrt{\kappa} \mid 0_A\rangle \mid \mathbf{1}_E\rangle. & (5.308) \end{aligned}$$

In plain language, if the observed qubit A is in the ground state, it stays in the ground state; nothing changes. But if it is in the higher energy state, $\mid 1_A\rangle$, then there is the probability κ that it is going to decay, transferring the energy to the environment that now flips from $\mid \mathbf{0}_E\rangle$ to $\mid \mathbf{1}_E\rangle$. But then, there is also the probability $1-\kappa$ that qubit A is going to remain in state $\mid 1_A\rangle$.

The process differs from depolarization. If this were depolarization, then we would also allow state $\mid 0_A\rangle$ to flip to $\mid 1_A\rangle$. We use only two states of the environment here: $\mid \mathbf{0}_E\rangle$ and $\mid \mathbf{1}_E\rangle$, so we are going to have only two operators $\boldsymbol{M}_i$ with $i = 0$ and 1. They are defined by

$$\begin{aligned} \boldsymbol{M}_0 &= \langle \mathbf{0}_E \mid \boldsymbol{U}_{AE} \mid \mathbf{0}_E\rangle, & (5.309)\\ \boldsymbol{M}_1 &= \langle \mathbf{1}_E \mid \boldsymbol{U}_{AE} \mid \mathbf{0}_E\rangle. & (5.310) \end{aligned}$$

To reconstruct matrices representing $\boldsymbol{M}_0$ and $\boldsymbol{M}_1$ in the space of qubit A, we need to figure out what the effects are of $\boldsymbol{M}_0$ and $\boldsymbol{M}_1$ acting on the basis vectors of qubit A. We find

$$\begin{aligned} \boldsymbol{M}_0 \mid 0_A\rangle &= \langle \mathbf{0}_E \mid \boldsymbol{U}_{AE} \mid 0_A\rangle \mid \mathbf{0}_E\rangle = \langle \mathbf{0}_E \left(\mid 0_A\rangle \mid \mathbf{0}_E\rangle\right) = \mid 0_A\rangle, & (5.311)\\ \boldsymbol{M}_0 \mid 1_A\rangle &= \langle \mathbf{0}_E \mid \boldsymbol{U}_{AE} \mid 1_A\rangle \mid \mathbf{0}_E\rangle = \langle \mathbf{0}_E \left(\sqrt{1-\kappa} \mid 1_A\rangle \mid \mathbf{0}_E\rangle + \sqrt{\kappa} \mid 0_A\rangle \mid \mathbf{1}_E\rangle\right) \\ &= \langle \mathbf{0}_E \sqrt{1-\kappa} \mid 1_A\rangle \mid \mathbf{0}_E\rangle = \sqrt{1-\kappa} \mid 1_A\rangle. & (5.312) \end{aligned}$$

These tell us that the matrix of $\boldsymbol{M}_0$ is diagonal and looks as follows.

$$\boldsymbol{M}_0 = \begin{pmatrix} 1 & 0 \\ 0 & \sqrt{1-\kappa} \end{pmatrix}. \qquad (5.313)$$

Similarly, for $\boldsymbol{M}_1$ we find

$$\begin{aligned}\boldsymbol{M}_1 \mid 0_A\rangle &= \langle \mathbf{1}_E \mid \boldsymbol{U}_{AE} \mid 0_a\rangle \mid \mathbf{0}_E\rangle = \langle \mathbf{1}_E \mid (\mid 0_A\rangle \mid \mathbf{0}_E\rangle) = 0, \quad (5.314)\\ \boldsymbol{M}_1 \mid 1_A\rangle &= \langle \mathbf{1}_E \mid \boldsymbol{U}_{AE} \mid 1_a\rangle \mid \mathbf{0}_E\rangle = \langle \mathbf{1}_E \mid \left(\sqrt{1-\kappa} \mid 1_A\rangle \mid \mathbf{0}_E\rangle + \sqrt{\kappa} \mid 0_A\rangle \mid \mathbf{1}_E\rangle\right)\\ &= \langle \mathbf{1}_E \mid \sqrt{\kappa} \mid 0_A\rangle \mid \mathbf{1}_E\rangle = \sqrt{\kappa} \mid 0_A\rangle. \quad (5.315)\end{aligned}$$

Hence, the resulting matrix of $\boldsymbol{M}_1$ is

$$\boldsymbol{M}_1 = \begin{pmatrix} 0 & \sqrt{\kappa} \\ 0 & 0 \end{pmatrix}. \quad (5.316)$$

The map $\mathfrak{A}$ looks as follows:

$$\begin{aligned}\mathfrak{A}[\boldsymbol{\rho}_A] &= \boldsymbol{M}_0\boldsymbol{\rho}_A\boldsymbol{M}_0^\dagger + \boldsymbol{M}_1\boldsymbol{\rho}_A\boldsymbol{M}_1^\dagger\\ &= \begin{pmatrix} 1 & 0 \\ 0 & \sqrt{1-\kappa} \end{pmatrix}\boldsymbol{\rho}_A\begin{pmatrix} 1 & 0 \\ 0 & \sqrt{1-\kappa} \end{pmatrix} + \begin{pmatrix} 0 & \sqrt{\kappa} \\ 0 & 0 \end{pmatrix}\boldsymbol{\rho}_A\begin{pmatrix} 0 & 0 \\ \sqrt{\kappa} & 0 \end{pmatrix}.\end{aligned} \quad (5.317)$$

For a general $\boldsymbol{\rho}_A = \begin{pmatrix} \rho_{00} & \rho_{01} \\ \rho_{10} & \rho_{11} \end{pmatrix}$ this becomes

Quantum operation

$$\begin{aligned}\mathfrak{A}[\boldsymbol{\rho}_A] &= \begin{pmatrix} \rho_{00} & \sqrt{1-\kappa}\rho_{01} \\ \sqrt{1-\kappa}\rho_{10} & (1-\kappa)\rho_{11} \end{pmatrix} + \begin{pmatrix} \kappa\rho_{11} & 0 \\ 0 & 0 \end{pmatrix}\\ &= \begin{pmatrix} \rho_{00}+\kappa\rho_{11} & \sqrt{1-\kappa}\rho_{01} \\ \sqrt{1-\kappa}\rho_{10} & (1-\kappa)\rho_{11} \end{pmatrix}.\end{aligned} \quad (5.318)$$

If we were to apply $\mathfrak{A}$ twice, the resulting new $\boldsymbol{\rho}$ would look as follows in terms of the original $\boldsymbol{\rho}$:

$$\mathfrak{A}[\mathfrak{A}[\boldsymbol{\rho}_A]] = \begin{pmatrix} \rho_{00}+\kappa\rho_{11}+\kappa(1-\kappa)\rho_{11} & \sqrt{1-\kappa}\sqrt{1-\kappa}\rho_{01} \\ \sqrt{1-\kappa}\sqrt{1-\kappa}\rho_{10} & (1-\kappa)(1-\kappa)\rho_{11} \end{pmatrix}. \quad (5.319)$$

As we apply $\mathfrak{A}$ n times, it is easy to see that the off-diagonal terms get multiplied by $(1-\kappa)^{n/2}$ and the ρ_{11} term gets multiplied by $(1-\kappa)^n$. The ρ_{00} term evolves as follows.

$$\begin{aligned}&\text{for } n=0: && \rho_{00},\\ &\text{for } n=1: && \rho_{00}+\kappa\rho_{11},\\ &\text{for } n=2: && \rho_{00}+\kappa\rho_{11}+\kappa(1-\kappa)\rho_{11},\\ &\text{for } n=3: && \rho_{00}+\kappa\rho_{11}+\kappa(1-\kappa)\rho_{11}+\kappa(1-\kappa)^2\rho_{11}, \quad \ldots\end{aligned}$$

So, for an arbitrary n this becomes

$$\begin{aligned}\rho_{00} + \kappa\rho_{11} + \kappa(1-\kappa)\rho_{11} + \ldots + \kappa(1-\kappa)^{n-1}\rho_{11} \\ = \rho_{00} + \kappa\rho_{11}\sum_{k=0}^{n-1}(1-\kappa)^k .\end{aligned} \tag{5.320}$$

This is the sum of a geometric series, which evaluates to

$$\sum_{k=0}^{n-1}(1-\kappa)^k = \frac{1-(1-\kappa)^n}{1-(1-\kappa)} = \frac{1-(1-\kappa)^n}{\kappa}. \tag{5.321}$$

Hence,

$$\kappa\sum_{k=0}^{n-1}(1-\kappa)^k = 1-(1-\kappa)^n, \tag{5.322}$$

and so the $(0,0)$ term of the density matrix becomes

$$\rho_{00} + \rho_{11}\left(1-(1-\kappa)^n\right). \tag{5.323}$$

Let us again exploit the trick that has served us so well in the preceding sections. We assume that $\kappa = \Gamma\Delta t$, where $n\Delta t = t$. Then

$$\mathfrak{A}^n\left[\boldsymbol{\rho}_A\right] = \begin{pmatrix} \rho_{00} + \rho_{11}\left(1-\left(1-\Gamma\frac{t}{n}\right)^n\right) & \left(1-\Gamma\frac{t}{n}\right)^{n/2}\rho_{01} \\ \left(1-\Gamma\frac{t}{n}\right)^{n/2}\rho_{10} & \left(1-\Gamma\frac{t}{n}\right)^n\rho_{11} \end{pmatrix}. \tag{5.324}$$

In the limit $n \to \infty$ and $\Delta t \to 0$, such that $n\Delta t = t$, we obtain

$$\boldsymbol{\rho}(t) = \begin{pmatrix} \rho_{00}(0) + \rho_{11}(0)\left(1-\mathrm{e}^{-\Gamma t}\right) & \mathrm{e}^{-\Gamma t/2}\rho_{01}(0) \\ \mathrm{e}^{-\Gamma t/2}\rho_{10}(0) & \mathrm{e}^{-\Gamma t}\rho_{11}(0) \end{pmatrix}. \tag{5.325}$$

We find that as time flows the system converges exponentially on a state that "points up," which here is the lower energy state, $|\, 0_A\rangle$.

Amplitude damping

If the original state $|\, \Psi(0)\rangle$ had a nonvanishing amplitude of being registered in the $|\, 1_A\rangle$ state, the amplitude would decay exponentially to zero, as would amplitudes of finding the system in any of the transverse states, $|{\rightarrow}\rangle$ or $|\, \otimes\rangle$. For this reason physicists call this process *amplitude damping.*

Spontanous emission has one beneficial side effect. We can use it to force a quantum system, which may have been in some thermal chaotic state initially, to cool down and become pure. This is how quantum computations often begin.

5.12 Schrödinger's Cat

Here is the sad story of Schrödinger's cat.

A quantum system, more specifically an atom of Rubidium, is put in a superposition of two *circular Rydberg* states with principal quantum numbers of 51 (we call this the $|\, e\rangle$ state) and 50 (we call this the $|\, g\rangle$ state).

Circular Rydberg states

> Circular Rydberg states are states of multi-electron atoms, of which Rubidium, which has 37 electrons normally, is one. They are characterized by large magnetic quantum numbers that derive from the orbital motion of many electrons around the nucleus. They are also characterized by large principal quantum numbers of their valence electrons, that is, electrons that are in the outer shells of the atom. They tend to respond strongly to electromagnetic stimulation and have long lifetimes. The circular Rydberg states of Rubidium mentioned here have lifetimes as long as 30 ms. For these reasons multi-electron atoms in circular Rydberg states are often used in quantum experiments, including quantum computing systems. The other reason is that the energy difference between $|\, e\rangle$ and $|\, g\rangle$ is in the microwave range, 51.099 GHz. Electromagnetic radiation in this range can be controlled with great precision.

The atom is observed with a detector that is connected to a vial filled with poisonous gas. The detector works by smashing the vial if it detects the atom in the $|\, e\rangle$ state, and not smashing it if it detects the atom in the $|\, g\rangle$ state.

Smearing out a cat

Now comes the cruelty of this needless experiment—*gedanken* or not. The contraption is put in a sealed box together with a live cat. And Dr. Schrödinger says [126], "If one has left this entire system to itself for an hour, one would say that the cat still lives if meanwhile no atom has decayed. The psi-function of the entire system would express this by having in it the living and dead cat (pardon the expression) mixed or smeared out in equal parts."[11]

The real purpose of this somewhat exaggerated *gedanken* experiment was to point out that we could not separate the logic of the microscopic, quantum world, from the logic of the macroscopic world, that we could always conceive of a situation in which the states of the microscopic world would have a direct bearing on macroscopic systems.

As is often the case with *gedanken* experiments, the conclusions drawn depend on a great many untold assumptions. *Gedanken* experiments became fashionable

[11] Translation by John D. Trimmer.

after Einstein's initial success with them. But Einstein's experiments dealt with macroscopic systems and with macroscopic physics, all of which we are familiar with. The situations were made strikingly simple, so as to expose a particular point Einstein was after. But there is nothing so simple when it comes to coupling between an atom and a macroscopic system. The difference between the two is described by the Avogadro number, $6.02252 \times 10^{23}/\text{mol}$. There is an Avogadro number of interacting atoms in the detector, versus only one atom the detector observes.

Unsuitability of gedanken experiments to quantum domain

We have seen in this chapter the degree of complexity that arises when we move from contemplating one qubit to contemplating two qubits. If we were to switch from one qubit to Avogadro number of qubits, the resulting complexity would be mind-boggling. What would a density operator of such a system be like? What would be the atom's evolution if we were to trace out the detector? And what would be the detector's evolution if we were to trace out the atom?

Release the cat!

As to the cat, we should really take it out of the box, because its presence only confuses the issue. The difference between $|\,\text{live}\rangle$ and $|\,\text{dead}\rangle$ is nowhere near as sharply defined as the difference between the $|\,e\rangle$ and $|\,g\rangle$ states of the atom. The cat may be sick, very sick, half dead, barely alive—or very angry. It may also rub against the detector and trigger the mechanism regardless of the state of the atom.

Rather than focusing on the cat, we could focus on the vial itself and contemplate a superposition of $|\,\text{smashed}\rangle$ and $|\,\text{whole}\rangle$. And we don't need to fill it with poison.

But the presence of the vial confuses the issue just as much as the presence of the cat. Clearly, it is the detector itself and its interaction with the atom that are of interest to us here. How the detector manifests the detected state of the atom, be it by smashing a vial, or beeping, or moving a pointer, is of secondary importance. What is important is that the detector entangles with the observed atom somehow—some physical interaction must be present—and eventually responds to the state of the atom by counting either "up" or "down," with the "up" response on the detector side leaving the atom in the $|\,e\rangle$ state and the "down" response on the detector side leaving the atom in the $|\,g\rangle$ state. The atom will *dephase,* and the detector will respond to the atom's dephasing by dephasing itself into a state that can be read by us.

This is how detectors work.

The superstition of observation

What greatly puzzled physicists in Schrödinger's days was that back then they believed that it was the "act of observation," however ill-defined, that was responsible for the observed quantum system flipping to one of its basis states. Some even believed that a conscious observer was needed. This belief resulted in one of the axioms of quantum mechanics stating that "upon observation a quantum system finds itself irreversibly in one of its basis states." For this reason physicists also believed that the cat would be fine as long as we wouldn't peek into the box. In other words, it would be our act of observing the cat that would throw it out of the superposition of $(|\text{ dead}\rangle + |\text{ alive}\rangle)/\sqrt{2}$.

The other issue here is what we referred to in Section 4.4, page 121, as "the superstition of superposition." There we argued that a state that is in a superposition of two basis states is not in both states at the same time, but, instead, it is in neither. It is in a third state that is altogether different. The physical meaning of this third state is not always easy to figure out, but it can be always identified by evaluating the full density operator and the full vector of probabilities for the state. Consequently, even if we could put the cat in the superposition of $|\text{ dead}\rangle$ and $|\text{ alive}\rangle$, it may not necessarily be something out of this world. It may, instead, be just $|\text{ sick}\rangle$.

Today we have a more sophisticated view of what goes on. We don't need an axiom. We can derive the dynamics of the combined system from the basic principles of quantum mechanics using just the Schrödinger equation, in principle and in practice even, as long as the detector itself is not too large. A detector comprising the Avogadro number of atoms is out of the question here. It is not computable. But a detector comprising no more than 10 qubits can be analyzed fully; and if the corresponding system can be implemented in a laboratory, the theoretical predictions can be compared against the actual measurements.

5.12.1 The Haroche-Ramsey Experiment

Just such an experiment was carried out by Haroche and his colaborators, Brune, Hagley, Dreyer, Maître, Maali, Wunderlich, and Raimond, at the Kastler Brossel Laboratory of the Paris l'Ecole Normale Supérieure in 1996 [17]. The experiment is similar to Ramsey measurement but with an important modification, hence the name.

Observing an atom with a handful of photons

Haroche's group created an "atom + measuring apparatus" system in which the measuring apparatus was a mesoscopic cavity holding up to 10 photons. They demonstrated that the photons in the cavity themselves were put in a superposition of states by their interaction with the atom. Then they observed their dephasing and transformation of the cavity state into a statistical mixture. The observed behavior of the system was contrasted with theoretical predictions [27] obtained by an analysis somewhat like what we have presented in Section 5.11—with the difference that instead of entangling a qubit with another qubit, here we entangle it with 10 photons—and a highly accurate match was demonstrated.

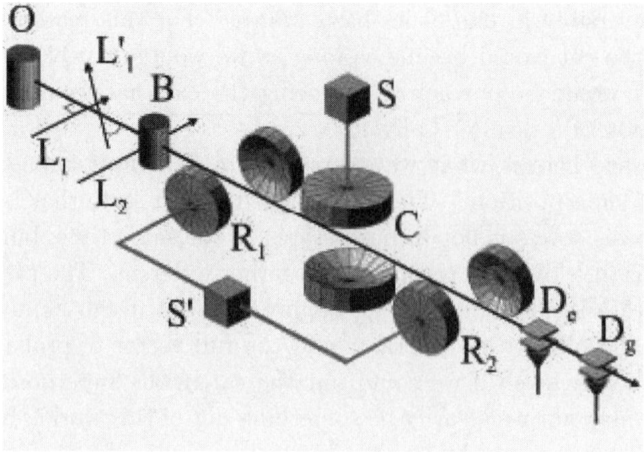

Figure 5.9: Setup for the Haroche-Ramsey experiment. Figure reprinted with permission from [17]. © 1996 by the American Physical Society.

The measurement setup for the experiment is shown in Figure 5.9. The apparatus is placed in vacuum and cooled to 0.6 K to reduce thermal radiation to a negligible level.

The cylindrical can labeled O is an oven that effuses Rubidium atoms. The atoms emerging from the oven are first conditioned by diode lasers L_1 and L_1' so that a certain subset of them—namely, the ones that move with velocity of $400 \pm 6\,\mathrm{m/s}$, and these atoms only—are in a state that is then pumped into a circular Rydberg state $\mid e\rangle$ in the box labeled B. The procedure prepares on average half an atom every 1.5 ms, and it takes about $2\,\mu$s in B to condition the atom. Other atoms will be naturally filtered away by the remaining part of the experiment, so we can forget about them.

As the selected atoms cross the cavity labeled R_1, in which they spend a precise amount of time on account of their selected velocity, their quantum state $\mid e\rangle$ is rotated by $\pi/2$ into $(\mid e\rangle + \mid g\rangle)/\sqrt{2}$. This is the Rabi rotation we had studied in Section 2.11, page 78.

Let us, for the moment, assume that there is no cavity labeled S. As the atoms fly from the microwave cavity R_1 to the microwave cavity R_2, they precess and are then rotated again by $\pi/2$ in R_2. So, we end up with a pure Ramsey experiment, like the one discussed in Section 2.11, that is here carried out on atoms of Rubidium in the two circular Rydberg states. The atoms have more states besides $\mid e\rangle$ and $\mid g\rangle$, but the dynamics of the experiment is confined to the two states only. Consequently, this is a qubit experiment, which is good, because qubits are simple and easy to understand.

Past the microwave cavity R_2 there are two field ionization detectors, D_e and D_g. The first one is tuned to detect atoms in state $| e\rangle$, and the second one is tuned to detect atoms in state $| g\rangle$. Both have a detection efficiency of $40 \pm 15\%$.

The frequency of the Rabi oscillations field in cavities R_1 and R_2 is varied slightly—by up to 10 kHz—around the resonance Rabi frequency for the $| e\rangle \rightarrow | g\rangle$ transition, $\nu_0 = 51.099\,\text{GHz}$. The variation is only two parts per 10 million. This has a similar effect to stretching or shrinking the free precession time between the two Ramsey pulses, so that when the atoms are finally detected either at D_e or D_g we observe Ramsey fringes as shown in Figure 5.10 (a). It's just that instead of observing the fringes in function of time elapsed between the two Ramsey pulses, here we observe them in function of the pulse frequency.

What is plotted in the diagrams of Figure 5.10 is the measured probability of detecting an atom in the $| g\rangle$ state, based on the exploration of a statistical ensemble of 1,000 events per each point of the graph and sampling for 50 discrete values of frequency ν [61]. To collect enough statistics for each graph takes about 10 minutes. The expected standard deviation for a count of $1,000$ is $\sqrt{1000} \approx 32$ (see, for example, [41]), which yields the relative error in the estimated probability of about 3%. The smooth lines are sinusoids fitted through the experimental data.

Even though the detector efficiencies are about 40% only, here we take a ratio of N_g to $N_g + N_e$, where N_g and N_e are the actually registered counts, so the curve in Figure 5.10 should vary between 0 and 1 in principle. Instead it is squashed to between 0.22 and 0.78, on average, or, to put it in another way, its *contrast* is reduced to $55 \pm 5\%$. This is, among other reasons, because of static and microwave field inhomogeneities over the diameter of the qubit beam, which is 0.7 mm. The finite lifetime of the $| e\rangle$ and $| g\rangle$ states contributes to this loss of contrast as well.

Now let us turn our attention to the cavity called C. C stands for "cat." This is the mesoscopic "detector" the atom interacts with. The cavity is made of two concave superconducting niobium mirrors separated by about 2.7 cm. The diameter of each mirror is 5 cm, and the curvature radius of their inner surface is 4 cm. The electromagnetic fields trapped between the mirrors are focused on a small region between them that is about 6 mm across. The Rubidium beam traverses the region in about 19 μs.

The field in the cavity is quantized and coherent. This had been demonstrated prior to the Haroche–Ramsey experiment by the Haroche group [18] and was one of the first such observations in history, even though the idea of the electromagnetic

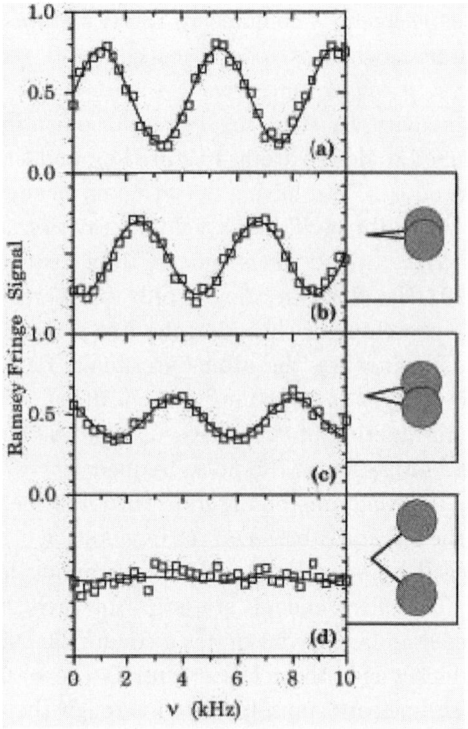

Figure 5.10: Ramsey fringes: (a) the microwave cavity C contains no photons, (b)–(d) the microwave cavity C contains photons; frequency detuning is (b) 712 kHz, (c) 347 kHz, (d) 104 kHz. Figure reprinted with permission from [17]. © 1996 by the American Physical Society.

field quantization goes back to the Einstein's photoelectric effect paper of 1905 [37].[12] The average number of photons in the cavity varies between 0 and 10, their lifetime is about 160 μs, and their collective state prior to the interaction with the atom can be described symbolically by $| \alpha\rangle$.

Jaynes-Cummings theory

Interaction between the qubit and this kind of a quantum field is not quite like interaction between the qubit and the *classical* Rabi field, which we discussed in Section 2.11. Here the interaction is with just a handful of photons, the number

[12] In Planck's original derivation of the black body radiation formula, it was the matter's ability to absorb and emit radiation that was quantized, not the electromagnetic field.

of which is quantumly uncertain, meaning that there is a certain distribution $P(n)$ that gives us the probability of there being n photons in the cavity. The problem was worked out by Jaynes and Cummings in 1963 [70], and the result is such that if the cavity is filled with photons of the $\mid e\rangle \rightarrow\mid g\rangle$ transition frequency, then the probability of the transition is

$$P_{\mid e\rangle\rightarrow\mid g\rangle}(t) = \sum_n P(n)\sin^2\left(\Omega\sqrt{n+1}t\right), \tag{5.326}$$

where t is the time of interaction and $\Omega = 2\pi \times 24\,\text{kHz}$ is the qubit-field coupling. This parameter plays here a role similar to the Rabi frequency.

The exact distance between the mirrors can be varied, which results in tuning the cavity. It can be tuned so that the field frequency in it matches the $\mid e\rangle \rightarrow\mid g\rangle$ transition frequency, but instead it is *detuned* by between $\delta = 2\pi \times 70\,\text{kHz}$ and $\delta = 2\pi \times 800\,\text{kHz}$. In effect, when the Rubidium atoms traverse the field region, the frequency of the field is too far from the Rabi frequency, and the atoms spend too little time in the field region for any energy exchange to take place between them. Instead the field in the cavity becomes coupled to the atom, which changes the phase of the field by

$$\phi = \frac{\Omega^2 t}{\delta}, \quad \text{for} \quad \Omega << \delta, \tag{5.327}$$

if the atom was in the $\mid e\rangle$ state prior to entering the cavity and by $-\phi$ if the atom was in the $\mid g\rangle$ state.

This is how the field in the cavity C becomes the "detector" of the atom's state.

After entanglement with the atom, the combined state of the photons+atom system becomes $\mid e\rangle \otimes \mid \alpha e^{i\phi}\rangle$ or $\mid g\rangle \otimes \mid \alpha e^{-i\phi}\rangle$; and if the atom was in the superposition state $\left(\mid e\rangle + \mid g\rangle\right)/\sqrt{2}$ prior to entering the cavity C, the combined state of the photons+atom system, after $19\,\mu$s of the interaction in the cavity, ends up in the superposition

$$\frac{1}{\sqrt{2}}\left(\mid e\rangle \otimes \mid \alpha e^{i\phi}\rangle + \mid g\rangle \otimes \mid \alpha e^{-i\phi}\rangle\right). \tag{5.328}$$

This has a profound effect on the Ramsey fringes that are detected by D_g.

Observing the states of the cat

Figure 5.10 (b) shows Ramsey fringes when the cavity C is filled with 9.5 photons on average, and is detuned from the Rabi frequency by 712 kHz. We can see that the contrast has diminished somewhat and the peaks have shifed to the right a little. The insert on the right-hand side of the graph shows the field phase shift ϕ, as phasors, for both the $\mid e\rangle$ and the $\mid g\rangle$ states.

Figure 5.10 (c) shows the same, but this time the cavity is detuned by 347 kHz. We are getting closer to the Rabi frequency, thus increasing the entanglement between the atom and the photons in the cavity. The response registered by D_g shows even more diminished contrast and even more shift in the location of the peaks. The accompanying cavity field phase shift ϕ is somewhat larger than in the previous case.

Figure 5.10 (d) shows Ramsey fringes for the cavity detuning of 104 kHz. This time we are quite close to the Rabi frequency, though still not close enough for the full interaction described by equation (5.326) to take place. The entanglement between the atom and the photons is stronger still, and the Ramsey fringes almost disappear. The cavity field phase shift ϕ is larger still.

The blue balls at the tips of the phasors in the inserts represent the uncertainty in the field's phase, which is due to quantum fluctuations of the field. We can see that in Figure 5.10 (d) the phase difference between the two field states is sufficiently large to be resolved even in the presence of the uncertainties. On the other hand, the separation in Figure 5.10 (b) is too small, and the two states are not resolvable.

The connection between our ability to resolve the two states of the "cat," $| \alpha e^{i\phi}\rangle$ and $| \alpha e^{-i\phi}\rangle$, and the disappearance of the fringe pattern, which in the unitary picture can be thought of as resulting from the interference of two amplitudes

$$\langle g \mid e \rangle = \langle g \mid R_1 \mid e \rangle + \langle g \mid R_2 \mid e \rangle, \tag{5.329}$$

is characteristic of quantum physics and shows up in many other situations. It is often explained by hand-waving arguments about the conspiracy of nature. However, here we can carry out detailed calculations, because the problem of coupling a qubit to 10 photons is still computable, and compare theoretical predictions to observed, experimental data.

The two graphs in Figure 5.11 show the results of theoretical calculations (smooth curves) and of the actual measurements (data points with error bars) for the Ramsey fringe contrast (a), and for the fringe shift (b) both in function of the cavity field phase shift ϕ. The agreement between theory and measurement is indeed striking, and it is here that the Haroche–Ramsey experiment is so remarkable.

Fiducial explanation of the loss of contrast

The almost total loss of contrast in Figure 5.10 (d) can be understood by invoking the Bloch sphere picture. Let us identify $| e\rangle$ with $|\uparrow\rangle$, and $| g\rangle$ with $|\downarrow\rangle$.

Just before the atom arrives at R_1 it is in the $|\uparrow\rangle$ state. The pulse in R_1 rotates it to $|\rightarrow\rangle$ or some other "equatorial" state. Then the atom enters the "cat" cavity

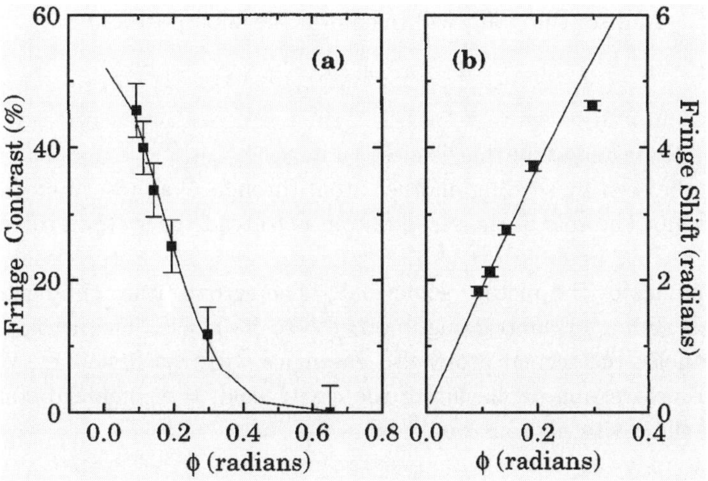

Figure 5.11: Ramsey fringe contrast (a) and shift (b). Figure reprinted with permission from [17]. © 1996 by the American Physical Society.

and becomes entangled with the photons in it. The combined superposition state that emerges from the interaction

$$\frac{1}{\sqrt{2}}\left(|\uparrow\rangle \mid \alpha e^{i\phi}\rangle + |\downarrow\rangle \mid \alpha e^{-i\phi}\rangle\right) \tag{5.330}$$

is pure, but if we were to trace the photons out, we would find that the qubit itself is no longer in a pure state. We would find that it has *dephased*, as we have seen in Section 5.11.3. The dephasing parameter Γ in equation (5.306) would be quite large on account of the detuning being relatively small. Consequently, the exponent $e^{-\Gamma t}$ would kill r^x and r^y of the qubit, leaving only r^z. But an "equatorial" state of a qubit does not have any r^z. So the qubit emerges from the cavity C in a completely chaotic state $\boldsymbol{r} = 0$. The application of the next Rabi pulse in R_2 does nothing to this state. It remains chaotic; and when it is finally measured by D_g and D_e, it returns a flat curve, as seen in Figure 5.10 (d).

Superposition of cat states

The act of measurement commited by the "cat" destroys the original state of the qubit, but the information about it survives in the "cat's" ϕ. What's more, tracing the qubit out from the combined state has much less effect on the "cat" than does tracing the "cat" out of the combined state had on the qubit, because there is more of the "cat." And so, the "cat" will remain in the superposition

$$\frac{1}{\sqrt{2}}\left(\mid \alpha e^{i\phi}\rangle + \mid \alpha e^{-i\phi}\rangle\right) \tag{5.331}$$

or in a slightly mixed state that is close to it.

This can be seen by sending another atom through C almost immediately after the first atom. The role of the second atom is to read the state of the field in the cavity.

This complicates the picture somewhat. The second atom entangles with the cavity field, adding or subtracting another ϕ to its phase. Furthermore, through the cavity field, the second atom also entangles with the first one. Were we to neglect the progressing dephasing of the cavity field, the combined state of both atoms and the cavity photons would be

$$\frac{1}{\sqrt{4}}\left(\mid e\rangle_2 \mid \alpha e^{i2\phi}\rangle \mid e\rangle_1 + \mid e\rangle_2 \mid \alpha\rangle \mid g\rangle_1 + \mid g\rangle_2 \mid \alpha\rangle \mid e\rangle_1 + \mid g\rangle_2 \mid \alpha e^{-i2\phi}\rangle \mid g\rangle_1\right). \tag{5.332}$$

Tracing the cavity field out lets us evaluate probabilities of correlated detections for both atoms, P_{ee}, P_{eg}, P_{ge} and P_{gg}. These depend on the frequency of the field applied to the two Ramsey cavities R_1 and R_2 and on the time lapse between the two atoms, τ. But it turns out that the following combination of the probabilities,

$$\eta = \frac{P_{ee}}{P_{ee} + P_{eg}} - \frac{P_{ge}}{P_{ge} + P_{gg}}, \tag{5.333}$$

is largely independent of the Ramsey (R_1 and R_2) pulse frequency. This quantity is our measure of the cavity field coherence. It should be 1/2 for very short times τ (it is actually less than this in Figure 5.12 because of the same experimental difficulties that reduce the expected contrast of Ramsey fringes), and we expect it to decay exponentially with τ.

Why should the "cat" itself dephase? It dephases because it is, in turn, entangled with the environment—for example, its power source marked S in Figure 5.9.

Figure 5.12 shows what happens for the two detunings of $\delta = 2\pi \times 170\,\mathrm{kHz}$ (circles and the dashed curve), and of $\delta = 2\pi \times 70\,\mathrm{kHz}$ (triangles and the smooth curve), with the cavity C filled by 3.3 photons on average. Circles and triangles correspond to the measured data, and curves correspond to theoretical predictions.

The field in the cavity indeed decoheres rapidly and on the time scale that is well short of the photon lifetime $T_r = 160\,\mu\mathrm{s}$. Furthermore, the field configuration that corresponds to stronger entanglement and yields a larger phase shift ϕ (triangles and the smooth line) decays faster.

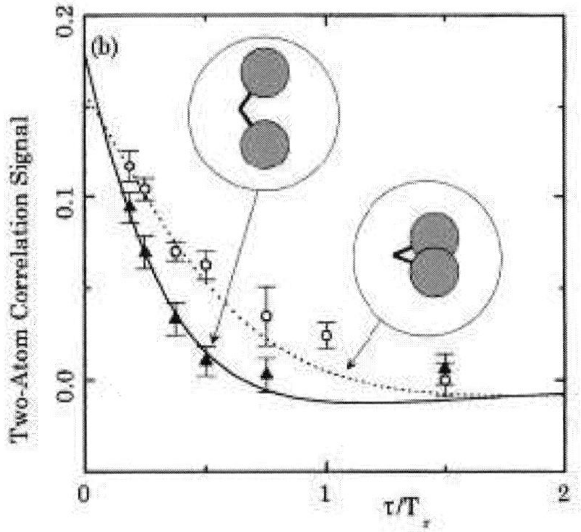

Figure 5.12: Two-atom correlation signal η as a function of τ/T_r, where τ is the delay between the two atoms and T_r is the cavity photon lifetime of 160 μs. Figure reprinted with permission from [17]. © 1996 by the American Physical Society.

The "cat" may remain longer in a superposition state if the two components of the state do not differ sufficiently to see the difference. The more they do, the faster the dephasing of the state.

6 The Controlled NOT Gate

6.1 The Quintessence of Quantum Computing

The controlled-NOT gate (CNOT for short) is a biqubit gate that is worshipped in quantum computing because every unitary operation on an n-qubit register can be implemented by combining controlled-NOT and single qubit unitary gates. In this the combination of controlled-NOT and unitary gates is similar to NAND or NOR gates known from classical computing, where every Boolean and arithmetic operation on an n-bit register can be implemented by a combination of NAND or NOR gates. We say that the controlled-NOT gate, in combination with single qubit unitary gates, is universal for quantum computation, as NOR and NAND gates are universal for classical ditigal computation.

Although the definition of the controlled-NOT gate is strikingly simple, the gate is rather hard to construct. Chapter 5 explains why: we have to perform a controlled operation on a biqubit. This is hard. Biqubits are hard to produce and hard to control, while maintaing their biqubitness, that is, entanglement, and the state's purity at the same time.

This chapter illustrates this point by discussing various implementations and simple uses of the controlled-NOT gate.

But first things first. What is a controlled-NOT gate? It is a single-qubit quantum NOT gate that can be activated or deactivated by the state of another qubit.

*Definition of controlled-*NOT

Let us consider a biqubit $| \psi\rangle \otimes | \eta\rangle$. Let the first qubit, the one on the left, be the control qubit and the one on the right the object qubit of the gate. We define the gate by saying that when the control qubit is in the $| 0\rangle$ state, then the gate is inactive. It leaves the object qubit as it is. But when the control qubit is in the $| 1\rangle$ state, then the gate flips the object qubit:

$$| 0\rangle | 0\rangle \rightarrow | 0\rangle | 0\rangle, \tag{6.1}$$
$$| 0\rangle | 1\rangle \rightarrow | 0\rangle | 1\rangle, \tag{6.2}$$
$$| 1\rangle | 0\rangle \rightarrow | 1\rangle | 1\rangle, \tag{6.3}$$
$$| 1\rangle | 1\rangle \rightarrow | 1\rangle | 0\rangle. \tag{6.4}$$

The definition is then extended onto linear combinations of biqubit basis states canonically, which makes the gate quantum. Equations (6.1)–(6.4) without this extension can be understood also as a definition of a classical controlled-NOT gate.

As we have done already once in Section 5.10, we replace the explicit tensor notation with the following binary and then decimal labeling of biqubit states:

$$| 0\rangle \otimes | 0\rangle \equiv | 00\rangle = | \mathbf{0}\rangle, \tag{6.5}$$

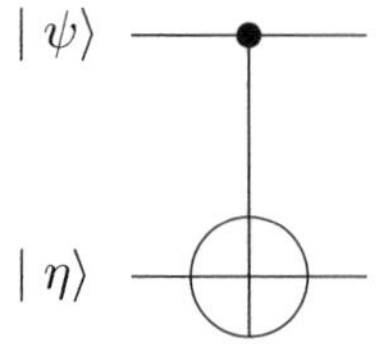

Figure 6.1: Diagrammatic representation of the controlled-NOT gate.

$$
\begin{aligned}
|\,0\rangle \otimes |\,1\rangle &\equiv |\,01\rangle =|\,\mathbf{1}\rangle, && (6.6)\\
|\,1\rangle \otimes |\,0\rangle &\equiv |\,10\rangle =|\,\mathbf{2}\rangle, && (6.7)\\
|\,1\rangle \otimes |\,1\rangle &\equiv |\,11\rangle =|\,\mathbf{3}\rangle. && (6.8)
\end{aligned}
$$

Using this notation, we can rewrite equations (6.1)–(6.4) in decimal form:

$$
\begin{aligned}
|\,\mathbf{0}\rangle &\rightarrow |\,\mathbf{0}\rangle, && (6.9)\\
|\,\mathbf{1}\rangle &\rightarrow |\,\mathbf{1}\rangle, && (6.10)\\
|\,\mathbf{2}\rangle &\rightarrow |\,\mathbf{3}\rangle, && (6.11)\\
|\,\mathbf{3}\rangle &\rightarrow |\,\mathbf{2}\rangle. && (6.12)
\end{aligned}
$$

*Controlled-*NOT *in matrix representation*

This leads to the following matrix representation of the gate:

$$
\text{CNOT} = \begin{pmatrix} 1 & 0 & 0 & 0 \\ 0 & 1 & 0 & 0 \\ 0 & 0 & 0 & 1 \\ 0 & 0 & 1 & 0 \end{pmatrix}. \tag{6.13}
$$

*Controlled-*NOT*, diagrammatic representation*

NOT *as modulo-2 addition*

A diagrammatic representation of the controlled-NOT gate is shown in Figure 6.1. The symbol used for the control connection is the little black dot on the $|\,\psi\rangle$ line and the vertical line that drops from it. The symbol used for NOT is a plus in a circle, $\oplus$, because it can be thought of as a modulo-2 addition of both qubits' values. If $|\,\psi\rangle$ is zero, then $|\,\eta\rangle$ remains unchanged. But if $|\,\psi\rangle$ is one, then when $|\,\eta\rangle$ is zero, we get $1 +_2 0 = 1$, and when $|\,\eta\rangle$ is one, we get $1 +_2 1 = 0$, so this is a bona fide controlled-NOT operation.

When the operation is used in equations describing algorithms, we may see a variety of notational devices, for example, $\bigoplus(|\,\psi\rangle\,|\,\eta\rangle)$ or, sometimes, $|\,\psi\rangle \oplus |\,\eta\rangle$. In order to avoid confusion, it is a good idea to clearly mark the control qubit, which does not change its value, and target qubit, which may or may not flip. This can be done by subscripting them, for example, $|\,\psi\rangle_c \oplus |\,\eta\rangle_t$, where c stands for "control" and t stands for "target."

The controlled-NOT gate is sometimes called the *measurement* gate because it can be used to measure one qubit by looking at the other one. For example, let $| \eta\rangle = | 0\rangle$ initially. By measuring it after the application of the gate we can find what $| \psi\rangle$ is. If $| \psi\rangle = | 0\rangle$, $| \eta\rangle$ remains zero, but if $| \psi\rangle = | 1\rangle$, $| \eta\rangle$ flips to $| 1\rangle$. So, in effect, after the application of the gate $| \eta\rangle$ is always the same as $| \psi\rangle$.

Controlled-NOT *as measurement gate*

Alas, the above is really classical reasoning. The controlled-NOT gate *is* a measurement gate in classical digital computing, but it does not let us carry out non-demolition measurements on the control qubit in the quantum domain.

Let us suppose that the control qubit, $| \psi\rangle$, is in the sideways state,

$$(| 0\rangle + | 1\rangle) / \sqrt{2}, \tag{6.14}$$

and the target qubit, $| \eta\rangle$, is in the $| 0\rangle$ state initially. What is going to be the final state of the biqubit after the operation has been completed?

This can be seen as follows.

$$\begin{aligned} &\bigoplus\left(\frac{1}{\sqrt{2}}(| 0\rangle + | 1\rangle)_c \otimes | 0\rangle_t\right) \\ &= \frac{1}{\sqrt{2}}\left(\bigoplus(| 0\rangle_c \otimes | 0\rangle_t) + \bigoplus(| 1\rangle_c \otimes | 0\rangle_t)\right) \\ &= \frac{1}{\sqrt{2}}(| 0\rangle_c \otimes | 0\rangle_t + | 1\rangle_c \otimes | 1\rangle_t) \end{aligned} \tag{6.15}$$

This is one of the four fully entangled Bell states, $| \Phi^+\rangle$, which we have encountered in Section 5.1; see, in particular, equation (5.53), page 200. Measuring the target qubit affects the control qubit. If the target qubit comes out in the $| 0\rangle$ state, the control qubit is in the $| 0\rangle$ state, too. If the target qubit comes out in the $| 1\rangle$ state, so does the control qubit. By measuring the target qubit we do not discover the initial state of the control qubit. Instead, we force it into the same state as the observed read-out of the target qubit. Furthermore, in this case the target qubit, when measured separately, will appear completely chaotic, as we have discovered in Section 5.8.

Controlled-NOT *creates entanglement.*

On the other hand, performing the measurements on both qubits will reveal the correlations: although totally random, both qubits always will show the same result. From this outcome, and knowing the initial state of the target qubit, we can reconstruct the initial state of the control qubit. The gate is reversible.

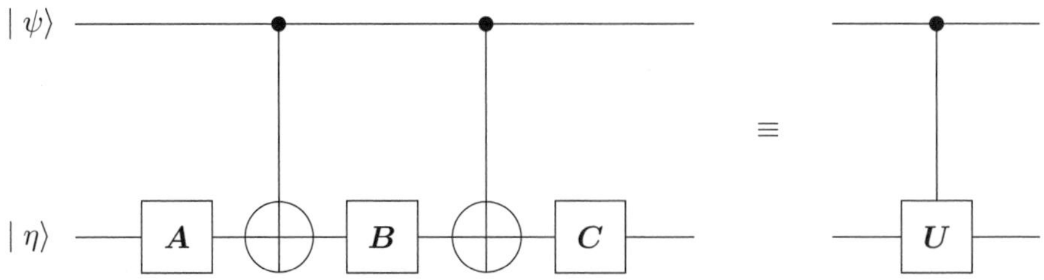

Figure 6.2: Decomposition of an arbitrary single-qubit $\boldsymbol{U}$ into $\boldsymbol{C\sigma_x B\sigma_x A}$, such that $\boldsymbol{CBA} = \mathbf{1}$.

6.2 Universal Gates

Although a controlled-NOT is not by itself a universal gate for quantum computation, in combination with single-qubit unitary gates it is—assuming that we restrict ourselves to unitary operations only. This can be ascertained in various ways and has been a subject of lively publishing since the late 1980s [33], with useful results published as recently as 2004 [140].

6.2.1 The ABC Decomposition and Controlled-U

One can easily show that a combination of controlled-NOT and single-qubit gates, as illustrated in Figure 6.2, can implement an arbitrary controlled-$\boldsymbol{U}$ gate, where $\boldsymbol{U}$ is a unitary single-qubit gate.

Gates $\boldsymbol{A}$, $\boldsymbol{B}$ and $\boldsymbol{C}$ are such that

$$\boldsymbol{CBA} = \mathbf{1} \tag{6.16}$$

and

$$\boldsymbol{C\sigma_x B\sigma_x A} = \boldsymbol{U}, \tag{6.17}$$

where $\boldsymbol{\sigma}_x$ are quantum NOTs. There is a $\boldsymbol{\sigma}_x$ sitting in every $\oplus$.

Clearly, if we can find such $\boldsymbol{A}$, $\boldsymbol{B}$, and $\boldsymbol{C}$ for any unitary $\boldsymbol{U}$, then the circuit on the left will be equivalent to the one on the right. When $|\psi\rangle$ is $|0\rangle$, then the two controlled-NOT gates on the $|\eta\rangle$ line remain inactive, $\boldsymbol{CBA}$ evaluates to $\mathbf{1}$, and $|\eta\rangle$ remains unchanged. But when $|\psi\rangle$ is $|1\rangle$, then the two controlled-NOT gates fire up, and the expression on the $|\eta\rangle$ line changes to $\boldsymbol{C\sigma_x B\sigma_x A}$, which evaluates to $\boldsymbol{U}$. So this is indeed a controlled-$\boldsymbol{U}$ gate.

The question is whether we can always find such $\boldsymbol{A}$, $\boldsymbol{B}$, and $\boldsymbol{C}$ that the equations (6.16) and (6.17) hold.

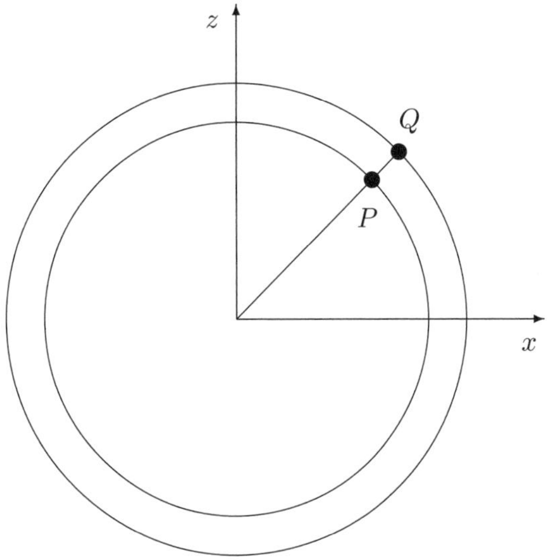

Figure 6.3: A Bloch sphere encased inside a reference sphere.

Revolutions of the Bloch sphere are trajectories on the reference sphere.

To see that we can and how, we need to examine revolutions of the Bloch sphere more closely. Unitary operations in the single-qubit space are just revolutions of the Bloch sphere, so we can switch between these two pictures as needed. To help us visualize the revolutions better, we consider the following contraption. We have a Bloch sphere surrounded by a reference sphere, as shown in Figure 6.3. The reference sphere is fixed, as are the x, y, and z axes, but the Bloch sphere inside it can rotate freely and about any axis. The line that connects the center of the Bloch sphere with point P on its surface is made of a stiff wire that having punctured its surface stretches all the way to the reference sphere, where it is attached to a little magnet that can move freely on the inner surface of the reference sphere. The location of the magnet is flagged as Q.

By moving another magnet on the outer surface of the reference sphere in the vicinity of Q, we can affect the inner magnet and, through its connection to the wire, move the Bloch sphere itself. The resulting movements of Q reflect the rotations of the Bloch sphere.

How can we move point Q? It can be moved from any location to any other location on the reference sphere by traveling in the south-north direction first, to get to the required latitude, and then by traveling in the east-west direction to reach the required longitude. The east-west travels are rotations about the z axis, but the south-north travels are rotations about an axis that is in the (x, y) plane and pointing in some direction that may be neither x nor y.

If we want to restrict south-north travels to a specific plane, for example, perpendicular to the y axis, we can do as follows. First we perform a rotation about the z axis that would get us to the (z, x) plane, which is perpendicular to the y axis; next we rotate in this plane about the y axis to get north or south to the required latitude; then we rotate again about the z axis to get to the required longitude.

Euler angles

We find that we can represent any rotation of the Bloch sphere by

$$\boldsymbol{R}_z(\gamma)\boldsymbol{R}_y(\beta)\boldsymbol{R}_z(\alpha), \tag{6.18}$$

where $\boldsymbol{R}_{y,z}$ are rotations about the y and z axes, respectively, and α, β, and γ are the rotation angles. The decomposition of arbitrary rotations into $\boldsymbol{R}_z$ and $\boldsymbol{R}_y$ is attributed to Euler, and the angles are referred to as Euler angles, but this was common knowledge among the navigators and early astronomers well before Euler.

We can think of these $\boldsymbol{R}_{y,z}$ as rotations or as single-qubit unitary operations. It's the same thing.

Next, we are going to figure out what $\boldsymbol{\sigma}_x$ does when it flanks a rotation on both sides, for example,

$$\boldsymbol{\sigma}_x \boldsymbol{R}_z \boldsymbol{\sigma}_x. \tag{6.19}$$

As we have seen in Section 5.11.2, equation (5.265), page 267, the meaning of $\boldsymbol{\sigma}_x$ is that it rotates the qubit polarization by 180° about the x axis, or, in other words, it flips the polarization about the x axis (but not about the center of the Bloch sphere!).

Let us assume that the point Q's starting position is in the northern hemisphere, closer to us, and in the (y, z) plane that is perpendicular to the x axis. This is shown in Figure 6.4, where the starting position of point Q is given number 1. Let us assume next that $\boldsymbol{R}_z(\alpha)$ rotates it to the right by a small angle so that 1 moves onto 2.

If instead of $\boldsymbol{R}_z$ our first move is $\boldsymbol{\sigma}_x$, then point 1 will not move onto point 2. Instead it'll flip onto point 3, which is on the other side of the sphere, but on the parallel of the same latitude, though in the south, not in the north. If we apply the same rotation $\boldsymbol{R}_z$, point 3 will move to the left, not to the right, because it's on the other side, but by exactly the same distance as the distance between 1 and 2, because the latitude is the same, and will eventually end up in the position marked 4.

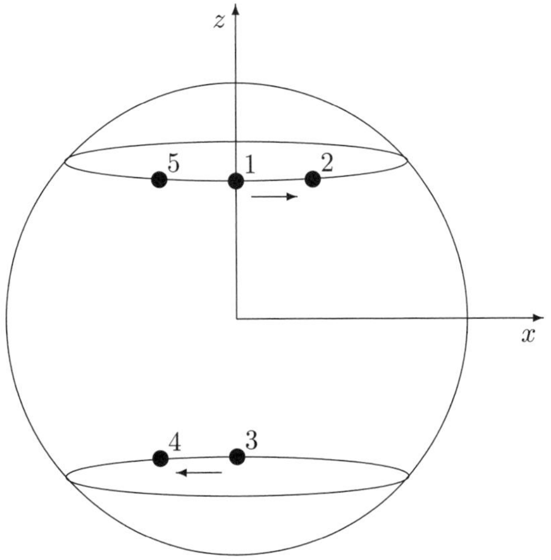

Figure 6.4: Flanking $\boldsymbol{R}_z$ with $\boldsymbol{\sigma}_x$ reverses the sign of rotation.

If we apply $\boldsymbol{\sigma}_x$ once more, point 4 will flip onto point 5, which is to the left of point 1, but exactly at the same distance from it as point 2.

In effect, we see that $\boldsymbol{\sigma}_x \boldsymbol{R}_z \boldsymbol{\sigma}_x$ rotates Q in the opposite direction. In other words,

Flanking $\boldsymbol{R}_{y,z}(\alpha)$ with $\boldsymbol{\sigma}_x$ reverses the rotation angle.

$$\boldsymbol{\sigma}_x \boldsymbol{R}_z(\alpha) \boldsymbol{\sigma}_x = \boldsymbol{R}_z(-\alpha). \tag{6.20}$$

Because y is just as perpendicular to x as z is, $\boldsymbol{\sigma}_x$ must have the same effect on $\boldsymbol{R}_y$:

$$\boldsymbol{\sigma}_x \boldsymbol{R}_y(\beta) \boldsymbol{\sigma}_x = \boldsymbol{R}_y(-\beta). \tag{6.21}$$

Now, let us make the following substitutions for $\boldsymbol{A}$, $\boldsymbol{B}$, and $\boldsymbol{C}$:

$$\boldsymbol{A} = \boldsymbol{R}_z\left(\frac{\alpha - \gamma}{2}\right), \tag{6.22}$$

$$\boldsymbol{B} = \boldsymbol{R}_y\left(-\frac{\beta}{2}\right) \boldsymbol{R}_z\left(-\frac{\alpha + \gamma}{2}\right), \tag{6.23}$$

$$\boldsymbol{C} = \boldsymbol{R}_z(\gamma)\, \boldsymbol{R}_y\left(\frac{\beta}{2}\right). \tag{6.24}$$

We find that

$$\begin{aligned} \boldsymbol{CBA} &= \boldsymbol{R}_z(\gamma)\,\boldsymbol{R}_y\left(\frac{\beta}{2}\right)\boldsymbol{R}_y\left(-\frac{\beta}{2}\right)\boldsymbol{R}_z\left(-\frac{\alpha+\gamma}{2}\right)\boldsymbol{R}_z\left(\frac{\alpha-\gamma}{2}\right) \\ &= \boldsymbol{R}_z(\gamma)\,\boldsymbol{R}_z(-\gamma) = \boldsymbol{1}, \end{aligned} \tag{6.25}$$

but

$$\begin{aligned} &\boldsymbol{C}\boldsymbol{\sigma}_x\boldsymbol{B}\boldsymbol{\sigma}_x\boldsymbol{A} \\ &= \boldsymbol{R}_z(\gamma)\,\boldsymbol{R}_y\left(\frac{\beta}{2}\right)\boldsymbol{\sigma}_x\boldsymbol{R}_y\left(-\frac{\beta}{2}\right)\boldsymbol{\sigma}_x\boldsymbol{\sigma}_x\boldsymbol{R}_z\left(-\frac{\alpha+\gamma}{2}\right)\boldsymbol{\sigma}_x\boldsymbol{R}_z\left(\frac{\alpha-\gamma}{2}\right) \\ &= \boldsymbol{R}_z(\gamma)\,\boldsymbol{R}_y\left(\frac{\beta}{2}\right)\boldsymbol{R}_y\left(\frac{\beta}{2}\right)\boldsymbol{R}_z\left(\frac{\alpha+\gamma}{2}\right)\boldsymbol{R}_z\left(\frac{\alpha-\gamma}{2}\right) \\ &= \boldsymbol{R}_z(\gamma)\,\boldsymbol{R}_y(\beta)\,\boldsymbol{R}_z(\alpha) = \boldsymbol{U}, \end{aligned} \tag{6.26}$$

where we have made use of $\boldsymbol{\sigma}_x\boldsymbol{\sigma}_x = \boldsymbol{1}$ and inserted it between $\boldsymbol{R}_y$ and $\boldsymbol{R}_z$ in the second line.

In summary, we can always express any given $\boldsymbol{U}$ in terms of Euler angles, and then we can construct $\boldsymbol{A}$, $\boldsymbol{B}$, and $\boldsymbol{C}$ such that a controlled-$\boldsymbol{U}$ can be constructed as shown in Figure 6.2.

Although this and similar results deliver much merriment to people who busy themselves drawing quantum circuits on paper and thinking of quantum algorithms, it is of less practical benefit than is commonly believed. It is often easier to implement a controlled-$\boldsymbol{U}$ gate directly. On the other hand, the controlled-$\boldsymbol{U}$ gate is useful and shows up in quantum algorithms frequently.

6.2.2 General Biqubit Unitary Gates

Vidal-Dawson decomposition

But a controlled-$\boldsymbol{U}$ gate is still not a general biqubit gate. It turns out that any biqubit gate can be expressed in terms of *three* controlled-NOT gates and eight single qubit unitary gates, as shown in Figure 6.5.

This follows from two observations.

Khaneja-Glaser decomposition

The first observation is that an arbitrary biqubit unitary operation $\boldsymbol{U}$ can be reduced to $e^{-\mathrm{i}\boldsymbol{H}}$, where

$$\boldsymbol{H} = h_x\boldsymbol{\sigma}_x \otimes \boldsymbol{\sigma}_x + h_y\boldsymbol{\sigma}_y \otimes \boldsymbol{\sigma}_y + h_z\boldsymbol{\sigma}_y \otimes \boldsymbol{\sigma}_y \tag{6.27}$$

by rotating the individual qubits before and after the application of $e^{-\mathrm{i}\boldsymbol{H}}$, as shown in Figure 6.6.

That it should be possible to do just this is obvious. The reasoning is similar to that used in Section 5.9, where we reduced terms proportional to $\boldsymbol{\sigma}_i \otimes \boldsymbol{\sigma}_j$ to a

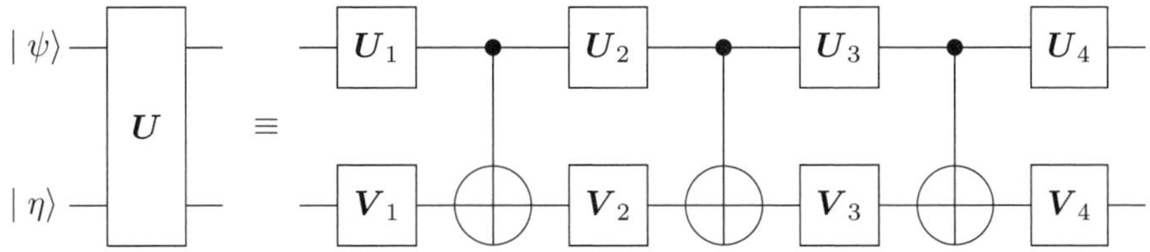

Figure 6.5: Decomposition of an arbitrary unitary biqubit operation $\boldsymbol{U}$ into a sequence of three controlled-NOTs and eight single qubit unitary operations, after Vidal and Dawson [140].

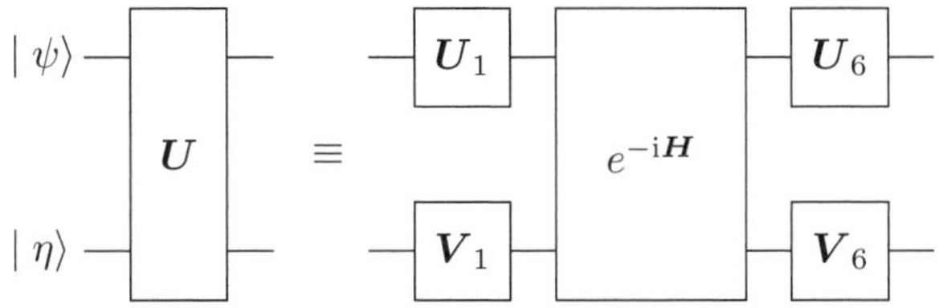

Figure 6.6: Reduction of an arbitrary biqubit unitary operation $\boldsymbol{U}$ to $\exp(-\mathrm{i}\boldsymbol{H})$, where $\boldsymbol{H}$ is given by equation (6.27).

diagonal sum such as (6.27). A more formal proof can be found in Kraus and Cirac [83] Appendix A, or in Khaneja, Brockett, and Glaser [77].

The decomposition of an arbitrary unitary biqubit gate into components shown in Figure 6.6 is a special case of the so-called Khaneja-Glaser decomposition [78]. The biqubit box labeled $e^{-\mathrm{i}\boldsymbol{H}}$ usually corresponds to free biqubit evolution in some background field, described by $\boldsymbol{H}$. When natural qubit-qubit couplings cannot be controlled easily, as is the case, for example, in molecular registers, this may be the only way to manipulate a biqubit: just do nothing and wait for it to evolve all by itself for a certain precisely measured amount of time. In such systems various biqubit gates are engineered by subjecting the component qubits to prescribed "local" (meaning "single-qubit") rotations before and after the free biqubit evolution. The controlled-NOT gate is constructed the same way, which, obviously, makes it less useful in real computations than the Khaneja-Glaser prescription. The application of controlled-NOT is more in quantum circuit design than in quantum circuit implementation.

Returning to our discussion of the universality of the controlled-NOT gate, the second observation is that $\exp(-\mathrm{i}\boldsymbol{H})$ can be decomposed, formally, into a sequence

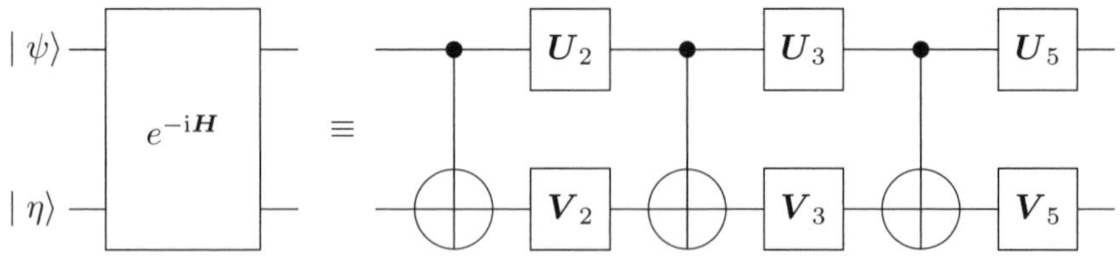

Figure 6.7: Decomposition of $\exp(-\mathrm{i}\boldsymbol{H})$ into three controlled-NOTs and single-qubit unitary gates, after Vidal and Dawson [140].

of three controlled-NOT gates each followed by single qubit unitary gates as shown in Figure 6.7.

Combining diagrams in Figures 6.6 and 6.7 gives us the diagram in Figure 6.5, with $\boldsymbol{U}_4 = \boldsymbol{U}_6\boldsymbol{U}_5$ and $\boldsymbol{V}_4 = \boldsymbol{V}_6\boldsymbol{V}_5$, which completes the proof that every biqubit unitary operation can be expressed in terms of controlled-NOT and single-qubit unitary gates.

The proof that the decomposition shown in Figure 6.7 is possible is as follows [140].

Khaneja-Glaser operator is diagonal in the Bell basis.

First we notice that $\exp(-\mathrm{i}\boldsymbol{H})$ can be diagonalized by rewriting equation (6.27) in the Bell basis. We are going to use a somewhat different notation for the Bell basis vectors this time, because it is more convenient than $|\,\Phi^{\pm}\rangle$ and $|\,\Psi^{\pm}\rangle$, namely,

$$
\begin{aligned}
|\,\Phi^{+}\rangle &\equiv |\,\bar{\mathbf{0}}\rangle, & (6.28)\\
|\,\Psi^{+}\rangle &\equiv |\,\bar{\mathbf{1}}\rangle, & (6.29)\\
|\,\Psi^{-}\rangle &\equiv |\,\bar{\mathbf{2}}\rangle, & (6.30)\\
|\,\Phi^{-}\rangle &\equiv |\,\bar{\mathbf{3}}\rangle. & (6.31)
\end{aligned}
$$

This notation has the advantage that the transformation from the computational basis $|\,\mathbf{0}\rangle \ldots |\,\mathbf{3}\rangle$ to the Bell basis and the reverse one look the same:

$$
\begin{aligned}
|\,\bar{\mathbf{0}}\rangle &= \frac{1}{\sqrt{2}}\left(|\,\mathbf{0}\rangle + |\,\mathbf{3}\rangle\right), & |\,\mathbf{0}\rangle &= \frac{1}{\sqrt{2}}\left(|\,\bar{\mathbf{0}}\rangle + |\,\bar{\mathbf{3}}\rangle\right), & (6.32)\\
|\,\bar{\mathbf{1}}\rangle &= \frac{1}{\sqrt{2}}\left(|\,\mathbf{1}\rangle + |\,\mathbf{2}\rangle\right), & |\,\mathbf{1}\rangle &= \frac{1}{\sqrt{2}}\left(|\,\bar{\mathbf{1}}\rangle + |\,\bar{\mathbf{2}}\rangle\right), & (6.33)\\
|\,\bar{\mathbf{2}}\rangle &= \frac{1}{\sqrt{2}}\left(|\,\mathbf{1}\rangle - |\,\mathbf{2}\rangle\right), & |\,\mathbf{2}\rangle &= \frac{1}{\sqrt{2}}\left(|\,\bar{\mathbf{1}}\rangle - |\,\bar{\mathbf{2}}\rangle\right), & (6.34)\\
|\,\bar{\mathbf{3}}\rangle &= \frac{1}{\sqrt{2}}\left(|\,\mathbf{0}\rangle - |\,\mathbf{3}\rangle\right), & |\,\mathbf{3}\rangle &= \frac{1}{\sqrt{2}}\left(|\,\bar{\mathbf{0}}\rangle - |\,\bar{\mathbf{3}}\rangle\right). & (6.35)
\end{aligned}
$$

Appendix C summarizes this and expresses the Pauli matrix tensor products $\boldsymbol{\sigma}_x \otimes \boldsymbol{\sigma}_x$, $\boldsymbol{\sigma}_y \otimes \boldsymbol{\sigma}_y$, and $\boldsymbol{\sigma}_z \otimes \boldsymbol{\sigma}_z$ in terms of both the computational and Bell bases. We find that the Bell basis diagonalizes tensor products of Pauli matrices as follows:

$$\boldsymbol{\sigma}_x \otimes \boldsymbol{\sigma}_x = |\,\bar{\mathbf{0}}\rangle\langle\bar{\mathbf{0}}\,| + |\,\bar{\mathbf{1}}\rangle\langle\bar{\mathbf{1}}\,| - |\,\bar{\mathbf{2}}\rangle\langle\bar{\mathbf{2}}\,| - |\,\bar{\mathbf{3}}\rangle\langle\bar{\mathbf{3}}\,|, \tag{6.36}$$

$$\boldsymbol{\sigma}_y \otimes \boldsymbol{\sigma}_y = -|\,\bar{\mathbf{0}}\rangle\langle\bar{\mathbf{0}}\,| + |\,\bar{\mathbf{1}}\rangle\langle\bar{\mathbf{1}}\,| - |\,\bar{\mathbf{2}}\rangle\langle\bar{\mathbf{2}}\,| + |\,\bar{\mathbf{3}}\rangle\langle\bar{\mathbf{3}}\,|, \tag{6.37}$$

$$\boldsymbol{\sigma}_z \otimes \boldsymbol{\sigma}_z = |\,\bar{\mathbf{0}}\rangle\langle\bar{\mathbf{0}}\,| - |\,\bar{\mathbf{1}}\rangle\langle\bar{\mathbf{1}}\,| - |\,\bar{\mathbf{2}}\rangle\langle\bar{\mathbf{2}}\,| + |\,\bar{\mathbf{3}}\rangle\langle\bar{\mathbf{3}}\,|\,. \tag{6.38}$$

This lets us rewrite (6.27):

$$\begin{aligned} & h_x\boldsymbol{\sigma}_x \otimes \boldsymbol{\sigma}_x + h_y\boldsymbol{\sigma}_y \otimes \boldsymbol{\sigma}_y + h_z\boldsymbol{\sigma}_z \otimes \boldsymbol{\sigma}_z \\ &\quad = (h_x - h_y + h_z)\,|\,\bar{\mathbf{0}}\rangle\langle\bar{\mathbf{0}}\,| + (h_x + h_y - h_z)\,|\,\bar{\mathbf{1}}\rangle\langle\bar{\mathbf{1}}\,| \\ &\qquad + (-h_x - h_y - h_z)\,|\,\bar{\mathbf{2}}\rangle\langle\bar{\mathbf{2}}\,| + (-h_x + h_y + h_z)\,|\,\bar{\mathbf{3}}\rangle\langle\bar{\mathbf{3}}\,|\,. \end{aligned} \tag{6.39}$$

By introducing

$$h_{\bar{\mathbf{0}}} = h_x - h_y + h_z, \tag{6.40}$$

$$h_{\bar{\mathbf{1}}} = h_x + h_y - h_z, \tag{6.41}$$

$$h_{\bar{\mathbf{2}}} = -h_x - h_y - h_z, \tag{6.42}$$

$$h_{\bar{\mathbf{3}}} = -h_x + h_y + h_z, \tag{6.43}$$

we can rewrite (6.39),

$$\boldsymbol{H} = h_{\bar{\mathbf{0}}}\,|\,\bar{\mathbf{0}}\rangle\langle\bar{\mathbf{0}}\,| + h_{\bar{\mathbf{1}}}\,|\,\bar{\mathbf{1}}\rangle\langle\bar{\mathbf{1}}\,| + h_{\bar{\mathbf{2}}}\,|\,\bar{\mathbf{2}}\rangle\langle\bar{\mathbf{2}}\,| + h_{\bar{\mathbf{3}}}\,|\,\bar{\mathbf{3}}\rangle\langle\bar{\mathbf{3}}\,|, \tag{6.44}$$

and now

$$e^{-\mathrm{i}\boldsymbol{H}} = e^{-\mathrm{i}h_{\bar{\mathbf{0}}}}\,|\,\bar{\mathbf{0}}\rangle\langle\bar{\mathbf{0}}\,| + e^{-\mathrm{i}h_{\bar{\mathbf{1}}}}\,|\,\bar{\mathbf{1}}\rangle\langle\bar{\mathbf{1}}\,| + e^{-\mathrm{i}h_{\bar{\mathbf{2}}}}\,|\,\bar{\mathbf{2}}\rangle\langle\bar{\mathbf{2}}\,| + e^{-\mathrm{i}h_{\bar{\mathbf{3}}}}\,|\,\bar{\mathbf{3}}\rangle\langle\bar{\mathbf{3}}\,|\,. \tag{6.45}$$

All that remains to be demonstrated is that the circuit on the right-hand side of Figure 6.7 does the same thing, meaning that for any $|\,\bar{\boldsymbol{k}}\rangle$, where $\bar{\boldsymbol{k}} = \bar{\mathbf{0}}\ldots\bar{\mathbf{3}}$, its action is to tranform it into $\exp(-\mathrm{i}h_{\bar{\boldsymbol{k}}})\,|\,\bar{\boldsymbol{k}}\rangle$.

This is not going to happen for arbitrary $\boldsymbol{U}_{2,3,5}$ and for arbitrary $\boldsymbol{V}_{2,3,5}$. We will need to choose them carefully. The following substitutions do the trick:

$$\boldsymbol{U}_2 = \frac{\mathrm{i}}{\sqrt{2}}(\boldsymbol{\sigma}_x + \boldsymbol{\sigma}_z)\,e^{-\mathrm{i}h_x\boldsymbol{\sigma}_x}, \tag{6.46}$$

$$\boldsymbol{V}_2 = e^{-\mathrm{i}h_z\boldsymbol{\sigma}_z}, \tag{6.47}$$

$$\boldsymbol{U}_3 = \frac{-\mathrm{i}}{\sqrt{2}}(\boldsymbol{\sigma}_x + \boldsymbol{\sigma}_z), \tag{6.48}$$

$$\boldsymbol{V}_3 = e^{\mathrm{i}h_y\boldsymbol{\sigma}_z}, \tag{6.49}$$

$$\boldsymbol{U}_5 = \frac{1}{\sqrt{2}}\left(\mathbf{1} - \mathrm{i}\boldsymbol{\sigma}_x\right), \tag{6.50}$$

$$\boldsymbol{V}_5 = \boldsymbol{U}_5^{-1}. \tag{6.51}$$

It helps to rewrite equations (6.46) to (6.51) by expressing the Pauli matrices in terms of the computational basis vectors and forms, as shown in Appendix A.3, equations (A.32)–A.35). Because $\boldsymbol{\sigma}_z$ is diagonal, we have

$$\boldsymbol{V}_2 = e^{-\mathrm{i}h_z\boldsymbol{\sigma}_z} = e^{-\mathrm{i}h_z} \mid 0\rangle\langle 0 \mid + e^{\mathrm{i}h_z} \mid 1\rangle\langle 1 \mid, \tag{6.52}$$

$$\boldsymbol{V}_3 = e^{\mathrm{i}h_y\boldsymbol{\sigma}_z} = e^{\mathrm{i}h_y} \mid 0\rangle\langle 0 \mid + e^{-\mathrm{i}h_y} \mid 1\rangle\langle 1 \mid. \tag{6.53}$$

$\boldsymbol{U}_3$, $\boldsymbol{U}_5$, and $\boldsymbol{V}_5$ are easy to evaluate:

$$\boldsymbol{U}_3 = -\frac{\mathrm{i}}{\sqrt{2}}\left(\mid 0\rangle + \mid 1\rangle\right)\langle 0 \mid - \frac{\mathrm{i}}{\sqrt{2}}\left(\mid 0\rangle - \mid 1\rangle\right)\langle 1 \mid, \tag{6.54}$$

$$\boldsymbol{U}_5 = \frac{1}{\sqrt{2}}\left(\mid 0\rangle - \mathrm{i} \mid 1\rangle\right)\langle 0 \mid - \frac{\mathrm{i}}{\sqrt{2}}\left(\mid 0\rangle + \mathrm{i} \mid 1\rangle\right)\langle 1 \mid, \tag{6.55}$$

$$\boldsymbol{V}_5 = \boldsymbol{U}_5^{-1} = \frac{1}{\sqrt{2}}\left(\mathbf{1} + \mathrm{i}\boldsymbol{\sigma}_x\right) \tag{6.56}$$

$$= \frac{1}{\sqrt{2}}\left(\mid 0\rangle + \mathrm{i} \mid 1\rangle\right)\langle 0 \mid + \frac{\mathrm{i}}{\sqrt{2}}\left(\mid 0\rangle - \mathrm{i} \mid 1\rangle\right)\langle 1 \mid. \tag{6.57}$$

It is easy to see that $\boldsymbol{U}_5\boldsymbol{V}_5 = \mathbf{1}$.

$\boldsymbol{U}_2$ is trickier because $\boldsymbol{\sigma}_x$ is not diagonal. But the following comes to the rescue:

$$\frac{1}{\sqrt{2}}\left(\boldsymbol{\sigma}_x + \boldsymbol{\sigma}_z\right)\boldsymbol{\sigma}_z\frac{1}{\sqrt{2}}\left(\boldsymbol{\sigma}_x + \boldsymbol{\sigma}_z\right) = \boldsymbol{\sigma}_x, \tag{6.58}$$

$$\frac{1}{\sqrt{2}}\left(\boldsymbol{\sigma}_x + \boldsymbol{\sigma}_z\right)\frac{1}{\sqrt{2}}\left(\boldsymbol{\sigma}_x + \boldsymbol{\sigma}_z\right) = \mathbf{1}, \tag{6.59}$$

which implies that, first, $\left(\boldsymbol{\sigma}_x + \boldsymbol{\sigma}_z\right)/\sqrt{2}$ is its own inverse, and, second, it diagonalizes $\boldsymbol{\sigma}_x$. Let us recall equations (4.300) and (4.301) (page 156 in Section 4.9.1). Combining them with (6.58) and (6.59) yields

$$e^{\boldsymbol{\sigma}_x} = \exp\left(\frac{1}{\sqrt{2}}\left(\boldsymbol{\sigma}_x + \boldsymbol{\sigma}_z\right)\boldsymbol{\sigma}_z\frac{1}{\sqrt{2}}\left(\boldsymbol{\sigma}_x + \boldsymbol{\sigma}_z\right)\right)$$

$$= \frac{1}{\sqrt{2}}\left(\boldsymbol{\sigma}_x + \boldsymbol{\sigma}_z\right)e^{\boldsymbol{\sigma}_z}\frac{1}{\sqrt{2}}\left(\boldsymbol{\sigma}_x + \boldsymbol{\sigma}_z\right), \tag{6.60}$$

and this we already know how to handle. In effect

$$\boldsymbol{U}_2 = \frac{\mathrm{i}}{\sqrt{2}}\left(\boldsymbol{\sigma}_x + \boldsymbol{\sigma}_z\right)e^{-\mathrm{i}h_x\boldsymbol{\sigma}_x}$$

$$\begin{aligned}
&= \frac{\mathrm{i}}{\sqrt{2}}\left(\boldsymbol{\sigma}_x + \boldsymbol{\sigma}_z\right)\frac{1}{\sqrt{2}}\left(\boldsymbol{\sigma}_x + \boldsymbol{\sigma}_z\right)e^{-\mathrm{i}h_x\boldsymbol{\sigma}_z}\frac{1}{\sqrt{2}}\left(\boldsymbol{\sigma}_x + \boldsymbol{\sigma}_z\right) \\
&= \mathrm{i}e^{-\mathrm{i}h_x\boldsymbol{\sigma}_z}\frac{1}{\sqrt{2}}\left(\boldsymbol{\sigma}_x + \boldsymbol{\sigma}_z\right) \\
&= \frac{\mathrm{i}}{\sqrt{2}}\left(e^{-\mathrm{i}h_x} \mid 0\rangle\langle 0 \mid + e^{\mathrm{i}h_x} \mid 1\rangle\langle 1 \mid\right) \\
&\quad \left(\mid 0\rangle\langle 1 \mid + \mid 1\rangle\langle 0 \mid + \mid 0\rangle\langle 0 \mid - \mid 1\rangle\langle 1 \mid\right) \\
&= \mathrm{i}e^{-\mathrm{i}h_x} \mid 0\rangle\frac{1}{\sqrt{2}}\left(\langle 0 \mid + \langle 1 \mid\right) \\
&\quad + \mathrm{i}e^{\mathrm{i}h_x} \mid 1\rangle\frac{1}{\sqrt{2}}\left(\langle 0 \mid - \langle 1 \mid\right).
\end{aligned} \tag{6.61}$$

Now we are ready to analyze the circuit shown in Figure 6.7. We move from the left to the right in steps.

1. Apply $\mid \bar{\mathbf{0}}\rangle$ to the input:

$$\mid \bar{\mathbf{0}}\rangle = \frac{1}{\sqrt{2}}\left(\mid \mathbf{0}\rangle + \mid \mathbf{3}\rangle\right) = \frac{1}{\sqrt{2}}\left(\mid 0\rangle \mid 0\rangle + \mid 1\rangle \mid 1\rangle\right).$$

2. Apply the first controlled-NOT gate to the input state:

$$\begin{aligned}
&\bigoplus\left(\frac{1}{\sqrt{2}}\left(\mid 0\rangle_c \mid 0\rangle_t + \mid 1\rangle_c \mid 1\rangle_t\right)\right) \\
&\quad = \frac{1}{\sqrt{2}}\left(\mid 0\rangle \mid 0\rangle + \mid 1\rangle \mid 0\rangle\right) = \frac{1}{\sqrt{2}}\left(\mid 0\rangle + \mid 1\rangle\right) \mid 0\rangle.
\end{aligned}$$

3. Apply $\boldsymbol{U}_2 \otimes \boldsymbol{V}_2$:

$$\begin{aligned}
&\left(\mathrm{i}e^{-\mathrm{i}h_x} \mid 0\rangle\frac{1}{\sqrt{2}}\left(\langle 0 \mid + \langle 1 \mid\right) + \mathrm{i}e^{\mathrm{i}h_x} \mid 1\rangle\frac{1}{\sqrt{2}}\left(\langle 0 \mid - \langle 1 \mid\right)\right) \\
&\otimes\left(e^{-\mathrm{i}h_z} \mid 0\rangle\langle 0 \mid + e^{\mathrm{i}h_z} \mid 1\rangle\langle 1 \mid\right)\left(\frac{1}{\sqrt{2}}\left(\mid 0\rangle + \mid 1\rangle\right) \mid 0\rangle\right) \\
&= \mathrm{i}e^{-\mathrm{i}h_x} \mid 0\rangle \otimes e^{-\mathrm{i}h_z} \mid 0\rangle = \mathrm{i}e^{-\mathrm{i}(h_x+h_z)} \mid 0\rangle \mid 0\rangle,
\end{aligned}$$

   where we have made use of perpendicularity of $(\mid 0\rangle \pm \mid 1\rangle)/\sqrt{2}$.

4. Apply the second controlled-NOT gate:

$$\bigoplus\left(\mathrm{i}e^{-\mathrm{i}(h_x+h_z)} \mid 0\rangle \mid 0\rangle\right) = \mathrm{i}e^{-\mathrm{i}(h_x+h_z)} \mid 0\rangle \mid 0\rangle.$$

   This gate doesn't change anything in this case.

5. Apply $\boldsymbol{U}_3 \otimes \boldsymbol{V}_3$:

$$\left(-\frac{\mathrm{i}}{\sqrt{2}}(\mid 0\rangle + \mid 1\rangle)\langle 0 \mid -\frac{\mathrm{i}}{\sqrt{2}}(\mid 0\rangle - \mid 1\rangle)\langle 1 \mid\right)$$
$$\otimes \left(e^{\mathrm{i}h_y} \mid 0\rangle\langle 0 \mid + e^{-\mathrm{i}h_y} \mid 1\rangle\langle 1 \mid\right)\left(\mathrm{i}e^{-\mathrm{i}(h_x+h_z)} \mid 0\rangle \mid 0\rangle\right)$$
$$= e^{-\mathrm{i}(h_x+h_z)}\frac{1}{\sqrt{2}}(\mid 0\rangle + \mid 1\rangle) \otimes e^{\mathrm{i}h_y} \mid 0\rangle$$
$$= e^{-\mathrm{i}(h_x-h_y+h_z)}\frac{1}{\sqrt{2}}(\mid 0\rangle + \mid 1\rangle) \mid 0\rangle.$$

6. Apply the third controlled-NOT gate:

$$\bigoplus\left(e^{-\mathrm{i}(h_x-h_y+h_z)}\frac{1}{\sqrt{2}}(\mid 0\rangle_c + \mid 1\rangle_c) \mid 0\rangle_t\right)$$
$$= e^{-\mathrm{i}(h_x-h_y+h_z)}\frac{1}{\sqrt{2}}(\mid 0\rangle \mid 0\rangle + \mid 1\rangle \mid 1\rangle).$$

7. Apply $\boldsymbol{U}_5 \otimes \boldsymbol{V}_5$:

$$\left(\frac{1}{\sqrt{2}}(\mid 0\rangle - \mathrm{i} \mid 1\rangle)\langle 0 \mid -\frac{\mathrm{i}}{\sqrt{2}}(\mid 0\rangle + \mathrm{i} \mid 1\rangle)\langle 1 \mid\right)$$
$$\otimes \left(\frac{1}{\sqrt{2}}(\mid 0\rangle + \mathrm{i} \mid 1\rangle)\langle 0 \mid + \frac{\mathrm{i}}{\sqrt{2}}(\mid 0\rangle - \mathrm{i} \mid 1\rangle)\langle 1 \mid\right)$$
$$e^{-\mathrm{i}(h_x-h_y+h_z)}\frac{1}{\sqrt{2}}(\mid 0\rangle \mid 0\rangle + \mid 1\rangle \mid 1\rangle)$$
$$= e^{-\mathrm{i}(h_x-h_y+h_z)}\frac{1}{\sqrt{2}}\left(\frac{(\mid 0\rangle - \mathrm{i} \mid 1\rangle) \otimes (\mid 0\rangle + \mathrm{i} \mid 1\rangle)}{2}\right.$$
$$\left.+\frac{-\mathrm{i}(\mid 0\rangle + \mathrm{i} \mid 1\rangle) \otimes \mathrm{i}(\mid 0\rangle - \mathrm{i} \mid 1\rangle)}{2}\right)$$
$$= e^{-\mathrm{i}(h_x-h_y+h_z)}\frac{1}{\sqrt{2}}(\mid 0\rangle \mid 0\rangle + \mid 1\rangle \mid 1\rangle) = e^{-\mathrm{i}h_{\bar{0}}} \mid \bar{\mathbf{0}}\rangle.$$

In seven similar steps we can demonstrate that the circuit transforms $\mid \bar{\mathbf{1}}\rangle$ into $e^{-\mathrm{i}h_{\bar{1}}} \mid \bar{\mathbf{1}}\rangle$, $\mid \bar{\mathbf{2}}\rangle$ into $e^{-\mathrm{i}h_{\bar{2}}} \mid \bar{\mathbf{2}}\rangle$, and $\mid \bar{\mathbf{3}}\rangle$ into $e^{-\mathrm{i}h_{\bar{3}}} \mid \bar{\mathbf{3}}\rangle$. It is a good exercise for the reader to do so, and it completes the proof that the circuits shown in Figure 6.7 are indeed equivalent.

The seven steps outlined above demonstrate how to analyze quantum circuits explicitly. But such explicit and manual analysis is not possible for circuits that may operate on, say, 20 qubits, because the number of possible input states is $2^{20} = 1,048,576$. Still, such circuits may be analyzed by a conventional computer. Quantum circuit analysis rules are straightforward and can be encoded in a computer program easily. But what if a circuit comprises 512 qubits? In this case the number of possible input states is $2^{512} \approx 1.34 \times 10^{154}$. This is an insanely huge number. And 512 qubits is not all that much. People do consider quantum computations carried out on thousands of qubits—there may be about 75,000 atoms in a very short viral DNA chain, and if every one of these atoms could be used as a qubit, we would have $2^{75,000}$ possible initial states for the circuit.

Quantum circuit analysis

For circuits with more than a handful of qubits, we have to resort to mathematical means of circuit analysis. This approach is possible for many circuits of interest and may be often coupled to explicit circuit simulation on small subcircuits that are then assembled into larger units—exactly as is done in classical circuit analysis.

But let us return to our discussion of the universality of the controlled-NOT and single-qubit unitary gates. We have demonstrated above that any unitary biqubit gate can be constructed, as shown in Figure 6.5, of three controlled-NOT gates and eight single-qubit gates. How about a triqubit gate?

6.2.3 Triqubit Gates

If biqubits are complicated, triqubits are even more so. Pure triqubits can be all separate or arranged into a single qubit and a biqubit (and there are three different ways in which this may happen) or all three may be entangled. And then there are numerous ways in which mixed qubits may enter a triqubit system: again, all three may be separate and mixed, one may be mixed and the remaining biqubit pure, and so on. A way to classify them, based on their separability, was presented by Acín, Bruß, Lewenstein, and Sanpera in 2001 [3], but we won't dwell on this complicated topic here.

Triqubit classification

In 2004 Vatan and Williams [139] published an explicit construction for an arbitrary triqubit gate implemented in terms of single-qubit and biqubit gates. Their solution derives from Khaneja-Glaser decomposition for an arbitrary triqubit gate [78] shown in Figure 6.8, where the triqubit gates $\boldsymbol{U}_1$, $\boldsymbol{V}$, and $\boldsymbol{U}_2$ are given by

Khaneja-Glaser decomposition of a triqubit gate

$$\boldsymbol{U}_1 = \exp\left(\mathrm{i}\left(a_1\boldsymbol{\sigma}_x \otimes \boldsymbol{\sigma}_x \otimes \boldsymbol{\sigma}_z + b_1\boldsymbol{\sigma}_y \otimes \boldsymbol{\sigma}_y \otimes \boldsymbol{\sigma}_z + c_1\boldsymbol{\sigma}_z \otimes \boldsymbol{\sigma}_z \otimes \boldsymbol{\sigma}_z\right)\right), \tag{6.62}$$

$$\boldsymbol{U}_2 = \exp\left(\mathrm{i}\left(a_2\boldsymbol{\sigma}_x \otimes \boldsymbol{\sigma}_x \otimes \boldsymbol{\sigma}_z + b_2\boldsymbol{\sigma}_y \otimes \boldsymbol{\sigma}_y \otimes \boldsymbol{\sigma}_z + c_2\boldsymbol{\sigma}_z \otimes \boldsymbol{\sigma}_z \otimes \boldsymbol{\sigma}_z\right)\right), \tag{6.63}$$

$$\boldsymbol{V} = \exp\left(\mathrm{i}\left(a\boldsymbol{\sigma}_x \otimes \boldsymbol{\sigma}_x \otimes \boldsymbol{\sigma}_x + b\boldsymbol{\sigma}_y \otimes \boldsymbol{\sigma}_y \otimes \boldsymbol{\sigma}_x\right.\right.$$

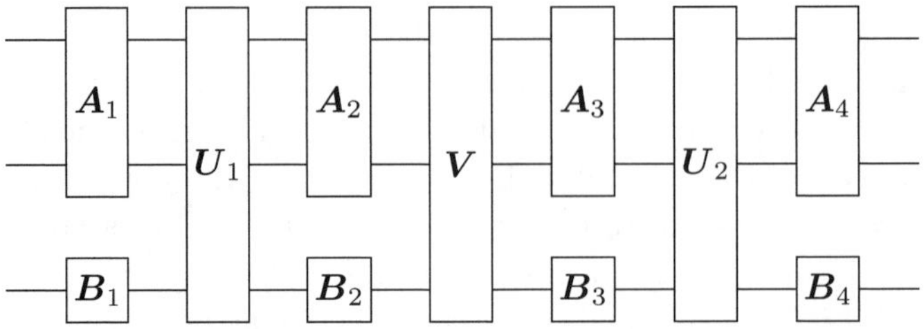

Figure 6.8: Khaneja-Glaser decomposition of an arbitrary unitary triqubit gate, after Vatan and Williams [139].

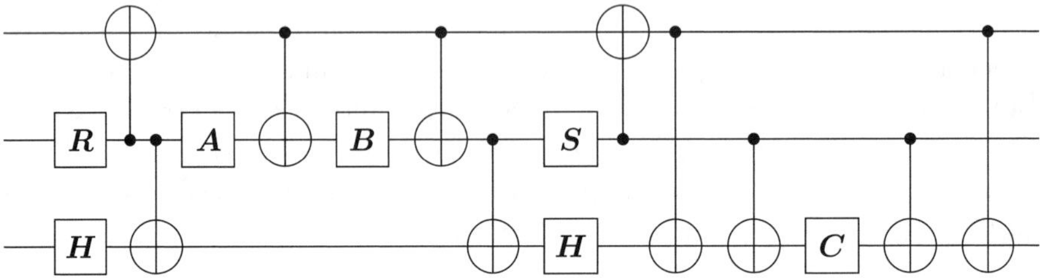

Figure 6.9: Decomposition of $U_{1,2}$ gates, after Vatan and Williams [139].

$$+c\boldsymbol{\sigma}_z \otimes \boldsymbol{\sigma}_z \otimes \boldsymbol{\sigma}_x + d\mathbf{1} \otimes \mathbf{1} \otimes \boldsymbol{\sigma}_x)). \tag{6.64}$$

The biqubit gates, $\boldsymbol{A}_1 \dots \boldsymbol{A}_4$, whatever they may be, we already know how to implement as a sequence of three controlled-NOT and eight single-qubit gates, as was demonstrated in Figure 6.5.

The $\boldsymbol{U}_1$ and $\boldsymbol{U}_2$ gates are much the same, the only difference being in parameters (a_i, b_i, c_i), where $i \in \{1, 2\}$. Figure 6.9 shows how these gates can be implemented as sequences of ten controlled-NOT gates and seven single-qubit gates, where

$$\boldsymbol{R} = \boldsymbol{R}_z\left(-\frac{\pi}{2}\right), \tag{6.65}$$

$$\boldsymbol{A} = \boldsymbol{R}_y\left(2a_i\right), \tag{6.66}$$

$$\boldsymbol{B} = \boldsymbol{R}_y\left(-2b_i\right), \tag{6.67}$$

$$\boldsymbol{S} = \boldsymbol{R}_z\left(\frac{\pi}{2}\right), \tag{6.68}$$

$$\boldsymbol{C} = \boldsymbol{R}_z\left(2c_i\right), \tag{6.69}$$

$$\boldsymbol{H} = \frac{1}{\sqrt{2}}\left(\boldsymbol{\sigma}_x + \boldsymbol{\sigma}_z\right), \tag{6.70}$$

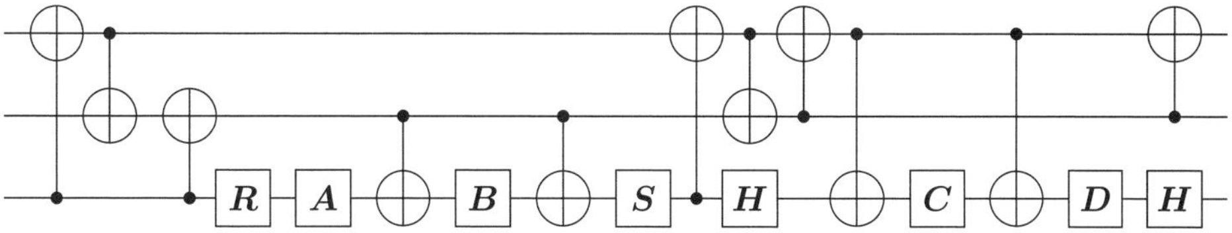

Figure 6.10: Decomposition of $\boldsymbol{V}$, after Vatan and Williams [139].

and $\boldsymbol{R}_{x,y,z}$ are qubit rotations about the x, y, and z axes, respectively.

Vatan-Williams decomposition of Khaneja-Glaser triqubit operators

The $\boldsymbol{V}$ gate can be implemented by eleven controlled-NOT gates and eight single-qubit gates, as shown in Figure 6.10, where $\boldsymbol{R}$, $\boldsymbol{A}$, $\boldsymbol{B}$, $\boldsymbol{S}$, and $\boldsymbol{C}$ are as above and

$$\boldsymbol{D} = \boldsymbol{R}_z\left(2d\right). \tag{6.71}$$

How many single-qubit and controlled-NOT gates do we need, then, to implement an arbitrary triqubit unitary operation? Looking at Figure 6.8 and then at Figures 6.9 and 6.10, we calculate 7 single-qubit gates and 10 controlled-NOT gates per each triqubit $\boldsymbol{U}$ gate, of which we have two in the Khaneja-Glaser circuit, then 8 single-qubit gates and 11 controlled-NOT gates per single triqubit $\boldsymbol{V}$ gate. So, the three triqubit gates require $2 \times 7 + 8 = 22$ single-qubit gates and $2 \times 10 + 11 = 31$ controlled-NOT gates. But then we also have 4 additional single-qubit gates and 4 additional biqubit gates in the Khaneja-Glaser circuit. Looking at Figure 6.5, we find that 8 single-qubit and 3 controlled-NOT gates implement any biqubit gate. The total number of single-qubit gates is $22 + 4 + 4 \times 8 = 58$, and the total number of controlled-NOT gates is $31 + 4 \times 3 = 43$.

These numbers can be further reduced because single-qubit gates on the joints between the two and three qubit boxes can be merged and some of the controlled-NOT gates on the boundaries can be moved from the triqubit gates to their biqubit neighbors. In effect, then, each $\boldsymbol{U}_i$ gate contributes 5 single qubit and 9 controlled-NOT gates and the $\boldsymbol{V}$ gate contributes 6 single-qubit and 10 controlled-NOT gates. In summary, the total number of gates may be reduced to $2 \times 5 + 6 + 4 + 4 \times 8 = 52$ single-qubit gates and $2 \times 9 + 10 + 4 \times 3 = 40$ controlled-NOT gates.

Maintaining the accuracy of the computation and preventing the triqubit system from leaking unitarity to its environment are going to be major problems, if we have to perform 40 controlled-NOTs, each of which involves numerous internal operations, and 52 single-qubit rotations on top, to reach a final state for an arbitrary triqubit computation.

On the other hand, the Khaneja-Glaser decomposition shown in Figure 6.8 provides us with a more economic implementation of an arbitrary triqubit gate in terms of biqubit gates and freely evolving triqubit gates given by $\boldsymbol{U}_1$, $\boldsymbol{U}_2$, and $\boldsymbol{V}$. If molecules are used as registers, then there is little to be gained, in practice, by decomposing these triqubit gates into controlled-NOTs.

6.2.4 Universality of the Deutsch Gate

How about 4-qubit gates and larger systems? Can n-qubit unitary transformations be implemented in terms of controlled-NOT and single-qubit unitary gates, too? Yes, they can.

A unitary transformation of a system of n-qubits can be thought of as a complex matrix $\boldsymbol{U}$, sized $2^n \times 2^n$, such that $\boldsymbol{U}^\dagger \boldsymbol{U} = \boldsymbol{U}\boldsymbol{U}^\dagger = \mathbf{1}$. The matrix can be decomposed into a product $\boldsymbol{U}_k \boldsymbol{U}_{k-1} \ldots \boldsymbol{U}_2 \boldsymbol{U}_1$ of simpler matrices—ultimately matrices that correspond to single-qubit and controlled-NOT operations. That this is possible for any n was first demonstrated by Deutsch in 1988 [29]. More precisely, Deutsch demonstrated that a certain triqubit gate of his invention (more about this below) was universal for quantum computation, but following our results presented in Sections 6.2.2 and 6.2.3 above, his demonstration reduces to the universality of controlled-NOT and single-qubit gates.

Deutsch gate

In the following we reproduce Deutsch reasoning with only a few additional explanations, because his demonstration is clear and does not require much knowledge of Lie group theory.[1]

We will focus first on triqubit systems and will demonstrate how various gates can be built from the Deutsch gate alone. Eventually we will show how *any* tri-, bi-, and single-qubit gates can be made of Deutsch gates. We will then extend these results to four-qubit registers and, while doing so, will point the way to extend it indefinitely to any number of qubits.

A triqubit system can be described in terms of the following eight basis states:

$$| 0\rangle \otimes | 0\rangle \otimes | 0\rangle = | 000\rangle \equiv | \mathbf{0}\rangle, \quad (6.72)$$

$$| 0\rangle \otimes | 0\rangle \otimes | 1\rangle = | 001\rangle \equiv | \mathbf{1}\rangle, \quad (6.73)$$

$$| 0\rangle \otimes | 1\rangle \otimes | 0\rangle = | 010\rangle \equiv | \mathbf{2}\rangle, \quad (6.74)$$

$$| 0\rangle \otimes | 1\rangle \otimes | 1\rangle = | 011\rangle \equiv | \mathbf{3}\rangle, \quad (6.75)$$

$$| 1\rangle \otimes | 0\rangle \otimes | 0\rangle = | 100\rangle \equiv | \mathbf{4}\rangle, \quad (6.76)$$

[1] Lie groups are groups that are also differential manifolds. Examples of Lie groups are rotations, Lorentz boosts, and unitary tranformations. Although we will not borrow from the theory explicitly, Deutsch's reasoning effectively reconstructs the required elements of it.

$$
\begin{aligned}
|\,1\rangle \otimes |\,0\rangle \otimes |\,1\rangle &= |\,101\rangle \equiv |\,\mathbf{5}\rangle, && (6.77)\\
|\,1\rangle \otimes |\,1\rangle \otimes |\,0\rangle &= |\,110\rangle \equiv |\,\mathbf{6}\rangle, && (6.78)\\
|\,1\rangle \otimes |\,1\rangle \otimes |\,1\rangle &= |\,111\rangle \equiv |\,\mathbf{7}\rangle. && (6.79)
\end{aligned}
$$

In this basis, a controlled-NOT triqubit gate that operates on qubits 2 (control) and 3 (target), as counted from the left, corresponds to the following matrix.

$$
\bigoplus(2_c, 3_t) = \begin{pmatrix}
1 & 0 & 0 & 0 & 0 & 0 & 0 & 0\\
0 & 1 & 0 & 0 & 0 & 0 & 0 & 0\\
0 & 0 & 0 & 1 & 0 & 0 & 0 & 0\\
0 & 0 & 1 & 0 & 0 & 0 & 0 & 0\\
0 & 0 & 0 & 0 & 1 & 0 & 0 & 0\\
0 & 0 & 0 & 0 & 0 & 1 & 0 & 0\\
0 & 0 & 0 & 0 & 0 & 0 & 0 & 1\\
0 & 0 & 0 & 0 & 0 & 0 & 1 & 0
\end{pmatrix}. \qquad (6.80)
$$

We have two occurrences of off-diagonal terms here. The first group swaps $|\,\mathbf{2}\rangle$ and $|\,\mathbf{3}\rangle$, and the second one swaps $|\,\mathbf{6}\rangle$ and $|\,\mathbf{7}\rangle$. They both correspond to the second qubit being set to $|\,1\rangle$, regardless of the value of the first qubit.

In turn, a single-qubit NOT gate applied to the third qubit in this context looks as follows.

$$
\neg\,(3) = \begin{pmatrix}
0 & 1 & 0 & 0 & 0 & 0 & 0 & 0\\
1 & 0 & 0 & 0 & 0 & 0 & 0 & 0\\
0 & 0 & 0 & 1 & 0 & 0 & 0 & 0\\
0 & 0 & 1 & 0 & 0 & 0 & 0 & 0\\
0 & 0 & 0 & 0 & 0 & 1 & 0 & 0\\
0 & 0 & 0 & 0 & 1 & 0 & 0 & 0\\
0 & 0 & 0 & 0 & 0 & 0 & 0 & 1\\
0 & 0 & 0 & 0 & 0 & 0 & 1 & 0
\end{pmatrix}. \qquad (6.81)
$$

Here we have four groups of off-diagonal terms. They correspond to $|\,x\rangle\,|\,y\rangle\,|\,0\rangle$ and $|\,x\rangle\,|\,y\rangle\,|\,1\rangle$ pairs. The pairs get swapped regardless of the values of the first and second qubit.

Such very simple operations do not necessarily have simple matrix representations in the triqubit universe. Instead, the simplest triqubit swap operation looks like *Toffoli gate*

Figure 6.11: Diagrammatic representation of the Toffoli gate.

this:

$$\boldsymbol{T} = \begin{pmatrix} 1 & 0 & 0 & 0 & 0 & 0 & 0 & 0 \\ 0 & 1 & 0 & 0 & 0 & 0 & 0 & 0 \\ 0 & 0 & 1 & 0 & 0 & 0 & 0 & 0 \\ 0 & 0 & 0 & 1 & 0 & 0 & 0 & 0 \\ 0 & 0 & 0 & 0 & 1 & 0 & 0 & 0 \\ 0 & 0 & 0 & 0 & 0 & 1 & 0 & 0 \\ 0 & 0 & 0 & 0 & 0 & 0 & 0 & 1 \\ 0 & 0 & 0 & 0 & 0 & 0 & 1 & 0 \end{pmatrix}. \tag{6.82}$$

Interpreting this operation in context of the triqubit basis states list (6.72–6.79), we arrive at the following understanding of the gate. If either of qubits one and two (counting from the left) is zero, and this covers all states from $|\,\mathbf{0}\rangle$ through $|\,\mathbf{5}\rangle$, the gate does nothing. If both are one, which holds for states $|\,\mathbf{6}\rangle$ and $|\,\mathbf{7}\rangle$, then the gate flips the state of the third qubit. It is a controlled-controlled-NOT gate, known as the Toffoli gate in classical computing.

Decomposition of controlled-controlled-$\boldsymbol{U}$

Square root of NOT

Its diagrammatic symbol is shown in Figure 6.11. As we may already suspect, although the Toffoli gate looks rather simple in its matrix and diagrammatic representations, it is not so simple to implement. Figure 6.12 shows a general prescription for a controlled-controlled-$\boldsymbol{U}$ gate, where $\boldsymbol{U}$ is an arbitrary single-qubit unitary operation. Figure 6.2 (page 292) shows us how to implement a controlled-$\sqrt{\boldsymbol{U}}$ gate, once we know how to make $\sqrt{\boldsymbol{U}}$. In this case we will need to implement $\sqrt{\text{NOT}}$. A quantum NOT is $\boldsymbol{\sigma}_x$. It is easy to see that

$$\sqrt{\boldsymbol{\sigma}_x} = \frac{1}{2}\left(\mathbf{1} + \boldsymbol{\sigma}_x + \mathrm{i}\left(\mathbf{1} - \boldsymbol{\sigma}_x\right)\right) \tag{6.83}$$

makes a good square root of $\boldsymbol{\sigma}_x$. Indeed

$$\sqrt{\boldsymbol{\sigma}_x}\sqrt{\boldsymbol{\sigma}_x}$$
$$= \frac{1}{4}\Big(\left(\mathbf{1} + \boldsymbol{\sigma}_x\right)^2 + \mathrm{i}\left(\mathbf{1} + \boldsymbol{\sigma}_x\right)\left(\mathbf{1} - \boldsymbol{\sigma}_x\right)$$

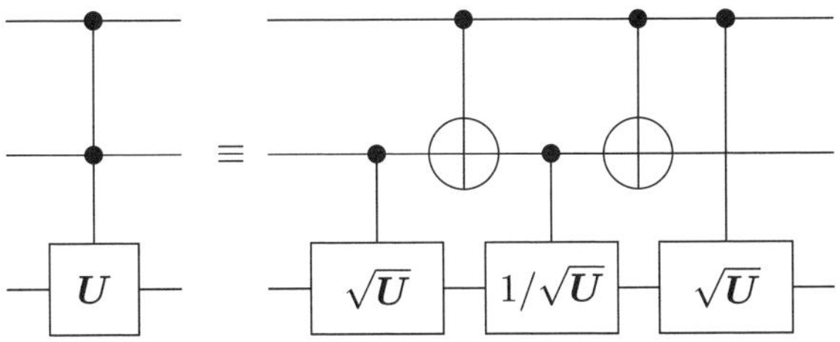

Figure 6.12: Decomposition of a controlled-controlled-$\boldsymbol{U}$ gate into biqubit gates.

$$
\begin{aligned}
&\quad + \mathrm{i}\,(\mathbf{1} - \boldsymbol{\sigma}_x)\,(\mathbf{1} + \boldsymbol{\sigma}_x) - (\mathbf{1} - \boldsymbol{\sigma}_x)^2 \Big) \\
&= \frac{1}{4}\,(\mathbf{1} + 2\boldsymbol{\sigma}_x + \mathbf{1} - \mathbf{1} + 2\boldsymbol{\sigma}_x - \mathbf{1}) \qquad (6.84) \\
&= \boldsymbol{\sigma}_x. \qquad (6.85)
\end{aligned}
$$

The imaginary terms have cancelled out because $(\mathbf{1} + \boldsymbol{\sigma}_x)\,(\mathbf{1} - \boldsymbol{\sigma}_x) = \mathbf{0}$.

Similarly, we can see that

$$\frac{1}{2}\,(\mathbf{1} + \boldsymbol{\sigma}_x - \mathrm{i}\,(\mathbf{1} - \boldsymbol{\sigma}_x)) \qquad (6.86)$$

is the inverse of $\sqrt{\boldsymbol{\sigma}_x}$, as defined by (6.83). This gives us an explicit prescription for the Toffoli gate in terms of controlled-NOT gates and single-qubit gates.

It also explains why, in the *classical* digital computing world, the Toffoli gate cannot be reduced to two-bit and single-bit gates. We cannot construct a square root of NOT gate in classical digital computing.

Why is the Toffoli gate so important?

Reversible computing

It turns out that the Toffoli gate is a universal gate for classical reversible computing—a method of computation that does not erase any data [45]. Reversible computation can be always carried out back to front, reaching the initial condition of the registers exactly as they were when the computation had unfolded in the original direction first. In reversible computing no data ever gets lost.

Von Neumann-Landauer limit

Reversible computing has an advantage over normal computing because, in principle at least, heat gets generated primarily when data is erased. According to the von Neumann-Landauer limit, single-bit erasure produces at least $kT \ln 2$ of heat [84, 9]. Thus, if we could carry out computations without erasing data, we would generate less heat, or no heat at all even, if we could only eliminate ohmic heat generation by the logic circuitry—and this could possibly be achieved in superconducting circuits.[2]

Classical reversible computing is a subset of quantum computing.

Quantum computing is also reversible, if we were to restrict ourselves to unitary operations only, because unitary operations are trivially reversible ($\boldsymbol{U}^{-1} = \boldsymbol{U}^{\dagger}$). One can think of quantum computing as a cocoon wrapped around classical reversible computing. The latter is a subset of the former. Many definitions of various quantum computing operations, such as controlled-NOT and controlled-controlled-NOT, if *not* extended linearly beyond the basis vectors to account for superposition, are just definitions of classical reversible computing. This is also why, when demonstrating a quantum gate, it is not enough to show its correct performance on basis states. To prove the gate's quantumness, we must demonstrate its performance on superpositions of basis states, as well.

Definition of Deutsch gate

A natural, and this time fully quantum, generalization of the Toffoli gate is the already mentioned Deutsch gate, which is a controlled-controlled-$\left(\mathrm{i}e^{-\mathrm{i}\pi\alpha/2}\boldsymbol{Q}(\alpha)\right)$ gate, where

$$\boldsymbol{Q}(\alpha) = \frac{1}{2}\begin{pmatrix} 1+e^{\mathrm{i}\pi\alpha} & 1-e^{\mathrm{i}\pi\alpha} \\ 1-e^{\mathrm{i}\pi\alpha} & 1+e^{\mathrm{i}\pi\alpha} \end{pmatrix} \tag{6.87}$$

and α is a fixed *irrational* number—any will do. Like the Toffoli gate, the Deutsch gate can be implemented by using controlled-NOT and single-qubit unitary gates, following the procedure shown in Figure 6.12. Hence, any universality results that we can demonstrate for this gate extend to the combination of controlled-NOT and single-qubit unitary gates.

To understand $\boldsymbol{Q}(\alpha)$ better, we assume for a moment that $\alpha = 1$. In this case $e^{\mathrm{i}\pi\alpha} = e^{\mathrm{i}\pi} = -1$ and

$$\boldsymbol{Q}(1) = \boldsymbol{\sigma}_x. \tag{6.88}$$

What if $\alpha = 2$? Then $e^{\mathrm{i}\pi\alpha} = e^{\mathrm{i}2\pi} = 1$, and

$$\boldsymbol{Q}(2) = \mathbf{1} = (\boldsymbol{\sigma}_x)^2. \tag{6.89}$$

What if $\alpha = 1/2$? In this case $e^{\mathrm{i}\pi\alpha} = e^{\mathrm{i}\pi/2} = \mathrm{i}$, so

$$\boldsymbol{Q}\left(\frac{1}{2}\right) = \frac{1}{2}\begin{pmatrix} 1+\mathrm{i} & 1-\mathrm{i} \\ 1-\mathrm{i} & 1+\mathrm{i} \end{pmatrix} = \frac{1}{2}\left(\mathbf{1} + \boldsymbol{\sigma}_x + \mathrm{i}\left(\mathbf{1} - \boldsymbol{\sigma}_x\right)\right) = \sqrt{\boldsymbol{\sigma}_x}, \tag{6.90}$$

[2] Electronic circuits also generate electromagnetic radiation. The only way to eliminate this would be to replace electrons with electrically neutral information carriers, such as photons, but such carriers are difficult to manipulate.

as we have already seen in equation (6.83). Also, for $\alpha = -1/2$, we get the same, but with $\mathrm{i} \to -\mathrm{i}$, and this, comparing with equation (6.86), gives us

$$\boldsymbol{Q}(-1/2) = \frac{1}{\sqrt{\boldsymbol{\sigma}_x}}. \tag{6.91}$$

A power of NOT

We find that $\boldsymbol{Q}(\alpha)$ evaluates to $(\boldsymbol{\sigma}_x)^\alpha$ in all cases we have investigated so far. For this reason the Deutsch operator $\boldsymbol{Q}(\alpha)$ is sometimes called *a power of* NOT, with square root of NOT, defined by equation (6.83), a special case.

Irrationality of α

What do we gain by making α be irrational? This ensures that consecutive applications of $\boldsymbol{Q}(\alpha)$ fill *densely* everything between $\boldsymbol{\sigma}_x$, which is $\boldsymbol{Q}(1)$, $\boldsymbol{Q}(3)$, $\boldsymbol{Q}(5)$, ..., and $\boldsymbol{1}$, which is $\boldsymbol{Q}(0)$, $\boldsymbol{Q}(2)$, $\boldsymbol{Q}(4)$, ..., allowing us to get *arbitrarily close* to both, or to any other operator in between, even if never hitting any that would correspond to a rational value of α exactly. But then, what is *exact?* Clearly, it is physically impossible to produce qubit rotation by *exactly* 180° about *exactly* the x axis anyway. There is always going to be some experimental error, however small, not to mention fundamental physics limitations due to vacuum fluctuations, vacuum polarization, and the uncertainty principle. The Deutsch operator lets us approach any $\boldsymbol{Q}(\alpha_0)$, for α_0 rational or irrational, with arbitrary precision, in a finite number of consecutive applications. It fills the continuum.[3]

This is exactly what we want if we are to approach arbitrarily closely every possible n-qubit unitary transformation, which constitute a certain continuum, by multiple applications of the Deutsch gate. The universality of the Deutsch gate is in this spirit. The gate is not universal in the classical sense, meaning that it cannot reproduce even the Toffoli gate exactly, but its consecutive applications can get arbitrarily close to it.

But what is "controlled-controlled" in the Deutsch gate is not exactly $\boldsymbol{Q}(\alpha)$, but $\boldsymbol{Q}(\alpha)$ multiplied by $\mathrm{i}\exp(-\mathrm{i}\pi\alpha/2)$, which evaluates to

$$\begin{aligned}
\tilde{\boldsymbol{D}}(\alpha) &= \mathrm{i}e^{-\mathrm{i}\pi\alpha/2}\boldsymbol{Q}(\alpha) = \mathrm{i}e^{-\mathrm{i}\pi\alpha/2}\frac{1}{2}\begin{pmatrix} 1+e^{\mathrm{i}\pi\alpha} & 1-e^{\mathrm{i}\pi\alpha} \\ 1-e^{\mathrm{i}\pi\alpha} & 1+e^{\mathrm{i}\pi\alpha} \end{pmatrix} \\
&= \begin{pmatrix} \mathrm{i}\frac{e^{\mathrm{i}\pi\alpha/2}+e^{-\mathrm{i}\pi\alpha/2}}{2} & \frac{e^{\mathrm{i}\pi\alpha/2}-e^{-\mathrm{i}\pi\alpha/2}}{2\mathrm{i}} \\ \frac{e^{\mathrm{i}\pi\alpha/2}-e^{-\mathrm{i}\pi\alpha/2}}{2\mathrm{i}} & \mathrm{i}\frac{e^{\mathrm{i}\pi\alpha/2}+e^{-\mathrm{i}\pi\alpha/2}}{2} \end{pmatrix} \\
&= \begin{pmatrix} \mathrm{i}\cos\frac{\pi\alpha}{2} & \sin\frac{\pi\alpha}{2} \\ \sin\frac{\pi\alpha}{2} & \mathrm{i}\cos\frac{\pi\alpha}{2} \end{pmatrix}.
\end{aligned} \tag{6.92}$$

[3] On the other hand, the bi- and triqubit decompositions we have discussed in Sections 6.2.2 and 6.2.3 were exact.

The full matrix of the Deutsch gate in the tri-qubit space now looks as follows:

$$\boldsymbol{D}(\alpha) = \begin{pmatrix} 1 & 0 & 0 & 0 & 0 & 0 & 0 & 0 \\ 0 & 1 & 0 & 0 & 0 & 0 & 0 & 0 \\ 0 & 0 & 1 & 0 & 0 & 0 & 0 & 0 \\ 0 & 0 & 0 & 1 & 0 & 0 & 0 & 0 \\ 0 & 0 & 0 & 0 & 1 & 0 & 0 & 0 \\ 0 & 0 & 0 & 0 & 0 & 1 & 0 & 0 \\ 0 & 0 & 0 & 0 & 0 & 0 & \mathrm{i}\cos\frac{\pi\alpha}{2} & \sin\frac{\pi\alpha}{2} \\ 0 & 0 & 0 & 0 & 0 & 0 & \sin\frac{\pi\alpha}{2} & \mathrm{i}\cos\frac{\pi\alpha}{2} \end{pmatrix}. \tag{6.93}$$

Matrix representation of the Deutsch gate

Matrix $\boldsymbol{D}(\alpha)$ has this nice property that on taking its powers the submatrix $\tilde{\boldsymbol{D}}(\alpha)$ in its lower-right corner stays there and is merely replaced by its own powers. And certain powers of the submatrix are quite easy to evaluate.

First, we note that

$$\begin{aligned} \left(\tilde{\boldsymbol{D}}(\alpha)\right)^2 &= \begin{pmatrix} -\cos^2\frac{\pi\alpha}{2} + \sin^2\frac{\pi\alpha}{2} & 2\mathrm{i}\cos\frac{\pi\alpha}{2}\sin\frac{\pi\alpha}{2} \\ 2\mathrm{i}\cos\frac{\pi\alpha}{2}\sin\frac{\pi\alpha}{2} & -\cos^2\frac{\pi\alpha}{2} + \sin^2\frac{\pi\alpha}{2} \end{pmatrix} \\ &= \begin{pmatrix} -\cos\pi\alpha & \mathrm{i}\sin\pi\alpha \\ \mathrm{i}\sin\pi\alpha & -\cos\pi\alpha \end{pmatrix}. \end{aligned} \tag{6.94}$$

Taking a square of this yields

$$\begin{aligned} \left(\tilde{\boldsymbol{D}}(\alpha)\right)^4 &= \begin{pmatrix} \cos^2\pi\alpha - \sin^2\pi\alpha & -2\mathrm{i}\cos\pi\alpha\sin\pi\alpha \\ -2\mathrm{i}\cos\pi\alpha\sin\pi\alpha & \cos^2\pi\alpha - \sin^2\pi\alpha \end{pmatrix} \\ &= \begin{pmatrix} \cos 2\pi\alpha & -\mathrm{i}\sin 2\pi\alpha \\ -\mathrm{i}\sin 2\pi\alpha & \cos 2\pi\alpha \end{pmatrix}. \end{aligned} \tag{6.95}$$

And taking another square give us

$$\begin{aligned} \left(\tilde{\boldsymbol{D}}(\alpha)\right)^8 &= \begin{pmatrix} \cos^2 2\pi\alpha - \sin^2 2\pi\alpha & -2\mathrm{i}\cos 2\pi\alpha\sin 2\pi\alpha \\ -2\mathrm{i}\cos 2\pi\alpha\sin 2\pi\alpha & \cos^2 2\pi\alpha - \sin^2 2\pi\alpha \end{pmatrix} \\ &= \begin{pmatrix} \cos 4\pi\alpha & -\mathrm{i}\sin 4\pi\alpha \\ -\mathrm{i}\sin 4\pi\alpha & \cos 4\pi\alpha \end{pmatrix}. \end{aligned} \tag{6.96}$$

This looks like a pattern, namely,

$$\left(\tilde{\boldsymbol{D}}(\alpha)\right)^{4n} = \begin{pmatrix} \cos 2n\pi\alpha & -\mathrm{i}\sin 2n\pi\alpha \\ -\mathrm{i}\sin 2n\pi\alpha & \cos 2n\pi\alpha \end{pmatrix}. \tag{6.97}$$

Another pattern can be discovered when $\left(\tilde{\boldsymbol{D}}(\alpha)\right)^4$ and $\left(\tilde{\boldsymbol{D}}(\alpha)\right)^8$ are muliplied by $\tilde{\boldsymbol{D}}(\alpha)$. This time we find that

$$\left(\tilde{\boldsymbol{D}}(\alpha)\right)^{4+1}$$

$$
\begin{aligned}
&= \begin{pmatrix} \mathrm{i}\left(\cos\frac{\pi\alpha}{2}\cos 2\pi\alpha - \sin\frac{\pi\alpha}{2}\sin 2\pi\alpha\right) & \cos\frac{\pi\alpha}{2}\sin 2\pi\alpha + \sin\frac{\pi\alpha}{2}\cos 2\pi\alpha \\ \cos\frac{\pi\alpha}{2}\sin 2\pi\alpha + \sin\frac{\pi\alpha}{2}\cos 2\pi\alpha & \mathrm{i}\left(\cos\frac{\pi\alpha}{2}\cos 2\pi\alpha - \sin\frac{\pi\alpha}{2}\sin 2\pi\alpha\right) \end{pmatrix} \\
&= \begin{pmatrix} \mathrm{i}\cos\left(2+\frac{1}{2}\right)\pi\alpha & \sin\left(2+\frac{1}{2}\right)\pi\alpha \\ \sin\left(2+\frac{1}{2}\right)\pi\alpha & \mathrm{i}\cos\left(2+\frac{1}{2}\right)\pi\alpha \end{pmatrix}
\end{aligned}
\tag{6.98}
$$

and

$$
\begin{aligned}
&\left(\tilde{\boldsymbol{D}}(\alpha)\right)^{8+1} \\
&= \begin{pmatrix} \mathrm{i}\left(\cos\frac{\pi\alpha}{2}\cos 4\pi\alpha - \sin\frac{\pi\alpha}{2}\sin 4\pi\alpha\right) & \cos\frac{\pi\alpha}{2}\sin 4\pi\alpha + \sin\frac{\pi\alpha}{2}\cos 4\pi\alpha \\ \cos\frac{\pi\alpha}{2}\sin 4\pi\alpha + \sin\frac{\pi\alpha}{2}\cos 4\pi\alpha & \mathrm{i}\left(\cos\frac{\pi\alpha}{2}\cos 4\pi\alpha - \sin\frac{\pi\alpha}{2}\sin 4\pi\alpha\right) \end{pmatrix} \\
&= \begin{pmatrix} \mathrm{i}\cos\left(4+\frac{1}{2}\right)\pi\alpha & \sin\left(4+\frac{1}{2}\right)\pi\alpha \\ \sin\left(4+\frac{1}{2}\right)\pi\alpha & \mathrm{i}\cos\left(4+\frac{1}{2}\right)\pi\alpha \end{pmatrix}.
\end{aligned}
\tag{6.99}
$$

And so, the new emerging pattern is

$$
\left(\tilde{\boldsymbol{D}}(\alpha)\right)^{4n+1} = \begin{pmatrix} \mathrm{i}\cos\left(2n+\frac{1}{2}\right)\pi\alpha & \sin\left(2n+\frac{1}{2}\right)\pi\alpha \\ \sin\left(2n+\frac{1}{2}\right)\pi\alpha & \mathrm{i}\cos\left(2n+\frac{1}{2}\right)\pi\alpha \end{pmatrix}. \tag{6.100}
$$

Equations (6.97) and (6.100) imply that by multiplying $\boldsymbol{D}(\alpha)$ by itself as many times as needed we can get arbitrarily close to

$$
\tilde{\boldsymbol{U}}_\lambda = \begin{pmatrix}
1 & 0 & 0 & 0 & 0 & 0 & 0 & 0 \\
0 & 1 & 0 & 0 & 0 & 0 & 0 & 0 \\
0 & 0 & 1 & 0 & 0 & 0 & 0 & 0 \\
0 & 0 & 0 & 1 & 0 & 0 & 0 & 0 \\
0 & 0 & 0 & 0 & 1 & 0 & 0 & 0 \\
0 & 0 & 0 & 0 & 0 & 1 & 0 & 0 \\
0 & 0 & 0 & 0 & 0 & 0 & \mathrm{i}\cos\lambda & \sin\lambda \\
0 & 0 & 0 & 0 & 0 & 0 & \sin\lambda & \mathrm{i}\cos\lambda
\end{pmatrix} \tag{6.101}
$$

and

$$
\boldsymbol{U}_\lambda = \begin{pmatrix}
1 & 0 & 0 & 0 & 0 & 0 & 0 & 0 \\
0 & 1 & 0 & 0 & 0 & 0 & 0 & 0 \\
0 & 0 & 1 & 0 & 0 & 0 & 0 & 0 \\
0 & 0 & 0 & 1 & 0 & 0 & 0 & 0 \\
0 & 0 & 0 & 0 & 1 & 0 & 0 & 0 \\
0 & 0 & 0 & 0 & 0 & 1 & 0 & 0 \\
0 & 0 & 0 & 0 & 0 & 0 & \cos\lambda & \mathrm{i}\sin\lambda \\
0 & 0 & 0 & 0 & 0 & 0 & \mathrm{i}\sin\lambda & \cos\lambda
\end{pmatrix} \tag{6.102}
$$

$$\xrightarrow[\lambda\to 0]{} \begin{pmatrix} 1 & 0 & 0 & 0 & 0 & 0 & 0 & 0 \\ 0 & 1 & 0 & 0 & 0 & 0 & 0 & 0 \\ 0 & 0 & 1 & 0 & 0 & 0 & 0 & 0 \\ 0 & 0 & 0 & 1 & 0 & 0 & 0 & 0 \\ 0 & 0 & 0 & 0 & 1 & 0 & 0 & 0 \\ 0 & 0 & 0 & 0 & 0 & 1 & 0 & 0 \\ 0 & 0 & 0 & 0 & 0 & 0 & 1-\frac{\lambda^2}{2} & \mathrm{i}\lambda \\ 0 & 0 & 0 & 0 & 0 & 0 & \mathrm{i}\lambda & 1-\frac{\lambda^2}{2} \end{pmatrix} + \mathcal{O}\left(\lambda^3\right), \quad (6.103)$$

for *any* real value of λ. We have also printed the last matrix in the limit $\lambda \to 0$, because we are going to need it in a moment.

Toffoli gate reproduced

One of these matrices is the Toffoli gate, which corresponds to $\tilde{\boldsymbol{U}}_{\pi/2}$. Hence, we have demonstrated that the Deutsch gate can be used to reproduce the Toffoli gate and thus the whole classical reversible computing framework.

All classical reversible computing operations are permutations of the register.

But what is this "classical reversible computing framework" exactly?

It is a set of all possible *permutations* of the register, because these are all and the only reversible register mappings. The Toffoli gate itself is a permutation, which when expressed by equation (6.82) swaps $|\,\mathbf{6}\rangle$ and $|\,\mathbf{7}\rangle$ and leaves all other states of the triqubit register unchanged. A permutation that just swaps two states is called a transposition. That every permutation can be made of successive applications of transpositions is one of the elementary theorems of algebra, so the universality of the Toffoli gate is simply an expression of the theorem.

Since the Toffoli gate is universal, we can approach any permutation with arbitrary accuracy, by applying $\boldsymbol{D}(\alpha)$, in various configurations, for example, upside down and as many times as needed. In particular, we can approach a transposition that swaps $|\,\mathbf{5}\rangle$ and $|\,\mathbf{6}\rangle$, as well as a transposition that swaps $|\,\mathbf{5}\rangle$ and $|\,\mathbf{7}\rangle$. Their matrix representations are

$$\boldsymbol{P}_{56} = \begin{pmatrix} 1 & 0 & 0 & 0 & 0 & 0 & 0 & 0 \\ 0 & 1 & 0 & 0 & 0 & 0 & 0 & 0 \\ 0 & 0 & 1 & 0 & 0 & 0 & 0 & 0 \\ 0 & 0 & 0 & 1 & 0 & 0 & 0 & 0 \\ 0 & 0 & 0 & 0 & 1 & 0 & 0 & 0 \\ 0 & 0 & 0 & 0 & 0 & 0 & 1 & 0 \\ 0 & 0 & 0 & 0 & 0 & 1 & 0 & 0 \\ 0 & 0 & 0 & 0 & 0 & 0 & 0 & 1 \end{pmatrix} \quad \text{and} \qquad (6.104)$$

$$\boldsymbol{P}_{57} = \begin{pmatrix} 1 & 0 & 0 & 0 & 0 & 0 & 0 & 0 \\ 0 & 1 & 0 & 0 & 0 & 0 & 0 & 0 \\ 0 & 0 & 1 & 0 & 0 & 0 & 0 & 0 \\ 0 & 0 & 0 & 1 & 0 & 0 & 0 & 0 \\ 0 & 0 & 0 & 0 & 1 & 0 & 0 & 0 \\ 0 & 0 & 0 & 0 & 0 & 0 & 0 & 1 \\ 0 & 0 & 0 & 0 & 0 & 0 & 1 & 0 \\ 0 & 0 & 0 & 0 & 0 & 1 & 0 & 0 \end{pmatrix}. \tag{6.105}$$

Let us now evaluate[4]

$$\boldsymbol{P}_{56} \left(\boldsymbol{U}_\lambda \boldsymbol{P}_{57}\right)^2 \left(\boldsymbol{U}_{-\lambda} \boldsymbol{P}_{57}\right)^2 \boldsymbol{P}_{56}, \tag{6.106}$$

in the $\lambda \to 0$ limit using equation (6.103). Truncating the polynomial terms at λ^3, we obtain the following:

$$\tilde{\boldsymbol{V}}(\lambda) = \begin{pmatrix} 1 & 0 & 0 & 0 & 0 & 0 & 0 & 0 \\ 0 & 1 & 0 & 0 & 0 & 0 & 0 & 0 \\ 0 & 0 & 1 & 0 & 0 & 0 & 0 & 0 \\ 0 & 0 & 0 & 1 & 0 & 0 & 0 & 0 \\ 0 & 0 & 0 & 0 & 1 & 0 & 0 & 0 \\ 0 & 0 & 0 & 0 & 0 & 1 & 0 & 0 \\ 0 & 0 & 0 & 0 & 0 & 0 & 1 & \lambda^2 \\ 0 & 0 & 0 & 0 & 0 & 0 & -\lambda^2 & 1 \end{pmatrix} + \mathcal{O}(\lambda^3). \tag{6.107}$$

The 2×2 lower-right corner submatrix of $\tilde{\boldsymbol{V}}(\lambda)$ is a generator of rotations. This can be seen as follows. Let us consider

$$\lim_{n\to\infty} \left(\tilde{\boldsymbol{V}}\left(\sqrt{\frac{\lambda}{n}}\right)\right)^n. \tag{6.108}$$

We can have a closer look at this expression by focusing on the resulting 2×2 corner submatrix

$$\lim_{n\to\infty} \begin{pmatrix} 1 & \lambda/n \\ -\lambda/n & 1 \end{pmatrix}^n. \tag{6.109}$$

That this must converge to

$$\begin{pmatrix} \cos\lambda & \sin\lambda \\ -\sin\lambda & \cos\lambda \end{pmatrix} \tag{6.110}$$

[4]It may help to use Mathematica or Maple at this stage.

is obvious. The latter is a rotation in a plane by angle λ. Instead of rotating by the whole λ, we can divide the rotation into n tiny chunks and rotate in step of λ/n. To reach λ in such small steps, we need to perform n of them. Therefore, we get

$$\begin{pmatrix} \cos\lambda & \sin\lambda \\ -\sin\lambda & \cos\lambda \end{pmatrix} = \begin{pmatrix} \cos\frac{\lambda}{n} & \sin\frac{\lambda}{n} \\ -\sin\frac{\lambda}{n} & \cos\frac{\lambda}{n} \end{pmatrix}^n. \tag{6.111}$$

For a very large n, λ/n is very small; hence, in the limit $n \to \infty$ we can replace cos and sin by their linear approximations around zero, $\cos(\lambda/n) \approx 1$, and $\sin(\lambda/n) \approx \lambda/n$. Hence

$$\begin{pmatrix} \cos\lambda & \sin\lambda \\ -\sin\lambda & \cos\lambda \end{pmatrix} \underset{n\to\infty}{=} \begin{pmatrix} 1 & \frac{\lambda}{n} \\ -\frac{\lambda}{n} & 1 \end{pmatrix}^n. \tag{6.112}$$

Rotations in the $(|\,\mathbf{6}\rangle, |\,\mathbf{7}\rangle)$ *plane*

In summary, we find that

$$\lim_{n\to\infty}\left(\tilde{\boldsymbol{V}}\left(\sqrt{\frac{\lambda}{n}}\right)\right)^n \stackrel{\text{def}}{=} \boldsymbol{V}(\lambda) = \begin{pmatrix} 1 & 0 & 0 & 0 & 0 & 0 & 0 & 0 \\ 0 & 1 & 0 & 0 & 0 & 0 & 0 & 0 \\ 0 & 0 & 1 & 0 & 0 & 0 & 0 & 0 \\ 0 & 0 & 0 & 1 & 0 & 0 & 0 & 0 \\ 0 & 0 & 0 & 0 & 1 & 0 & 0 & 0 \\ 0 & 0 & 0 & 0 & 0 & 1 & 0 & 0 \\ 0 & 0 & 0 & 0 & 0 & 0 & \cos\lambda & \sin\lambda \\ 0 & 0 & 0 & 0 & 0 & 0 & -\sin\lambda & \cos\lambda \end{pmatrix}. \tag{6.113}$$

Now we can take another limit of an expression that combines $\boldsymbol{U}_\lambda$ and $\boldsymbol{V}_\lambda$ operators. Let us consider

$$\lim_{n\to\infty}\left(\boldsymbol{U}_{\sqrt{\frac{\lambda}{2n}}}\boldsymbol{V}_{\sqrt{\frac{\lambda}{2n}}}\boldsymbol{U}_{-\sqrt{\frac{\lambda}{2n}}}\boldsymbol{V}_{-\sqrt{\frac{\lambda}{2n}}}\right)^n. \tag{6.114}$$

This we can figure out by looking just at the corner submatrices of $\boldsymbol{U}_x$ and $\boldsymbol{V}_x$ and working with $x \to 0$ limits of these. In this approximation we have to evaluate the following:

$$\begin{aligned} &\begin{pmatrix} 1-\frac{x^2}{2} & ix \\ ix & 1-\frac{x^2}{2} \end{pmatrix}\begin{pmatrix} 1-\frac{x^2}{2} & x \\ -x & 1-\frac{x^2}{2} \end{pmatrix} \\ &\quad\times\begin{pmatrix} 1-\frac{x^2}{2} & -ix \\ -ix & 1-\frac{x^2}{2} \end{pmatrix}\begin{pmatrix} 1-\frac{x^2}{2} & -x \\ x & 1-\frac{x^2}{2} \end{pmatrix} \\ &= \begin{pmatrix} 1-2ix^2 & 0 \\ 0 & 1+2ix^2 \end{pmatrix} + \mathcal{O}\left(x^3\right). \end{aligned} \tag{6.115}$$

We then substitute $x = \sqrt{\lambda/(2n)}$, which yields

$$\begin{pmatrix} 1 - \mathrm{i}\frac{\lambda}{n} & 0 \\ 0 & 1 + \mathrm{i}\frac{\lambda}{n} \end{pmatrix}. \tag{6.116}$$

When the diagonal terms are raised to the power of n, we obtain[5]

$$\left(1 - \mathrm{i}\frac{\lambda}{n}\right)^n \xrightarrow[n\to\infty]{} e^{-\mathrm{i}\lambda} \quad \text{and} \tag{6.117}$$

$$\left(1 + \mathrm{i}\frac{\lambda}{n}\right)^n \xrightarrow[n\to\infty]{} e^{\mathrm{i}\lambda}. \tag{6.118}$$

Diagonal operators

And so, we find that equation (6.114) gives us

$$\boldsymbol{W}_\lambda = \begin{pmatrix} 1 & 0 & 0 & 0 & 0 & 0 & 0 & 0 \\ 0 & 1 & 0 & 0 & 0 & 0 & 0 & 0 \\ 0 & 0 & 1 & 0 & 0 & 0 & 0 & 0 \\ 0 & 0 & 0 & 1 & 0 & 0 & 0 & 0 \\ 0 & 0 & 0 & 0 & 1 & 0 & 0 & 0 \\ 0 & 0 & 0 & 0 & 0 & 1 & 0 & 0 \\ 0 & 0 & 0 & 0 & 0 & 0 & e^{-\mathrm{i}\lambda} & 0 \\ 0 & 0 & 0 & 0 & 0 & 0 & 0 & e^{\mathrm{i}\lambda} \end{pmatrix}. \tag{6.119}$$

Subjecting a diagonal operator, such as $\boldsymbol{W}_\lambda$ to transpositions, $\boldsymbol{P}_{ab}\boldsymbol{W}_\lambda\boldsymbol{P}_{ab}$, does not change the diagonality of the operator; it merely reorders the terms on the diagonal. In turn, by multiplying such permuted $\boldsymbol{W}_\lambda$ by each other, and for various values of λ, we can reproduce just about any diagonal operator, including

$$\boldsymbol{X}_\lambda = \begin{pmatrix} 1 & 0 & 0 & 0 & 0 & 0 & 0 & 0 \\ 0 & 1 & 0 & 0 & 0 & 0 & 0 & 0 \\ 0 & 0 & 1 & 0 & 0 & 0 & 0 & 0 \\ 0 & 0 & 0 & 1 & 0 & 0 & 0 & 0 \\ 0 & 0 & 0 & 0 & 1 & 0 & 0 & 0 \\ 0 & 0 & 0 & 0 & 0 & 1 & 0 & 0 \\ 0 & 0 & 0 & 0 & 0 & 0 & 1 & 0 \\ 0 & 0 & 0 & 0 & 0 & 0 & 0 & e^{\mathrm{i}\lambda} \end{pmatrix}. \tag{6.120}$$

Collection of gates produced by application of the Deutsch gate

In summary, by manipulating the Deutsch gate we have reproduced the Toffoli gate, $\boldsymbol{T}$, and with it the whole body of classical reversible computing on three

[5] We again apply the trick we had figured out when looking for a general solution to the Schrödinger equation in Section 4.9.1, page 153.

bits, that is, all possible permutations of a three bit register, including all possible transpositions $\boldsymbol{P}_{ab}$. We have reproduced all rotations $\boldsymbol{V}_\lambda$ in the $(|\,\mathbf{6}\rangle, |\,\mathbf{7}\rangle)$ plane, equation (6.113), as well as the $\boldsymbol{U}_\lambda$ operator (equation (6.102)), and their respective inifitesimal versions given by equations (6.107) and (6.103), respectively. Moreover, we have reproduced diagonal operators $\boldsymbol{W}_\lambda$, equation (6.119), and $\boldsymbol{X}_\lambda$, equation (6.120), and with these and classical transpositions every other diagonal operator.

This is a formidable arsenal of operations, but is it enough to reproduce all other triqubit unitary operations? We are going to show this with the help of a yet another special gate that is made of $\boldsymbol{X}_\lambda$, $\boldsymbol{V}_\lambda$, and $\boldsymbol{W}_\lambda$ and of a selected triqubit state $|\,\Psi\rangle$.

Zeroing gate

Let

$$|\,\Psi\rangle = \sum_{\boldsymbol{k}=\mathbf{0}}^{\boldsymbol{k}=\mathbf{7}} C_{\boldsymbol{k}}\,|\,\boldsymbol{k}\rangle. \tag{6.121}$$

Let us apply the following operation, which is made of the components of the $|\,\Psi\rangle$ state,

$$\boldsymbol{Z}_{\mathbf{6}}(|\,\Psi\rangle) = \boldsymbol{X}_{-\frac{1}{2}\arg(C_{\mathbf{6}}C_{\mathbf{7}})}\,\boldsymbol{V}_{-\arctan|C_{\mathbf{6}}/C_{\mathbf{7}}|}\,\boldsymbol{W}_{-\frac{1}{2}\arg(C_{\mathbf{7}}/C_{\mathbf{6}})}, \tag{6.122}$$

to $|\,\Psi\rangle$ itself. Because all three operators are restricted to the $(|\,\mathbf{6}\rangle, |\,\mathbf{7}\rangle)$ subspace, leaving the remaining basis vectors unchanged, we can analyze this operation using 2×2 matrices only. Let

$$C_{\mathbf{6}} = |C_{\mathbf{6}}|\,e^{\mathrm{i}\phi_6} \quad \text{and} \quad C_{\mathbf{7}} = |C_{\mathbf{7}}|\,e^{\mathrm{i}\phi_7}. \tag{6.123}$$

Then

$$\frac{C_{\mathbf{7}}}{C_{\mathbf{6}}} = \frac{|C_{\mathbf{7}}|}{|C_{\mathbf{6}}|}e^{\mathrm{i}(\phi_7-\phi_6)}. \tag{6.124}$$

The arg operation retrieves the phase angle, so

$$-\frac{1}{2}\arg\left(\frac{C_{\mathbf{7}}}{C_{\mathbf{6}}}\right) = -\frac{1}{2}\left(\phi_7 - \phi_6\right), \tag{6.125}$$

and the $\boldsymbol{W}$ suboperator now looks as follows:

$$\tilde{\boldsymbol{W}}_{-\frac{1}{2}\arg(C_{\mathbf{7}}/C_{\mathbf{6}})} = \begin{pmatrix} e^{\mathrm{i}(\phi_7-\phi_6)/2} & 0 \\ 0 & e^{-\mathrm{i}(\phi_7-\phi_6)/2} \end{pmatrix}. \tag{6.126}$$

When this is applied to the sixth and seventh components of $|\,\Psi\rangle$, it yields

$$\begin{pmatrix} e^{\mathrm{i}(\phi_7-\phi_6)/2} & 0 \\ 0 & e^{-\mathrm{i}(\phi_7-\phi_6)/2} \end{pmatrix}\begin{pmatrix} |C_{\mathbf{6}}|\,e^{\mathrm{i}\phi_6} \\ |C_{\mathbf{7}}|\,e^{\mathrm{i}\phi_7} \end{pmatrix} = \begin{pmatrix} |C_{\mathbf{6}}|\,e^{\mathrm{i}(\phi_6+\phi_7)/2} \\ |C_{\mathbf{7}}|\,e^{\mathrm{i}(\phi_6+\phi_7)/2} \end{pmatrix}. \tag{6.127}$$

The effect here is that both components have acquired the same phase.

Now, let us turn to the $\boldsymbol{V}_\lambda$ operator. First we find that

$$\frac{C_{\mathbf{6}}}{C_{\mathbf{7}}} = \frac{|C_{\mathbf{6}}|}{|C_{\mathbf{7}}|} e^{\mathrm{i}(\phi_6 - \phi_7)}. \tag{6.128}$$

The absolute value of this is the fraction in front of the exp, that is, $|C_{\mathbf{6}}| \,/\, |C_{\mathbf{6}}|$, and the arctan of this is whatever it is. But when used as an argument of cos and sin in the $\boldsymbol{V}_\lambda$ matrix, it turns into the following:

$$\sin\left(-\arctan\left|\frac{C_{\mathbf{6}}}{C_{\mathbf{7}}}\right|\right) = \frac{-|C_{\mathbf{6}}|}{\sqrt{|C_{\mathbf{6}}|^2 + |C_{\mathbf{7}}|^2}}, \tag{6.129}$$

$$\cos\left(-\arctan\left|\frac{C_{\mathbf{6}}}{C_{\mathbf{7}}}\right|\right) = \frac{|C_{\mathbf{7}}|}{\sqrt{|C_{\mathbf{6}}|^2 + |C_{\mathbf{7}}|^2}}. \tag{6.130}$$

Consequently, the effect of the $\boldsymbol{V}_\lambda$ submatrix acting on the $(C_{\mathbf{6}}, C_{\mathbf{7}})$ vector is

$$\begin{aligned} &\frac{e^{\mathrm{i}(\phi_6+\phi_7)/2}}{\sqrt{|C_{\mathbf{6}}|^2 + |C_{\mathbf{7}}|^2}} \begin{pmatrix} |C_{\mathbf{7}}| & -|C_{\mathbf{6}}| \\ |C_{\mathbf{6}}| & |C_{\mathbf{7}}| \end{pmatrix} \begin{pmatrix} |C_{\mathbf{6}}| \\ |C_{\mathbf{7}}| \end{pmatrix} \\ &= \frac{e^{\mathrm{i}(\phi_6+\phi_7)/2}}{\sqrt{|C_{\mathbf{6}}|^2 + |C_{\mathbf{7}}|^2}} \begin{pmatrix} 0 \\ |C_{\mathbf{6}}|^2 + |C_{\mathbf{7}}|^2 \end{pmatrix} \\ &= e^{\mathrm{i}(\phi_6+\phi_7)/2} \begin{pmatrix} 0 \\ \sqrt{|C_{\mathbf{6}}|^2 + |C_{\mathbf{7}}|^2} \end{pmatrix}. \end{aligned} \tag{6.131}$$

Finally, the $\boldsymbol{X}_\lambda$ matrix multiplies the seventh component of $|\,\Psi\rangle$ by a phase factor $\exp(\mathrm{i}\lambda)$. Here the phase factor is

$$-\frac{1}{2}\arg(C_{\mathbf{6}}C_{\mathbf{7}}) = -\frac{1}{2}(\phi_6 + \phi_7), \tag{6.132}$$

which cancels the phase factor we find in equation (6.131). The final effect of our peregrinations is

$$\boldsymbol{Z}_{\mathbf{6}}(|\,\Psi\rangle)\,|\,\Psi\rangle = \sum_{\boldsymbol{k}=\mathbf{0}}^{\boldsymbol{k}=\mathbf{5}} C_{\boldsymbol{k}}\,|\,\boldsymbol{k}\rangle + \sqrt{|C_{\mathbf{6}}|^2 + |C_{\mathbf{7}}|^2}\,|\,\mathbf{7}\rangle. \tag{6.133}$$

We see here that the $\boldsymbol{Z}_{\mathbf{6}}(|\,\Psi\rangle)$ operator "rotates" state $|\,\Psi\rangle$ *away* from the $|\,\mathbf{6}\rangle$ direction and *onto* the $|\,\mathbf{7}\rangle$ direction.

In a similar way, we can construct $\boldsymbol{Z_k}\left(\mid\Psi\rangle\right)$ operators for other directions. Applying them in turn, we can rotate $\mid\Psi\rangle$ onto the $\mid\mathbf{7}\rangle$ direction altogether, if we so desire:

$$\left(\prod_{\boldsymbol{k}=\mathbf{0}}^{\boldsymbol{k}=\mathbf{6}} \boldsymbol{Z_k}\left(\mid\Psi\rangle\right)\right)\mid\Psi\rangle = \mid\mathbf{7}\rangle. \tag{6.134}$$

Let us call this operator $\boldsymbol{R_7}(\mid\Psi\rangle)$.

Every triqubit operation implemented

And this, at long last, is how we are going to implement every possible unitary operation on a triqubit. For an arbitrary $\boldsymbol{U}$ we have that

$$\boldsymbol{U} = \sum_{n=0}^{n=7} \mid\Psi_n\rangle e^{\mathrm{i}\omega_n}\langle\Psi_n\mid, \tag{6.135}$$

where $\mid\Psi_n\rangle$ are eigenstates of $\boldsymbol{U}$ and $e^{\mathrm{i}\omega_n}$ are its eigenvalues. So, what we are going to do to imitate $\boldsymbol{U}$ is to rotate its every eigenvector $\mid\Psi_n\rangle$ on $\mid\mathbf{7}\rangle$ first, then subject it to $\boldsymbol{X}_{\omega_n}$, and then rotate it back in place. Repeating this for all eigenvectors reproduces $\boldsymbol{U}$:

$$\boldsymbol{U} = \prod_{n=0}^{n=7} \boldsymbol{R_7}^{-1}(\mid\Psi_n\rangle)\boldsymbol{X}_{\omega_n}\boldsymbol{R_7}(\mid\Psi_n\rangle). \tag{6.136}$$

Some of the $\boldsymbol{U}$ operations act on biqubit components of the triqubit, or on individual qubits only, and these are also covered by the above formula. All triqubit operations can be implemented this way. The Deutsch gate is then indeed universal within the triqubit world.

Extension to an arbitrary number of qubits

The reasoning presented so far extends trivially to any number of qubits. Let us consider a quadqubit (or a four-qubit) Deutsch gate presented in Figure 6.13.

The gate can be made of a triqubit Deutsch gate and two Toffoli gates, with one auxiliary line in the middle preset to $\mid 0\rangle$ and not participating in computations otherwise, as shown in the figure. All reasoning steps we have employed in demonstrating that a triqubit Deutsch gate is universal for triqubit computations will run the same way for the quadqubit Deutsch gate, because they are concerned mostly with operations on the 2×2 submatrix in the lower right corner of the 8×8 triqubit matrix, which was identity-diagonal for the first six basis states. Where we have gone beyond that corner, it was to generate $\boldsymbol{V}_\lambda$ out of $\boldsymbol{U}_\lambda$ and two transpositions, and the same trick will work here with the transpositions renumbered to $\boldsymbol{P}_{14,15}$ and $\boldsymbol{P}_{13,15}$. All else works the same way as before, thereby proving that the quadqubit Deutsch gate is universal for quadqubit computing.

And so on, we can extend this reasoning indefinitely to any number of qubits, thereby proving the universality of the Deutsch gate and hence also the universality

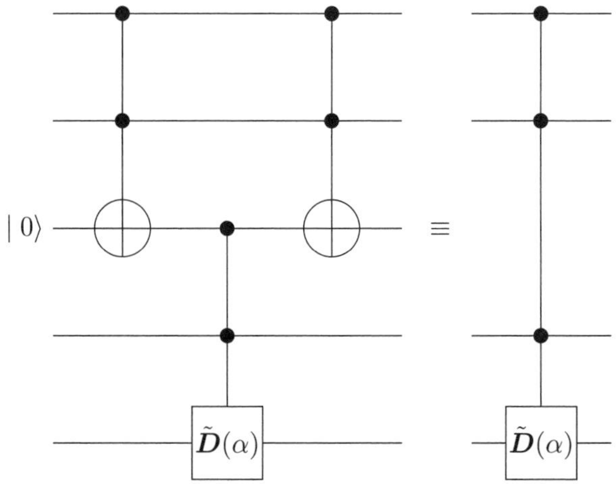

Figure 6.13: A four-qubit Deutsch gate.

of controlled-NOT and single qubit unitary gates for quantum computing on an arbitrary number of qubits.

Quantum computing programs

This is a classic result of quantum computing theory. It *defines* quantum computing as a sequence of logically simple steps that can produce any quantum outcome eventually. The sequence constitutes a quantum computing program. The final unitary transformation is the answer.

Khaneja-Glaser decomposition

In practical terms, as we have already remarked, neither the Deutsch gate nor the controlled-NOT gate is very useful outside of the algorithm design domain. What is useful instead is the Khaneja-Glaser decomposition, special cases of which we have discussed in Sections 6.2.2 and 6.2.3. The Khaneja-Glaser decomposition provides us with a more practical method of manipulating n-qubit systems to obtain any desired outcome, including the Deutsch gate and the controlled-NOT gate, applicable especially to situations where we cannot easily control the couplings—this is the case, for example, when molecules are used as registers. The idea derives from nuclear magnetic resonance (NMR) procedures, which have been most successful in demonstrating various aspects of quantum computation so far.

Nevertheless, because of its algorithmic importance, the demonstration of a working quantum controlled-NOT gate has become something of a Holy Grail in quantum computing. It is noteworthy that, as of the time of this writing, there have been few such demonstrations outside the NMR field. The reason is that, whereas theorizing about quantum computing is somewhat trivial, and so enormous body of literature

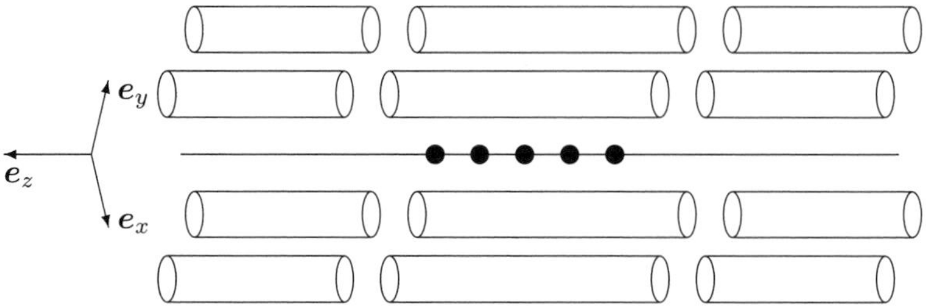

Figure 6.14: Linear Paul trap.

exists here already, doing it for real is hard. The next two sections will illustrate this by discussing two most beautiful experiments.

6.3 The Cirac-Zoller Gate

Among the first proposals for the implementation of the controlled-NOT gate and other quantum circuits was that by Cirac and Zoller in 1995 [22], who suggested using cold ions in a linear trap. A qubit would be associated with two distinct states of each ion, whereas the collective quantized motion of ion groups, which could be excited with lasers, would provide a *bus* connection by means of which various qubits could be coupled to form controlled-NOT or other multiqubit gates.

A linear Paul trap, which was used in the realization of the Cirac-Zoller idea discussed here, was invented by Raizen, Gilligan, Bergquist, Itano, and Wineland of the National Institute of Standards in 1991 [118]. It is a modification of the original idea by Wolfgang Paul, for which he received the 1989 Nobel Prize in physics, together with Hans Dehmelt and Norman F. Ramsey. A schematic representation of the linear Paul trap is shown in Figure 6.14. The device, which is enclosed in vacuum, is made of twelve metal rods that form three groups: the shorter eight rods at both ends and the longer four rods in the middle. Normally a radio frequency potential is applied to the central four rods, so as to create a quadrupole potential field $V(x, y, t)$ in the vicinity of the line that runs through the center of the trap—we will refer to this line as the *axis* of the device:

Linear Paul trap

$$V(x, y, t) = \frac{1}{2R^2} V_0 \left(x^2 - y^2\right) \cos \omega_{RF} t, \tag{6.137}$$

where x and y are coordinates in the plane perpendicular to the axis, which crosses the plane at $x = y = 0$, and R is the distance from the axis to the surface of the rods. If the radio frequency ω_{RF} is sufficiently high, the effective pseudopotential harmonic well forms around the axis of the device of the form

$$V_p = \frac{m}{2q}\omega_p^2\left(x^2 + y^2\right), \tag{6.138}$$

where m and q are the mass and charge of trapped ions and

$$\omega_p = \frac{qV_0}{m\omega_{RF}R^2\sqrt{2}}. \tag{6.139}$$

And this is how the ions get trapped. Static voltage is then applied to the eight short rods at the ends of the trap to prevent the ions from leaking out of the trap in the $\boldsymbol{e}_z$ direction. The short rods work like caps. Sometimes two rings perpendicular to the axis of the device are used for this purpose instead of the eight short rods.

Earnshaw's theorem

What is so ingenious about this and similar contraptions is that Samuel Earnshaw proved in 1842 that no static configuration of electric or magnetic fields can confine electric charges. But his theorem said nothing about varying electromagnetic fields, and so here we confine electric charges with a combination of RF and static fields.

By injecting a minute amount of preconditioned ions into the trap, we can produce chains of equidistant and isolated ions like the ones shown symbolically in Figure 6.14 as fat black dots. The trap pseudopotential and the end caps force them to align on the axis of the device, whereas their own electrostatic repulsion keeps them away from each other.

$^{40}Ca^+$ ions in a trap

Into such a trap Schmidt-Kaler and his colleagues from the Institute for Experimental Physics of the University of Innsbruck—Häffner, Riebe, Gulde, Lancaster, Deuschle, Becher, Roos, Eschner, and Blatt—loaded just two $^{40}\mathrm{Ca}^+$ ions [125]. The two specific $^{40}\mathrm{Ca}^+$ states that were used to encode qubits were the ground state $|\,\mathrm{S}_{1/2}\rangle$, which was interpreted as $|\,0\rangle$, and a metastable state $|\,\mathrm{D}_{5/2}\rangle$, the lifetime of which is nearly one second and which was interpreted as $|\,1\rangle$.

Qubit encoding

Manipulating qubits

Transitions between the $|\,\mathrm{S}_{1/2}\rangle$ and $|\,\mathrm{D}_{5/2}\rangle$ states were induced by shining a narrow-band titanium-sapphire laser beam of 729 nm wavelength onto the locations of the ions, which were separated by a distance of 5.3 μm. Because the laser beam diameter was 2.5 μm, Schmidt-Kaler and his colleagues could control each ion separately, without affecting the other one at the same time. The beam could be moved from one ion to another in 15 μs by an electro-optical deflector. They could also control the phase of the light field by tweaking it with an *acousto-optical modulator* with an accuracy of 0.06 radians. Unlike spin qubits, energy level qubits do not have a direction in space. Yet their mathematics is the same as that of spin qubits, which we covered in depth in Chapter 2. Qubit rotations about various directions in space map in this case on qubit transformations induced by Rabi pulses with various phases between the incident radiation and the atomic polarization of the ion.

Electron shelving

The states of the qubits were measured by resorting to the so-called electron shelving technique. The shelf in this case was a $|\,\mathrm{P}_{1/2}\rangle$ state to which $|\,\mathrm{S}_{1/2}\rangle$ was excited by irradiation with a laser beam of 397 nm wavelength. An electron that's put on "the shelf" drops off it eventually, and this event can be detected by observing the atom's fluorescence with a CCD camera, separately for each ion. If there is fluorescence, it means that the atom was in the $|\,\mathrm{S}_{1/2}\rangle$ state when it had been buzzed with the detection beam. If there is no fluorescence, it means the atom was in the $|\,\mathrm{D}_{5/2}\rangle$ state. The CCD exposure times used by the Innsbruck group were 10 ms or 23 ms depending on the measurement. Light collected by the CCD pixels was integrated within a $3\,\mu\mathrm{m} \times 3\,\mu\mathrm{m}$ area around each ion. The detection efficiency of this system is very high, about 98%, the 2% error deriving from occasional registration of spurious fluorescence from the adjacent ion, or, for longer exposure times, from spontaneous decay.

Dynamics of trapped ions

To understand what they did next, we need to look more closely at the dynamics of the two ions in the trap. The ions have their internal electronic degrees of freedom—these are the ones we use to encode qubits—but they also have external degrees of freedom, associated with their physical motion within the trap. Because of their electrostatic repulsion, the movement of one ion transfers to the other one, so their motion is collective. Given the one-dimensional dynamics of the trap, they can move in two basic modes: in the same direction, in which case their center of mass moves, too—this is called the CM mode, or in opposite directions, in which case their center of mass does not move. They can also move in a combination of the two modes.

Phonons

A general feature of quantum mechanics is that enclosed systems have their energy levels quantized, which basically derives from the way waves can slosh inside a box. This is also the case here. The vibrations of both modes are quantized. We call these quantized vibrations *phonons.* A full description of the system must therefore include not only the electron states of the two ions but also the phonon states of their collective vibrational motions.

Detuning

If we shine the laser beam onto one ion only, we affect the ion's internal electron state, given that the beam wavelength corresponds exactly to its internal transition, for example, from $|\,\mathrm{S}_{1/2}\rangle$ to $|\,\mathrm{D}_{5/2}\rangle$. If the beam is well focused, so that its diameter is far less than the distance between the ions, the other ion is not affected. But if we detune the laser so that its frequency becomes

$$\omega = \omega_{\text{ion}} \pm \omega_{\text{phonon}}, \tag{6.140}$$

where ω_{ion} is the electronic transition frequency, ω_{phonon} is the phonon frequency, and $\omega_{\text{phonon}} << \omega_{\text{ion}}$, so that the resulting radiation is still within the resonance peak of the electron transition of the ion, then we will trigger the electronic transition of the ion and at the same time excite the phonon state of the "bus."

This kind of a pulse, called a *sideband* pulse, can be produced with the acousto-optical modulator. Sideband pulses can be *blue detuned,* if $\omega = \omega_{\text{ion}} + \omega_{\text{phonon}}$, or *red detuned,* if $\omega = \omega_{\text{ion}} - \omega_{\text{phonon}}$.

Let $| \, 1\rangle_{\text{bus}}$ be a bus state without any phonons, and let $| \, 0\rangle_{\text{bus}}$ be a bus state with one CM-mode phonon—this is the lowest energy bus excitation. Thus, the bus itself constitutes a *third qubit* in the system, and all following manipulations are really triqubit operations. But the bus qubit is auxiliary and does not participate in the logic of the computation, even if it participates in its physics.

The sequence of operations is as follows.

1. We cool the system to force all three qubits into their ground state

   $$| \, 0\rangle_c \, | \, 1\rangle_{\text{bus}} \, | \, 0\rangle_t, \tag{6.141}$$

   where c stands for "control" and t stands for "target." This is a critical part of the experiment without which none of the following steps would be possible. Rapidly moving ions would not respond properly to the highly tuned and focused laser beams, first, because they would not be where we'd expect them—they can move with velocities as high as 1 km/s—and second, because the resulting Doppler shift would move the incident laser beam frequency in their reference frame away from the resonance.

   Hence, we have to immobilize them completely. We do so with lasers. The discovery and development of the technique are due to Steven Chu of Stanford University, Claude Cohen-Tannoudji of the École Normale Supérieure, and William D. Phillips of the National Institute of Standards and Technology, who all received the 1997 Nobel Prize in physics for this work. *Laser cooling*

   When ions absorb photons, they absorb their energy and their momentum. Having absorbed a photon, they usually re-emit it within a few nanoseconds and then absorb another one. In this way a single ion can absorb and re-emit up to a hundred million photons per second. Even though a velocity kick that results from a single absorption is very small, only about 3 cm/s for a calcium ion, when multiplied by a hundred million of such scattering events, this yields enormous acceleration of a thousand kilometers per second in a second. This is 100,000 times more than standard gravity, which is slightly less than 10 m/s in a second.

Doppler cooling

Let us consider two high-intensity superimposed laser beams that propagate in opposite directions and are identically detuned toward longer wavelengths away from the resonant transition of an ion. If the ion moves against one of the beams, the beam gets Doppler shifted in the ion's own frame of reference back toward the resonance, while the other beam redshifts even farther from the resonance. In effect the ion absorbs more photons from the beam against which it is moving, and so is slowed. If the ion overshoots, the same thing happens against the other beam, until the ion stops altogether. This procedure is called *Doppler cooling.* Normally we would have three pairs of laser beams focused on the trap region from three orthogonal directions, so as to kill the x, y, and z components of the ion's velocity.

In the Innsbruck experiment only 2 ms of such Doppler cooling were needed to slow both ions. This was then followed by 8 ms of *sideband cooling,* a similar though more subtle procedure that detunes the ion transition additionally by the bus phonon frequencies, in order to kill bus excitations.

The final result of this preparation procedure was that all components of the quantum register, the ions and the bus, were forced into their ground states with 99% probability.

Range of initial conditions explored

2. The next step was to prepare an initial condition for the gate operation. This was accomplished by buzzing a selected ion with a π pulse, a Rabi pulse needed to flip its qubit.

   The functioning of the gate for all four initial conditions,

$$
\begin{aligned}
&|\, 0\rangle_c \,|\, 1\rangle_{\text{bus}} \,|\, 0\rangle_t, \\
&|\, 0\rangle_c \,|\, 1\rangle_{\text{bus}} \,|\, 1\rangle_t, \\
&|\, 1\rangle_c \,|\, 1\rangle_{\text{bus}} \,|\, 0\rangle_t, \qquad \text{and} \\
&|\, 1\rangle_c \,|\, 1\rangle_{\text{bus}} \,|\, 1\rangle_t,
\end{aligned}
$$

   was explored. Additionally the experimenters tested the quantumness of the gate against the following initial condition:

$$
\frac{1}{\sqrt{2}} \left(|\, 0\rangle_c + |\, 1\rangle_c\right) |\, 1\rangle_{\text{bus}} \,|\, 0\rangle_t,
$$

which, as we have seen (see equation (6.15) on page 291), should have produced a fully entangled Bell state $|\,\Phi^+\rangle$,

$$\frac{1}{\sqrt{2}}\left(|\,0\rangle_c\,|\,x\rangle_{\text{bus}}\,|\,0\rangle_t + |\,1\rangle_c\,|\,x\rangle_{\text{bus}}\,|\,1\rangle_t\right),$$

where $|\,x\rangle_{\text{bus}}$ is whatever state the bus ends up in after the operation is over—since the bus qubit is auxiliary, we don't care.

3. Next, the experimenters commenced the operations that constituted the gate. *Gate execution*

   The first step here was to submit the control qubit to the blue sideband rotation $\boldsymbol{R}^+(\pi, 0)$, where + marks that the rotation is blue sideband, π is the angle of the rotation (here it ends up being a full flip), and 0 is the phase between the incident radiation and the atomic polarization of the ion, which is mathematically equivalent to choosing the rotation axis.

   This operation can be described formally for arbitrary angles θ and ϕ as follows [58]:

$$\boldsymbol{R}^+(\theta, \phi) = \exp\left[\mathrm{i}\frac{\theta}{2}\left(e^{\mathrm{i}\phi}\left(|\,1\rangle\langle 0\,|\right)_{\text{ion}}\boldsymbol{a}^\dagger + e^{-\mathrm{i}\phi}\left(|\,0\rangle\langle 1\,|\right)_{\text{ion}}\boldsymbol{a}\right)\right], \qquad (6.142)$$

   where $\boldsymbol{a}^\dagger$ is an operator that creates a phonon on the bus and $\boldsymbol{a}$ is an operator that annihilates a phonon on the bus.

   The transformation keeps the state $|\,1\rangle_c\,|\,1\rangle_{\text{bus}}$ unaltered, whereas at the same time, it flips [22]

$$|\,0\rangle_c\,|\,1\rangle_{\text{bus}} \leftrightarrow |\,1\rangle_c\,|\,0\rangle_{\text{bus}}. \qquad (6.143)$$

   This way we map the state of the control qubit onto the bus: if the control qubit is $|\,1\rangle_c$, the bus remains vibration-free, $|\,1\rangle_{\text{bus}}$; if the control qubit is $|\,0\rangle_c$, the bus becomes excited by one CM-mode phonon and switches to the $|\,0\rangle_{\text{bus}}$ state.

   In order to flip the target qubit, by the means of operations that are discussed below, the bus must be vibration free ($|\,1\rangle_{\text{bus}}$), meaning that the control qubit must be $|\,1\rangle_c$. If there is a phonon on the bus, the target qubit will not flip. So this is how the two-ion controlled-NOT gate is implemented.

Triqubit gate reduced to a biqubit one

Alas! An astute reader will notice that we have merely postponed the problem, because we still have to implement a full controlled-NOT gate, this time on a biqubit system—without the auxiliary third qubit—with the bus in the control role. This is why we have made this counterintuitive assigment, calling the vibration *free* state of the bus $|\,1\rangle_{\text{bus}}$.

Single ion and bus gate

4. How to implement a single-ion-and-bus controlled-NOT gate?

The following six rotations were all performed on the second ion, the target qubit. The train of operations, with no pauses in between, because the qubit would evolve out of control during such pauses, was as follows (from right to left).

$$\boldsymbol{R}\left(\frac{\pi}{2},\pi\right)\boldsymbol{R}^{+}\left(\frac{\pi}{\sqrt{2}},\frac{\pi}{2}\right)\boldsymbol{R}^{+}(\pi,0)\,\boldsymbol{R}^{+}\left(\frac{\pi}{\sqrt{2}},\frac{\pi}{2}\right)\boldsymbol{R}^{+}(\pi,0)\,\boldsymbol{R}\left(\frac{\pi}{2},0\right). \tag{6.144}$$

The first and the last rotation can be easily recognized as two bracketing Ramseys. The mathematical formula that describes them is [58]

$$\boldsymbol{R}(\theta,\phi)) = \exp\left[\mathrm{i}\frac{\theta}{2}\left(e^{\mathrm{i}\phi}\,|\,1\rangle\langle 0\,| + e^{-\mathrm{i}\phi}\,|\,0\rangle\langle 1\,|\right)\right]. \tag{6.145}$$

The second Ramsey pulse is, in fact, an inverse of the first one. It rotates the target vector by $\pi/2$ in the opposite direction, because $e^{\mathrm{i}\pi} = e^{-\mathrm{i}\pi} = -1$.

The four blue sideband rotations in the middle implement a composite single ion phase gate with the following truth table.[6]

	$\langle 0\,\|_{\text{bus}}\,\langle 0\,\|_t$	$\langle 0\,\|_{\text{bus}}\,\langle 1\,\|_t$	$\langle 1\,\|_{\text{bus}}\,\langle 0\,\|_t$	$\langle 1\,\|_{\text{bus}}\,\langle 1\,\|_t$
$\|\,0\rangle_{\text{bus}}\,\|\,0\rangle_t$	−1	0	0	0
$\|\,0\rangle_{\text{bus}}\,\|\,1\rangle_t$	0	−1	0	0
$\|\,1\rangle_{\text{bus}}\,\|\,0\rangle_t$	0	0	−1	0
$\|\,1\rangle_{\text{bus}}\,\|\,1\rangle_t$	0	0	0	1

[6]To show how this comes about, we substitute $\begin{pmatrix} 0 & 0 \\ 1 & 0 \end{pmatrix}$ in place of $(|\,1\rangle\langle 0\,|)_{\text{ion}}$ and $\begin{pmatrix} 0 & 1 \\ 0 & 0 \end{pmatrix}$ in place of $(|\,0\rangle\langle 1\,|)_{\text{ion}}$. Then we use $\begin{pmatrix} 0 & 0 & 0 \\ 1 & 0 & 0 \\ 0 & \sqrt{2} & 0 \end{pmatrix}$ for $\boldsymbol{a}^{\dagger}$ and $\begin{pmatrix} 0 & 1 & 0 \\ 0 & 0 & \sqrt{2} \\ 0 & 0 & 0 \end{pmatrix}$ for $\boldsymbol{a}$ in equation (6.142). When performing these computations in Matlab or Octave, we define $\boldsymbol{R}^{+}(\theta,\phi)$ as a function—remembering that matrix exponential is `expm`, i and π are `i` and `pi`, a $\sqrt{}$ is `sqrt()`, and a tensor product of two matrices is provided by the `kron()` function—and multiply the four blue sideband rotations specified here, in such order as they're written. The table can be read from the resulting matrix for places that correspond to zero and one phonons only. The reason the $\boldsymbol{a}$ and $\boldsymbol{a}^{\dagger}$ matrices have to be 3×3 in this computation is that the single ion phase gate moves the bus through a state with two phonons temporarily.

Ramsey pulses rotate a qubit from its base position ($| 1\rangle$ or $| 0\rangle$) halfway to, say, $(| 0\rangle + | 1\rangle)/\sqrt{2}$, possibly multiplied by some phase factor. If the bus is in the $| 0\rangle_{\text{bus}}$ state, then the whole rotated target state is multiplied by -1. This doesn't really change anything, so when the second reverse Ramsey pulse comes along, the state is rotated back to where it was in the beginning.

But when the bus state is $| 1\rangle_{\text{bus}}$, then only the $| 0\rangle_t$ component gets multiplied by -1, and the $| 1\rangle_t$ component stays as it was. In this case when

$$(| 0\rangle + | 1\rangle)/\sqrt{2}$$

is mapped onto

$$(- | 0\rangle + | 1\rangle)/\sqrt{2},$$

and when the reverse Ramsey pulse comes to it, it doesn't rotate the qubit back. Instead it continues to rotate the qubit, until the qubit has been flipped over completely.

We can just as easily see the above by performing a simple direct computation. The two Ramsey pulses evaluate to

$$\boldsymbol{R}\left(\frac{\pi}{2}, 0\right) = e^{\mathrm{i}\pi\boldsymbol{\sigma}_x/4}, \qquad (6.146)$$

$$\boldsymbol{R}\left(\frac{\pi}{2}, \pi\right) = e^{-\mathrm{i}\pi\boldsymbol{\sigma}_x/4}. \qquad (6.147)$$

If the bus is in the $| 0\rangle_{\text{bus}}$ state, the action of the phase gate on the target qubit is given by $-\mathbf{1}$. Hence, in this case the full action of the gate, including the Ramsey pulses, is

$$e^{\mathrm{i}\pi\boldsymbol{\sigma}_x/4}(-\mathbf{1})e^{-\mathrm{i}\pi\boldsymbol{\sigma}_x/4} = -\mathbf{1}. \qquad (6.148)$$

The phase of the qubit changes, but otherwise the qubit stays put. On the other hand, if the bus is in the $| 1\rangle_{\text{bus}}$ state, the action of the phase gate on the target qubit is given by $-\boldsymbol{\sigma}_{\boldsymbol{z}}$. So, in this case the full action of the gate, including the Ramsey pulses, is[7]

$$e^{\mathrm{i}\pi\boldsymbol{\sigma}_x/4}(-\boldsymbol{\sigma}_z)e^{-\mathrm{i}\pi\boldsymbol{\sigma}_x/4} = \begin{pmatrix} 0 & \mathrm{i} \\ -\mathrm{i} & 0 \end{pmatrix}. \qquad (6.149)$$

This is an almost pure NOT. To make it into $\boldsymbol{\sigma}_x$, we need only to apply $\exp(-\mathrm{i}\pi\boldsymbol{\sigma}_z/2)$ to the target qubit, that is, to rotate it around the z axis by π.

[7]This can be also checked easily with Matlab or Octave. Alternatively, equation (6.60), page 300, can be used to evaluate $\exp(\mathrm{i}\pi\boldsymbol{\sigma}_x/4)$.

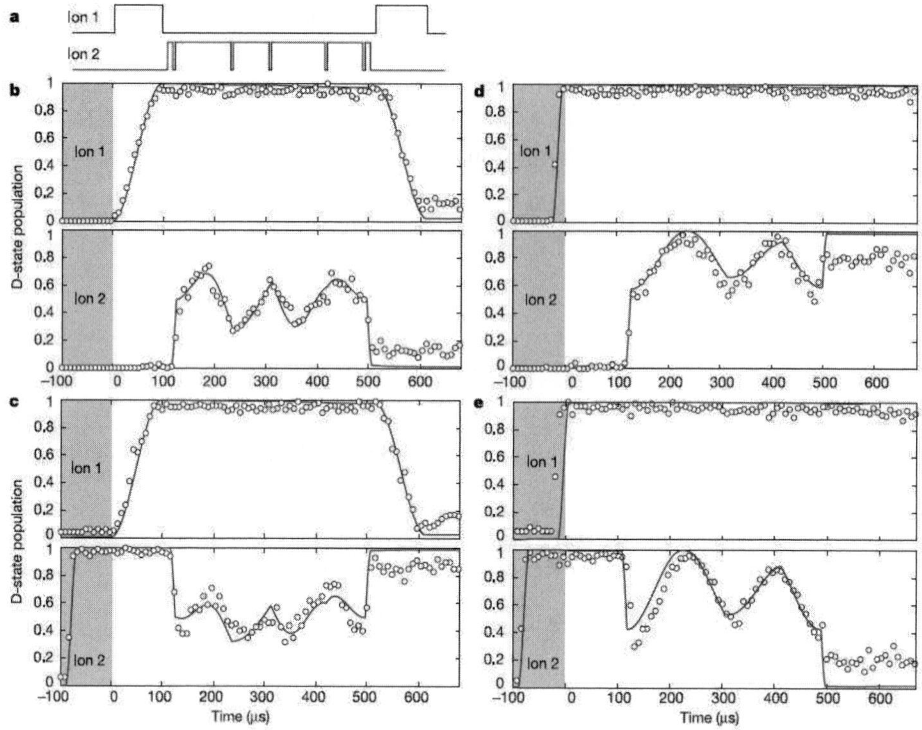

Figure 6.15: Time evolution of qubit states during execution of the controlled-NOT gate. Figure reprinted by permission from Macmillan Publishers Ltd: Nature [125], © 2003.

Restoration of the control qubit state

5. In the last step we perform a sideband rotation on the first—control—qubit, this time it is $\boldsymbol{R}^+(\pi, \pi)$, which rotates the qubit back to its original position, if it was moved from it at all, and removes the phonon from the bus, if it's been put there.

This completes the controlled-NOT gate.

What is most interesting about the Innsbruck experiment is that the state of both qubits is closely monitored as the computation unfolds.

State tomography

This is shown in Figure 6.15. Inset (a) shows the pulse sequence. Ion 1 is the control qubit and ion 2 the target qubit. The shaded area corresponds to the preparation part. Once the qubits' initial condition is prepared, the execution of

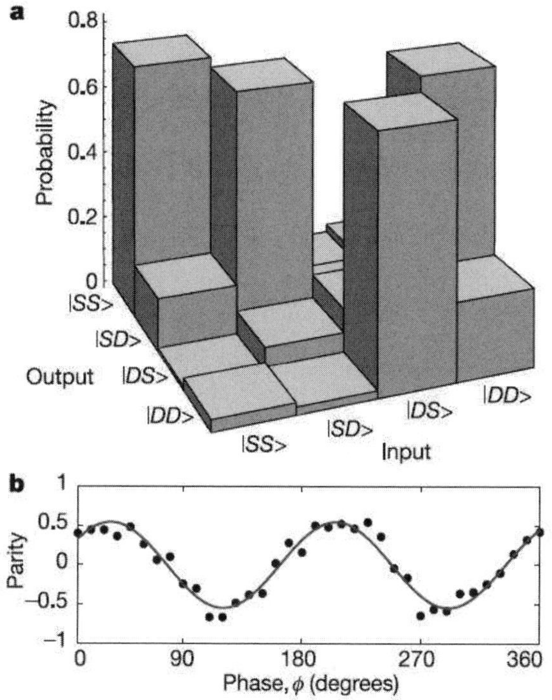

Figure 6.16: (a) Experimentally obtained truth table for the gate. (b) The biqubit *parity* $P = P_{|00\rangle} + P_{|11\rangle} - \left(P_{|01\rangle} + P_{|10\rangle}\right)$ in function of phase ϕ. Figure reprinted by permission from Macmillan Publishers Ltd: Nature [125], © 2003.

the gate commences at time $t = 0$. Experimental points in the graphs are marked with open dots, and the solid lines correspond to theoretically predicted evolution. Plotted in all eight diagrams are the probabilities of finding an ion (1 or 2) in the $| D_{5/2}\rangle$ state. On every single inspection, of course, each qubit is found to be either in the $| S_{1/2}\rangle$ or in the $| D_{5/2}\rangle$ state.

The first pulse is the sideband rotation of the control qubit, which takes $95\,\mu$s. We can see that the qubit indeed flips from $| 0\rangle_c$ to $| 1\rangle_c$—panels (b) and (c). The qubit does not flip when its initial condition is $| 1\rangle_c$—panels (d) and (e). In each case, the target qubit responds as the theory predicts.

The flips are not clean enough for serious multigate computation; and after the gate has been completed, both qubits end up in mixed states that are close to $| 0\rangle$

and $|\,1\rangle$, but not quite there. Hence, when the target qubit is expected to be, say, $|\,1\rangle_t$ after the gate operations have been completed, it will be there most of the time, but in up to 30% of cases, we will see $|\,0\rangle_t$ instead, and vice versa.

Statistical ensemble of the device

Every experimental point of the graph, which is marked by an open dot, was obtained by allowing the system to evolve up to this point in time, then interrupting the evolution and measuring both qubits. This was repeated 100 times for each point in order to collect enough statistics to estimate probabilities. The resulting statistical ensemble for each point is rather small, and the standard deviation error on these estimates is about 5%.

Gate quantumness

Figure 6.16 shows the measured truth-table for the gate in panel (a). Ideally, the high columns should all reach one, and the low background should all be zero. In reality the high peaks vary between 0.71 and 0.77, and the low squares vary between 0.01 and 0.22. Panel (b) shows the result of a measurement that proves the gate's quantumness, that is, the linearity with respect to the input states' superposition. This was tested by applying a $\pi/2$ pulse with phase ϕ to both ions after the gate had been completed and measuring the biqubit's parity given by $P = P_{|00\rangle} + P_{|11\rangle} - \left(P_{|01\rangle} + P_{|10\rangle}\right)$, where $P_{|xy\rangle}$ are probabilities of finding the biqubit in the $|\,xy\rangle$ state.

If the control qubit is in the superposition $(|\,0\rangle_c + |\,1\rangle_c)/\sqrt{2}$ and the target qubit is $|\,0\rangle_t$ initially, the gate should produce the entangled Bell state $|\,\Phi^+\rangle$, for which the observed variation of $P(\phi)$ should be $\cos(2\phi)$—which it is—whereas if the biqubit was not entangled, the variation would be $\cos\phi$.

Gate fidelity

In the final account, the gate has the fidelity F of 80%, and the entanglement is created with the fidelity of 71%—fidelity being a measure of discrepancy between the ideal and the observed behavior of the gate given by

$$F = \frac{1}{4}\mathrm{Tr}\left(\boldsymbol{M}_{\mathrm{exp}}\boldsymbol{M}^T_{\mathrm{cnot}}\right), \tag{6.150}$$

where $\boldsymbol{M}_{\mathrm{exp}}$ and $\boldsymbol{M}^T_{\mathrm{cnot}}$ are measured and theoretically expected controlled-NOT matrices.

Where do the discrepancies, 20% on average, come from?

Most errors derive from the laser equipment used. Laser frequency noise contributes about 10%, and laser intensity fluctuations contribute between 1% and 3%. Laser detuning error contributes some 2%, and ion addressing errors contribute between 3% and 5%. Off-resonant excitations account for 4% and further small errors are caused by residual thermal excitations.

Further improvements in the equipment and techniques are likely to yield more accurate controlled-NOT and other multiqubit gates in the future. Very long lifetimes of metastable ion states, highly accurate read-out methodology, ease of addressing individual qubits and of manipulating the bus—all speak in its favor.

On the other hand, the setup is clumsy. The need for vacuum equipment, ion preparation equipment, lasers, detectors, acousto-optical modulators, at least ten highly trained physicists, auxiliary personnel, and an endless supply of coffee are all a far cry from what the generation of laptop and iPod users expects their future computers to look like.

So how about a superconducting controlled-NOT gate?

6.4 The Superconducting Gate

NEC/RIKEN device

One of the first demonstrations of a superconducting controlled-NOT gate was by Yamamoto, Pashkin, Astafiev, Nakamura, and Tsai of NEC and RIKEN laboratories in Japan [150]. The group had considerable experience developing superconducting devices for quantum computing. Theirs was the first demonstration of macroscopic quantum-coherent states in a Cooper pair box in 1999 [97] and a later demonstration of quantum oscillations in a coupled charge superconducting biqubit in 2003 [106]. The controlled-NOT gate followed the biqubit demonstration by a few months only. However, the authors had to resort to a numerical simulation of their device in order to recover the truth table for the operation.

For a more complete demonstration we had to wait another four years.

Delft device

It wasn't until June 2007 that Plantenberg, de Groot, Harmans, and Mooij of Delft University of Technology in the Netherlands published their results in *Nature* [114]. The Delft group used SQUID detectors (which we encountered in a similar context in the preceding chapter) to measure the output state of the biqubit, including biqubit correlations. All biqubit computational basis states, as well as their superpositions, were used as inputs. The gate's truth table was determined experimentally and, after correcting for poor contrast, was found to have fidelity of 0.4. The device was noisy, but all the right features were there and accessible experimentally.

Three-junction flux qubit

The device is shown in Figure 6.17. The qubits used were of a type we haven't discussed yet. They are called "three-junction flux qubits" and were discussed in depth by Robertson, Plourde, Reichardt, Hime, Wu, and Clarke of the University of California, Berkeley, in two papers: the first one published with Linzen in 2005 [115], and the second one published in 2006 [121].

A three-junction flux qubit is made of three Josephson junctions connected in series on a superconducting loop. A close inspection of Figure 6.17 may show—to a reader with keen eyesight—the three junctions on the middle branch of each of the

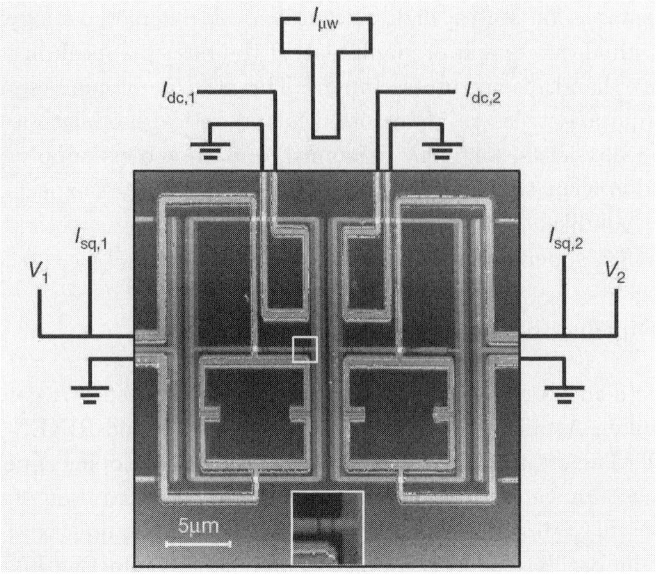

Figure 6.17: An atomic force micrograph of the Delft biqubit. Figure reprinted by permission from Macmillan Publishers Ltd: Nature [114], © 2007.

two figure "8" patterns made by aluminum tracks (they actually look more like two ⊟ figures side by side). The rightmost junction of the left "8" is framed in a little white square and expanded at the bottom of the figure. The two qubits are covered with a thin insulating layer, on top of which two SQUIDs are defined, one above the bottom half of each 8, with currents labeled $I_{\mathrm{sq},1}$ and $I_{\mathrm{sq},2}$. The SQUIDs are used as state detectors, similar to how they were also used in the superconducting biqubit, discussed in Section 5.3. The qubits themselves are biased through the small coils that appear in the upper part of each 8, also in orange, and with currents labeled $I_{\mathrm{dc},1}$ and $I_{\mathrm{dc},2}$.

How do they work?

Flux qubit

It is well known that magnetic field flux through a closed superconducting loop, without any junctions on it, is quantized. This derives basically from the observation that the superconducting wave function on the loop must fold onto itself on circumambulating the loop. Otherwise it would not be single-valued. This observation, in combination with the Kelvin-Stokes theorem, and a simple proportionality relation between the magnetic potential $\boldsymbol{A}$ and the gradient of the wave function phase within the superconductor, implies that the magnetic flux Φ through the loop must be

$$\Phi = n\frac{h}{2q_e} = n\Phi_0, \tag{6.151}$$

where n is an integer number, h is the Planck constant (without it being divided by 2π), q_e is the electron charge, and Φ_0 is the flux quantum.

A flux qubit builds on this idea, but it inserts a Josephson junction into the ring. This has the effect of a persistent current being built in the ring whenever an external flux is applied through it. The current may flow clockwise or counterclockwise or both at the same time. The counteraction of the currents on the applied flux is such as to force the total flux resulting to comply with the quantization condition (6.151).

With each of the two currents, we can associate a collective quantum state, $|\circlearrowleft\rangle$ and $|\circlearrowright\rangle$. The energy levels of the two states get closer as the applied flux approaches a half-integer multiples of flux quantum, that is,

$$\frac{1}{2}\Phi_0, \frac{3}{2}\Phi_0, \frac{5}{2}\Phi_0, \ldots, \tag{6.152}$$

becoming degenerate, meaning "of the same energy," at half-integer multiples of flux quantum exactly. By approaching the degeneracy, but not reaching it exactly, we can make the two energy levels associated with $|\circlearrowleft\rangle$ and $|\circlearrowright\rangle$ sufficiently close to make the device a useful qubit that can be manipulated by pulsing it with microwaves. In particular, the state of the loop can be put in the superposition

$$\frac{1}{\sqrt{2}}\left(|\circlearrowleft\rangle + |\circlearrowright\rangle\right). \tag{6.153}$$

Another Schrödinger's cat

Such a superposition was observed for the first time in 2000 by Friedman, Patel, Chen, Tolpygo, and Lukens of the State University of New York [47]. Because the superconducting loop used in the experiments is a macroscopic device, the experiment was pronounced to be a practical demonstration of Schrödinger's cat.

The junction critical current in the New York experiment was adjusted with a second magnetic flux, and the tunnel coupling between $|\circlearrowleft\rangle$ and $|\circlearrowright\rangle$ was extremely small, which made manipulation of the loop's quantum state difficult.

Inserting three junctions in the loop, instead of one, can produce much stronger coupling, which is determined by junction sizes when the device is fabricated [121].

Back to the Delft device.

The two qubits weren't exactly identical. The persistent current in the first one (the one on the left) was 450 nA, and 480 nA in the second. Similarly, their energy gaps were $h \times 2.6\,\text{GHz}$ for the first one, and $h \times 2.2\,\text{GHz}$ for the second.

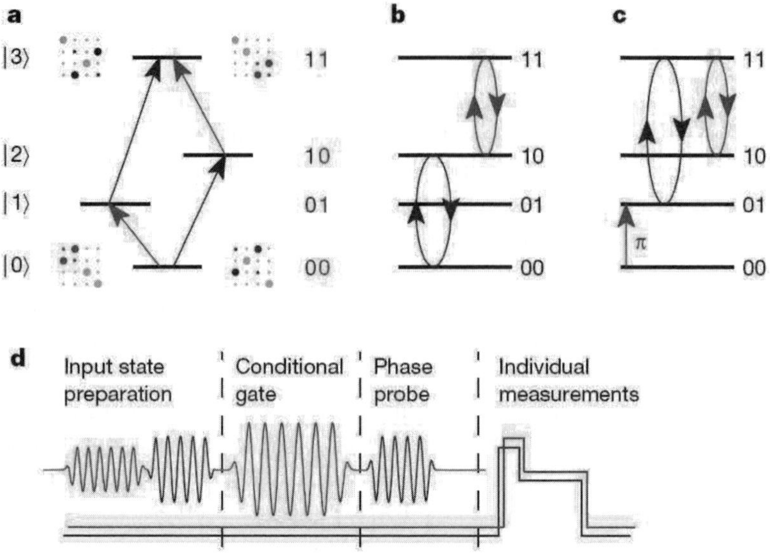

Figure 6.18: Energy level diagram of the Delft biqubit. Figure reprinted by permission from Macmillan Publishers Ltd: Nature [114], © 2007.

State measurement

The two SQUIDs measured qubit states by sending pulses from current sources $I_{\mathrm{sq},1}$ and $I_{\mathrm{sq},2}$, which would trigger voltage pulses V_1 and V_2 dependent on the states of the qubits. These, in turn, would be detected with the help of amplifiers and threshold detectors.

Under the bias resulting from applying $I_{\mathrm{dc},1}$ and $I_{\mathrm{dc},2}$ of about $10\,\mu$A to the two smaller loops at the top of Figure 6.17, the energy level separations for both qubits would widen to $h \times 7\,$GHz for the left qubit and $h \times 5\,$GHz for the right one.

Driving the biqubit

The qubits were irradiated through an on-chip antenna, schematically drawn on top of Figure 6.17 and labeled $I_{\mu\mathrm{w}}$.

In effect, the two qubits formed a tunable four-level quantum system described by the following Hamiltonian:

$$\boldsymbol{H} = -\frac{1}{2}\left(\epsilon_1 \boldsymbol{\sigma}_{z1} + \Delta_1 \boldsymbol{\sigma}_{x1}\right) \otimes \mathbf{1}_2 - \frac{1}{2}\mathbf{1}_1 \otimes \left(\epsilon_2 \boldsymbol{\sigma}_{z2} + \Delta_1 \boldsymbol{\sigma}_{x2}\right) + J\boldsymbol{\sigma}_{z1} \otimes \boldsymbol{\sigma}_{z2}, \quad (6.154)$$

where $J = \frac{1}{2} \times h \times 400\,$MHz was a measure of coupling between the qubits.

Energy levels

Figure 6.18 shows the energy level diagram of the Delft biqubit. Because the qubits differ, the energy levels of the $|\,0\rangle \otimes |\,1\rangle \equiv |\,\mathbf{1}\rangle$ and $|\,1\rangle \otimes |\,0\rangle \equiv |\,\mathbf{2}\rangle$ states differ, too. Consequently, we end up with a system of four different energy levels. The arrows in Figure 6.18 (a) show possible transitions within the system resulting from a single qubit flip. The pale arrows correspond to the $| \, 00\rangle \rightarrow | \, 01\rangle$ and $| \, 10\rangle \rightarrow | \, 11\rangle$ transitions, which result from the second qubit flip, and the darker arrows correspond to the $| \, 00\rangle \rightarrow | \, 10\rangle$ and $| \, 01\rangle \rightarrow | \, 11\rangle$ transitions, which result from the first qubit flip. Each of these can be thought of as a two-state system and understood in terms of qubit dynamics.

Let us recall equation (4.194), which we derived in Section 4.6. It described a transformation of spinor coefficients under a frame rotation in a plane defined by $\phi = \text{constant}$. The resulting transformation matrix was

$$\begin{pmatrix} \cos\frac{\theta}{2} & \mathrm{i}e^{-\mathrm{i}\phi}\sin\frac{\theta}{2} \\ \mathrm{i}e^{\mathrm{i}\phi}\sin\frac{\theta}{2} & \cos\frac{\theta}{2} \end{pmatrix}. \tag{6.155}$$

As we saw in Section 2.11.1, equation (2.156), subjecting a spinor to a Rabi pulse with the resonance frequency of $\omega_L = 2\mu B_{\|}/\hbar$ changed θ linearly, $\theta = \omega_R t$, where $\omega_R = 2\mu B_{\perp}/\hbar$ was the Rabi (not resonance) frequency. For each of the transitions shown in Figure 6.18, there is a specific resonance frequency that corresponds to the energy gap between the levels.

Rabi rotations in a 4-level system

Let us then focus on the possible Rabi rotations of such a four-level system while affixing ϕ at 90° so as to kill $\sin\phi$. This reduces matrix (6.155) to

$$\begin{pmatrix} \cos\frac{\omega t}{2} & \mathrm{i}\sin\frac{\omega t}{2} \\ \mathrm{i}\sin\frac{\omega t}{2} & \cos\frac{\omega t}{2} \end{pmatrix}. \tag{6.156}$$

The first pale arrow operation rotates the four-state system between $| \, \mathbf{0}\rangle$ and $| \, \mathbf{1}\rangle$. Its 4×4 matrix fills the upper left corner with (6.156) and leaves the other states, $| \, \mathbf{2}\rangle$ and $| \, \mathbf{3}\rangle$, unchanged, so its rotation matrix looks as follows.

$$\boldsymbol{R}(\omega_{01}, t) = \begin{pmatrix} \cos\frac{\omega_{01}t}{2} & \mathrm{i}\sin\frac{\omega_{01}t}{2} & 0 & 0 \\ \mathrm{i}\sin\frac{\omega_{01}t}{2} & \cos\frac{\omega_{01}t}{2} & 0 & 0 \\ 0 & 0 & 1 & 0 \\ 0 & 0 & 0 & 1 \end{pmatrix}. \tag{6.157}$$

For the second pale arrow, which connects $| \, \mathbf{2}\rangle$ and $| \, \mathbf{3}\rangle$ we will have the rotation 2×2 matrix in the lower right corner, and the identity will be in the upper left corner. The resulting 4×4 matrix is as follows.

$$\boldsymbol{R}(\omega_{23}, t) = \begin{pmatrix} 1 & 0 & 0 & 0 \\ 0 & 1 & 0 & 0 \\ 0 & 0 & \cos\frac{\omega_{23}t}{2} & \mathrm{i}\sin\frac{\omega_{23}t}{2} \\ 0 & 0 & \mathrm{i}\sin\frac{\omega_{23}t}{2} & \cos\frac{\omega_{23}t}{2} \end{pmatrix}. \tag{6.158}$$

The first dark arrow transition connects $| \mathbf{0}\rangle$ and $| \mathbf{2}\rangle$. So its matrix will have rotation terms in the $(0,2)$ and $(2,0)$ locations, as well as $(0,0)$ and $(2,2)$, and identity in the $(1,3)$, $(3,1)$, $(1,1)$, and $(3,3)$ locations:

$$\boldsymbol{R}(\omega_{02}, t) = \begin{pmatrix} \cos\frac{\omega_{02}t}{2} & 0 & \mathrm{i}\sin\frac{\omega_{02}t}{2} & 0 \\ 0 & 1 & 0 & 0 \\ \mathrm{i}\sin\frac{\omega_{02}t}{2} & 0 & \cos\frac{\omega_{02}t}{2} & 0 \\ 0 & 0 & 0 & 1 \end{pmatrix}. \tag{6.159}$$

The second dark arrow transition connects $| \mathbf{1}\rangle$ with $| \mathbf{3}\rangle$, so its matrix looks as follows.

$$\boldsymbol{R}(\omega_{13}, t) = \begin{pmatrix} 1 & 0 & 0 & 0 \\ 0 & \cos\frac{\omega_{13}t}{2} & 0 & \mathrm{i}\sin\frac{\omega_{13}t}{2} \\ 0 & 0 & 1 & 0 \\ 0 & \mathrm{i}\sin\frac{\omega_{13}t}{2} & 0 & \cos\frac{\omega_{13}t}{2} \end{pmatrix}. \tag{6.160}$$

The little inserts in the corners of Figure 6.18 (a) show matrices $\boldsymbol{R}(\omega_{01}, t)$, $\boldsymbol{R}(\omega_{23}, t)$, $\boldsymbol{R}(\omega_{02}, t)$, and $\boldsymbol{R}(\omega_{13}, t)$ schematically. For ω outside of ω_{01}, ω_{23}, ω_{02}, and ω_{13} resonance peaks, $\boldsymbol{R}(\omega, t)$ is an identity.

Single-qubit operations can be generated by combining $\boldsymbol{R}(\omega_{01}, t)$ and $\boldsymbol{R}(\omega_{23}, t)$. Figure 6.18 (b) shows rotations that produce superpositions of $| 00\rangle$ and $| 10\rangle$, and of $| 10\rangle$ and $| 11\rangle$. In turn, a rotation that would form a superposition of $| 01\rangle$ and $| 11\rangle$ is shown in Figure 6.18 (c).

*Generating the controlled-*NOT *gate*

An operation that is controlled by the first qubit (from the left) and that looks almost like a controlled-NOT gate is obtained from $\boldsymbol{R}(\omega_{23}, t)$ for $\omega_{23}t/2 = \pi/2$:

$$\boldsymbol{R}\left(\omega_{23}, \frac{\pi}{\omega_{23}}\right) = \begin{pmatrix} 1 & 0 & 0 & 0 \\ 0 & 1 & 0 & 0 \\ 0 & 0 & 0 & \mathrm{i} \\ 0 & 0 & \mathrm{i} & 0 \end{pmatrix}. \tag{6.161}$$

We call this operation a π_{23} pulse.

We can convert gate (6.161) to a full controlled-NOT gate, up to a phase constant, by combining it with a single qubit rotation about the z axis. In the two-dimensional Hilbert space of a single qubit, such an operation was described by (see equation (4.192) in Section 4.6)

$$\begin{pmatrix} e^{\mathrm{i}\phi/2} & 0 \\ 0 & e^{-\mathrm{i}\phi/2} \end{pmatrix}. \tag{6.162}$$

If the operation was to be performed on the left qubit, while ignoring the state of the right one, the corresponding matrix would be

$$\begin{pmatrix} e^{i\phi/2} & 0 & 0 & 0 \\ 0 & 1 & 0 & 0 \\ 0 & 0 & e^{-i\phi/2} & 0 \\ 0 & 0 & 0 & 1 \end{pmatrix} \begin{pmatrix} 1 & 0 & 0 & 0 \\ 0 & e^{i\phi/2} & 0 & 0 \\ 0 & 0 & 1 & 0 \\ 0 & 0 & 0 & e^{-i\phi/2} \end{pmatrix}$$

$$= \begin{pmatrix} e^{i\phi/2} & 0 & 0 & 0 \\ 0 & e^{i\phi/2} & 0 & 0 \\ 0 & 0 & e^{-i\phi/2} & 0 \\ 0 & 0 & 0 & e^{-i\phi/2} \end{pmatrix}. \tag{6.163}$$

For $\phi = \pi/2$ this becomes

$$\frac{1}{\sqrt{2}} \begin{pmatrix} 1+i & 0 & 0 & 0 \\ 0 & 1+i & 0 & 0 \\ 0 & 0 & 1-i & 0 \\ 0 & 0 & 0 & 1-i \end{pmatrix}, \tag{6.164}$$

because $\cos\frac{\pi}{4} = \sin\frac{\pi}{4} = 1/\sqrt{2}$. In combination with $\boldsymbol{R}(\omega_{23}, \pi/\omega_{23})$ the operation yields the controlled-NOT gate as follows:

$$\frac{1}{\sqrt{2}} \begin{pmatrix} 1+i & 0 & 0 & 0 \\ 0 & 1+i & 0 & 0 \\ 0 & 0 & 1-i & 0 \\ 0 & 0 & 0 & 1-i \end{pmatrix} \begin{pmatrix} 1 & 0 & 0 & 0 \\ 0 & 1 & 0 & 0 \\ 0 & 0 & 0 & i \\ 0 & 0 & i & 0 \end{pmatrix} = \frac{1+i}{\sqrt{2}} \begin{pmatrix} 1 & 0 & 0 & 0 \\ 0 & 1 & 0 & 0 \\ 0 & 0 & 0 & 1 \\ 0 & 0 & 1 & 0 \end{pmatrix}. \tag{6.165}$$

How can we implement a rotation about the z axis? The easiest way is to just wait and do nothing for a specific amount of time. The system should perform a Larmor precession about the z axis during this time all by itself. In practice, Larmor precession is so much faster than Rabi oscillations that we do not need to wait at all, and we couldn't even do so with sufficient precision anyway. Instead we just change the phase of the incident signal, the phase mapping directly on the ϕ angle.

Controlled phase gate

Applying the π_{23} pulse twice yields another operation, called a controlled-phase gate:

$$\left(\boldsymbol{R}\left(\omega_{23}, \frac{\pi}{\omega_{23}}\right)\right)^2 = \begin{pmatrix} 1 & 0 & 0 & 0 \\ 0 & 1 & 0 & 0 \\ 0 & 0 & 0 & i \\ 0 & 0 & i & 0 \end{pmatrix}^2 = \begin{pmatrix} 1 & 0 & 0 & 0 \\ 0 & 1 & 0 & 0 \\ 0 & 0 & -1 & 0 \\ 0 & 0 & 0 & -1 \end{pmatrix}. \tag{6.166}$$

Bell states

All four Bell states (see Section 5.1) can be prepared by applying two pulses to $|\,00\rangle$:

$$|\,\Phi^+\rangle = \frac{|\,00\rangle + |\,11\rangle}{\sqrt{2}} = \boldsymbol{R}\left(\omega_{23}, \frac{-\pi}{\omega_{23}}\right)\boldsymbol{R}\left(\omega_{02}, \frac{\pi}{2\omega_{02}}\right)|\,00\rangle \quad (6.167)$$

$$|\,\Phi^-\rangle = \frac{|\,00\rangle - |\,11\rangle}{\sqrt{2}} = \boldsymbol{R}\left(\omega_{23}, \frac{\pi}{\omega_{23}}\right)\boldsymbol{R}\left(\omega_{02}, \frac{\pi}{2\omega_{02}}\right)|\,00\rangle \quad (6.168)$$

$$|\,\Psi^+\rangle = \frac{|\,01\rangle + |\,10\rangle}{\sqrt{2}} = \boldsymbol{R}\left(\omega_{02}, \frac{\pi}{\omega_{02}}\right)\boldsymbol{R}\left(\omega_{01}, \frac{\pi}{2\omega_{01}}\right)|\,00\rangle \quad (6.169)$$

$$|\,\Psi^-\rangle = \frac{|\,01\rangle - |\,10\rangle}{\sqrt{2}} = \boldsymbol{R}\left(\omega_{02}, \frac{-\pi}{\omega_{02}}\right)\boldsymbol{R}\left(\omega_{01}, \frac{\pi}{2\omega_{01}}\right)|\,00\rangle. \quad (6.170)$$

The above is illustrated in Figure 6.18 (d), where the input state is prepared by applying two pulses to the $|\,00\rangle$ state, into which the system relaxes naturally, when left alone, upon having radiated excess energy away.

After the initial state of the circuit is prepared, it is operated by applying a gate rotation to the register, for example, as described by equation (6.165). The resulting density matrix of the final state is read by applying probe pulses. The readings are taken for both qubits simultaneously and independently. For each initial state, the whole operation is repeated 8,192 times (which is 2^{13}) in order to collect sufficient statistics—the average deviation from the expected probability on the sample of this size being about 1%. In the process the truth table amplitudes for the controlled-NOT gate are evaluated as well, because the initial state is always known.

Statistical ensemble of the device

Figure 6.19 (a) shows the directly measured statistics for the gate, the Manhattan-like towers that correspond to the number of counts for which a specific transition between input and output states was observed. Although the controlled-NOT matrix pattern can be discerned, there is a great deal of other high-rise in this Manhattan. The sensitivity of the read-out system deployed in the device is only 40%. If the measurement is performed so as to compensate for it, the contrast between the highest and the lowest towers in the controlled-NOT Manhattan improves, as shown in Figure 6.19 (b). But even with this improvement, the estimated fidelity of the device (see equation (6.150, page 332) is ≈ 0.4 only.

Yale and NIST devices

Even more sophisticated biqubits have been demonstrated by other groups since the Delft result was published in the June 2007 issue of *Nature* [114]. Most notable is a device demonstrated by the Schoelkopf's group of Yale University, described in a letter published in the late September 2007, that used a virtual photon cavity to couple the qubits [91], and a similar device made by the Simmond's group of

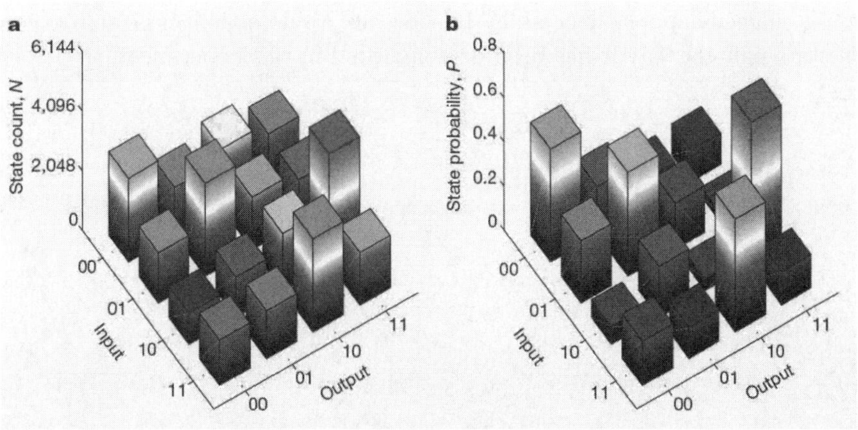

Figure 6.19: Experimentally obtained truth table for the controlled-NOT operation: (a) directly measured statistics, (b) conditional spectroscopy that corrects for the limited sensitivity of the measuring system. Figure reprinted by permission from Macmillan Publishers Ltd: Nature [114], © 2007.

the National Institute of Standards and Technology [129]. There have been no other demonstrations of a superconducting controlled-NOT gate by the time of this writing. The device presented in this section, noisy as it is, represents the state of the art in solid-state quantum computing as of fall 2007.

Resolving transition lines

An important feature of the Delft device is that it is biased so that ω_{01}, ω_{23}, ω_{02}, and ω_{13} are all different and sufficiently separated. Consequently, the corresponding transitions can be activated and controlled in separation from one another. But this feature is also the undoing of the approach. Let us suppose that instead of working with a 2-qubit register, we have a modest size, by classical computing standards, 32-bit register. The number of computational basis states of the register is $2^{32} = 4,294,967,296$, and the number of computational basis state pairs, between which we might want to perform the rotations is

$$\binom{2^{32}}{2} = \frac{2^{32} \times \left(2^{32} - 1\right)}{1 \times 2} = \frac{2^{64} - 2^{32}}{2} = 2^{63} - 2^{31} \approx 9.2 \times 10^{18}. \qquad (6.171)$$

This is a huge number of transition frequencies—so huge that, if they were all made to be different (which is probably impossible anyway), yet restricted to some sensible finite range $[\omega_{\min}, \omega_{\max}]$, and each with its own finite resonance width $\Delta\omega$, they would fill the range continuously forming a band rather than a set of isolated lines—as indeed happens in crystals. In effect, we would no longer be able to control each state pair the way it has been demonstrated in this experiment.

7 Yes, It Can Be Done with Cogwheels

7.1 The Deutsch Oracle

At last we have arrived at the very heart of quantum computing—the controlled-NOT gate, which we have scrutinized both theoretically and experimentally. In this final chapter we use examples to illustrate what we can do with it.

Among the first quantum algorithms that attracted attention was the Deutsch oracle [28] and its generalization to an arbitrary number of qubits by Deutsch and Jozsa [30]. Figure 7.1 shows the circuit for the original biqubit oracle.

Hadamard gates rotate qubits sideways

The device is constructed as follows. The two boxes labeled $\boldsymbol{H}$ represent Hadamard rotations, which we have seen in Section 4.10, page 162. Their role is to rotate the qubit's polarization sideways (the polarization is expected to be up or down initially).

$$\boldsymbol{H} \mid 0\rangle = \frac{1}{\sqrt{2}} (\mid 0\rangle + \mid 1\rangle), \tag{7.1}$$

$$\boldsymbol{H} \mid 1\rangle = \frac{1}{\sqrt{2}} (\mid 0\rangle - \mid 1\rangle). \tag{7.2}$$

When applied again, it restores the original polarization, since $\boldsymbol{H}^2 = \mathbf{1}$. But this is not what is going to happen in this circuit, because the controlled-$\boldsymbol{U}_f$ gate couples both qubits. So, something is going to happen to the top qubit between the two applications of $\boldsymbol{H}$.

$\boldsymbol{U}_f$ is defined as

$$\boldsymbol{U}_f \mid x\rangle \mid y\rangle = \mid x\rangle \mid y \oplus f(x)\rangle, \tag{7.3}$$

where $\oplus$ is modulo-2 addition, $\oplus = +_2$. It is a controlled-by-$f(x)$-NOT gate, rather than a controlled-by-x-NOT gate. But if $f(x) = x$, then this is a normal controlled-NOT gate.

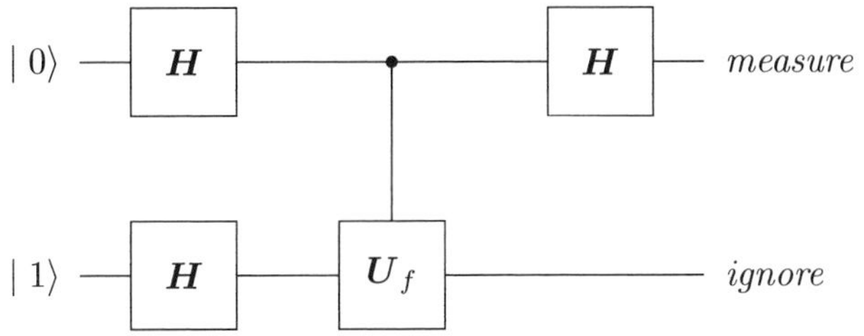

Figure 7.1: The Deutsch oracle circuit.

Function $f(x)$ is either constant or balanced.

Here $f(x)$ is an arbitrary function of the top qubit state that maps $\{0,1\}$ onto $\{0,1\}$. Alas, $\{0,1\}$ is a small universe; hence $f(x)$ can be one of only two things: either *constant* or *balanced,*. The latter means that it's either $f(0)=0$ and $f(1)=1$, or $f(0)=1$ and $f(1)=0$.

The oracle as a computer program

The purpose of the device is to tell us whether $f(x)$ is constant or balanced—if we don't know the answer a priori, which is rather silly because we need to know what it is in order to implement it in the first place, as we're going to see soon enough. The circuit is called an "oracle" because it answers a profound question: "Is this function balanced?" The oracle says "yes" or "no," as the case may be.

Step-by-step analysis of the Deutsch oracle

Let us analyze how the device proceeds, step by step.

1. The first pair of Hadamard rotations converts $\mid 0\rangle\otimes\mid 1\rangle$ into

$$\frac{1}{2}\left(\mid 0\rangle+\mid 1\rangle\right)\otimes\left(\mid 0\rangle-\mid 1\rangle\right). \tag{7.4}$$

2. We apply controlled-$\boldsymbol{U}_f$ to this state. Now, $f(x)$ can be either 0 or 1. If it is 0, then

$$\left(\mid 0\rangle-\mid 1\rangle\right)\oplus 0=\left(\mid 0\rangle-\mid 1\rangle\right)=(-1)^0\left(\mid 0\rangle-\mid 1\rangle\right). \tag{7.5}$$

If it is 1, then

$$\begin{aligned}\left(\mid 0\rangle-\mid 1\rangle\right)\oplus 1 &= \neg\left(\mid 0\rangle-\mid 1\rangle\right)=\left(\mid 1\rangle-\mid 0\rangle\right)\\ &= -\left(\mid 0\rangle-\mid 1\rangle\right)=(-1)^1\left(\mid 0\rangle-\mid 1\rangle\right).\end{aligned} \tag{7.6}$$

In summary,

$$\left(\mid 0\rangle-\mid 1\rangle\right)\oplus f(x)=(-1)^{f(x)}\left(\mid 0\rangle-\mid 1\rangle\right), \tag{7.7}$$

and

$$\boldsymbol{U}_f\mid x\rangle\otimes\left(\mid 0\rangle-\mid 1\rangle\right)=(-1)^{f(x)}\mid x\rangle\otimes\left(\mid 0\rangle-\mid 1\rangle\right). \tag{7.8}$$

Because $\boldsymbol{U}_f$ is linear,

$$\begin{aligned}&\boldsymbol{U}_f\frac{1}{2}\left(\mid 0\rangle+\mid 1\rangle\right)\otimes\left(\mid 0\rangle-\mid 1\rangle\right)\\ &\quad=\frac{1}{2}\left((-1)^{f(0)}\mid 0\rangle+(-1)^{f(1)}\mid 1\rangle\right)\otimes\left(\mid 0\rangle-\mid 1\rangle\right).\end{aligned} \tag{7.9}$$

3. Next, we apply the Hadamard gate to the top qubit. We get

$$\frac{1}{2}\left((-1)^{f(0)}\boldsymbol{H}\mid 0\rangle+(-1)^{f(1)}\boldsymbol{H}\mid 1\rangle\right)\otimes\left(\mid 0\rangle-\mid 1\rangle\right)$$

$$= \frac{1}{2}\left((-1)^{f(0)}\frac{1}{\sqrt{2}}(|\,0\rangle + |\,1\rangle) + (-1)^{f(1)}\frac{1}{\sqrt{2}}(|\,0\rangle - |\,1\rangle)\right) \otimes (|\,0\rangle - |\,1\rangle)$$
$$= \frac{1}{2}\left(|\,0\rangle\left((-1)^{f(0)} + (-1)^{f(1)}\right) + |\,1\rangle\left((-1)^{f(0)} - (-1)^{f(1)}\right)\right)$$
$$\otimes \frac{1}{\sqrt{2}}(|\,0\rangle - |\,1\rangle). \tag{7.10}$$

4. Finally, we measure the top qubit. If $f(x)$ is constant, then $(-1)^{f(0)} - (-1)^{f(1)} = 0$, and so the top qubit must be in the $|\,0\rangle$ state. On the other hand, if $f(x)$ is balanced, then $(-1)^{f(0)} + (-1)^{f(1)} = 0$, and then the top qubit must be in the $|\,1\rangle$ state.

   Therefore, by measuring the top qubit just once, we can answer whether function $f(x)$ is constant or balanced.

This is sometimes quoted as a great triumph of quantum computing because in the classical digital computing world we would have to evaluate $f(x)$ at two points, 0 and 1, in order to ascertain whether it is constant or balanced.

Another way of describing the problem is in terms of a coin. What does it take to check whether a coin has its sides identical or different? Well, normally we have to look at one side first, then at the other, and then we'll know. Yet here, because of the magic of quantum mechanics, just one look at the top qubit lets us answer the question.

Why does it work? It works because we have put both qubits in the superpositions of $|\,0\rangle$ and $|\,1\rangle$, so the inspection occurs here for both values simultaneously. A clever focusing procedure then projects the answer onto the top qubit.

Experimental demonstration of the Deutsch oracle

How does it work in practice? The same group of researchers from the University of Innsbruck, who demonstrated the Cirac-Zoller controlled-NOT gate in 2003, which we had discussed in Section 6.3, page 322, also implemented the Deutsch algorithm using a similar experimental framework [58]. The difference in this case was twofold: the system had only one calcium ion, which was made the control qubit; and the second qubit, the target one, was encoded on the phonon.

The exact sequence of operations is shown in Figure 7.2 and in Table 7.1, which shows pulses that were used to implement the $\boldsymbol{R}_{\bar{y}}\boldsymbol{U}_f\boldsymbol{R}_y$ sequence (the ordering is from right to left) on the lower qubit, depending on what the function $f(x)$ was meant to be. The single-qubit rotation $\boldsymbol{R}(\theta, \phi)$ and the blue sideband rotation $\boldsymbol{R}^+(\theta, \phi)$ operators are as in Section 6.3, page 322, with θ defining the length and ϕ the phase of the pulse.

True computational cost of the Deutsch oracle

In the simplest case of $f(x) = 0$, we end up with two pulses only for the whole algorithm, namely, the two Hadamard operations that the top qubit is subjected

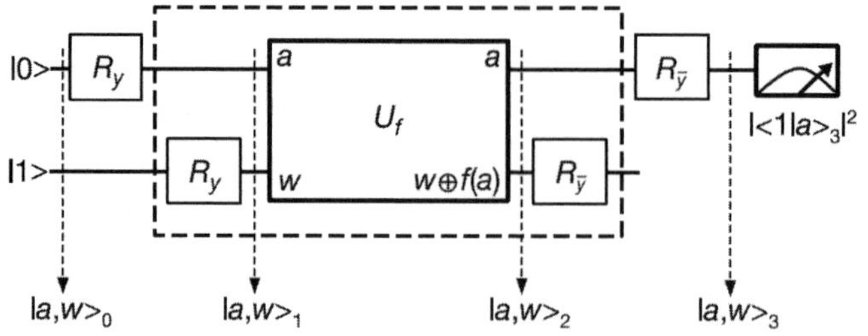

Figure 7.2: Implementation of the Deutsch algorithm on an ion-trap system. The top (control) qubit, $|\, a\rangle$, is encoded on a $^{40}Ca^{+}$ ion, and the bottom (target) qubit, $|\, w\rangle$, is encoded on a phonon—the mechanical vibrations of the ion in the trap. The rotations are $\boldsymbol{R}_y = \boldsymbol{R}\left(\frac{\pi}{2}, 0\right)$ and $\boldsymbol{R}_{\bar{y}} = \boldsymbol{R}\left(\frac{\pi}{2}, \pi\right)$. Figure reprinted by permission from Macmillan Publishers Ltd: Nature [58], © 2003.

to. But in the most complex case, that of $f(x) = 1$, the total number of gates is 11. In other words, we may have to perform up to 11 atomic operations on the two qubits in order to find our answer, a calculation that classically requires only two operations: $f(0)$ and $f(1)$.

Figure 7.3 shows the evolution of the upper qubit during the computation. The total duration of all the pulses, including the pauses, is about 250 μs. The measured probabilities of detecting the upper qubit in the $|\, 1\rangle$ state at the end of the computation were 0.019 ± 0.006 for $f(x) = 0$, 0.087 ± 0.006 for $f(x) = 1$, 0.975 ± 0.004 for $f(x) = x$, and 0.975 ± 0.002 for $f(x) = \neg x$. Every dot in Figure 7.3 is the result of repeating the computation up to this point 100 times in order to build a sufficient statistical ensemble to estimate the probabilities.

From the physics point of view this is a tremendously impressive result.

From the computational point of view, though, this result is disappointing. Because on every repeat of the operations the ion and the bus have to be cooled and then manipulated to restore the initial condition, we should probably double the time per computation to, say, 500 μs. The computation has to be repeated at least 100 times for each $f(x)$, which is going to take $50{,}000\,\mu\text{s} = 50\,\text{ms}$. And this is far slower than what it would take on a typical 3 GHz Intel chip.

Quantum computing wins when applied to large problems only.

The argument, of course, is that quantum computing, like massively parallel computing, is heavy and therefore should be applied only to very large problems, where it is expected to shine. The Deutsch oracle is certainly not a very large problem, and so more lightweight computational methods win easily.

Table 7.1: Sequences of pulses implementing $\boldsymbol{R}_{\bar{y}}\boldsymbol{U}_f\boldsymbol{R}_y$ in the Innsbruck implementation of the Deutsch algorithm. Here $\phi_{\text{swap}} = \arccos\left(\cot^2\left(\pi/\sqrt{2}\right)\right)$. We note that for f_3 the $\boldsymbol{U}_f$ operation is just the controlled-NOT gate. Table reprinted by permission from Macmillan Publishers Ltd: Nature [58], © 2003.

$f(x)$	Implementation of $\boldsymbol{R}_{\bar{y}}\boldsymbol{U}_f\boldsymbol{R}_y$
$f_1 : f(x) = 0$	no pulses
$f_2 : f(x) = 1$	$\boldsymbol{R}^+\left(\frac{\pi}{\sqrt{2}},0\right)\boldsymbol{R}^+\left(\frac{2\pi}{\sqrt{2}},\phi_{\text{swap}}\right)\boldsymbol{R}^+\left(\frac{\pi}{\sqrt{2}},0\right)$
	$\boldsymbol{R}\left(\frac{\pi}{2},0\right)\boldsymbol{R}\left(\pi,\frac{\pi}{2}\right)\boldsymbol{R}\left(\frac{\pi}{2},\pi\right)$
	$\boldsymbol{R}^+\left(\frac{\pi}{\sqrt{2}},\pi\right)\boldsymbol{R}^+\left(\frac{2\pi}{\sqrt{2}},\pi+\phi_{\text{swap}}\right)\boldsymbol{R}^+\left(\frac{\pi}{\sqrt{2}},\pi\right)$
$f_3 : f(x) = x$	$\boldsymbol{R}^+\left(\frac{\pi}{\sqrt{2}},0\right)\boldsymbol{R}^+\left(\pi,\frac{\pi}{2}\right)\boldsymbol{R}^+\left(\frac{\pi}{\sqrt{2}},0\right)\boldsymbol{R}^+\left(\pi,\frac{\pi}{2}\right)$
$f_4 : f(x) = \neg x$	$\boldsymbol{R}(\pi,0)\boldsymbol{R}^+\left(\pi,\frac{\pi}{\sqrt{2}}\right)\boldsymbol{R}^+\left(\pi,\frac{\pi}{2}\right)\boldsymbol{R}^+\left(\frac{\pi}{\sqrt{2}},0\right)\boldsymbol{R}^+\left(\pi,\frac{\pi}{2}\right)\boldsymbol{R}(\pi,0)$

The core of the argument in favor of quantum computing is that a single quantum mechanical measurement suffices where two classical measurements are needed to find the answer. This is certainly so when comparisons are made with classical digital computing. But let us consider a simple contraption made of two perpendicular mirrors, illustrated in Figure 7.4, that let us see both sides of a coin at the same time. To view the coin, one should place it symmetrically between the two mirrors subtending a 45° angle with each. The reflected images show both sides at the same time. With additional mirrors and lenses, we can merge both images and then filter them so that only differences are shown. If nothing gets shown, both sides of the coin are identical.

The two-mirror contraption is an example of a classical analog computer that, in many cases, can match a quantum computer in algorithmic efficiency.[1]

But let us consider a more elaborate example, a much larger version of the Deutch oracle—the Deutsch-Jozsa oracle. Figure 7.5 shows a relatively small 4-qubit version of it, but in principle the oracle may grow to an arbitrary number of $n+1$ qubits. The Hadamard gates in the oracle work as before, and the $\boldsymbol{U}_f$ gate is now controlled not by one but by n lines. The function f maps from $\{0,1\}^n$ to $\{0,1\}$. It may take various shapes, but we still restrict it to being either constant or balanced. But what does *balanced* mean for the n control qubits? It means that the function

Deutsch-Jozsa oracle

[1]Both sides of the coin can be seen by using just one mirror, though not with equal resolution, because the image of one side is more removed than the image of the other; also, the far side is reversed, whereas the near side isn't. Our simple contraption addresses both shortcomings, reversing both images the same way and placing them side by side.

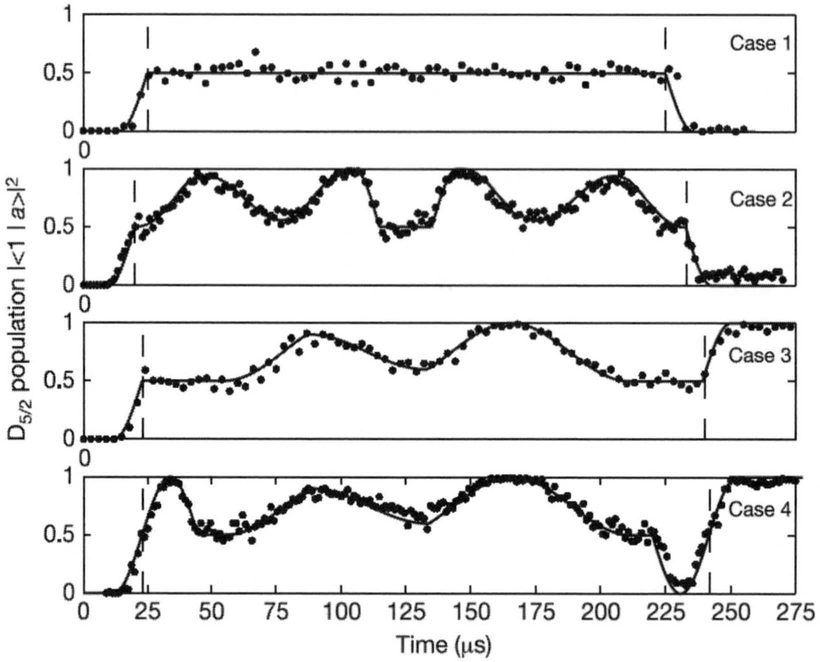

Figure 7.3: Probability of detecting the $^{40}Ca^{+}$ ion in the $\mid D_{5/2}\rangle \equiv \mid 1\rangle$ state, for the sequences of operations that correspond to f_1, f_2, f_3, and f_4. Each dot is a probability estimated from 100 repeated measurements. The continuous lines correspond to theoretically predicted evolutions of the ion. They are *not* fits to experimental data. Figure reprinted by permission from Macmillan Publishers Ltd: Nature [58], © 2003.

value is zero for half of the possible values of its arguments and one for the other half. The oracle lets us determine whether the function is constant or balanced. To do so, we have to measure all the output lines on the top. If we find $\mid 0\rangle$ on *every* line, the function is constant; otherwise, it is balanced.

A classical oracle would require 2^n measurements, one for each value of the argument, to ascertain that f is constant. For $n = 512$ this converts to $\approx 1.34 \times 10^{154}$ versus 512 measurements only. The saving returned by the quantum version of the oracle is enormous. Even if the gates are slow and many, we'll end up ahead of the classical digital device.

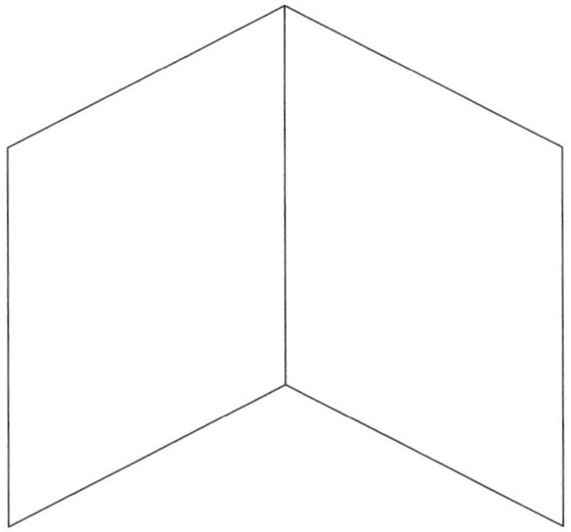

Figure 7.4: Two perpendicular mirrors let us observe both sides of a coin at the same time.

Here is how the device works.

Step-by-step analysis of Deutsch-Jozsa oracle

1. First we need to apply $\boldsymbol{H}$ to n qubits that all start in the $\mid 0\rangle$ state. This, as it's shown below, produces a superposition of all numbers between 0 and $2^n - 1$.

$$\begin{aligned}
&\boldsymbol{H} \mid 0\rangle \boldsymbol{H} \mid 0\rangle \cdots \boldsymbol{H} \mid 0\rangle \\
&= \frac{1}{\sqrt{2}}(\mid 0\rangle + \mid 1\rangle) \frac{1}{\sqrt{2}}(\mid 0\rangle + \mid 1\rangle) \cdots \frac{1}{\sqrt{2}}(\mid 0\rangle + \mid 1\rangle) \\
&= \frac{1}{2^{n/2}}(\mid 00\ldots 0\rangle + \mid 00\ldots 1\rangle + \ldots + \mid 11\ldots 1\rangle) \\
&= \frac{1}{2^{n/2}}(\mid \mathbf{0}\rangle + \mid \mathbf{1}\rangle + \mid \mathbf{2}\rangle + \ldots + \mid 2^n - 1\rangle) \\
&= \frac{1}{2^{n/2}} \sum_{\boldsymbol{x}=0}^{2^n-1} \mid \boldsymbol{x}\rangle. \qquad (7.11)
\end{aligned}$$

This is the source of the quantum computer's power: the ability to run operations simultaneously on 2^n different values. Although we can carry out various tasks in parallel on present-day supercomputers, deploying tens of thousands of CPUs sometimes, we may never be able to put together 1.34×10^{154} CPUs to match quantum computers.

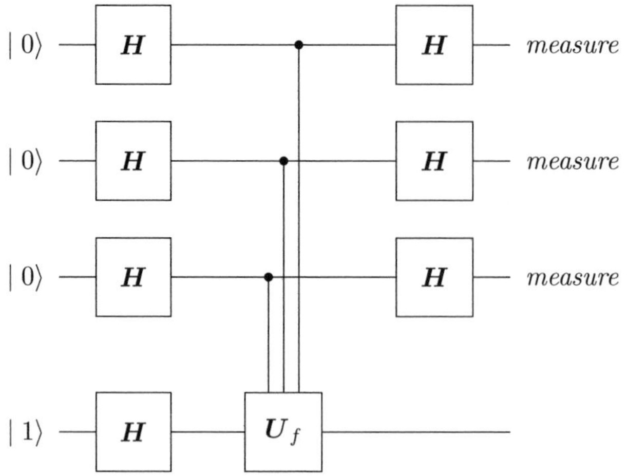

Figure 7.5: A 4-qubit Deutsch-Jozsa oracle.

Applying $\boldsymbol{H}$ to the bottom line yields

$$\frac{1}{\sqrt{2}}\left(\mid 0\rangle - \mid 1\rangle\right). \tag{7.12}$$

So the state of the whole computer becomes

$$\frac{1}{2^{n/2}}\left(\sum_{\boldsymbol{x}=0}^{2^n-1} \mid \boldsymbol{x}\rangle\right) \otimes \frac{1}{\sqrt{2}}\left(\mid 0\rangle - \mid 1\rangle\right). \tag{7.13}$$

2. Now the n-line controlled $\boldsymbol{U}_f$ is applied, and, by extending our result from the Deutsch oracle, equation (7.8), we obtain

$$\frac{1}{2^{n/2}}\left(\sum_{\boldsymbol{x}=0}^{2^n-1} (-1)^{f(\boldsymbol{x})} \mid \boldsymbol{x}\rangle\right) \otimes \frac{1}{\sqrt{2}}\left(\mid 0\rangle - \mid 1\rangle\right). \tag{7.14}$$

3. Finally, we have to apply the Hadamard transform to the top n lines again. But the top lines are no longer just $\mid 0\rangle$, so here we have to do some more work.

   Let us observe that the basic definition for the Hadamard transform can be rewritten as follows:

$$\boldsymbol{H} \mid 0\rangle \quad = \quad \frac{1}{\sqrt{2}}\left(\mid 0\rangle + \mid 1\rangle\right) = \frac{1}{\sqrt{2}}\left((-1)^{0\cdot 0} \mid 0\rangle + (-1)^{0\cdot 1} \mid 1\rangle\right), \tag{7.15}$$

$$\boldsymbol{H} \mid 1\rangle \quad = \quad \frac{1}{\sqrt{2}}(\mid 0\rangle - \mid 1\rangle) = \frac{1}{\sqrt{2}}\left((-1)^{1\cdot 0} \mid 0\rangle + (-1)^{1\cdot 1} \mid 1\rangle\right). \quad (7.16)$$

In summary,

$$\boldsymbol{H} \mid x\rangle = \frac{1}{\sqrt{2}} \sum_{y=0}^{1} (-1)^{x\cdot y} \mid y\rangle. \quad (7.17)$$

To use this formula, we must figure out how to apply it not to an individual qubit but to a tensor product of n qubits. So here is how we go about this task:

$$\begin{aligned} & \boldsymbol{H} \mid x_1\rangle \otimes \boldsymbol{H} \mid x_2\rangle \otimes \ldots \otimes \boldsymbol{H} \mid x_n\rangle \\ &= \left(\frac{1}{\sqrt{2}} \sum_{y_1=0}^{1} (-1)^{x_1\cdot y_1} \mid y_1\rangle\right) \otimes \left(\frac{1}{\sqrt{2}} \sum_{y_2=0}^{1} (-1)^{x_2\cdot y_2} \mid y_2\rangle\right) \otimes \ldots \\ & \quad \otimes \left(\frac{1}{\sqrt{2}} \sum_{y_n=0}^{1} (-1)^{x_n\cdot y_n} \mid y_n\rangle\right) \\ &= \frac{1}{2^{n/2}} \sum_{y_1 y_2 \ldots y_n} (-1)^{x_1\cdot y_1} (-1)^{x_2\cdot y_2} \ldots (-1)^{x_n\cdot y_n} \mid y_1 y_2 \ldots y_n\rangle \\ &= \frac{1}{2^{n/2}} \sum_{\boldsymbol{y}=\boldsymbol{0}}^{2^n-1} (-1)^{\boldsymbol{x}\cdot\boldsymbol{y}} \mid \boldsymbol{y}\rangle, \end{aligned} \quad (7.18)$$

where

$$\boldsymbol{x} \cdot \boldsymbol{y} = x_1 \cdot y_1 +_2 x_2 \cdot y_2 +_2 \ldots +_2 x_n \cdot y_n. \quad (7.19)$$

Now we combine the result with the outcome of step 2, given by equation (7.14), and obtain

$$\begin{aligned} & \frac{1}{2^{n/2}} \left(\sum_{\boldsymbol{x}=\boldsymbol{0}}^{2^n-1} (-1)^{f(\boldsymbol{x})} \bigotimes^{n} \boldsymbol{H} \mid \boldsymbol{x}\rangle\right) \otimes \frac{1}{\sqrt{2}}(\mid 0\rangle - \mid 1\rangle) \\ &= \frac{1}{2^{n/2}} \left(\sum_{\boldsymbol{x}=\boldsymbol{0}}^{2^n-1} (-1)^{f(\boldsymbol{x})} \frac{1}{2^{n/2}} \sum_{\boldsymbol{y}=\boldsymbol{0}}^{2^n-1} (-1)^{\boldsymbol{x}\cdot\boldsymbol{y}} \mid \boldsymbol{y}\rangle\right) \otimes \frac{1}{\sqrt{2}}(\mid 0\rangle - \mid 1\rangle) \\ &= \frac{1}{2^{n}} \left(\sum_{\boldsymbol{x}=\boldsymbol{0}}^{2^n-1} \sum_{\boldsymbol{y}=\boldsymbol{0}}^{2^n-1} (-1)^{f(\boldsymbol{x})} (-1)^{\boldsymbol{x}\cdot\boldsymbol{y}} \mid \boldsymbol{y}\rangle\right) \otimes \frac{1}{\sqrt{2}}(\mid 0\rangle - \mid 1\rangle). \end{aligned} \quad (7.20)$$

4. Finally, we come to the measurement. If $f(x)$ is constant, the $(-1)^{f(x)}$ factor can be moved in front of the sums, leaving

$$\sum_{\boldsymbol{x}=\boldsymbol{0}}^{2^n-1}\sum_{\boldsymbol{y}=\boldsymbol{0}}^{2^n-1}(-1)^{\boldsymbol{x}\cdot\boldsymbol{y}}\mid\boldsymbol{y}\rangle \tag{7.21}$$

inside. Let us fix $\boldsymbol{y}$ at some value. If $\boldsymbol{y}\neq\boldsymbol{0}$, then

$$\sum_{\boldsymbol{x}=\boldsymbol{0}}^{2^n-1}(-1)^{\boldsymbol{x}\cdot\boldsymbol{y}} \tag{7.22}$$

must be zero, because $\boldsymbol{x}\cdot\boldsymbol{y}$ will "push as often to the right as to the left." Hence, the only term that is going to survive in this case is for $\boldsymbol{y}=\boldsymbol{0}$. Consequently, in this case the final state of the oracle is going to be

$$\begin{aligned}&\frac{1}{2^n}(-1)^{f(\boldsymbol{x})}\left(\sum_{\boldsymbol{x}=\boldsymbol{0}}^{2^n-1}(-1)^{\boldsymbol{x}\cdot\boldsymbol{0}}\mid\boldsymbol{0}\rangle\right)\otimes\frac{1}{\sqrt{2}}(\mid 0\rangle-\mid 1\rangle)\\&=(-1)^{f(\boldsymbol{x})}\mid\boldsymbol{0}\rangle\otimes\frac{1}{\sqrt{2}}(\mid 0\rangle-\mid 1\rangle)\,.\end{aligned} \tag{7.23}$$

On the other hand, if $f(\boldsymbol{x})$ is balanced, then for $\mid\boldsymbol{y}\rangle=\mid\boldsymbol{0}\rangle$ we get

$$\sum_{\boldsymbol{x}=\boldsymbol{0}}^{2^n-1}(-1)^{f(\boldsymbol{x})}(-1)^{\boldsymbol{x}\cdot\boldsymbol{0}}\mid\boldsymbol{0}\rangle=\sum_{\boldsymbol{x}=\boldsymbol{0}}^{2^n-1}(-1)^{f(\boldsymbol{x})}\mid\boldsymbol{0}\rangle=0 \tag{7.24}$$

because $f(\boldsymbol{x})$ pushes as often to the right as it pushes to the left, on account of being balanced. Here, then, the probability amplitude of finding $\mid\boldsymbol{y}\rangle$ in the $\mid\boldsymbol{0}\rangle$ state is zero.

In summary, if $f(\boldsymbol{x})$ is constant, then measuring control lines on exit *must* return $\mid 0\rangle$ on every line. If this is not the case, then $f(\boldsymbol{x})$ is balanced.

7.2 NMR Computing

Another way of looking at $\sum_{\boldsymbol{x}=0}^{2^n-1}\mid\boldsymbol{x}\rangle$

At this juncture, an astute reader who took to heart our admonitions about the superstition of superposition and our comments on individual qubits being computationally equivalent to a point within a ball of radius 1 may well ask the question, "What's all this business of

$$\boldsymbol{H}\mid 0\rangle\otimes\boldsymbol{H}\mid 0\rangle\otimes\cdots\otimes\boldsymbol{H}\mid 0\rangle=\frac{1}{2^{n/2}}\sum_{\boldsymbol{x}=0}^{2^n-1}\mid\boldsymbol{x}\rangle \tag{7.25}$$

about? Isn't $\boldsymbol{H} \mid 0\rangle$ just a qubit with its polarization turned sideways, $|\rightarrow\rangle$? And if so, isn't the above just $|\rightarrow\rangle\otimes |\rightarrow\rangle \otimes \cdots \otimes |\rightarrow\rangle$?"

Of course, it is. But, as we have seen in Chapter 4, a qubit can be described in various ways, by using various languages, all of them to a certain degree equivalent. The unitary description in terms of basis states $\mid 0\rangle$ and $\mid 1\rangle$ and their normalized complex-valued superpositions conveys exactly the same amount of information about the state of a qubit as its fiducial description in terms of a four-component real-valued vector of probabilities, assuming that the qubit is in a pure state. If not, then, sadly, the unitary formalism becomes speechless, unless we entangle the qubit with its environment.

Quantum computing as an artifact of unitary notation

This leads us to the startling conclusion that the idea of quantum computation may be an artifact of the notation commonly used in quantum physics. But this is fine. A great many human insights and discoveries are artifacts of formalisms and notations that people use. This is what makes some notations better than others.

Can a Babbage machine for quantum computing be constructed?

Still, holding on to the above, one could ask whether the Deutsch-Jozsa algorithm could not be implemented by using, say, marked ping-pong balls instead of qubits, and some clever mechanism that would rotate them in various ways, reflecting the actions of quantum gates. At the end we could look at the ping-pong balls that correspond to the control qubits in the Deutsch-Jozsa circuit, and if they all have their marks pointing, say, downwards, then the mysterious function $f(x)$ would be constant. This would be a kind of a Babbage machine [6] for quantum computing.[2]

A natural machine of this kind exists, called a nuclear magnetic resonance (NMR) computer. It is shown in Figure 7.6.

Success of NMR computing

Nuclear magnetic resonance experiments have been in the forefront of quantum computing ever since the idea of a *qubit* was first proposed. NMR practitioners immediately recognized that manipulating qubits was what they'd been doing for years and years—it was just that no one had called it "quantum computing." By now, several well-known quantum algorithms, including the famous Brassard teleportation circuit [10, 15], about which more in the next section, and the Shor algorithm for factoring integers [127], have been demonstrated by using NMR computers. In 1998 Nielsen, Knill, and Laflamme used an NMR computer to teleport a

[2]Not all quantum algorithms make use of qubit entanglement. Where it is present, its classical implementation may have to be more involved. But it is possible to fake entanglement as well, as we are going to discuss in Section 7.5. It is also possible to model the effect of a local measurement made on an entangled system of qubits, which is computationally more important, by transmitting information to the remaining ping-pong balls and repositioning them according to what the entangled state requires. Although this is not instantaneous in the macroscopic world, it may be fast enough within a sufficiently small system, especially given that quantum gates and quantum measurements are not in themselves instantaneous processes either.

Figure 7.6: An 18.8 T, 800 MHz nuclear magnetic resonance spectrometer at the Pacific Northwest National Laboratory in Kennewick, Washington. A public domain photograph from Wikimedia Commons.

quantum state from a ^{13}C nucleus to the hydrogen nucleus in molecules of labeled trichloroethylene [98], and Vandersypen, Steffen, Bryta, Yannoni, Sherwood, and Chuang used an NMR computer to factorize the number 15 [138]—and it did turn out to be 3×5! In turn, the new field of quantum computing helped redesign and improve various NMR procedures with broad applications beyond the field of quantum information processing. So this has been a happy and fruitful marriage for both.

Is NMR computing quantum enough?

Alas, such happiness often attracts spoilers, who in this case accused NMR computing of not being "quantum" enough. Some even went as far as to state that "if something can be understood in simple terms like spinning tops, it cannot be quantum." (They, of course, are great devotees of Magic.) The distinguished NMR practitioners Milburn, Laflamme, Sanders, and Knill responded by scrutinizing NMR procedures and demonstrated that the dynamics of NMR systems was genuinely "quantum," meaning "not derivable from principles of classical mechanics," even if the actual measurement and handling of NMR samples were macroscopic [93]. This is no different from, for example, ion beam experiments that produce a typical quantum diffraction image. The beam is macroscopic, the fringes are macroscopic, and the double slit is macroscopic, whereas the dynamics that produces the fringes is "quantum."

To show how NMR computing can be both "quantum" and "classical analog" at the same time, we now take a closer look at NMR procedures. An NMR experiment involves a very large number of magnetically active molecules. It is a macroscopic sample that is normally dissolved in a suitable magnetically inactive solvent, such as deuterated tetrahydrofuran (C_4D_8O) or deuterated acetone (CD_3COCD_3)—about a dozen solvents are used commonly. The dilution must be high in order to break molecule-molecule interactions, but the resulting number may be still about 100 million molecules in a sample of one cubic centimeter volume. For smaller concentrations the produced signal may be too weak to detect.

NMR solvents

A typical molecule used in an NMR measurement may comprise a number of protons, all of which are magnetically active (which is why the solvents are deuterated) and produce an NMR signal at about 500 MHz in a magnetic field of about 12 T, although in most quantum computing experiments active nuclei other than protons are used. The reason is that there are normally just a few (or one) of ^{13}C, ^{19}F, ^{15}N, or ^{31}P in a molecule of interest, and it is easier to separate them from one another. Qubits are associated with these nuclei, more precisely, with their magnetic moments, each molecule constituting a separate quantum register, and the whole sample representing a statistical ensemble of quantum registers. The size of the register depends on the specific molecule chosen to carry out the computation and on the number of magnetically active nuclei within it that can be manipulated individually.

Magnetically active nuclei

For example, the molecule that was used by Nielsen, Knill, and Laflamme in their teleportation experiment [98] was trichloroethylene, $^{13}C_2HCl_3$, which has six atoms. The Brassard teleportation circuit [15] deployed in the computation used only three qubits. The nuclei chosen to construct the three-qubit register were the two carbon-13 nuclei, resonating at 125.772580 MHz and 125.771669 MHz, respectively, and the hydrogen nucleus, resonating at 500.133491 MHz, the three chlorine nuclei remaining unused—but not useless.

Magnetic resonance frequencies corresponding to even identical nuclei in a molecule, for example, the two carbon atoms in trichloroethylene, may differ by between a few kilohertz to a few hundred kilohertz depending on their position within the molecule. The differences, called *chemical shifts,* are caused by the presence of local magnetic fields generated by electron shells within the molecule. The fields vary from place to place, which is how inferences can be made about the structure of a molecule by looking at its NMR spectrum. The same chemical shifts help us separate messages transmitted to individual qubits. In trichloroethylene one of the carbon atoms is connected to two chlorine atoms, and the other one is con-

Chemical shifts

nected to one chlorine atom and one hydrogen atom. This configuration is enough to produce the 0.911 kHz difference between their nuclear resonances—sufficient to separate messages directed to the two carbon nuclei from one another.

The heart of an NMR computer is a superconducting magnet that generates a highly uniform magnetic field within a small region of about $1\,\mathrm{cm}^3$. The sample must fit within this space to ensure that all molecules are placed within the field of the same direction and strength. The uniformity of the field can be ensured to within 1 part per billion. Nuclei in the molecules respond by Larmor-precessing their spin polarizations about the direction of the magnetic field, exactly as we saw in Chapter 2.

NMR agitation

Helmholtz coils are then used to generate small oscillating magnetic fields perpendicular to the direction of the background magnetic field. As a result, the nuclear magnetic moments flip between the *up* and *down* configurations, exactly as we saw in Section 2.11. The fields can be pulsed rapidly. But ensuring the homogeneity of this radio frequency field is extremely difficult because the Helmholtz coils are much smaller than the superconducting coils that generate the background field.

NMR read-out

The same Helmoltz coils are also used to pick up radio frequency fields generated by the precessing nuclei, which is how the system is read. The coils convert the radio frequency field to a decaying voltage signal, which is recorded and then analyzed by using Fourier transform. The decay of the signal is related to the decoherence of nuclear spin states, and frequency peaks read from the transform correspond directly to spin resonances of activated nuclei, with areas under the peaks related to spin states.

The first complication of NMR is that the hundreds of millions of molecules of the target substance all float around and tumble in the solvent pretty much chaotically. The ensemble is not in a pure state. It is not like a fully polarized neutron beam.

Thermal equilibrium state

At any given temperature T—and these measurements are normally carried out at room temperature in order to keep the solvent liquid—the density operator of the ensemble in the thermal equilibrium state is given by

$$\boldsymbol{\rho} = \frac{e^{-\boldsymbol{H}/(kT)}}{\mathrm{Tr}\left(e^{-\boldsymbol{H}/(kT)}\right)}, \tag{7.26}$$

where $k \approx 8.617 \times 10^{-5}\,\mathrm{eV/K}$ is the Boltzmann constant and $\boldsymbol{H}$ is the system's Hamiltonian. At high temperature, where "high" means that every eigenvalue of $\boldsymbol{H}$ is much smaller than kT, we can use the approximation

$$e^{-\boldsymbol{H}/(kT)} \approx \mathbf{1} - \frac{\boldsymbol{H}}{kT}. \tag{7.27}$$

Since the trace is a matrix invariant, we can evaluate it in the eigenbasis of $\boldsymbol{H}$, in which case it is going to be

$$\begin{aligned} &\operatorname{Tr}\left(e^{-\boldsymbol{H}/(kT)}\right) \\ &\approx \operatorname{Tr}\left(\mathbf{1} - \frac{\boldsymbol{H}}{kT}\right) = \left(1 - \frac{E_1}{kT}\right) + \left(1 - \frac{E_2}{kT}\right) + \cdots + \left(1 - \frac{E_n}{kT}\right) \\ &\approx N, \end{aligned} \tag{7.28}$$

where E_1, E_2, ..., and E_n are the eigenvalues of $\boldsymbol{H}$ and N is the total dimension of the system, which for an n-qubit register is 2^n. We have made use here of $E_i/(kT) \ll 1$.

In summary,

$$\boldsymbol{\rho} \approx \frac{1}{2^n}\left(\mathbf{1} - \frac{\boldsymbol{H}}{kT}\right). \tag{7.29}$$

NMR Hamiltonian

The second complication is the effective form of the Hamiltonian. The real and full quantum Hamiltonian that describes the dynamics of nuclear magnetic moments in a molecule arises from the sum of all interactions within the molecule between nuclear spins and the interactions of the spins with externally applied magnetic fields. The nuclear spins are coupled with one another directly, each exerting its own dipolar magentic field on all other nuclear spins within the molecule and, indirectly, through the electron cloud that envelops and binds the molecule. The former averages away as the molecules tumble in the liquid, so it is not seen in the (macroscopic) NMR measurements. The latter survives and assumes the following form for a 2-qubit interaction:

$$\boldsymbol{H}_{AB} = \frac{\hbar J_{AB}}{4}\left(\boldsymbol{\sigma}_{xA} \otimes \boldsymbol{\sigma}_{xB} + \boldsymbol{\sigma}_{yA} \otimes \boldsymbol{\sigma}_{yB} + \boldsymbol{\sigma}_{zA} \otimes \boldsymbol{\sigma}_{zB}\right), \tag{7.30}$$

where J_{AB} is a coupling coefficient, which is a tensor in general. This Hamiltonian should be compared with the Khaneja-Glaser Hamiltonian (6.27), page 296. Indeed, their choice in computations was to use free evolution of NMR registers, as given in part by (7.30).

Hence, we end up with an impure statistical ensemble of hundreds of millions of tumbling molecular registers and with a simplified "NMR" Hamiltonian that is the product of macroscopic averaging. Yet, this is quite enough to demonstrate a broad range of quantum computations. The simplified Hamiltonian is a blessing in disguise, because it facilitates operations on the registers, while preserving enough quantumness to let us implement any quantum algorithm. And the large number of the registers delivers excellent statistics, which would be hard to generate in single-atom systems such as the Cirac-Zoller gate.

Pseudo-pure states

Pure states are generated by post factum data manipulation. The physical operations themselves are always carried out by taking (7.26) as the starting point, but a simple procedure exists that lets us extract from the data obtained what the result *would be* if the operations were performed on a pure state.

The nuclear spin coupling term in the NMR Hamiltonian, given by (7.30), is small in comparison with the $\sum_i \boldsymbol{\mu}_i \boldsymbol{B}$ term. The resulting Hamiltonian and the density operator $\boldsymbol{\rho}$ are almost diagonal. It is therefore a fair approximation to assume that the initial density matrix of the ensemble is given by

$$\boldsymbol{\rho}(0) = \begin{pmatrix} \rho_{00} & 0 & 0 & 0 & \cdots \\ 0 & \rho_{11} & 0 & 0 & \cdots \\ 0 & 0 & \rho_{22} & 0 & \cdots \\ 0 & 0 & 0 & \rho_{33} & \cdots \\ \vdots & \vdots & \vdots & \vdots & \ddots \end{pmatrix}. \tag{7.31}$$

In Section 6.2.4, we saw how various permutations of the register would arise naturally by application of the Toffoli gate, which in turn could be implemented by application of controlled-NOT and controlled-$\sqrt{\neg}$ (square root of NOT) gates, where the latter could be expressed in terms of NOTs, too (see equations (6.83) and (6.86)). When a permutation is applied to $\boldsymbol{\rho}(0)$, the result is a permuted population.

Let us consider, for simplicity, a biqubit system. The density matrix is only 4×4. We can generate the following new state $\boldsymbol{\rho}_1$ by applying a cyclic permutation that replaces $(|\,1\rangle, |\,2\rangle, |\,3\rangle)$ with $(|\,2\rangle, |\,3\rangle, |\,1\rangle)$:

$$\boldsymbol{\rho}_1 = \boldsymbol{P}\boldsymbol{\rho}(0)\boldsymbol{P}^\dagger = \begin{pmatrix} \rho_{00} & 0 & 0 & 0 \\ 0 & \rho_{22} & 0 & 0 \\ 0 & 0 & \rho_{33} & 0 \\ 0 & 0 & 0 & \rho_{11} \end{pmatrix}. \tag{7.32}$$

We can also generate another state, $\boldsymbol{\rho}_2$, by applying a reverse permutation, as follows:

$$\boldsymbol{\rho}_2 = \boldsymbol{P}^{-1}\boldsymbol{\rho}(0)\left(\boldsymbol{P}^{-1}\right)^\dagger = \boldsymbol{P}^\dagger\boldsymbol{\rho}(0)\boldsymbol{P} = \begin{pmatrix} \rho_{00} & 0 & 0 & 0 \\ 0 & \rho_{33} & 0 & 0 \\ 0 & 0 & \rho_{11} & 0 \\ 0 & 0 & 0 & \rho_{22} \end{pmatrix}. \tag{7.33}$$

A quantum computing program $\boldsymbol{U}$ executed on $\boldsymbol{\rho}(0)$ generates $\boldsymbol{U}\boldsymbol{\rho}(0)\boldsymbol{U}^\dagger$. Let us execute it also on $\boldsymbol{\rho}_1$ and then on $\boldsymbol{\rho}_2$ and add the results on paper:

$$\boldsymbol{U}\boldsymbol{\rho}(0)\boldsymbol{U}^\dagger + \boldsymbol{U}\boldsymbol{\rho}_1\boldsymbol{U}^\dagger + \boldsymbol{U}\boldsymbol{\rho}_2\boldsymbol{U}^\dagger$$

$$
\begin{aligned}
&= \boldsymbol{U}\left(\boldsymbol{\rho}(0) + \boldsymbol{\rho}_1 + \boldsymbol{\rho}_2\right)\boldsymbol{U}^\dagger \\
&= \boldsymbol{U}\begin{pmatrix} 3\rho_{00} & 0 & 0 & 0 \\ 0 & \rho_{11}+\rho_{22}+\rho_{33} & 0 & 0 \\ 0 & 0 & \rho_{11}+\rho_{22}+\rho_{33} & 0 \\ 0 & 0 & 0 & \rho_{11}+\rho_{22}+\rho_{33} \end{pmatrix}\boldsymbol{U}^\dagger .
\end{aligned}
\tag{7.34}
$$

Ah, but $\rho_{11} + \rho_{22} + \rho_{33} = 1 - \rho_{00}$, because $\boldsymbol{\rho}(0)$ is a density matrix, so its trace must be 1. Consequently, the result is

$$
\boldsymbol{U}\begin{pmatrix} 3\rho_{00} & 0 & 0 & 0 \\ 0 & 1-\rho_{00} & 0 & 0 \\ 0 & 0 & 1-\rho_{00} & 0 \\ 0 & 0 & 0 & 1-\rho_{00} \end{pmatrix}\boldsymbol{U}. \tag{7.35}
$$

Observing that $3\rho_{00} = 1 - \rho_{00} + 4\rho_{00} - 1$ and that $\boldsymbol{U}\boldsymbol{1}\boldsymbol{U}^\dagger = \boldsymbol{1}$, we can wrap the above into

$$
\left(4\rho_{00} - 1\right)\boldsymbol{U}\begin{pmatrix} 1 & 0 & 0 & 0 \\ 0 & 0 & 0 & 0 \\ 0 & 0 & 0 & 0 \\ 0 & 0 & 0 & 0 \end{pmatrix}\boldsymbol{U}^\dagger + \left(1 - \rho_{00}\right)\boldsymbol{1}. \tag{7.36}
$$

The second term here is not observable in the NMR systems because the Helmholtz coils pick up magnetization signals only in the x and y directions.[3] What is therefore going to be read is

$$
(4\rho_{00} - 1)\boldsymbol{U} \mid 00\rangle\langle 00 \mid \boldsymbol{U}^\dagger . \tag{7.37}
$$

And so, we have our program $\boldsymbol{U}$ executed on a pure state, even if this result has been concocted by juggling data generated by several computational procedures after their completion. For larger registers, more juggling is needed, but the procedure is much the same, just longer.

Purists discombobulated

This impels wrath in some quantum computing purists, who exclaim against such trickery. But all that we really do here is to reconstruct a specific, unquestionably quantum, computational procedure. It does not matter that the samples and measurements are all macroscopic (carried out at room temperature), in the same way that it does not matter if the Babbage machine computes by using metal gears instead of transistor switches. The logic of the computation is the same.

[3]What is eventually seen on the output voltage of an NMR system is $V_0 \mathrm{Tr}\left(\boldsymbol{\rho}(t)\boldsymbol{M}_k\right)$, where the kth nucleus observable $\boldsymbol{M}_k$ is $\boldsymbol{\sigma}_x$ or $\boldsymbol{\sigma}_y$ or a combination thereof and $\boldsymbol{\rho}(t)$ is the density operator of the system evolving freely (and decohering) after the Helmholtz coils have been switched from agitation to listening. Obviously, $\mathrm{Tr}\left(\boldsymbol{1}\boldsymbol{\sigma}_x\right) = \mathrm{Tr}\left(\boldsymbol{1}\boldsymbol{\sigma}_y\right) = 0$, which is why the coils don't see the second term.

Refocusing

The third complication of NMR is that nuclear spins in the NMR registers evolve freely all the time, as described by their free Hamiltonians, that is, Hamiltonians without the Helmholtz coil forcing, whether it is desirable, as is used by Khaneja-Glaser decompositions, or not. An additional procedure is therefore required that would halt or reverse this evolution when needed.

The situation can be compared to that of a class full of unruly children. Whenever an activity slows or stops, children's attention inevitably begins to drift, whereupon the firm action of a teacher is needed to bring them back into focus. In the days of my distant youth one of my best teachers used to bang her fist on the table and shout, "Ich bitte um Ruhe," whereupon all children would hush, their attention perked up.

And we do exactly the same to refocus NMR systems.

What does the trick this time is the commutation relation

$$\boldsymbol{\sigma}_x \boldsymbol{\sigma}_z = -\boldsymbol{\sigma}_z \boldsymbol{\sigma}_x. \tag{7.38}$$

Multiplying it by $\boldsymbol{\sigma}_x$ from the right yields

$$\boldsymbol{\sigma}_x \boldsymbol{\sigma}_z \boldsymbol{\sigma}_x = -\boldsymbol{\sigma}_z. \tag{7.39}$$

How does this work? With Helmholtz coil activation switched off, the individual spin Hamiltonian is $-\mu B_z \boldsymbol{\sigma}_z$, and the imparted evolution pushed by

$$\boldsymbol{U} = e^{-\mathrm{i}\boldsymbol{H}t/\hbar} = e^{\mathrm{i}\mu B_z t \boldsymbol{\sigma}_z/\hbar}. \tag{7.40}$$

Let us recall equations (4.300) and (4.301), page 156, from Chapter 4. When applied to equation (7.39), they imply that

$$\boldsymbol{\sigma}_x e^{\mathrm{i}\mu B_z t \boldsymbol{\sigma}_z/\hbar} \boldsymbol{\sigma}_x = e^{-\mathrm{i}\mu B_z t \boldsymbol{\sigma}_z/\hbar}. \tag{7.41}$$

This lets us reverse the free evolution of every spin by sending the NOT signals to it. These are the "Ich bitte um Ruhe" requests that refocus the NMR registers.

Decoupling

The fourth complication of NMR is that nuclear spins in a molecular register are always coupled. Yet, most operations deployed by quantum algorithms are performed on individual qubits, only occasionally invoking a coupling when a controlled-NOT gate is required. Again a tricky NMR procedure comes to the rescue. The procedure derives from refocusing.

Let us suppose that we operate on a biqubit molecular register. We can let both qubits evolve for a certain amount of time, Δt. Then we can send a NOT gate to one of the qubits and let both evolve for time Δt again. Finally, we send another

NOT to the same qubit we have shouted at already. The result is such that the undisturbed qubit will have evolved by $\exp\left(\mathrm{i}\mu B_z 2\Delta t \boldsymbol{\sigma}_z/\hbar\right)$, whereas the disturbed one will have evolved by

$$e^{\mathrm{i}\mu B_z \Delta t \boldsymbol{\sigma}_z/\hbar} e^{\mathrm{i}\mu B_z(-\Delta t)\boldsymbol{\sigma}_z/\hbar} = \mathbf{1}, \tag{7.42}$$

meaning that it will not have evolved at all. By doing this we have decoupled the qubits from one another. One evolves; the other one doesn't.

NMR is an old experimental technique that dates back to the work of Felix Bloch and Edward Mills Purcell in 1946. The work was a by-product of various war-time research activities, including microwave radars and atomic energy, and delivered a veritable miracle of immense usefulness. For this accomplishment Bloch and Purcell were awarded the 1952 Nobel Prize in physics. *Nobel Prize for NMR*

A nicer way to say *old* is to say *mature* and *experienced*. By now, sixty years later, the method is so sophisticated that a trick like any of the above exists for everything. No wonder, then, that NMR has been more successful than just about any other approach in demonstrating quantum computing algorithms. NMR implementation of a controlled-NOT gate between qubits A and B goes back to 1990s. The following was proposed by Leung, Chuang, Yamaguchi, and Yamamoto [87]: *controlled-NOT*

$$\begin{aligned}
&\text{controlled-NOT}_{AB} = \\
&\quad \exp\left(-\mathrm{i}\frac{\pi}{4}\boldsymbol{\sigma}_{yA}\right)\exp\left(\mathrm{i}\frac{\pi}{4}\boldsymbol{\sigma}_{xA}\right)\exp\left(\mathrm{i}\frac{\pi}{4}\boldsymbol{\sigma}_{yA}\right) \\
&\quad \times \exp\left(-\mathrm{i}\frac{\pi}{4}\boldsymbol{\sigma}_{xB}\right)\exp\left(\mathrm{i}\frac{\pi}{4}\boldsymbol{\sigma}_{yB}\right) \\
&\quad \times \exp\left(-\mathrm{i}\frac{\pi}{4}\boldsymbol{\sigma}_{zA}\otimes\boldsymbol{\sigma}_{zB}\right) \quad \leftarrow \quad \text{both qubits evolve freely} \\
&\quad \times \exp\left(-\mathrm{i}\frac{\pi}{4}\boldsymbol{\sigma}_{yB}\right).
\end{aligned} \tag{7.43}$$

Just such a collection of controlled-NOT gates, in combination with many other NMR techniques, was used by Vandersypen, Steffen, Breyta, Yannoni, Sherwood, and Chuang in their remarkable demonstration of the Shor factoring algorithm in 2001, which we have mentioned above briefly [138]. *The molecule used to factor number 15*

In order to factor the number 15, a seven-qubit quantum register was needed. A new molecule, shown in Figure 7.7, a pentafluorobutadienyl cyclopentadienyldicarbonyl-iron complex ($C_{11}H_5F_5O_2Fe$) with the inner two carbons ^{13}C-labeled, was designed and synthesized specially for this purpose at IBM. The vial sample contained 10^{18} molecules of it. To minimize the impact of decoherence, the experimenters implemented and used special quantum error correction routines,

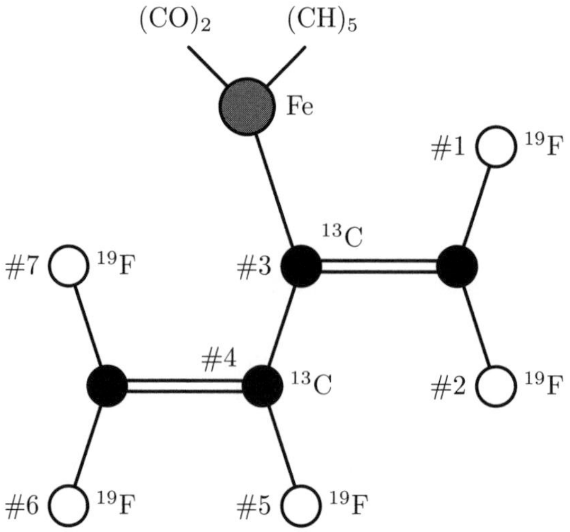

Figure 7.7: The 7-qubit molecule of a pentafluorobutadienyl cyclopentadienyl-dicarbonyl-iron complex specially designed for the quantum factoring of number 15. Magnetically active nuclei used in the computation are labeled #1 through #7.

which represented, in the words of Isaac Chuang, the project's most important accomplishment [68].

This remains the most elaborate quantum computation demonstrated so far.

7.3 Brassard Teleportation Circuit

Before we progress, it is instructive to see how NMR works, by discussing a relatively simple example, the already mentioned Brassard teleportation circuit.

No-cloning theorem

One of the first facts of life that quantum computing students learn is that quantum information cannot be copied. This is the subject of the much celebrated, although somewhat misunderstood, no cloning theorem of Wootters, Zurek, and Dieks [148, 32]. The theorem is stated as follows. Let us suppose we have a combined state, not necessarily a biqubit one, $\mid \psi\rangle_A \otimes \mid \eta\rangle_B$. Is there a unitary operation that would convert it into $\mid \psi\rangle_A \otimes \mid \psi\rangle_B$ for an arbitrary $\mid \psi\rangle_A$, thus copying ψ from A to B? We can immediately see, even without attempting any computation, that such an operation cannot be procured because it would destroy information stored on $| \eta\rangle_B$. Consequently it would not be a reversible operation, and all unitary operations must be reversible.

A simple proof for pure states loooks as follows. Let us assume that such an operation does exist, and let us call it $\boldsymbol{U}_C$. Because it must work for an arbitrary $| \psi\rangle_A$, we can replace ψ with some other ϕ and write that

$$\begin{aligned} \boldsymbol{U}_C \,| \psi\rangle_A \otimes | \eta\rangle_B &= | \psi\rangle_A \otimes | \psi\rangle_B \quad \text{and} && (7.44)\\ \boldsymbol{U}_C \,| \phi\rangle_A \otimes | \eta\rangle_B &= | \phi\rangle_A \otimes | \phi\rangle_B . && (7.45)\end{aligned}$$

Taking a dual of the second equation and contracting it with the first one, we obtain

$$\langle \eta \,|_B \,\langle \phi \,|_A \,\boldsymbol{U}_C^\dagger \boldsymbol{U}_C \,| \psi\rangle_A \,| \eta\rangle_B = \left(\langle \phi \,|_B \,\langle \phi \,|_A\right)\left(| \psi\rangle_A \,| \psi\rangle_B\right). \qquad (7.46)$$

Hence, contracting A kets with A bras and B kets with B bras,

$$\langle \phi \mid \psi\rangle_A = \langle \phi \mid \psi\rangle_A \langle \phi \mid \psi\rangle_B . \qquad (7.47)$$

For this to be true for arbitrary ψ and ϕ, we must have that $\langle \phi \mid \psi\rangle_B = 1$ or that $\langle \phi \mid \psi\rangle_A = 0$, which is not going to be the case in general, and hence implies that the $\boldsymbol{U}_C$ sought does not exist.

Extension to arbitrary mixtures

It is often argued, with much hand-waving, that the result extends trivially to mixtures and nonunitary quantum operations by the virtue of purification. But this is not true, and only in November 2007 was a proof of nonclonability of quantum information demonstrated for general density matrices [74]. The proof by Kalev and Hen is based on entropic considerations, which is a fundamental principle of information theory, and so it extends to the classical analog of the no-cloning theorem, which states that "an arbitrary probability distribution associated with a given source system cannot be copied onto another target system while leaving the original distribution of the source system unperturbed" [26].

Classical no-cloning theorem

How, then, can we learn and reproduce items such as cars, pictures, or, for that matter, this very book? I can easily imagine a smart-aleck student approaching her Latin teacher and saying, "Madam, because of the classical no cloning theorem of Daffertshofer and the Plastinos it was utterly impossible for me to memorize the first paragraph of Caesar's *De Bello Gallico*." Is the student right?

As every teacher knows, a student with this kind of excuse is never right.

Although my car may be an exact replica of the one in my neighbor's garage, it is not identical down to every atom and its quantum state. I know this because my driver seat squeaks and his doesn't. What we commonly perceive as macroscopic information is not dependent on the exact atomic configuration of the system it

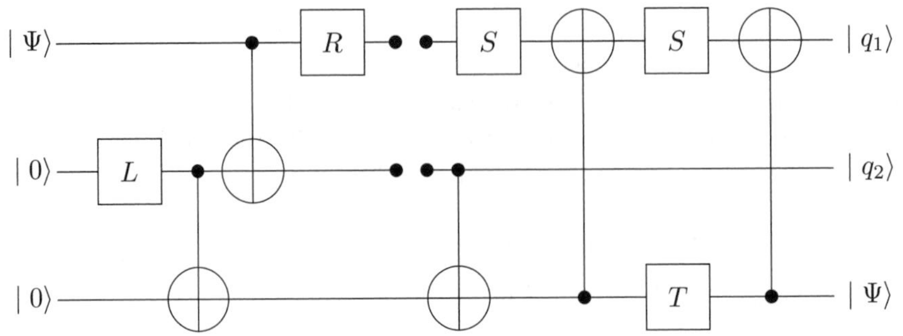

Figure 7.8: The Brassard teleportation circuit.

is written on. The student, with some effort, may learn Caesar's *Gallia est omnis divisa in partes tres...* [20], but the neural connections in the student's brain, on which these words are encoded, will look quite different from the ones in mine. Hers will be younger, for starters.[4]

Imperfect cloning

These and similar observations have been quantified recently by Walker and Braunstein, who demonstrated that probability distributions can be copied classically with arbitrarily high fidelities for any *finite* resolution, this being a measure of difference between the original and its slightly distorted copy [144]. Art forgers may breathe a sigh of relief. When the notions of fidelity and resolution are brought into play in the quantum world, quantum information can be copied, too [19].

Teleportation circuit

But in the teleportation circuit shown in Figure 7.8 we do not copy information. Instead, we *move* it from the first qubit, the one on top, to the third qubit, the one on the bottom. In the process, information that was originally stored on the first qubit gets wiped out. This is allowed, and it does not violate the no cloning theorem.

The single-qubit gates $\boldsymbol{L}$, $\boldsymbol{R}$, $\boldsymbol{S}$, and $\boldsymbol{T}$ used in the circuit have the following matrix definitions in the computational basis $|\,0\rangle = \binom{1}{0}$ and $|\,1\rangle = \binom{0}{1}$:

$$\boldsymbol{L} = \frac{1}{\sqrt{2}}\begin{pmatrix} 1 & -1 \\ 1 & 1 \end{pmatrix}, \tag{7.48}$$

$$\boldsymbol{R} = \frac{1}{\sqrt{2}}\begin{pmatrix} 1 & 1 \\ -1 & 1 \end{pmatrix}, \tag{7.49}$$

[4]The abstraction of information from the physical carriers on which it is conveyed lies behind the strategy of encoding bit and qubit values in order to protect them from corruption brought about by the environment.

$$\boldsymbol{S} = \begin{pmatrix} \mathrm{i} & 0 \\ 0 & 1 \end{pmatrix}, \tag{7.50}$$

$$\boldsymbol{T} = \begin{pmatrix} -1 & 0 \\ 0 & -\mathrm{i} \end{pmatrix}. \tag{7.51}$$

We can think of the gate $\boldsymbol{L}$ as implementing a 90° qubit rotation on the Bloch sphere "to the left"; then the gate $\boldsymbol{R}$ implements a 90° rotation "to the right." It is easy to see that $\boldsymbol{L} \cdot \boldsymbol{R} = \mathbf{1}$. Gates $\boldsymbol{S}$ and $\boldsymbol{T}$ represent a combination of rotations about the z axis with a multiplication by a fixed global phase-shift:

$$\boldsymbol{S} = \begin{pmatrix} \mathrm{i} & 0 \\ 0 & 1 \end{pmatrix} = e^{\mathrm{i}\pi/4} \begin{pmatrix} e^{\mathrm{i}\pi/4} & 0 \\ 0 & e^{-\mathrm{i}\pi/4} \end{pmatrix}, \tag{7.52}$$

$$\boldsymbol{T} = \begin{pmatrix} -1 & 0 \\ 0 & -\mathrm{i} \end{pmatrix} = e^{\mathrm{i}\pi/4} \begin{pmatrix} e^{\mathrm{i}3\pi/4} & 0 \\ 0 & e^{-\mathrm{i}3\pi/4} \end{pmatrix}. \tag{7.53}$$

Circuit analysis

Below, we analyze the functioning of the circuit step by step.

1. Inputs applied to the circuit are $|\,0\rangle$ to the two bottom lines and an arbitrary state $|\,\Psi\rangle = \binom{a}{b}$ applied to the top line. This is the state that is going to be teleported to the bottom line.

2. The first gate, $\boldsymbol{L}$, converts the input $|\,0\rangle$ to $\frac{1}{\sqrt{2}}(|\,0\rangle + |\,1\rangle)$. After the gate has been applied, the states on the three lines of the circuit are

$$|\,\Psi\rangle \otimes \frac{1}{\sqrt{2}}(|\,0\rangle + |\,1\rangle) \otimes |\,0\rangle. \tag{7.54}$$

3. The next step entangles the two bottom lines using the controlled-NOT gate. Here we use a subscript c to point to the control line and subscript t to point to the target line:

$$\begin{aligned} &\bigoplus \frac{1}{\sqrt{2}}(|\,0\rangle_c + |\,1\rangle_c)\,|\,0\rangle_t \\ &= \frac{1}{\sqrt{2}}\left(\bigoplus |\,0\rangle_c\,|\,0\rangle_t + \bigoplus |\,1\rangle_c\,|\,0\rangle_t\right) \\ &= \frac{1}{\sqrt{2}}(|\,0\rangle_c\,|\,0\rangle_t + |\,1\rangle_c\,|\,1\rangle_t). \end{aligned} \tag{7.55}$$

This is a new bipartite state, which now binds the two bottom lines together. Their state is maximally entangled. The state of the circuit at this stage is

$$|\,\Psi\rangle \otimes \frac{1}{\sqrt{2}}(|\,0\rangle\,|\,0\rangle + |\,1\rangle\,|\,1\rangle). \tag{7.56}$$

4. The next operation entangles $\mid \Psi\rangle$ with the middle qubit, which is already entangled with the bottom qubit. We continue with our convention, where the control qubit is marked with subscript c and the target qubit is marked with subscript t:

$$\begin{aligned}
&\bigoplus \mid \Psi\rangle_c \frac{1}{\sqrt{2}} (\mid 0\rangle_t \mid 0\rangle + \mid 1\rangle_t \mid 1\rangle) \\
&= \frac{1}{\sqrt{2}} \bigoplus (a \mid 0\rangle_c + b \mid 1\rangle_c) (\mid 0\rangle_t \mid 0\rangle + \mid 1\rangle_t \mid 1\rangle) \\
&= \frac{1}{\sqrt{2}} \Big(a \Big(\bigoplus (\mid 0\rangle_c \mid 0\rangle_t) \mid 0\rangle + \bigoplus (\mid 0\rangle_c \mid 1\rangle_t) \mid 1\rangle \Big) \\
&\quad + b \Big(\bigoplus (\mid 1\rangle_c \mid 0\rangle_t) \mid 0\rangle + \bigoplus (\mid 1\rangle_c \mid 1\rangle_t) \mid 1\rangle \Big) \Big) \\
&= \frac{1}{\sqrt{2}} \Big(a (\mid 0\rangle_c \mid 0\rangle_t \mid 0\rangle + \mid 0\rangle_c \mid 1\rangle_t \mid 1\rangle) \\
&\quad + b (\mid 1\rangle_c \mid 1\rangle_t \mid 0\rangle + \mid 1\rangle_c \mid 0\rangle_t \mid 1\rangle) \Big).
\end{aligned} \tag{7.57}$$

Now all three lines are entangled, and the computer is no longer in a state that will let us isolate any of the lines.

5. The black dots that interrupt the top and middle qubit lines are "projective measurements." The gap indicates "classical information transmission," and the following black dot indicates resetting the qubit according to the information received.

   The last operation before the measurement rotates the upper qubit "to the right." For ease of reading we mark the target qubit of this operation with subscript R:

$$\begin{aligned}
&\boldsymbol{R} \frac{1}{\sqrt{2}} \Big(a (\mid 0\rangle_R \mid 0\rangle \mid 0\rangle + \mid 0\rangle_R \mid 1\rangle \mid 1\rangle) \\
&\quad + b (\mid 1\rangle_R \mid 1\rangle \mid 0\rangle + \mid 1\rangle_R \mid 0\rangle \mid 1\rangle) \Big) \\
&= \frac{1}{\sqrt{2}} \Big(a (\boldsymbol{R} \mid 0\rangle_R \mid 0\rangle \mid 0\rangle + \boldsymbol{R} \mid 0\rangle_R \mid 1\rangle \mid 1\rangle) \\
&\quad + b (\boldsymbol{R} \mid 1\rangle_R \mid 1\rangle \mid 0\rangle + \boldsymbol{R} \mid 1\rangle_R \mid 0\rangle \mid 1\rangle) \Big) \\
&= \frac{1}{\sqrt{2}} \left(a \left(\frac{1}{\sqrt{2}} (\mid 0\rangle_R - \mid 1\rangle_R) \mid 0\rangle \mid 0\rangle + \frac{1}{\sqrt{2}} (\mid 0\rangle_R - \mid 1\rangle_R) \mid 1\rangle \mid 1\rangle \right) \right.
\end{aligned}$$

$$+b\left(\frac{1}{\sqrt{2}}(|\,0\rangle_R+|\,1\rangle_R)\,|\,1\rangle\,|\,0\rangle+\frac{1}{\sqrt{2}}(|\,0\rangle_R+|\,1\rangle_R)\,|\,0\rangle\,|\,1\rangle\right)\Bigg)$$

$$=\frac{1}{2}\Big(a\,(|\,0_R00\rangle-|\,1_R00\rangle+|\,0_R11\rangle-|\,1_R11\rangle)$$

$$+b\,(|\,0_R10\rangle+|\,1_R10\rangle+|\,0_R01\rangle+|\,1_R01\rangle)\Big). \tag{7.58}$$

6. Now we reach the "projective measurement." At this point the upper two qubits either are measured or are allowed to decohere naturally, as will be the case in the NMR experiment with trichloroethylene, with the result that they collapse jointly onto one of the following biqubit states: $|\,00\rangle$, $|\,01\rangle$, $|\,10\rangle$, or $|\,11\rangle$.

   This process forces the bottom qubit into a state that is commensurate with whatever the upper qubits become and with the original quantum state of all three qubits.

   To see what happens next, we need to carry out our analysis for all possible outcomes of the measurement. This is going to be tedious, so we choose just one possible outcome and go ahead with analyzing what happens in this case, leaving the analysis of the remaining three channels to the reader as an exercise.

   Let us suppose the upper two qubits decohere to $|\,01\rangle$. This process filters the state of the system into

$$a\,|\,011\rangle+b\,|\,010\rangle. \tag{7.59}$$

   Let us observe that every other outcome of the measurement on the upper two wires will produce a similar result, namely,

$$a\,|\,\text{something}\rangle+b\,|\,\text{something else}\rangle. \tag{7.60}$$

   No possible measurement outcome results in information loss about either a or b.

7. The first operation on the right-hand side of the gap applies $\bigoplus$ to the second and third qubit. We continue with our convention of marking control and target qubits with c and t:

$$a\,|\,0\rangle\bigoplus(|\,1\rangle_c\,|\,1\rangle_t)+b\,|\,0\rangle\bigoplus(|\,1\rangle_c\,|\,0\rangle_t)$$
$$=a\,|\,0\rangle\,|\,1\rangle_c\,|\,0\rangle_t+b\,|\,0\rangle\,|\,1\rangle_c\,|\,1\rangle_t. \tag{7.61}$$

8. Next, we pass the upper line through the $\boldsymbol{S}$ gate, which in this case just multiplies $\mid 0\rangle$ by i, so that the state of the system becomes

$$\mathrm{i}a \mid 0\rangle \mid 1\rangle \mid 0\rangle + \mathrm{i}b \mid 0\rangle \mid 1\rangle \mid 1\rangle. \tag{7.62}$$

9. The next operation is difficult to write down symbolically because it couples the top and the bottom line of the circuit and it is the bottom line that controls the gate. Again our subscript convention helps.

$$\begin{aligned} &\bigoplus (\mathrm{i}a \mid 0\rangle_t \mid 1\rangle \mid 0\rangle_c + \mathrm{i}b \mid 0\rangle_t \mid 1\rangle \mid 1\rangle_c) \\ &= \mathrm{i}a \mid 0\rangle_t \mid 1\rangle \mid 0\rangle_c + \mathrm{i}b \mid 1\rangle_t \mid 1\rangle \mid 1\rangle_c. \end{aligned} \tag{7.63}$$

10. Now we apply the $\boldsymbol{S}$ gate to the top qubit (labeled with S) and the $\boldsymbol{T}$ gate to the bottom one (labeled with T):

$$\begin{aligned} &\mathrm{i}a\boldsymbol{S} \mid 0\rangle_S \mid 1\rangle \boldsymbol{T} \mid 0\rangle_T + \mathrm{i}b\boldsymbol{S} \mid 1\rangle_S \mid 1\rangle \boldsymbol{T} \mid 1\rangle_T \\ &= \mathrm{i}a\mathrm{i} \mid 0\rangle_S \mid 1\rangle(-1) \mid 0\rangle_T + \mathrm{i}b(1) \mid 1\rangle_S \mid 1\rangle(-\mathrm{i}) \mid 1\rangle_T \\ &= a \mid 0\rangle \mid 1\rangle \mid 0\rangle + b \mid 1\rangle \mid 1\rangle \mid 1\rangle. \end{aligned} \tag{7.64}$$

11. Finally we apply a yet another upside down controlled-NOT gate:

$$\begin{aligned} &\bigoplus (a \mid 0\rangle_t \mid 1\rangle \mid 0\rangle_c + b \mid 1\rangle_t \mid 1\rangle \mid 1\rangle_c) \\ &= a \mid 0\rangle_t \mid 1\rangle \mid 0\rangle_c + b \mid 0\rangle_t \mid 1\rangle \mid 1\rangle_c \\ &= \mid 0\rangle \mid 1\rangle \begin{pmatrix} a \\ b \end{pmatrix} = \mid 0\rangle \mid 1\rangle \mid \Psi\rangle. \end{aligned} \tag{7.65}$$

We find the bottom qubit emerging in the same state that the top qubit entered the computation in.

NMR teleportation

Now, let us see how this computation was carried out by Nielsen, Knill, and Laflamme in their 1998 experiment. As we remarked earlier, the liquid sample holder in their NMR machine was filled with labeled trichloroethylene, $^{13}C_2HCl_3$. The structure of the molecule is shown in Figure 7.9.

Three nuclei are used in the computation: the two ^{13}C nuclei, which are labeled in Figure 7.9 A and B, and the hydrogen nucleus. The state is teleported from the $^{13}C_B$ nucleus to the hydrogen nucleus, with $^{13}C_A$ corresponding to the middle line in Figure 7.8. Because of chemical shifts, each nucleus that participates in the computation has its own Larmor frequency, namely,

$$\omega_H \approx 500.133491\,\mathrm{MHz},$$

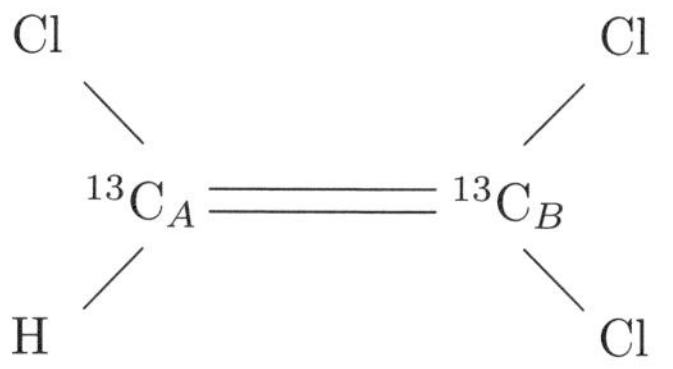

Figure 7.9: A molecule of labeled trichloroethylene used in the teleportation experiment.

$$
\begin{aligned}
\omega_{C_A} &\approx 125.772580\,\text{MHz}, \\
\omega_{C_B} &\approx \omega_{C_A} - 911\,\text{Hz}.
\end{aligned}
$$

The double junction that connects the two carbon nuclei has a resonance frequency of 103 Hz, and the junction that connects $^{13}C_A$ with the hydrogen nucleus has resonance frequency of 201 Hz.

We should remember that in a molecule all components are coupled to each other. So, we also have the hydrogen nucleus coupled to $^{13}C_B$, and the three nuclei of chlorine are all coupled to the three computational nuclei as well. But the resonance frequencies for these couplings are about 10 Hz and 1 Hz for the chlorine bonds. By manipulating the registers with MHz signals we stay safely away from any frequencies that may activate the bonds.

Decoherence properties of the molecule

The molecule was chosen for its convenient decoherence properties. The two carbon nuclei have phase decoherence times of 0.4 s and 0.3 s, whereas their relaxation times are between 20 s and 30 s. The phase decoherence time for the hydrogen nucleus is about 3 s, and the relaxation time for it is 5 s. We can therefore implement the "act of measurement" in the middle of the Brassard circuit, which is performed in this case on the two carbon nuclei, by waiting for about a second before attempting further operations. While we're waiting, the quantum states of the carbon nuclei decohere. The original information stored on the $^{13}C_A$ nucleus is wiped out, but it is preserved in the entanglement of all three qubits and can be retrieved by manipulating them, so as to flush it onto the hydrogen nucleus in the end.

Readout

While performing the computation the experimenters varied the delay between steps (6) and (7)—see the algorithm on page 367. The readout was then performed on the hydrogen nucleus and the entanglement fidelity was calculated in function of the delay. The results are shown in Figure 7.10, the upper curve marked with dots.

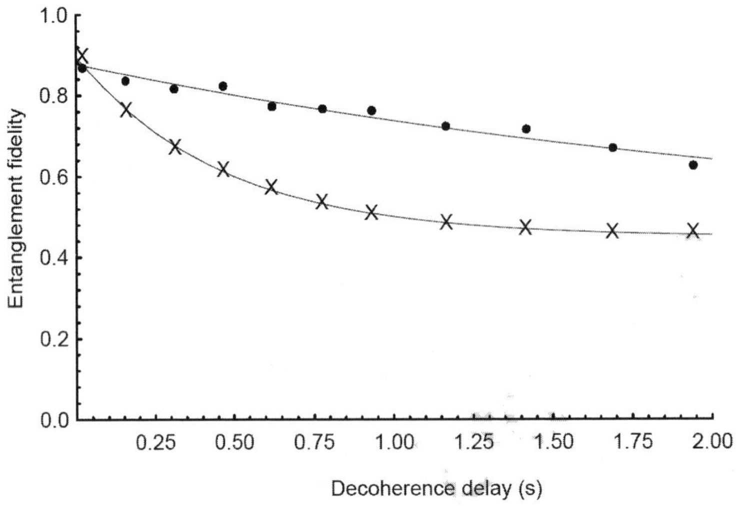

Figure 7.10: Entanglement fidelity in function of decoherence delay for the teleportation experiment. Figure reprinted by permission from Macmillan Publishers Ltd: Nature [98], © 1998.

The lower curve in Figure 7.10, the one marked with crosses, corresponds to a similar measurement, but this time no operations other than the final state tomography were carried out following the decoherence step. Fidelity of 1 corresponds to perfect transmission. Fidelity of 0.25 corresponds to a total information loss. Fidelity of 0.5 corresponds to a "perfect classical transmission."

Interpretation

The graph shows us that quantum information is transferred to the hydrogen nucleus, albeit with some loss that increases with the decoherence delay. The reason for the loss is the decoherence of the hydrogen nucleus state, which, although slower than that of the carbon nuclei, is not slow enough and cannot be totally eliminated. The higher fidelity of the upper curve attests to the need for the six additional operations that follow step (7) in the Brassard algorithm.

The NMR experiment by Nielsen, Knill, and Laflamme was the first *complete* demonstration of quantum teleportation. Prior to the experiment only partial teleportations were performed, meaning that the final state of the third qubit was not recovered completely.

There haven't been many such demonstrations since, as teleporting quantum states is not easily done. An interesting experiment was described in July 2007 by a group of researchers from the University of Innsbruck, who used an ion trap system to teleport quantum states with an average fidelity of 0.83 ± 0.01 [120]—only slightly better than what we have seen in the 1998 NMR experiment.

7.4 The Grover Search Algorithm

NMR's heart is quantum.

Although NMR computing is carried out by using macroscopic samples at room temperature and macroscopic measuring devices, just ordinary Helmholtz coils, there is a quantum heart beating there somewhere. NMR dynamics and analysis of NMR results are expressed in terms of quantum mechanics, which shows up in the NMR spectra as nicely separated resonance peaks, attributable to quantum transitions. This is no different, as we have pointed out above, from many other experiments that probe the world of quantum phenomena.

Evidently, whatever quantum physics can be still seen through the haze of high-temperature environment and macroscopic samples is enough to demonstrate even most elaborate quantum computations, such as the Shor factoring algorithm.

Can a system that is entirely classical (but not restricted to digital), with no reference to quantum mechanics whatsoever, be used to implement a quantum algorithm?

The answer depends on the algorithm. In some cases, the answer is *yes*. In other cases, the answer is not known.

Grover search algorithm

One of the most famous cases, for which we know the answer, is the Grover search algorithm [56, 55]. This algorithm finds, in the words of Grover himself, "a needle in a haystack." Such a situation arises often. For example, we may have a mysterious telephone number, and our task is to find it in a telephone book and identify the person it belongs to. This is a difficult task. If we are unlucky and begin our search from the wrong end, we may have to search through the whole book until we find the owner under "ZWIEG William J." Of course, a telephone company may have a directory where entries are ordered by number, rather than by name, but such a directory may not be available to the public. In other cases, there may not be a directory at all, just items thrown into a data base at random, and not ordered by the particular key we have in hand.

By classical "digital" reasoning, if the data base contains N items, then we have to carry out of the order of $\mathcal{O}(N)$ inspections to locate the item that matches the key.

Efficiency of Grover search algorithm

Grover's algorithm can do this in $\mathcal{O}(\sqrt{N})$ steps only, seemingly by virtue of quantum magic. The saving is enormous. If a data base contains 300,000,000 items, roughly the population of the United States today, Grover's algorithm lets

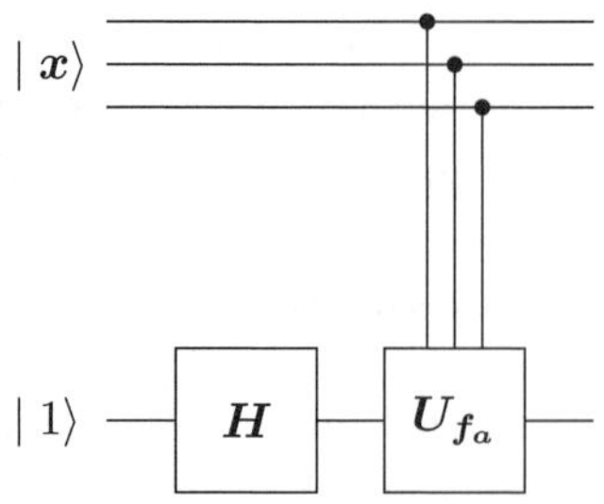

Figure 7.11: State marker for Grover's search algorithm.

us find the item pointed to by the key in about 17,321 steps on average. If it takes one second per iteration, we'll be done in about five hours. To search through the whole set of 300,000,000 items classically, at the same pace of one item per second, would take one person 43 years, assuming 8 hours of work per day, 5 days per week, 48 weeks per year, and 4 weeks leave.

State marker

To formalize and simplify our discussion of Grover's algorithm, we assume that our items are numbers from 0 through $N-1$. The key will be represented by a function that is zero on all N numbers, with the exception of one, for which it is 1:

$$f_a(x) : x \in \{0, 1, \ldots, N-1\} \mapsto \{0, 1\}, \tag{7.66}$$

$$\exists_{a\in\{0,1,\ldots,N-1\}} f_a(a) = 1 \quad \text{and} \quad \forall_{\{0,1,\ldots,N-1\}\ni x\neq a} f_a(x) = 0. \tag{7.67}$$

Here $f_a(x)$ is a *characteristic function* of the search.

To carry out the search using a quantum apparatus, we define a state marker for the function first. It is a unitary operation implemented on an *auxiliary qubit* and controlled by the register that holds $\boldsymbol{x}$ (a quantum representation of number x). The operation implements a controlled-by-$f_a(x)$-NOT gate as follows:

$$\boldsymbol{U}_{f_a} (|\,\boldsymbol{x}\rangle \otimes |\,y\rangle) = |\,y \oplus_2 f_a(x)\rangle. \tag{7.68}$$

Figure 7.11 shows how the gate is deployed in the Grover circuit. $\boldsymbol{U}_{f_a}$ activates a NOT operation on a Hadamarded auxiliary line when the state of the register $|\,\boldsymbol{x}\rangle = |\,\boldsymbol{a}\rangle$ and remains inactive otherwise.

We have seen this type of operation before, in the Deutsch oracle, so without much ado we can quote the result here:

$$\boldsymbol{U}_{f_a} |\,\boldsymbol{x}\rangle \otimes \frac{1}{\sqrt{2}} (|\,0\rangle - |\,1\rangle) = (-1)^{f_a(x)} |\,\boldsymbol{x}\rangle \otimes \frac{1}{\sqrt{2}} (|\,0\rangle - |\,1\rangle). \tag{7.69}$$

But because we know what f_a is, we can capitalize on it and take the above expression further. For $x \neq a$ we have that

$$(-1)^{f_a(x)} \mid \boldsymbol{x}\rangle = (-1)^0 \mid \boldsymbol{x}\rangle = \mid \boldsymbol{x}\rangle. \tag{7.70}$$

But for the single $x = a$

$$(-1)^{f_a(a)} \mid \boldsymbol{a}\rangle = (-1)^1 \mid \boldsymbol{a}\rangle = - \mid \boldsymbol{a}\rangle. \tag{7.71}$$

We can sum this up by combining both equations as follows:

$$(-1)^{f_a(x)} \mid \boldsymbol{x}\rangle = (\boldsymbol{1} - 2 \mid \boldsymbol{a}\rangle\langle \boldsymbol{a} \mid) \mid \boldsymbol{x}\rangle. \tag{7.72}$$

In summary, the output state in Figure 7.11 is

$$(\boldsymbol{1} - 2 \mid \boldsymbol{a}\rangle\langle \boldsymbol{a} \mid) \mid \boldsymbol{x}\rangle \otimes \frac{1}{\sqrt{2}} (\mid 0\rangle - \mid 1\rangle). \tag{7.73}$$

The operator $\boldsymbol{1} - 2 \mid \boldsymbol{a}\rangle\langle \boldsymbol{a} \mid$ is a reflection. It reflects the a component of a register state, changing its sign, much in the way a mirror reflects the component of a vector that is perpendicular to its surface. Let us call it $\boldsymbol{P}_a$, as it is going to be useful in what follows.

Now, let us make another reflection. It is constructed of a superposition of all possible register states from $\mid \boldsymbol{0}\rangle$ through $\mid (N-1)\rangle$. This, as we know already, can be generated by setting each register qubit to $\mid 0\rangle$, and Hadamarding it:

$$\mid \boldsymbol{x}\rangle = \boldsymbol{H} \mid 0\rangle \otimes \boldsymbol{H} \mid 0\rangle \otimes \cdots \otimes \boldsymbol{H} \mid 0\rangle = \frac{1}{\sqrt{N}} \sum_{n=0}^{N-1} \mid \boldsymbol{n}\rangle. \tag{7.74}$$

Using $\mid \boldsymbol{x}\rangle$, we define the reflection by

$$\boldsymbol{P}_x = 2 \mid \boldsymbol{x}\rangle\langle \boldsymbol{x} \mid - \boldsymbol{1}. \tag{7.75}$$

This time it is the part of the argument vector that is perpendicular to $\mid \boldsymbol{x}\rangle$ that is reflected, whereas the component parallel to $\mid \boldsymbol{x}\rangle$ is left alone.

The two reflections, when combined, constitute what is called a *Grover iteration:* *Grover iteration*

$$\boldsymbol{R}_G = \boldsymbol{P}_x \boldsymbol{P}_a = (2 \mid \boldsymbol{x}\rangle\langle \boldsymbol{x} \mid - \boldsymbol{1}) (\boldsymbol{1} - 2 \mid \boldsymbol{a}\rangle\langle \boldsymbol{a} \mid). \tag{7.76}$$

The Grover iteration is also a rotation in the $\mid \boldsymbol{a}\rangle \wedge \mid \boldsymbol{x}\rangle$ plane.

To see how this comes about, we now apply it to two perpendicular vectors in the $\mid \boldsymbol{a}\rangle \wedge \mid \boldsymbol{x}\rangle$ plane that can be used as the two-dimensional basis of the corresponding vector space. Let the vectors be $|\,\boldsymbol{a}\rangle$ and $|\,\boldsymbol{b}\rangle$, where $|\,\boldsymbol{b}\rangle \perp |\,\boldsymbol{a}\rangle$. We cannot use $|\,\boldsymbol{x}\rangle$ for $|\,\boldsymbol{b}\rangle$ because $|\,\boldsymbol{x}\rangle$ is not perpendicular to $|\,\boldsymbol{a}\rangle$. The subtended angle between the two has a cosine of $1/\sqrt{N}$, as can be read from equation (7.74), because $|\,\boldsymbol{a}\rangle$ is one of the $|\,\boldsymbol{n}\rangle$. Although this is close to zero for large N, and we intend to make use of it below, it is not zero exactly.

The action of $\boldsymbol{R}_G$ on $|\,\boldsymbol{a}\rangle$ returns

$$\begin{aligned} \boldsymbol{R}_G\,|\,\boldsymbol{a}\rangle &= (2\,|\,\boldsymbol{x}\rangle\langle\boldsymbol{x}\,| - \mathbf{1})\,(\mathbf{1} - 2\,|\,\boldsymbol{a}\rangle\langle\boldsymbol{a}\,|)\,|\,\boldsymbol{a}\rangle \\ &= -\,(2\,|\,\boldsymbol{x}\rangle\langle\boldsymbol{x}\,| - \mathbf{1})\,|\,\boldsymbol{a}\rangle \\ &= |\,\boldsymbol{a}\rangle - 2\,|\,\boldsymbol{x}\rangle\langle\boldsymbol{x}\,|\,\boldsymbol{a}\rangle \end{aligned} \tag{7.77}$$

and then

$$\begin{aligned} \boldsymbol{R}_G\,|\,\boldsymbol{b}\rangle &= (2\,|\,\boldsymbol{x}\rangle\langle\boldsymbol{x}\,| - \mathbf{1})\,(\mathbf{1} - 2\,|\,\boldsymbol{a}\rangle\langle\boldsymbol{a}\,|)\,|\,\boldsymbol{b}\rangle \\ &= (2\,|\,\boldsymbol{x}\rangle\langle\boldsymbol{x}\,| - \mathbf{1})\,|\,\boldsymbol{b}\rangle \\ &= 2\,|\,\boldsymbol{x}\rangle\langle\boldsymbol{x}\,|\,\boldsymbol{b}\rangle - |\,\boldsymbol{b}\rangle. \end{aligned} \tag{7.78}$$

To complete these two equations, we need to express $|\,\boldsymbol{x}\rangle$ in terms of $|\,\boldsymbol{a}\rangle$ and $|\,\boldsymbol{b}\rangle$. Let θ be the angle between $|\,\boldsymbol{a}\rangle$ and $|\,\boldsymbol{x}\rangle$. We know that $\cos\theta = 1/\sqrt{N}$. The angle between $|\,\boldsymbol{b}\rangle$ and $|\,\boldsymbol{x}\rangle$ is $90^\circ - \theta = \delta$, and

$$|\,\boldsymbol{x}\rangle = |\,\boldsymbol{a}\rangle\cos\theta + |\,\boldsymbol{b}\rangle\cos\delta. \tag{7.79}$$

Substituting this into $\boldsymbol{R}_G\,|\,\boldsymbol{a}\rangle$ and $\boldsymbol{R}_G\,|\,\boldsymbol{b}\rangle$ yields

$$\begin{aligned} \boldsymbol{R}_G\,|\,\boldsymbol{a}\rangle &= |\,\boldsymbol{a}\rangle - 2\,(|\,\boldsymbol{a}\rangle\cos\theta + |\,\boldsymbol{b}\rangle\cos\delta)\cos\theta \\ &= |\,\boldsymbol{a}\rangle\left(1 - 2\cos^2\theta\right) - |\,\boldsymbol{b}\rangle\,(2\cos\delta\cos\theta)\,, \qquad (7.80) \\ \boldsymbol{R}_G\,|\,\boldsymbol{b}\rangle &= 2\,(|\,\boldsymbol{a}\rangle\cos\theta + |\,\boldsymbol{b}\rangle\cos\delta)\cos\delta - |\,\boldsymbol{b}\rangle \\ &= |\,\boldsymbol{a}\rangle\,(2\cos\theta\cos\delta) + |\,\boldsymbol{b}\rangle\left(2\cos^2\delta - 1\right). \qquad (7.81) \end{aligned}$$

Grover iteration is a rotation.

We are going to replace θ with $90^\circ - \delta$ in these equations, remembering that $\cos(90^\circ - \delta) = \sin\delta$. In effect we obtain

$$\begin{aligned} \boldsymbol{R}_G\,|\,\boldsymbol{a}\rangle &= |\,\boldsymbol{a}\rangle\left(1 - 2\sin^2\delta\right) - |\,\boldsymbol{b}\rangle\,(2\cos\delta\sin\delta) \\ &= |\,\boldsymbol{a}\rangle\cos 2\delta - |\,\boldsymbol{b}\rangle\sin 2\delta, \qquad (7.82) \\ \boldsymbol{R}_G\,|\,\boldsymbol{b}\rangle &= |\,\boldsymbol{a}\rangle\,(2\sin\delta\cos\delta) + |\,\boldsymbol{b}\rangle\left(2\cos^2\delta - 1\right) \\ &= |\,\boldsymbol{a}\rangle\sin 2\delta + |\,\boldsymbol{b}\rangle\cos 2\delta. \qquad (7.83) \end{aligned}$$

This is indeed a clean rotation in the $|\,\boldsymbol{a}\rangle \wedge |\,\boldsymbol{b}\rangle$ plane by 2δ.

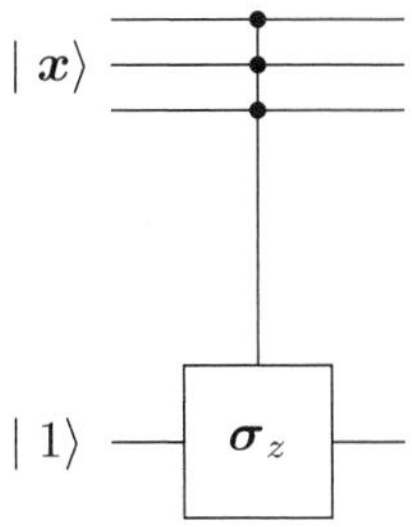

Figure 7.12: In this circuit, $\boldsymbol{\sigma}_z$ inverts the sign of $| 1\rangle \otimes | 1\rangle \otimes | 1\rangle$ but leaves all other states of the register unchanged.

If we make the initial state of the register equal $| \boldsymbol{x}\rangle$ and apply $\boldsymbol{R}_G$ to it, the new $| \boldsymbol{x}'\rangle$ will be rotated by 2δ toward $| \boldsymbol{a}\rangle$. Applying $\boldsymbol{R}_G$ again to $| \boldsymbol{x}'\rangle$ will produce $| \boldsymbol{x}''\rangle$, which will be rotated by another 2δ toward $| \boldsymbol{a}\rangle$.[5]

How many steps to solution?

How many Grover iterations are needed to rotate $| \boldsymbol{x}\rangle$ onto $| \boldsymbol{a}\rangle$? The question is easy to answer. The original angle subtended between $| \boldsymbol{x}\rangle$ and $| \boldsymbol{a}\rangle$ is $\theta = 90^\circ - \delta$, such that $\cos\theta = \sin\delta = 1/\sqrt{N} \approx \delta$, because for small angles $\delta \approx \sin\delta$. Since θ is almost 90°, we can say that in

$$\frac{90^\circ}{2\delta} \approx \frac{\pi/2}{2 \times 1/\sqrt{N}} = \frac{\pi\sqrt{N}}{4} \tag{7.84}$$

steps, we should come sufficiently close to $| \boldsymbol{a}\rangle$ to have it returned in the register on its measurement with probability close to one.

The way to implement $\boldsymbol{P}_x$ is as follows. Figure 7.12 shows a circuit with the $\boldsymbol{\sigma}_z$ gate on an auxiliary line controlled by lines of a 3-qubit register. For all $| \boldsymbol{x}\rangle$ but $| 1\rangle \otimes | 1\rangle \otimes | 1\rangle$, the $\boldsymbol{\sigma}_z$ remains inactive, so the output state of the system is unchanged, $| \boldsymbol{x}\rangle \otimes | 1\rangle$. But for $x = 7$ the sigma converts $| 1\rangle$ in the auxiliary line to $-| 1\rangle$, so the output state this time is $(-1) | \boldsymbol{7}\rangle \otimes | 1\rangle$.

How does this help? Let us observe that

$$-\boldsymbol{P}_x = \boldsymbol{1} - 2 | \boldsymbol{x}\rangle\langle \boldsymbol{x} | = \boldsymbol{1} - 2 \left(\bigotimes_{n=0}^{N-1} \boldsymbol{H} \right) | \boldsymbol{0}\rangle\langle \boldsymbol{0} | \left(\bigotimes_{n=0}^{N-1} \boldsymbol{H} \right)$$

[5] We must remember that $\boldsymbol{R}_G$ on consequtive applications stays defined in terms of the original $| \boldsymbol{x}\rangle$, not in terms of the new $| \boldsymbol{x}'\rangle$ and $| \boldsymbol{x}''\rangle$. These are the arguments only of $\boldsymbol{R}_G$, and they change as the iterations proceed, whereas $\boldsymbol{R}_G$ does not.

$$= \left(\bigotimes_{n=0}^{N-1} \boldsymbol{H}\right) (\mathbf{1} - 2 \mid \mathbf{0}\rangle\langle\mathbf{0} \mid) \left(\bigotimes_{n=0}^{N-1} \boldsymbol{H}\right). \tag{7.85}$$

The operator in the middle, $\mathbf{1} - 2 \mid \mathbf{0}\rangle\langle\mathbf{0} \mid$, does almost exactly what the circuit in Figure 7.12 does, but it triggers the sign change on $\mid 0\rangle \otimes \mid 0\rangle \otimes \mid 0\rangle$ instead of on $\mid 1\rangle\otimes \mid 1\rangle\otimes \mid 1\rangle$. We can fix this by negating every register line in Figure 7.12 before and after the control connection to $\boldsymbol{\sigma}_z$. In effect we find that

$$-\boldsymbol{P}_x = \left(\bigotimes_{n=0}^{N-1} \boldsymbol{H}\right) \left(\bigotimes_{n=0}^{N-1} \boldsymbol{\sigma}_x\right) \text{controlled-}\boldsymbol{\sigma}_z \left(\bigotimes_{n=0}^{N-1} \boldsymbol{\sigma}_x\right) \left(\bigotimes_{n=0}^{N-1} \boldsymbol{H}\right). \tag{7.86}$$

The full Grover circuit is more complicated. The reason is that we want to apply $\boldsymbol{R}_G$ to the register successively, without affecting $\boldsymbol{R}_G$ itself and without entangling any auxiliary lines used in the $\boldsymbol{R}_G$ implementation with the register.

Experimental demonstration

The first experimental demonstration of the Grover iteration technique was by Chuang, Gershenfeld, and Kubinec in 1998 [21]. It was an NMR computation carried out on chloroform molecules. It was a very small 2-qubit computation, but also the first complete demonstration of a quantum computing process. For a 2-qubit register, the Grover algorithm returns a correct answer in just one iteration, so it was not an involved computation.

Protecting qubits against decoherence

Because Grover's is an iterative algorithm, decoherence is its deadly enemy for queries against larger registers. In 2003, Ollerenshaw, Lidar, and Kay of the University of Toronto demonstrated a still 2-qubit Grover computation, in which logical qubits were quantum error avoidance encoded in two qubits each [102]. Hence, they actually had four physical qubits. The computation was carried out by using NMR as well and compared against computation that used "unprotected" raw qubits. The NMR computer operating on encoded qubits successfully executed the search algorithm in the presence of engineered decoherence, however strong, whereas the raw-qubit computer failed consistently under the same conditions.

But what interests us here is that a similar $\mathcal{O}(\sqrt{N})$ computation can be implemented by using a completely classical low-technology system of coupled pendula.

Yes, it can be done with cogwheels.

7.5 Cogwheels

Coupled pendula solve a Grover search in the same number of steps.

The coupled pendula of Grover and Sengupta [57] are illustrated in Figure 7.13. It is a system with two tiers of pendula. The first tier is the support pendulum of mass M and length L that is attached to the ceiling, which is here drawn with

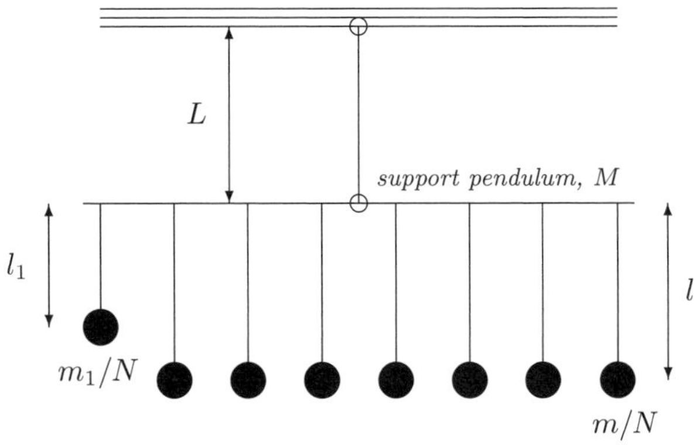

Figure 7.13: Coupled pendula of Grover and Sengupta. Figure redrawn with permission from [57]. © 2002 by the American Physical Society.

a triple horizontal line. The second tier comprises eight pendula; here we assume that $N = 8$, of which seven are of mass m/N and length l, and the first is of mass m_1/N and length l_1. The second-tier pendula all hang off the support pendulum. The coupling here is of a more subtle nature, not through explicit springs that are often used to connect coupled pendula, but through the support pendulum instead. We assume a small motion of the pendula in the x direction only, so that they don't hit each other, and so that the well-known oscillator approximation applies.

System Lagrangian

The system's Lagrangian is given by

$$\begin{aligned} 2L &= M\dot{X}^2 - KX^2 + \frac{1}{N}\left(m_1\dot{x}_1^2 - k_1(x_1 - X)^2\right) \\ &\quad + \frac{1}{N}\sum_{j=2}^{N}\left(m\dot{x}_j^2 - k(x_j - X)^2\right), \end{aligned} \tag{7.87}$$

where X is the center mass position of the support pendulum and $\dot{X}$ is its time derivative. Similarly, x_1, $\dot{x}_1$, x_j, and $\dot{x}_j$ are positions of the second-tier pendula and their time derivatives. The coefficients K, k_1, and k are given by

$$K = \left(M + \frac{m}{N}\right)\frac{g}{L}, \tag{7.88}$$

$$k = m\frac{g}{l}, \tag{7.89}$$

$$k_1 = m_1\frac{g}{l_1}, \tag{7.90}$$

where g is the acceleration due to gravity.

Extract center of mass mode

It'll help us to rewrite the Lagrangian by extracting the center of mass mode of the longer tier-two pendula, in terms of x_{cm}, and replace x_j for $j \geq 3$ with excitation modes that are perpendicular to the center-of-mass mode. We'll call them y_j for $j \in [3, N]$. We skip $j = 2$ because it is taken by the center-of-mass mode now. The y_j variables become decoupled from the other ones, as can be seen from the Lagrangian:

$$\begin{aligned} 2L &= M\dot{X}^2 - KX^2 + \frac{1}{N}\left(m_1\dot{x}_1^2 - k_1\left(x_1 - X\right)^2\right) \\ &\quad + \left(1 - \frac{1}{N}\right)\left(m\dot{x}_{\mathrm{cm}}^2 - k\left(x_{\mathrm{cm}} - x\right)^2\right) + \frac{1}{N}\sum_{j=3}^{N}\left(m\dot{y}_j^2 - ky_j^2\right). \quad (7.91) \end{aligned}$$

Because y_j and $\dot{y}_j$ are decoupled, if they are zero initially, they remain zero. Therefore we can drop them from the Lagrangian. We are left with N in the x_1 and $\dot{x}_1$ terms only. We can absorb $\sqrt{N}$ into x_1 by defining

$$\xi = \frac{1}{\sqrt{N}} x_1, \quad (7.92)$$

Simplified Lagrangian

which simplifies the Lagrangian as follows:

$$\begin{aligned} 2L &\approx M\dot{X}^2 - KX^2 + m_1\xi^2 - k_1\left(\xi - \frac{1}{\sqrt{N}}X\right)^2 \\ &\quad + m\dot{x}_{\mathrm{cm}}^2 - k\left(x_{\mathrm{cm}} - X\right)^2 + \mathcal{O}\left(\frac{1}{N}\right). \quad (7.93) \end{aligned}$$

Energy transfer to the shorter pendulum occurs in $\mathcal{O}(\sqrt{N})$ steps.

We see here that X and x_{cm} are strongly coupled through the $(x_{\mathrm{cm}} - X)$ term, but ξ coupling with X is $\sqrt{N}$ weaker. Hence we can get to understand the system by analyzing the (X, x_{cm}) component in the absence of the ξ term first, which should yield two resonance modes with frequencies ω_a and ω_b. The resonance frequency of the ξ mode, in the absence of the coupling term, is $\omega_1 = \sqrt{k_1/m_1}$; but when the coupling is activated and ω_1 made close to either ω_a or ω_b, then a resonant energy transfer between the (X, x_{cm}) and ξ components occurs, with the number of cycles required to transfer energy from the (X, x_{cm}) component to the ξ component inversely proportional to the coupling constant, that is, $\mathcal{O}(\sqrt{N})$.[6]

[6] Readers acquainted with classical mechanics can easily extract equations of motion from the simplified Lagrangian (7.93) and analyze the system in terms of small vibrations. But such analysis does not add anything new to what we can read from the Lagrangian directly.

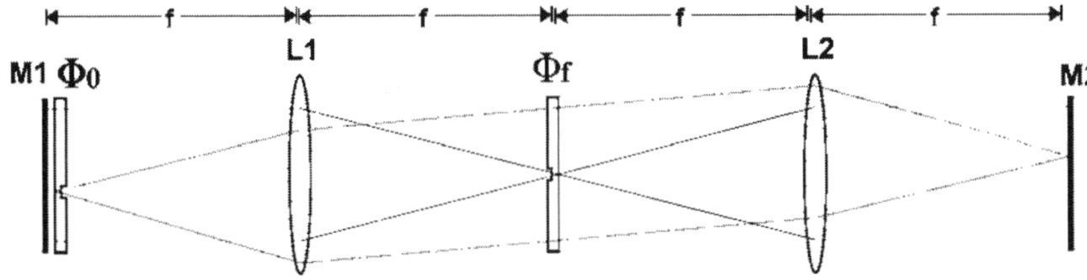

Figure 7.14: Optical cavity implementation of the Grover iterator. Figure reprinted from [64] with permission of the Optical Society of America.

And this is how the coupled pendula solve the Grover's needle in a haystack problem in the same number of steps as his quantum search algorithm. One aspect of the solution, however, needs to be pointed out. In this classical analog system, we must have eight tier-two pendula to solve a problem of size eight. But in the quantum case, we could have solved this problem with only a 3-qubit register (perhaps doubled, to account for auxiliary inputs). It is here that a difference shows up. Classical computers, even analog ones, seem to consume resources at a linear rate, linear in the size of a problem to be tackled, whereas quantum computers consume resources at a logarithmic rate. The Grover-Sengupta analog computer can be implemented electronically, rather than mechanically, using the well-known analogy between harmonic oscillators and LCR circuits, but this does not change the number of resources required.

The number of resources required by the classical system grows linearly with N.

The device is not the only example of a classical-analog competitor of a similar quantum system. In 2001, Dorrer, Londero, Anderson, Wallentowitz, and Walmsley of Rochester University demonstrated a Grover-like query against a 50-element data base using a classical optical interference setup [34]. More recently, Hijmans, Huussen, and Spreeuw of the University of Amsterdam demonstrated a similar classical optical arrangment that was used to search through the data base of up to 1,000 items [64] and produced improvements that would be difficult or even impossible to implement in the quantum version of the algorithm. Although a classical optical computer suffers from the same limitation of having to operate on N items—versus $\log N$ for a quantum computer—they are easier to provide than is the case for the Grover-Sengupta pendula.

Demonstration of Grover search with a classical optical computer

Figure 7.14 shows a schematic diagram of a classical optical cavity implementation of the Grover algorithm [64]. The two 90% reflectivity mirrors, M_1 and M_2, form an optical cavity. Inside the cavity are two lenses, L_1 and L_2, whose focal

Cavity configuration

Table 7.2: Correspondence between quantum and classical operators. Table reprinted from [64] with permission of the Optical Society of America.

Quantum	Classical
$\boldsymbol{P}_a$	Φ_0^2
$\bigotimes \boldsymbol{H}$	$\mathcal{F}$
$\boldsymbol{1} - 2 \mid \boldsymbol{0}\rangle\langle\boldsymbol{0} \mid$	Φ_f^2
$\bigotimes \boldsymbol{H}$	$\mathcal{F}^{-1}$

length f equals $1/4^{\text{th}}$ of the distance between the mirrors and which are positioned at $1f$ distance from each mirror within the cavity. Also inside the cavity are two phase plates, Φ_0, located on top of mirror M_1, and Φ_f, located in the center.

The Φ_0 phase plate has a small circular spot that changes the phase of the beam passing through it by 90°. This is the marker that corresponds to the Grover projector $\boldsymbol{P}_a$. The Φ_f phase plate has a small spot that changes the phase of the beam passing through it by 90°. The size of both spots in Φ_0 and Φ_f is equal to the waist of the light beam in the center of the cavity, at the focal point of both lenses. The Φ_f phase plate corresponds to the $\boldsymbol{1} - 2 \mid \boldsymbol{0}\rangle\langle\boldsymbol{0} \mid$ operator, which is a part of $\boldsymbol{P}_x$.

Correspondence between quantum and classical operations

The lenses and the mirrors are positioned so that the light profile at M_1, and the light profile at Φ_f, are Fourier transforms, $\mathcal{F}$, of each other. The Fourier transforms play the role of Hadamard operators in $\boldsymbol{P}_x$ that flank the $\boldsymbol{1} - 2 \mid \boldsymbol{0}\rangle\langle\boldsymbol{0} \mid$ operator. The correspondence is summarized in Table 7.2.

The equivalent of Grover iteration $\boldsymbol{R}_G = \boldsymbol{P}_x\boldsymbol{P}_a$ here is the passage of light beam from M_1, through Φ_0, L_1, Φ_f, L_2, to M_2 and then back through L_2, Φ_f, L_1, and Φ_0 to M_1. Because on every such passage each phase plate is traversed twice, we get Φ_0 and Φ_f squared in the mathematical formula that describes the operation:

$$\tilde{\boldsymbol{R}}_G = \Phi_0^2\mathcal{F}^{-1}\Phi_f^2\mathcal{F}. \tag{7.94}$$

What corresponds to N in this experiment? It is the ratio of the beam cross-section area to the area of the Φ_0's circular phase shifting spot. We can think of the beam as filled with patches of light of equal size, of which one has been labeled by having its phase shifted by 90°.

The system is measured by inserting a CCD detector behind M_2. Because M_2 leaks 10% of incident light, if we time the measurements just the right way, we can capture the transmitted image after $\frac{1}{2}$ iteration, then $1\frac{1}{2}$, $2\frac{1}{2}$, $3\frac{1}{2}$ iterations, and so on. On the other hand, because light leaks from the system at a considerable

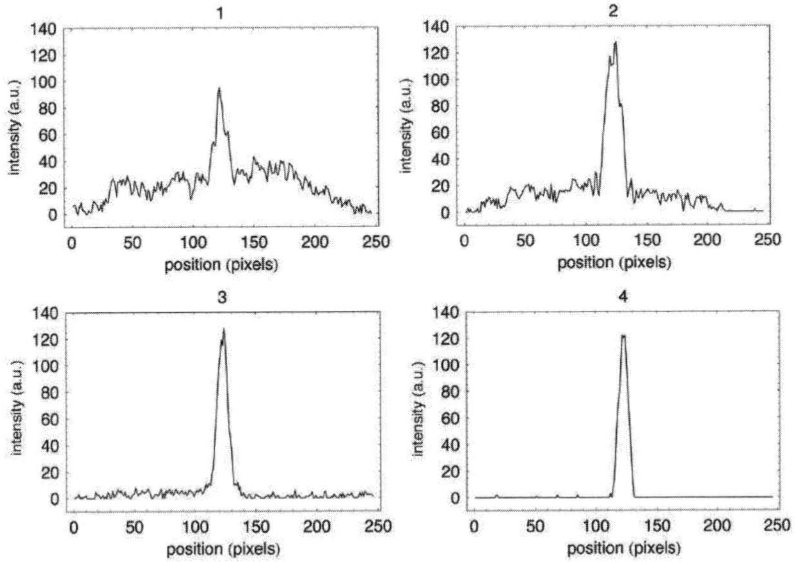

Figure 7.15: Cross section through the flat-fielded CCD image after $\frac{1}{2}$, $1\frac{1}{2}$, $2\frac{1}{2}$, and $3\frac{1}{2}$ iterations. Figure reprinted from [64] with permission of the Optical Society of America.

rate—between 30% and 50% on each full iteration—we cannot really perform a large number of iterations this way. Still, as shown in Figure 7.15, the image of the Φ_0's spot quickly improves with the number of iterations performed.

Grover search as a phase-contrast microscope

Hijmans, Huussen, and Spreeuw describe an interesting practical application of this technique, in which the system effectively becomes a "phase-contrast microscope operated in a multipass fashion," with the Grover algorithm enhancing the contrast by a large factor. It is remarkable that what was conceived to be a quantum algorithm has produced such an unexpected yet beneficial technological development in the domain of classical physics and technologies based on it.

But why does Grover's algorithm work in classical physics in the first place? And, can the same be done with other quantum algorithms? Do we really need quantum systems for "quantum computing" at all?

The answer is "not necessarily," although the final answer is not yet known with certainty.

The Grover algorithm does not work exactly the same way in the classical world as it does in the quantum world. Although, the search completes in $\mathcal{O}(\sqrt{N})$ steps in both cases, we need to have all N objects in the classical case, whereas we can do with $\log N$ objects only in the quantum case. This situation, however, may not necessarily be bad. One case might be if we have a classical data base with N classical items we want to search through. Furthermore, if we have to operate on a very large number of items, it is not so hard to generate all of them. For example, the still anticipated 45 nm technology Intel 80986 processor is going to have 1 billion transistors on a die, and the IBM Power6 has nearly as many—700 million transistors on a die. And as the transistor size continues to shrink—nanotube transistors have been demonstrated—and microprocessor architectures begin to dig into the third dimension, we may expect the progress to continue for a long time, with transistor counts eventually reaching trillions.

Classical implementations of Grover search require more resources.

Grover search does not require entanglement.

But let us get back to our fundamental question. Only two years after Grover's publication of his original algorithm, Seth Lloyd pointed out that the Grover search can be implemented without quantum entanglement if bivalued qubits were to be replaced with single-valued quantum objects; the only difference in this case would be the amount of resources required [90]. Classical demonstrations of the Grover search all derive ultimately from the Lloyd's paper.

Can one say, then, that whenever entanglement is present, the algorithm is truly quantum and not implementable classically?

Classical imitation of entanglement is possible

Not so fast.

Entanglement can be imitated classically as well. This was observed by Robert J. C. Spreeuw of the University of Amsterdam in 1997, who demonstrated a full classical analog of entanglement using classical light beams, their polarization and their splitting [133]. Where the similarity broke was in the nonlocality of quantum entanglement. The classical light beam entanglement corresponded closely to the entanglement of different degrees of freedom of the same atom, as we saw exploited in the first 1995 demonstration of the controlled-NOT gate [96] and in the 2003 demonstration of the Deutsch algorithm discussed in this chapter [58]. Following Spreeuw's result, one should be able to demonstrate similar "quantum logic" elements using classical polarized light beams.

Secret classical correlations

In 2002 Collins and Popescu discussed another classical analogy of entanglement [24]. They compared entanglement to secret classical correlations. In this context they discussed analogs of teleportation, pure states, entanglement concentration and dilution, and entanglement manipulations. Teleportation, for example, was shown to be equivalent to a one-time pad. Entanglement dilution and concentration was shown to be equivalent to classical secrecy manipulations. Entanglement purification turned out to be equivalent to classical privacy amplification. The most profound question that arose from their discussion was "What part of quantum entanglement was genuinely quantum, meaning, such that no classical analog could be found for it?"

Also in 2002, Yuri Orlov of Cornell University showed "how a classical system of numbered linear oscillators can possess such quantum properties as indeterminism, interference of probabilities, unitary transformations, wave functions, and noncommuting operators, and be used in quantumlike computations" [105]. Orlov's system comprised seven sets of special devices, all classical, which could be implemented as a microelectronic processor. The sets were as follows:

Orlov's classical computer imitates many aspects of quantum physics.

1. Numbered linear oscillators;
2. Perturbation devices to change and exchange oscillator amplitudes and phases;
3. Device that "multiplies" the oscillators in a way that imitates the tensor product operation;
4. Random number generators biased by relative energies of the oscillators, used for decision making;
5. Devices for quenching (to "$\mid 0\rangle$") and full activation (to "$\mid 1\rangle$") of the oscillators;
6. Devices to measure amplitudes, phases, and energies of the oscillators;
7. Auxiliary digital circuitry for managing logic and for digital computations.

The basic operational unit of the Orlov's computer is a classical *Qbit* (as opposed to a quantum *qubit*), which is made of *two* oscillators, numbered by $k \in \{0, 1\}$, each vibrating (classically), as in

Qbit

$$q_k = A\left(c_k e^{\mathrm{i}\omega t} + c_k^* e^{-\mathrm{i}\omega t}\right) = 2A\left|c_k\right|\cos\left(\omega t + \phi\right), \tag{7.95}$$

where $c_k = |c_k| \exp(\mathrm{i}\phi)$. The oscillators that constitute a Qbit are constructed so that

$$|c_0|^2 + |c_1|^2 = 1. \tag{7.96}$$

Because the (classical) energy of both oscillators is

$$E = 2m\omega^2 A^1 \left(|c_0|^2 + |c_1|^2\right), \tag{7.97}$$

where m is the mass of the oscillator (we assume it to be the same for all oscillators in the system), coefficients $|c_0|^2$ and $|c_1|^2$ determine how energy is partitioned between both components of a Qbit.

Operations on Qbits

Orlov's Qbits can be acted upon by perturbing their constituent amplitudes c_k, but in such a way as to preserve condition (7.96), which implies that the energy of a Qbit is conserved. The resulting transformation of Qbit's complex coefficients is a unitary transformation $\boldsymbol{U}(t, t')$,

$$c'_k(t') = \sum_{l=0}^{1} U_{kl}(t, t') c_l(t), \quad k \in \{0, 1\}, \tag{7.98}$$

where $U_{kl} = U^*_{lk}$.

The Qbit's state is read by a random number generator that reads the amplitudes first and then generates 0 or 1 with probability weights w_k given by

$$w_k = |c_k|^2, \quad k \in \{0, 1\}. \tag{7.99}$$

This turns out to be the only way to assign readable values to a Qbit that is consistent with the properties of unitary transformations [103, 104]. The assignment makes c_k into probability amplitudes.

Qbit unitary operations $\boldsymbol{U}$ map onto the following linear operations $\boldsymbol{M}$ that manipulate Qbit internal degrees of freedom: q_0, $\dot{q}_0$, q_1, and $\dot{q}_1$. Assuming that $A = m = \omega = 1$,

$$\begin{pmatrix} q'_0 \\ \dot{q}'_0 \\ q'_1 \\ \dot{q}'_1 \end{pmatrix} = \begin{pmatrix} M_{+00} & M_{-00} & M_{+01} & M_{-01} \\ -M_{-00} & M_{+00} & -M_{-01} & M_{+01} \\ M_{+10} & M_{-10} & M_{+11} & M_{-11} \\ -M_{-10} & M_{+10} & -M_{-11} & M_{+11} \end{pmatrix} \begin{pmatrix} q_0 \\ \dot{q}_0 \\ q_1 \\ \dot{q}_1 \end{pmatrix}, \tag{7.100}$$

where the eight coefficients $M_{\pm kl}$ are given by the following.

$$M_{+00} = \left(U_{00} + U^*_{00}\right)/2 \tag{7.101}$$
$$M_{-00} = -\mathrm{i}\left(U_{00} - U^*_{00}\right)/2 \tag{7.102}$$
$$M_{+01} = \left(U_{01} + U^*_{01}\right)/2 \tag{7.103}$$
$$M_{-01} = -\mathrm{i}\left(U_{01} - U^*_{01}\right)/2 \tag{7.104}$$
$$M_{+10} = \left(U_{10} + U^*_{10}\right)/2 \tag{7.105}$$
$$M_{-10} = -\mathrm{i}\left(U_{10} - U^*_{10}\right)/2 \tag{7.106}$$
$$M_{+11} = \left(U_{11} + U^*_{11}\right)/2 \tag{7.107}$$
$$M_{-11} = -\mathrm{i}\left(U_{11} - U^*_{11}\right)/2 \tag{7.108}$$

For noninstantaneous perturbations that take some time Δt to apply, an additional phase shift $U_{kl} \to U_{kl} \exp(\mathrm{i}\omega\Delta t)$ must be included in the above equations, but not in (7.98).

A single oscillator k within a Qbit can be subjected to a phase rotation $\boldsymbol{R}_k(\phi)$ by slightly altering its frequency by $\delta\omega$ for a certain duration Δt. The resulting transformation is described by the oscillator's equation of motion:

$$\ddot{q}_k = -\left(\omega^2 + \delta\omega^2(t)\right) q_k, \tag{7.109}$$

which, after time Δt, produces the U_{kk} term equal to

$$U_{kk} = \mathrm{e}^{\mathrm{i}\phi}, \tag{7.110}$$

where

$$\phi = \int_{\Delta t} \frac{\delta\omega^2(t)}{2\omega}\,\mathrm{d}t. \tag{7.111}$$

The other U_{ij} remain δ_{ij}, where δ_{ij} is Kronecker delta.

Two oscillators, k and l, not necessarily of the same Qbit, can be coupled by applying an analogous perturbation $\boldsymbol{C}_{kl}(\phi)$ that rotates both oscillators according to

$$U_{kk} = U_{ll} = \cos\phi, \tag{7.112}$$

$$U_{kl} = U_{lk} = \mathrm{i}\sin\phi. \tag{7.113}$$

The rotation is accomplished by applying the following coupling:

$$\ddot{q}_k = -\omega^2 q_k - \delta\omega^2(t) q_l, \tag{7.114}$$

$$\ddot{q}_l = -\omega_2 q_l - \delta\omega^2(t) q_k. \tag{7.115}$$

The accumulated angle ϕ is given by the same integral (7.111) as for the single oscillator case.

With this machinery in place, we obtain the following prescriptions for Pauli matrices and for the Hadamard operator $\frac{1}{\sqrt{2}}\begin{pmatrix} 1 & 1 \\ 1 & -1 \end{pmatrix}$:

$$\boldsymbol{\sigma}_x = \boldsymbol{R}_1\left(-\frac{\pi}{2}\right)\boldsymbol{R}_0\left(-\frac{\pi}{2}\right)\boldsymbol{C}_{01}\left(\frac{\pi}{2}\right), \tag{7.116}$$

$$\boldsymbol{\sigma}_y = \boldsymbol{R}_1(\pi)\,\boldsymbol{R}_0(0)\,\boldsymbol{C}_{01}\left(\frac{\pi}{2}\right) = \boldsymbol{\sigma}_z\boldsymbol{C}_{01}\left(\frac{\pi}{2}\right), \tag{7.117}$$

$$\boldsymbol{\sigma}_z = \boldsymbol{R}_1(\pi)\,\boldsymbol{R}_0(0), \tag{7.118}$$

$$\boldsymbol{H} = \boldsymbol{R}_1\left(\frac{\pi}{2}\right)\boldsymbol{\sigma}_z\boldsymbol{C}_{01}\left(\frac{\pi}{4}\right)\boldsymbol{\sigma}_z\boldsymbol{R}_1\left(\frac{\pi}{2}\right). \tag{7.119}$$

An entangled state of two Qbits A and B is made by connecting all four oscillators that constitute both Qbits and activating the tensor product device. The device creates a new object, a bi-Qbit, that is made of four new oscillators, the amplitudes of which are

Bi-Qbit and entanglement

$$c_0 = c_{0A}c_{0B}, \quad (7.120)$$
$$c_1 = c_{0A}c_{1B}, \quad (7.121)$$
$$c_2 = c_{1A}c_{0B}, \quad (7.122)$$
$$c_3 = c_{1A}c_{1B}. \quad (7.123)$$

The perturbation devices that operate on the bi-Qbit work the same as before, but all unitary rotations are performed on the new four oscillator complex, while preserving its normalization. For example, the prescription for controlled-NOT gate is

$$\text{controlled-NOT} = \boldsymbol{R}_2\left(-\frac{\pi}{2}\right)\boldsymbol{R}_3\left(-\frac{\pi}{2}\right)\boldsymbol{C}_{23}\left(\frac{\pi}{2}\right). \quad (7.124)$$

This operation entangles the two Qbits, in analogy to what happens in the qubit world.

Shor factorization

The proposed system proved powerful enough to implement core elements of the Shor factorization algorithm [127], step by step. The only thing it could not do was to imitate long-distance quantum communication. But quantum computing itself does not make explicit use of this property. Computational qubits can all live on the same single atom, if possible, for example, as its various degrees of freedom, and as we saw demonstrated earlier.

Orlov's computer brings us back to the first chapter of this book, where we talked about classical randomly fluctuating registers, while illustrating the basic concepts of probability calculus. With a few more refinements, as it turns out, we could have continued with this model much of the way into the quantum domain.

Particle diffraction in classical physics

That it is possible to imitate a quantum computing system with a classical analog computer so closely should not be surprising in view of other recent demonstrations in this area. In 2006 Couder and Fort of the University of Paris presented a classical physical system that closely imitated the famous quantum particle diffraction experiment, similar in its idea to the de Broglie's pilot wave theory [25]. They used a droplet of silicon oil that bounced on the surface of vertically vibrating silicon oil bath. The shaking of the droplet generated its own wave, to which the droplet's trajectory coupled. When confronted with a slit in a screen, limiting the transverse extent of its wave, the droplet scattered randomly. However, a histogram of the deviations of many successive droplets (successive, so that they would not interact with one other) prepared in the same way and pushed toward the screen showed diffraction or interference patterns, illustrated here in Figure 7.16, page 391, much like the ones observed in quantum mechanical scattering experiments of similar geometry.

Much of quantum physics is not unique to quantum world.

This is not to say that quantum physics is merely classical physics in disguise. Far from it. Nevertheless, it appears that many phenomena thought of as quantum are not really unique to quantum physics and have close classical analogs. It may well be that anything expressed in terms of differential equations can be mapped onto classical analog circuitry, including qubits and quantum computers, as has been demonstrated by Orlov.

Classicalization of quantum computing

The problem of quantum versus classical analog computing can be approached from the other direction as well, which yields some tantalizing results. In his 2002 paper, David Poulin of the University of Montreal made the following observation. Let us suppose we have a quantum computation that unfolds on a certain number of genuinely quantum qubits. It turns out that under certain conditions we can force some of the qubits into classical states *without* affecting the outcome of the computation [116]. The obvious question that arises is, "How much 'quantum' do we really need to have in a quantum computer to make it compute?" A precise answer to this question could greatly help us realize practical quantum computers.

7.6 The Crossroad

And so our peregrinations bring us to the crossroad, from which we can continue in several directions.

Abundant literature on the subject

Quantum computing has been around at least since the seminal 1985 paper by Deutsch about the universal quantum computer [28], which makes the discipline nearly a quarter of a century old. Thousands of papers on the subject have been published; and a simple query on `Amazon.com` returns 48 titles, to which more will be added each year. A newcomer may well receive the impression that quantum computers are all around us and that students frequently busy themselves by programming them while lounging in front of the TV. As this text explains, this is not so, and won't be so for years to come.

Benefits of quantum computing

Nevertheless, the adventure of quantum computing has opened our eyes to modes of computation other than the classical Turing machine on which present-day digital computers are based. Furthermore, the discipline may stimulate development of unexpected applications deriving one way or another from concepts of quantum computing, as we have seen in Section 7.5. So, it is well worth the effort of further exploration.

Quantum algorithms

Throughout this book we have been avoiding the topic of quantum algorithms, discussing them only when that was needed for our understanding of quantum devices that were the focus of our interest. Readers who would like to learn more about the algorithms we have mentioned but not really discussed—and especially about the Shor factoring algorithm—can peruse *Quantum Computer Science* by N. David Mermin (Cambridge University Press, 2007) [92]. This accessible text covers the subject from a purely algorithmic point of view. The breaking of the RSA encryption key, of which the Shor algorithm is a core component, is covered, as is quantum error correction (as well as quantum search and oracles, which we have discussed already). The book discusses all fundamental quantum algorithms in only 150 pages (about half of which are filled with illustrations).

Lack of progress

Quantum computing, in spite of being a quarter-century old, has progressed little beyond these algorithms. This state of affairs is the subject of a 2003 article by Shor himself, who pondered on why more quantum algorithms have not been found [128]. Two possible reasons have been raised. The first is that "quantum computers operate in a manner so different from classical computers that our techniques for designing algorithms and our intuitions for understanding the process of computation no longer work." As we have seen in the preceding section, this is not necessarily so, but it is the analog intuitions and analog computing techniques that we should rely on, more so than the digital ones. Alas, analog computing techniques are largely forgotten nowadays, as the discipline is out of fashion; hence, this pool of inspiration is mostly dry. More's the pity.

The second possible reason noted by Shor is that "there really might be relatively few problems for which quantum computers can offer a substantial speed-up over classical computers, and we may have already discovered many or all of the important techniques for constructing quantum algorithms." It may well be so regarding algorithms that, like the factorization algorithm, deal with simple classical world objects, integer numbers in this case, with which we are so familiar. Feynman, on the other hand, thought of quantum computers as devices that would help us simulate quantum phenomena, some of which—for example, those related to quantum chromodynamics—call for enormous computing resources, which back in Feynman's days were in short supply.[7]

Quantum computer simulation of quantum systems

Seth Lloyd and Christof Zalka noticed in 1996 [89, 152] that quantum computers could efficiently simulate quantum systems driven by local interactions, delivering

[7]The situation has changed so dramatically in recent years that one of the recent TeraGrid announcements lamented the oversupply of computer power on the network, with the effect that much of it went undersubscribed. On the other hand, this does not translate into oversupply of capability systems. Most of what is available are microprocessor clusters.

exponential speed-up compared to classical digital computers. This observation was further elaborated upon by Somaroo, Tseng, Havel, Laflamme, and Cory in 1999, who produced a general scheme for simulating one quantum system on another [131], and then by Somma, Ortiz, Gubernatis, Knill, and Laflamme in 2002 [132].

These and other papers on the subject spurred increased activity in the field that produced quantum algorithms with exponential speed-up for such activities as evaluating partition functions [88]; solving eigenproblems [2]; and simulating fermionic systems [1], quantum chaos [50], quantum many-body systems [141], and pair-interaction Hamiltonians [147]. Among the first practical demonstrations of the approach was the simulation in 2000 of a three-body quantum Hamiltonian interaction using an NMR quantum computer [136]. A year later, Khitrin and Fung simulated propagation of excitation along a one-dimensional chain of atoms using NMR [79]. But we have not yet seen many other such demonstrations.

David Deutsch himself [100] has stated the following:

> The most important application of quantum computing in the future is likely to be a computer simulation of quantum systems, because that's an application where we know for sure that quantum systems in general cannot be efficiently simulated on a classical computer. This is an application were the quantum computer is ideally suited.
>
> Perhaps in the long run, as nanotechnology becomes quantum technology, that will be a very important generic application.

This is the field to watch.

The D-Wave computer

In February 2007 a Canadian company D-Wave Systems demonstrated the first 16-qubit commercial quantum computer, much to the disbelief and distress of many research laboratories. The machine solved a Sudoku puzzle, a seating arrangements problem, and searched for molecules similar to omeprazole (otherwise known as Prilosec) in a data base.

Adiabatic quantum computer

The D-Wave machine represents a different approach to quantum computation that was first proposed by Farhi, Goldstone, and Sipser, all of MIT, and Gutmann of Northeastern University in Boston in 2000 and that is referred to as *adiabatic quantum computation* [40]. The idea here is that a given problem is defined in terms of a Hamiltonian, which may be quite intractable in general, and the solution to the problem is represented by the ground state of the Hamiltonian. If the Hamiltonian in question can be transformed smoothly into another Hamiltonian that is simpler and for which a ground state solution is known, then we can start our computation from that solution and from the simple Hamiltonian. Then, ever so gently, that is, *adiabatically,* we can drift the simple Hamiltonian into the original, hairy one, hoping all the time that the ground state solution originally configured would slide into the ground state of the hairy Hamiltonian. This procedure, if successful, provides us with a solution to the problem originally posed.

The process is similar to our discussion of the Berry phase in Section 4.11, where we did much the same, albeit mathematically only. The Berry phase manipulations discussed in Section 4.11.4 can be thought of as examples of single-qubit adiabatic computations. The difference is that in our discussion of Berry phase we did not restrict ourselves to ground states only, and we were interested in what would happen to superpositions of states of different energies—the energy difference between the levels appeared in the Berry phase equations. But in the computational model employed by D-Wave, the qubits are always in the ground state, which makes the task easier because we do not have to worry about superpositions with higher energy states. The initial and the final ground states are not entangled, although the state of the system becomes entangled in the middle of the evolution.

Figure 7.17, page 392, shows a processor that was at the core of the company's first 16-qubit version of the computer. A 28-qubit system was later presented at SC07 in November.

In a brilliant burst of lateral thinking, the company exploited both classical and quantum analog computing paradigms, to deliver a system that not only works but also (it is hoped) can be sold. And, whatever the objections raised in on-line blogs and popular articles, in the end "it doesn't matter if a cat is black or white, so long as it catches mice."[8]

[8]This memorable quote is attributed to many people, among them a Chinese statesman Deng Xiaoping (1904–1997), a British Prime Minister Alec Douglas-Home (1903–1995), an English poet Sir William Watson (1858–1935), and another Sir William Watson (1715–1787), who was an English physicist, physician, and botanist and the father of the electric charge conservation principle. According to the philosophy espoused in this book, the cat is *neither* black *nor* white. It is of a different color altogether, contrary to the popular assertion that the cat is *both* black *and* white at the same time.

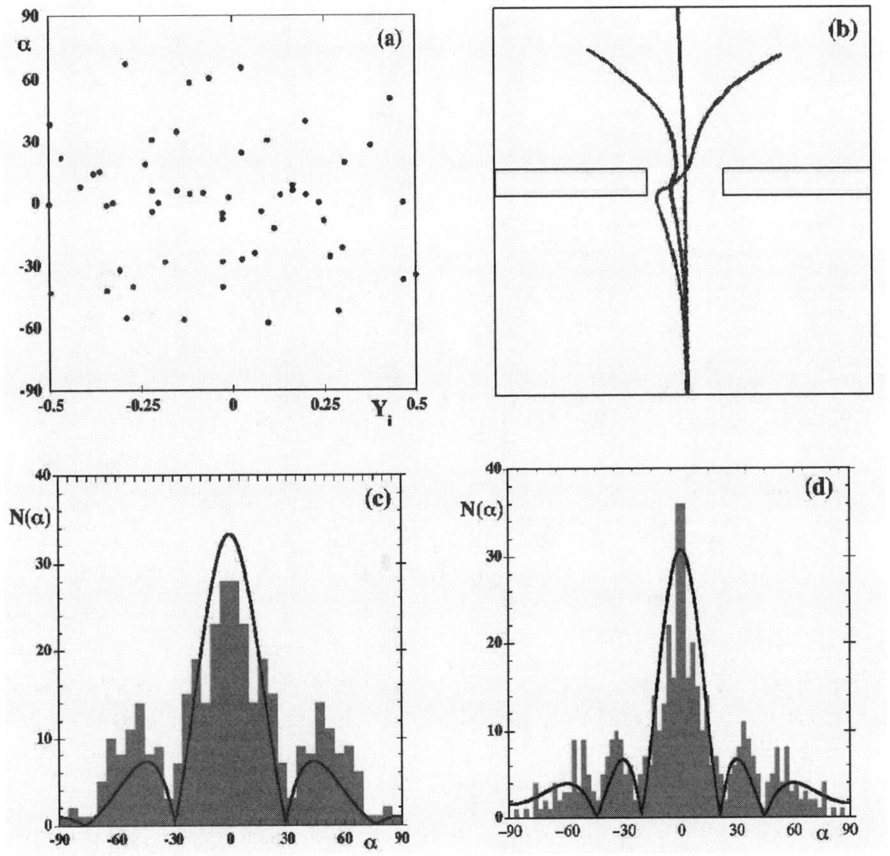

Figure 7.16: (a) Observed locations, seemingly random, of the oil droplet's impact on the detection screen in function of the impact parameter $Y_i = y_i/L$, where L is the width of the slit and y_i is measured from the slit center. (b) Example trajectories of the oil droplet on passing through the slit. (c) Histogram obtained by sending 250 identically prepared droplets through the slit. (d) Same for a changed L/λ geometry, where λ is the measured wavelength of standing waves that form on the the silicon oil surface in response to agitation. Figure reprinted with permission from [25]. © 2006 by the American Physical Society.

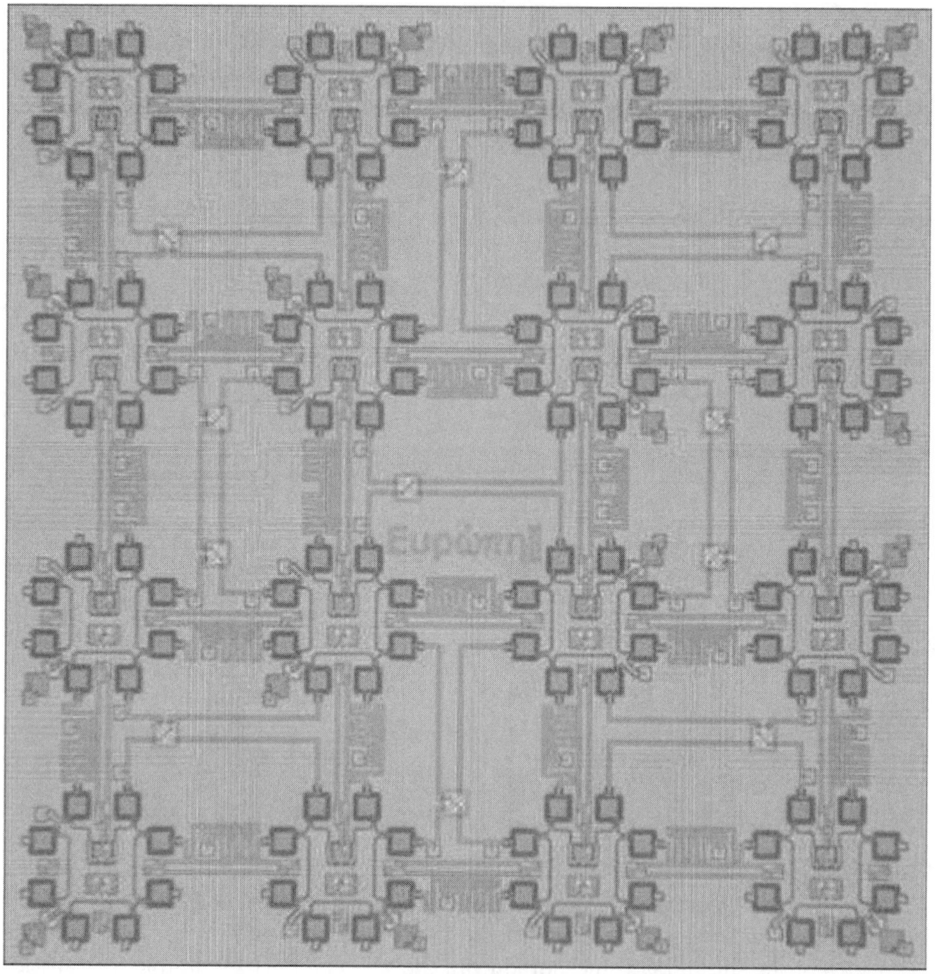

Figure 7.17: D-Wave Orion chip. Courtesy of D-Wave Systems Inc.

A Quaternions and Pauli Matrices

A.1 Hamilton Quaternions

Hamilton quaternions are numbers of the form

$$\boldsymbol{q} = a + b\boldsymbol{i} + c\boldsymbol{j} + d\boldsymbol{k}, \tag{A.1}$$

where

$$\boldsymbol{i}\boldsymbol{i} = \boldsymbol{j}\boldsymbol{j} = \boldsymbol{k}\boldsymbol{k} = -1, \tag{A.2}$$

and

$$\boldsymbol{i}\boldsymbol{j}\boldsymbol{k} = -1, \tag{A.3}$$

from which it follows that

$$\boldsymbol{i}\boldsymbol{j} = -\boldsymbol{j}\boldsymbol{i} = \boldsymbol{k}, \tag{A.4}$$

$$\boldsymbol{j}\boldsymbol{k} = -\boldsymbol{k}\boldsymbol{j} = \boldsymbol{i}, \tag{A.5}$$

$$\boldsymbol{k}\boldsymbol{i} = -\boldsymbol{i}\boldsymbol{k} = \boldsymbol{j}. \tag{A.6}$$

Extraction of quaternion components is accomplished by the following operators, similar to $\Re$ and $\Im$ used with complex numbers:

$$a = \Re(\boldsymbol{q}), \tag{A.7}$$

$$b = \Im_i(\boldsymbol{q}) = -\Re(\boldsymbol{i}\boldsymbol{q}), \tag{A.8}$$

$$c = \Im_j(\boldsymbol{q}) = -\Re(\boldsymbol{j}\boldsymbol{q}), \tag{A.9}$$

$$d = \Im_k(\boldsymbol{q}) = -\Re(\boldsymbol{k}\boldsymbol{q}). \tag{A.10}$$

For two quaternions

$$\boldsymbol{a} = a^x\boldsymbol{i} + a^y\boldsymbol{j} + a^z\boldsymbol{k}, \quad \text{and} \tag{A.11}$$

$$\boldsymbol{b} = b^x\boldsymbol{i} + b^y\boldsymbol{j} + b^z\boldsymbol{k}, \tag{A.12}$$

the following holds:

$$\boldsymbol{a}\boldsymbol{b} = -\vec{a}\cdot\vec{b} + \left(\vec{a}\times\vec{b}\right)^x \boldsymbol{i} + \left(\vec{a}\times\vec{b}\right)^y \boldsymbol{j} + \left(\vec{a}\times\vec{b}\right)^z \boldsymbol{k}. \tag{A.13}$$

A.2 Pauli Quaternions

Pauli quaternions are numbers of the form

$$\boldsymbol{q} = a + b\boldsymbol{\sigma}_x + c\boldsymbol{\sigma}_y + d\boldsymbol{\sigma}_z, \tag{A.14}$$

where

$$\begin{aligned}\boldsymbol{\sigma}_x &= i\boldsymbol{i}, &\text{(A.15)}\\ \boldsymbol{\sigma}_y &= i\boldsymbol{j}, &\text{(A.16)}\\ \boldsymbol{\sigma}_z &= i\boldsymbol{k}, &\text{(A.17)}\end{aligned}$$

and

$$\boldsymbol{\sigma}_x\boldsymbol{\sigma}_x = \boldsymbol{\sigma}_y\boldsymbol{\sigma}_y = \boldsymbol{\sigma}_z\boldsymbol{\sigma}_z = 1, \tag{A.18}$$

and

$$\boldsymbol{\sigma}_x\boldsymbol{\sigma}_y\boldsymbol{\sigma}_z = i. \tag{A.19}$$

From which it follows that

$$\begin{aligned}\boldsymbol{\sigma}_x\boldsymbol{\sigma}_y &= -\boldsymbol{\sigma}_y\boldsymbol{\sigma}_x = i\boldsymbol{\sigma}_z, &\text{(A.20)}\\ \boldsymbol{\sigma}_y\boldsymbol{\sigma}_z &= -\boldsymbol{\sigma}_z\boldsymbol{\sigma}_y = i\boldsymbol{\sigma}_x, &\text{(A.21)}\\ \boldsymbol{\sigma}_z\boldsymbol{\sigma}_x &= -\boldsymbol{\sigma}_x\boldsymbol{\sigma}_z = i\boldsymbol{\sigma}_y. &\text{(A.22)}\end{aligned}$$

These can be encapsulated into

$$\boldsymbol{\sigma}_i\boldsymbol{\sigma}_j = \delta_{ij}\mathbf{1} + i\sum_k \epsilon_{ijk}\boldsymbol{\sigma}_k. \tag{A.23}$$

Extraction of quaternion components is accomplished by the following four operators:

$$\begin{aligned}a &= \Re(\boldsymbol{q}), &\text{(A.24)}\\ b &= \Im_x(\boldsymbol{q}) = \Re(\boldsymbol{\sigma}_x\boldsymbol{q}), &\text{(A.25)}\\ c &= \Im_y(\boldsymbol{q}) = \Re(\boldsymbol{\sigma}_y\boldsymbol{q}), &\text{(A.26)}\\ d &= \Im_z(\boldsymbol{q}) = \Re(\boldsymbol{\sigma}_z\boldsymbol{q}). &\text{(A.27)}\end{aligned}$$

For

$$\begin{aligned}\boldsymbol{a} &= a^x\boldsymbol{\sigma}_x + a^y\boldsymbol{\sigma}_y + a^z\boldsymbol{\sigma}_z, \quad \text{and} &\text{(A.28)}\\ \boldsymbol{b} &= b^x\boldsymbol{\sigma}_x + b^y\boldsymbol{\sigma}_y + b^z\boldsymbol{\sigma}_z, &\text{(A.29)}\end{aligned}$$

the following holds.

$$\boldsymbol{ab} = \vec{a}\cdot\vec{b} + i\left(\left(\vec{a}\times\vec{b}\right)^x\boldsymbol{\sigma}_x + \left(\vec{a}\times\vec{b}\right)^y\boldsymbol{\sigma}_y + \left(\vec{a}\times\vec{b}\right)^z\boldsymbol{\sigma}_z\right). \tag{A.30}$$

A.3 Pauli Matrices

Complex 2×2 matrices can be parametrized by

$$\boldsymbol{q} = a\mathbf{1} + b\boldsymbol{\sigma}_x + c\boldsymbol{\sigma}_y + d\boldsymbol{\sigma}_z, \tag{A.31}$$

where

$$\mathbf{1} = \begin{pmatrix} 1 & 0 \\ 0 & 1 \end{pmatrix} = |\,0\rangle\langle 0\,| + |\,1\rangle\langle 1\,|, \tag{A.32}$$

$$\boldsymbol{\sigma}_x = \begin{pmatrix} 0 & 1 \\ 1 & 0 \end{pmatrix} = |\,0\rangle\langle 1\,| + |\,1\rangle\langle 0\,|, \tag{A.33}$$

$$\boldsymbol{\sigma}_y = \begin{pmatrix} 0 & -i \\ i & 0 \end{pmatrix} = -\mathrm{i}\,|\,0\rangle\langle 1\,| + \mathrm{i}\,|\,1\rangle\langle 0\,|, \tag{A.34}$$

$$\boldsymbol{\sigma}_z = \begin{pmatrix} 1 & 0 \\ 0 & -1 \end{pmatrix} = |\,0\rangle\langle 0\,| - |\,1\rangle\langle 1\,|, \tag{A.35}$$

and where $\boldsymbol{\sigma}_x$, $\boldsymbol{\sigma}_y$ and $\boldsymbol{\sigma}_z$ all have commutation properties identical with Pauli quaternions. Pauli matrices are a *representation* of Pauli quaternions.

For all Pauli quaternions and matrices

$$2\Re = \mathrm{Tr}. \tag{A.36}$$

Parameters a, b, c and d in (A.31) can be extracted with

$$a = \frac{1}{2}\mathrm{Tr}\,(\boldsymbol{q}), \tag{A.37}$$

$$b = \frac{1}{2}\mathrm{Tr}\,(\boldsymbol{\sigma}_x\boldsymbol{q}), \tag{A.38}$$

$$c = \frac{1}{2}\mathrm{Tr}\,(\boldsymbol{\sigma}_y\boldsymbol{q}), \tag{A.39}$$

$$d = \frac{1}{2}\mathrm{Tr}\,(\boldsymbol{\sigma}_z\boldsymbol{q}). \tag{A.40}$$

The canonical basis in the space of 2×2 matrices can be expressed in terms of Pauli matrices as follows.

$$\boldsymbol{M}_0 = \begin{pmatrix} 1 & 0 \\ 0 & 0 \end{pmatrix} = \frac{1}{2}(\mathbf{1} + \boldsymbol{\sigma}_z), \tag{A.41}$$

$$\boldsymbol{M}_1 = \begin{pmatrix} 0 & 1 \\ 0 & 0 \end{pmatrix} = \frac{1}{2}(\boldsymbol{\sigma}_x + i\boldsymbol{\sigma}_y), \tag{A.42}$$

$$\boldsymbol{M}_2 = \begin{pmatrix} 0 & 0 \\ 1 & 0 \end{pmatrix} = \frac{1}{2}(\boldsymbol{\sigma}_x - i\boldsymbol{\sigma}_y), \tag{A.43}$$

$$\boldsymbol{M}_3 = \begin{pmatrix} 0 & 0 \\ 0 & 1 \end{pmatrix} = \frac{1}{2}\left(\mathbf{1} - \boldsymbol{\sigma}_z\right). \tag{A.44}$$

B Biqubit Probability Matrices

Pauli vectors can be expressed in terms of canonical vectors as follows.

$$\begin{aligned} \varsigma_1 &= \boldsymbol{e}_0 + \boldsymbol{e}_1 + \boldsymbol{e}_2 + \boldsymbol{e}_3, && \text{(B.1)} \\ \varsigma_x &= \boldsymbol{e}_2, && \text{(B.2)} \\ \varsigma_y &= \boldsymbol{e}_3, && \text{(B.3)} \\ \varsigma_z &= \boldsymbol{e}_0 - \boldsymbol{e}_1. && \text{(B.4)} \end{aligned}$$

Some of their tensor products, the ones used in calculations in Chapter 5, have the following matrix representation in the canonical basis.

$$\varsigma_{1A} \otimes \varsigma_{1B} = (\boldsymbol{e}_{0A} + \boldsymbol{e}_{1A} + \boldsymbol{e}_{2A} + \boldsymbol{e}_{3A}) \otimes (\boldsymbol{e}_{0B} + \boldsymbol{e}_{1B} + \boldsymbol{e}_{2B} + \boldsymbol{e}_{3B}) \quad \text{(B.5)}$$

$$= \begin{pmatrix} 1 & 1 & 1 & 1 \\ 1 & 1 & 1 & 1 \\ 1 & 1 & 1 & 1 \\ 1 & 1 & 1 & 1 \end{pmatrix}, \quad \text{(B.6)}$$

$$\varsigma_{xA} \otimes \varsigma_{1B} = \boldsymbol{e}_{2A} \otimes (\boldsymbol{e}_{0B} + \boldsymbol{e}_{1B} + \boldsymbol{e}_{2B} + \boldsymbol{e}_{3B}) \quad \text{(B.7)}$$

$$= \begin{pmatrix} 0 & 0 & 0 & 0 \\ 0 & 0 & 0 & 0 \\ 1 & 1 & 1 & 1 \\ 0 & 0 & 0 & 0 \end{pmatrix}, \quad \text{(B.8)}$$

$$\varsigma_{1A} \otimes \varsigma_{xB} = (\boldsymbol{e}_{0A} + \boldsymbol{e}_{1A} + \boldsymbol{e}_{2A} + \boldsymbol{e}_{3A}) \otimes \boldsymbol{e}_{2B} \quad \text{(B.9)}$$

$$= \begin{pmatrix} 0 & 0 & 1 & 0 \\ 0 & 0 & 1 & 0 \\ 0 & 0 & 1 & 0 \\ 0 & 0 & 1 & 0 \end{pmatrix}, \quad \text{(B.10)}$$

$$\varsigma_{yA} \otimes \varsigma_{1B} = \boldsymbol{e}_{3A} \otimes (\boldsymbol{e}_{0B} + \boldsymbol{e}_{1B} + \boldsymbol{e}_{2B} + \boldsymbol{e}_{3B}) \quad \text{(B.11)}$$

$$= \begin{pmatrix} 0 & 0 & 0 & 0 \\ 0 & 0 & 0 & 0 \\ 0 & 0 & 0 & 0 \\ 1 & 1 & 1 & 1 \end{pmatrix}, \quad \text{(B.12)}$$

$$\varsigma_{1A} \otimes \varsigma_{yB} = (\boldsymbol{e}_{0A} + \boldsymbol{e}_{1A} + \boldsymbol{e}_{2A} + \boldsymbol{e}_{3A}) \otimes \boldsymbol{e}_{3B} \quad \text{(B.13)}$$

$$= \begin{pmatrix} 0 & 0 & 0 & 1 \\ 0 & 0 & 0 & 1 \\ 0 & 0 & 0 & 1 \\ 0 & 0 & 0 & 1 \end{pmatrix}, \quad \text{(B.14)}$$

$$\varsigma_{zA} \otimes \varsigma_{1B} = (\boldsymbol{e}_{0A} - \boldsymbol{e}_{1A}) \otimes (\boldsymbol{e}_{0B} + \boldsymbol{e}_{1B} + \boldsymbol{e}_{2B} + \boldsymbol{e}_{3B}) \quad \text{(B.15)}$$

$$= \begin{pmatrix} 1 & 1 & 1 & 1 \\ -1 & -1 & -1 & -1 \\ 0 & 0 & 0 & 0 \\ 0 & 0 & 0 & 0 \end{pmatrix}, \tag{B.16}$$

$$\varsigma_{1A} \otimes \varsigma_{zB} = (\boldsymbol{e}_{0A} + \boldsymbol{e}_{1A} + \boldsymbol{e}_{2A} + \boldsymbol{e}_{3A}) \otimes (\boldsymbol{e}_{0B} - \boldsymbol{e}_{1B}) \tag{B.17}$$

$$= \begin{pmatrix} 1 & -1 & 0 & 0 \\ 1 & -1 & 0 & 0 \\ 1 & -1 & 0 & 0 \\ 1 & -1 & 0 & 0 \end{pmatrix}, \tag{B.18}$$

$$\varsigma_{xA} \otimes \varsigma_{xB} = \boldsymbol{e}_{2A} \otimes \boldsymbol{e}_{2B} \tag{B.19}$$

$$= \begin{pmatrix} 0 & 0 & 0 & 0 \\ 0 & 0 & 0 & 0 \\ 0 & 0 & 1 & 0 \\ 0 & 0 & 0 & 0 \end{pmatrix}, \tag{B.20}$$

$$\varsigma_{yA} \otimes \varsigma_{yB} = \boldsymbol{e}_{3A} \otimes \boldsymbol{e}_{3B} \tag{B.21}$$

$$= \begin{pmatrix} 0 & 0 & 0 & 0 \\ 0 & 0 & 0 & 0 \\ 0 & 0 & 0 & 0 \\ 0 & 0 & 0 & 1 \end{pmatrix}, \tag{B.22}$$

$$\varsigma_{zA} \otimes \varsigma_{zB} = (\boldsymbol{e}_{0A} - \boldsymbol{e}_{1A}) \otimes (\boldsymbol{e}_{0B} - \boldsymbol{e}_{1B}) \tag{B.23}$$

$$= \begin{pmatrix} 1 & -1 & 0 & 0 \\ -1 & 1 & 0 & 0 \\ 0 & 0 & 0 & 0 \\ 0 & 0 & 0 & 0 \end{pmatrix}. \tag{B.24}$$

C Tensor Products of Pauli Matrices

In the biqubit computational basis,

$$|\,0\rangle \otimes |\,0\rangle \;=\; |\,00\rangle = |\,\mathbf{0}\rangle, \tag{C.1}$$

$$|\,0\rangle \otimes |\,1\rangle \;=\; |\,01\rangle = |\,\mathbf{1}\rangle, \tag{C.2}$$

$$|\,1\rangle \otimes |\,0\rangle \;=\; |\,10\rangle = |\,\mathbf{2}\rangle, \tag{C.3}$$

$$|\,1\rangle \otimes |\,1\rangle \;=\; |\,11\rangle = |\,\mathbf{3}\rangle, \tag{C.4}$$

it is easy to encode Bell states,

$$|\,\Phi^+\rangle \;=\; |\,\bar{\mathbf{0}}\rangle = \frac{1}{\sqrt{2}}\left(|\,\mathbf{0}\rangle + |\,\mathbf{3}\rangle\right), \tag{C.5}$$

$$|\,\Psi^+\rangle \;=\; |\,\bar{\mathbf{1}}\rangle = \frac{1}{\sqrt{2}}\left(|\,\mathbf{1}\rangle + |\,\mathbf{2}\rangle\right), \tag{C.6}$$

$$|\,\Psi^-\rangle \;=\; |\,\bar{\mathbf{2}}\rangle = \frac{1}{\sqrt{2}}\left(|\,\mathbf{1}\rangle - |\,\mathbf{2}\rangle\right), \tag{C.7}$$

$$|\,\Phi^-\rangle \;=\; |\,\bar{\mathbf{3}}\rangle = \frac{1}{\sqrt{2}}\left(|\,\mathbf{0}\rangle - |\,\mathbf{3}\rangle\right), \tag{C.8}$$

so that the reverse encoding looks the same,

$$|\,\mathbf{0}\rangle \;=\; \frac{1}{\sqrt{2}}\left(|\,\Phi^+\rangle + |\,\Phi^-\rangle\right) = \frac{1}{\sqrt{2}}\left(|\,\bar{\mathbf{0}}\rangle + |\,\bar{\mathbf{3}}\rangle\right), \tag{C.9}$$

$$|\,\mathbf{1}\rangle \;=\; \frac{1}{\sqrt{2}}\left(|\,\Psi^+\rangle + |\,\Psi^-\rangle\right) = \frac{1}{\sqrt{2}}\left(|\,\bar{\mathbf{1}}\rangle + |\,\bar{\mathbf{2}}\rangle\right), \tag{C.10}$$

$$|\,\mathbf{2}\rangle \;=\; \frac{1}{\sqrt{2}}\left(|\,\Psi^+\rangle - |\,\Psi^-\rangle\right) = \frac{1}{\sqrt{2}}\left(|\,\bar{\mathbf{1}}\rangle - |\,\bar{\mathbf{2}}\rangle\right), \tag{C.11}$$

$$|\,\mathbf{3}\rangle \;=\; \frac{1}{\sqrt{2}}\left(|\,\Phi^+\rangle - |\,\Phi^-\rangle\right) = \frac{1}{\sqrt{2}}\left(|\,\bar{\mathbf{0}}\rangle - |\,\bar{\mathbf{3}}\rangle\right). \tag{C.12}$$

We use this encoding, and the following tensor products, in Chapter 6.1.

$$\begin{aligned}
\mathbf{1}\otimes\mathbf{1} &= \begin{pmatrix} 1 & 0 \\ 0 & 1 \end{pmatrix} \otimes \begin{pmatrix} 1 & 0 \\ 0 & 1 \end{pmatrix} \\
&\equiv \begin{pmatrix} 1\begin{pmatrix} 1 & 0 \\ 0 & 1 \end{pmatrix} & 0\begin{pmatrix} 1 & 0 \\ 0 & 1 \end{pmatrix} \\ 0\begin{pmatrix} 1 & 0 \\ 0 & 1 \end{pmatrix} & 1\begin{pmatrix} 1 & 0 \\ 0 & 1 \end{pmatrix} \end{pmatrix} \\
&= \begin{pmatrix} 1 & 0 & 0 & 0 \\ 0 & 1 & 0 & 0 \\ 0 & 0 & 1 & 0 \\ 0 & 0 & 0 & 1 \end{pmatrix}
\end{aligned} \tag{C.13}$$

$$= \mid \mathbf{0}\rangle\langle\mathbf{0}\mid + \mid\mathbf{1}\rangle\langle\mathbf{1}\mid + \mid\mathbf{2}\rangle\langle\mathbf{2}\mid + \mid\mathbf{3}\rangle\langle\mathbf{3}\mid \tag{C.14}$$

$$= \mid \bar{\mathbf{0}}\rangle\langle\bar{\mathbf{0}}\mid + \mid\bar{\mathbf{1}}\rangle\langle\bar{\mathbf{1}}\mid + \mid\bar{\mathbf{2}}\rangle\langle\bar{\mathbf{2}}\mid + \mid\bar{\mathbf{3}}\rangle\langle\bar{\mathbf{3}}\mid, \tag{C.15}$$

$$\begin{aligned}
\boldsymbol{\sigma}_x \otimes \boldsymbol{\sigma}_x &= \begin{pmatrix} 0 & 1 \\ 1 & 0 \end{pmatrix} \otimes \begin{pmatrix} 0 & 1 \\ 1 & 0 \end{pmatrix} \\
&\equiv \begin{pmatrix} 0\begin{pmatrix} 0 & 1 \\ 1 & 0 \end{pmatrix} & 1\begin{pmatrix} 0 & 1 \\ 1 & 0 \end{pmatrix} \\ 1\begin{pmatrix} 0 & 1 \\ 1 & 0 \end{pmatrix} & 0\begin{pmatrix} 0 & 1 \\ 1 & 0 \end{pmatrix} \end{pmatrix} \\
&= \begin{pmatrix} 0 & 0 & 0 & 1 \\ 0 & 0 & 1 & 0 \\ 0 & 1 & 0 & 0 \\ 1 & 0 & 0 & 0 \end{pmatrix} && \text{(C.16)} \\
&= \mid\mathbf{0}\rangle\langle\mathbf{3}\mid + \mid\mathbf{1}\rangle\langle\mathbf{2}\mid + \mid\mathbf{2}\rangle\langle\mathbf{1}\mid + \mid\mathbf{3}\rangle\langle\mathbf{0}\mid && \text{(C.17)} \\
&= \mid\bar{\mathbf{0}}\rangle\langle\bar{\mathbf{0}}\mid + \mid\bar{\mathbf{1}}\rangle\langle\bar{\mathbf{1}}\mid - \mid\bar{\mathbf{2}}\rangle\langle\bar{\mathbf{2}}\mid - \mid\bar{\mathbf{3}}\rangle\langle\bar{\mathbf{3}}\mid, && \text{(C.18)}
\end{aligned}$$

$$\begin{aligned}
\boldsymbol{\sigma}_y \otimes \boldsymbol{\sigma}_y &= \begin{pmatrix} 0 & -i \\ i & 0 \end{pmatrix} \otimes \begin{pmatrix} 0 & -i \\ i & 0 \end{pmatrix} \\
&\equiv \begin{pmatrix} 0\begin{pmatrix} 0 & -i \\ i & 0 \end{pmatrix} & -i\begin{pmatrix} 0 & -i \\ i & 0 \end{pmatrix} \\ i\begin{pmatrix} 0 & -i \\ i & 0 \end{pmatrix} & 0\begin{pmatrix} 0 & -i \\ i & 0 \end{pmatrix} \end{pmatrix} \\
&= \begin{pmatrix} 0 & 0 & 0 & -1 \\ 0 & 0 & 1 & 0 \\ 0 & 1 & 0 & 0 \\ -1 & 0 & 0 & 0 \end{pmatrix} && \text{(C.19)} \\
&= -\mid\mathbf{0}\rangle\langle\mathbf{3}\mid + \mid\mathbf{1}\rangle\langle\mathbf{2}\mid + \mid\mathbf{2}\rangle\langle\mathbf{1}\mid - \mid\mathbf{3}\rangle\langle\mathbf{0}\mid && \text{(C.20)} \\
&= -\mid\bar{\mathbf{0}}\rangle\langle\bar{\mathbf{0}}\mid + \mid\bar{\mathbf{1}}\rangle\langle\bar{\mathbf{1}}\mid - \mid\bar{\mathbf{2}}\rangle\langle\bar{\mathbf{2}}\mid + \mid\bar{\mathbf{3}}\rangle\langle\bar{\mathbf{3}}\mid, && \text{(C.21)}
\end{aligned}$$

$$\begin{aligned}
\boldsymbol{\sigma}_z \otimes \boldsymbol{\sigma}_z &= \begin{pmatrix} 1 & 0 \\ 0 & -1 \end{pmatrix} \otimes \begin{pmatrix} 1 & 0 \\ 0 & -1 \end{pmatrix} \\
&\equiv \begin{pmatrix} 1\begin{pmatrix} 1 & 0 \\ 0 & -1 \end{pmatrix} & 0\begin{pmatrix} 1 & 0 \\ 0 & -1 \end{pmatrix} \\ 0\begin{pmatrix} 1 & 0 \\ 0 & -1 \end{pmatrix} & -1\begin{pmatrix} 1 & 0 \\ 0 & -1 \end{pmatrix} \end{pmatrix}
\end{aligned}$$

$$= \begin{pmatrix} 1 & 0 & 0 & 0 \\ 0 & -1 & 0 & 0 \\ 0 & 0 & -1 & 0 \\ 0 & 0 & 0 & 1 \end{pmatrix} \tag{C.22}$$

$$= |\,\mathbf{0}\rangle\langle\mathbf{0}\,| - |\,\mathbf{1}\rangle\langle\mathbf{1}\,| - |\,\mathbf{2}\rangle\langle\mathbf{2}\,| + |\,\mathbf{3}\rangle\langle\mathbf{3}\,| \tag{C.23}$$

$$= |\,\bar{\mathbf{0}}\rangle\langle\bar{\mathbf{0}}\,| - |\,\bar{\mathbf{1}}\rangle\langle\bar{\mathbf{1}}\,| - |\,\bar{\mathbf{2}}\rangle\langle\bar{\mathbf{2}}\,| + |\,\bar{\mathbf{3}}\rangle\langle\bar{\mathbf{3}}\,|\,. \tag{C.24}$$

References

[1] Daniel S. Abrams and Seth Lloyd. Simulation of many-body Fermi systems on a universal quantum computer. *Physical Review Letters*, 79(13):2586–2589, September 1997.

[2] Daniel S. Abrams and Seth Lloyd. Quantum algorithm providing exponential speed increase for finding eigenvalues and eigenvectors. *Physical Review Letters*, 83:5162–5165, December 1999.

[3] A. Acín, D. Bruß, M. Lewenstein, and A. Sanpera. Classification of mixed three-qubit states. *Physical Review Letters*, 87(4):040401(4), July 2001.

[4] A. Aspect, P. Grangier, and G. Roger. Experimental tests of realistic local theories via Bell's theorem. *Physical Review Letters*, 47(7):460–463, August 1981.

[5] Markus Aspelmeyer, Hannes R. Böhm, Tsewang Gyatso, Thomas Jennewein, Rainer Kaltenbaek, Michael Lindenthal, Gabriel Molina-Terriza, Andreas Poppe, Kevin Resch, Michael Taraba, Rupert Ursin, Philip Walther, and Anton Zeilinger. Long-distance free-space distribution of quantum entanglement. *Science*, 301:621–623, August 2003.

[6] Francis Baily. On Mr. Babbage's new machine for calculating and printing mathematical and astronomical tables. *Astronomische Nachrichten*, 2:407–408, 1824.

[7] J. S. Bell. On the Einstein Podolsky Rosen paradox. *Physics*, 1:195–200, 1964.

[8] J. S. Bell. *Speakable and Unspeakable in Quantum Mechanics.* Cambridge University Press, Cambridge, U.K., 1987.

[9] C. H. Bennett. The thermodynamics of computation—a review. *International Journal of Theoretical Physics*, 21(12):905–940, 1982.

[10] C. H. Bennett, G. Brassard, C. Crepeau, R. Jozsa, A. Peres, and W. Wooters. Teleporting an unknown quantum state via dual classical and EPR channels. *Physical Review Letters*, 70:1895–1899, 1993.

[11] A. J. Berkley, H. Xu, R. C. Ramos, M. A. Gubrud, F. W. Strauch, P. R. Johnson, J. R. Anderson, A. J. Dragt, C. J. Lobb, and F. C. Wellstood. Entangled macroscopic quantum states in two superconducting qubits. *Science*, 300:1548–1550, June 2003.

[12] M. V. Berry. Quantal phase factors accompanying adiabatic changes. *Proceedings of the Royal Society of London*, A 392:45–57, 1984.

[13] B. B. Blinov, D. L. Moehring, L.-M. Duan, and C. Monroe. Observation of entanglement between a single trapped atom and a single photon. *Nature*, 428:153–157, March 2004.

[14] D. Bohm and B. J. Hiley. *The Undivided Universe.* Routledge, London and New York, 1996.

[15] G. Brassard. Teleportation as a quantum computation. *Physica D*, 120:43–47, 1998. arXiv:quant-ph/9605035v1.

[16] J. R. Brownstein and J. W. Moffat. The Bullet Cluster 1E0657-558 evidence shows modified gravity in the absence of dark matter. *Monthly Notices of the Royal Astronomical Society*, 382(1):29–47, November 2007.

[17] M. Brune, E. Hagley, J. Dreyer, X. Maître, A. Maali, C. Wunderlich, J. M. Raimond, and S. Haroche. Observing the progressive decoherence of the "meter" in a quantum measurement. *Physical Review Letters*, 77(24):4887–4890, December 1996.

[18] M. Brune, F. Schmidt-Kaler, A. Maali, J. Dreyer, E. Hagley, J. M. Raimond, and S. Haroche. Quantum Rabi oscillation: A direct test of field quantization in a cavity. *Physical Review Letters*, 76(11):1800–1803, March 1996.

[19] V. Buzek and M. Hillery. Quantum cloning. *Physics World*, 14(11):25–29, 2001.

[20] Julius Caesar. *The Gallic War.* Harvard University Press, Cambridge, Massachusetts, 2006.

[21] Isaac L. Chuang, Neil Gershenfeld, and Mark Kubinec. Experimental implementation of fast quantum searching. *Physical Review Letters*, 80(15):3408–3411, April 1998.

[22] J. I. Cirac and P. Zoller. Quantum computation with cold trapped ions. *Physical Review Letters*, 74(20):4091–4094, May 1995.

[23] J. F. Clauser, M. A. Horne nd A. Shimony, and R. A. Hold. Proposed experiment to test local hidden-variable theories. *Physical Review Letters*, 23(15):880–884, October 1969.

[24] Daniel Collins and Sandu Popescu. Classical analog of entanglement. *Physical Review A*, 65(3):032321(11), 2002.

[25] Yves Couder and Emmanuel Fort. Single-particle diffraction and interference at a macroscopic scale. *Physical Review Letters*, 97(15):154101(4), October 2006.

[26] A. Daffertshofer, A. R. Plastino, and A. Plastino. Classical no-cloning theorem. *Physical Review Letters*, 88(21):210601(4), May 2002.

[27] L. Davidovich, M. Brune, J. M. Raimond, and S. Haroche. Mesoscopic quantum coherences in cavity QED: Preparation and decoherence monitoring schemes. *Physical Review A*, 53(3):1295–1309, March 1996.

[28] D. Deutsch. Quantum theory, the Church-Turing principle and the universal quantum computer. *Proceedings of the Royal Society of London A*, 400:97–117, 1985.

[29] D. Deutsch. Quantum computational networks. *Proceedings of the Royal Society of London*, 425(1868):73–90, September 1989.

[30] D. Deutsch and R. Jozsa. Rapid solution of problems by quantum computation. *Proceedings of the Royal Society of London A*, 439:553–558, 1992.

[31] B. S. DeWitt and N. Graham, editors. *The Many-Worlds Interpretation of Quantum Mechanics*. Princeton University Press, Princeton, New Jersey, 1973.

[32] D. Dieks. Communication by EPR devices. *Physics Letters A*, 92(6):271–272, 1982.

[33] David P. DiVincenzo. Two-bit gates are universal for quantum computation. *Physical Review A*, 51(2):1015–1022, February 1995.

[34] C. Dorrer, M. Anderson, P. Londero, S. Wallentowitz, K. Banaszek, and I. A. Walmsley. Computing with waves: All-optical single-query 50-element database search. In *Proceedings of the Conference on Lasers and Electro-Optics*. Optical Society of America, 2001.

[35] Thomas Durt. Quantum entanglement, interaction, and the classical limit. *Zeitschrift für Naturforschung A*, 59a(7/8):425–436, 2004. arXiv:quant-ph/0401121v1.

[36] Thomas Durt and Yves Pierseaux. Bohm's interpretation and maximally entangled states. *Physical Review A*, 66(5):052109(11), November 2002.

[37] A. Einstein. Über einen die Erzeugung und verwandlung des Lichtes betreffenden heuristischen Gesichtspunkt. *Annalen der Physik*, 17:132–148, 1905.

[38] A. Einstein, B. Podolsky, and N. Rosen. Can quantum-mechanical description of physical reality be considered complete? *Physical Review*, 47:777–780, 1935.

[39] Berthold-Georg Englert and Nasser Metwally. Remarks on 2-q-bit states. *Applied Physics B*, 72(1):35–42, January 2001. arXiv:quant-ph/0007053v1.

[40] Edward Farhi, Jeffrey Goldstone, Sam Gutmann, and Michael Sipser. Quantum computation by adiabatic evolution. January 2000. arXiv:quant-ph/0001106v1.

[41] Richard P. Feynman, Robert B. Leighton, and Matthew Sands. *The Feynman Lectures on Physics*, volume 1. Addison-Wesley Publishing Company, Reading, Massachusetts, 1975.

[42] Richard P. Feynman, Robert B. Leighton, and Matthew Sands. *The Feynman Lectures on Physics*, volume 3. Addison-Wesley Publishing Company, Reading, Massachusetts, 1979.

[43] Richard P. Feynman, Robert B. Leighton, and Matthew Sands. *The Feynman Lectures on Physics*, volume 2. Addison-Wesley Publishing Company, Reading, Massachusetts, 1981.

[44] M. Fitting, editor. *Beyond Two: Theory and Applications of Multiple Valued Logic*. Physica Verlag, December 2002.

[45] E. Fredkin and T. Toffoli. Conservative logic. *International Journal of Theoretical Physics*, 21:219–253, 1982.

[46] Michael Freedman, Michael Larsen, and Zhenghan Wang. A modular functor which is universal for quantum computation. February 2000. arXiv:quant-ph/0001108v2.

[47] Jonathan R. Friedman, Vijay Patel, W. Chen, S. K. Tolpygo, and J. E. Lukens. Quantum superposition of distinct macroscopic states. *Nature*, 406:43–46, July 2000.

[48] Frank Gaitan. Controlling qubit transitions during non-adiabatic rapid passage through quantum interference. 2004. arXiv:quant-ph/0402108v1.

[49] Julio Gea-Banacloche and Laszlo B. Kish. Comparison of energy requirements for classical and quantum information processing. *Fluctuation and Noise Letters*, 3(3):C3–C7, 2003.

[50] B. Georgeot and D. L. Shepelyansky. Exponential gain in quantum computing of quantum chaos and localization. *Physical Review Letters*, 86:2890–2893, March 2001.

[51] Walther Gerland and Otto Stern. Das magnetische Moment des Silberatoms. *Zeitschrift für Physik A, Hadrons and Nuclei*, 9(1):353–355, December 1922.

[52] Robert Percival Graves. *Life of Sir William Rowan Hamilton: Including Selections from His Poems, Correspondence, and Miscellaneous Writings*. Ayer Company Publishers, Doublin and London, April 1975. Reprinted from the 1882–1889 three-volume original.

[53] D. M. Greenberger, M. A. Horne, A. Shimony, and A. Zeilinger. Bell's theorem without inequalities. *American Journal of Physics*, 58:1131–1143, 1990.

[54] D. M. Greenberger, M. A. Horne, and A. Zeilinger. *Bell's Theorem, Quantum Theory, and Conception of the Universe*, pages 73–76. Kluwer Academic, Dordrecht, 1989.

[55] Lov K. Grover. Quantum computers can search arbitrarily large databases by a single query. *Physical Review Letters*, 79(23):4709–4712, December 1997.

[56] Lov K. Grover. Quantum mechanics helps in search for a needle in a haystack. *Physical Review Letters*, 79(2):325–328, July 1997.

[57] Lov K. Grover and Anirvan M. Sengupta. Classical analog of quantum search. *Physical Review A*, 65(3):032319(5), 2002.

[58] Stephan Gulde, Mark Riebe, Gavin P. T. Lancaster, Christoph Becher, Jürgen Eschner, Hartmut Häffner, Ferdinand Schmidt-Kaler, Isaac L. Chuang, and Rainer Blatt. Implementation of the Deutsch-Jozsa algorithm on an ion-trap quantum computer. *Nature*, 421:48–50, January 2003.

[59] Lucien Hardy. Quantum mechanics, local realistic theories, and Lorentz-invariant realistic theories. *Physical Review Letters*, 68(20):2981–2984, May 1992.

[60] Lucien Hardy. Quantum theory from five reasonable axioms. 2001. arXiv:quant-ph/0101012v4.

[61] S. Haroche. Personal communication, August 2006.

[62] S. Haroche, M. Brune, and J. M. Raimond. Experiments with single atoms in a cavity: Entanglement, Schrödinger's cats and decoherence. *Philosophical Transactions of the Royal Society*, 355:2367–2380, 1997.

[63] Tony Hey and Patrick Walters. *Einstein's Mirror.* Cambridge University Press, Cambridge, U.K., 1997.

[64] Tom W. Hijmans, Tycho N. Huussen, and Robert J. C. Spreeuw. Time- and frequency-domain solutions in an optical analogue of Grover's search algorithm. *Journal of the Optical Society of America B*, 24(2):214–220, February 2007.

[65] E. D. Hirsch, Joseph F. Kett, and James Trefil. *The New Dictionary of Cultural Literacy: What Every American Needs to Know.* Houghton Mifflin Company, New York, third edition, 2002.

[66] Michał Horodecki, Paweł Horodecki, and Ryszard Horodecki. Separability of mixed states: Necessary and sufficient conditions. *Physics Letters A*, 223:1–8, November 1996.

[67] Paweł Horodecki. Separability criterion and inseparable mixed states with positive partial transposition. *Physics Letters A*, 232:333–339, August 1997.

[68] IBM. IBM's test-tube quantum computer makes history. IBM Press Release, December 2001.

[69] William T. M. Irvine, Juan F. Hodelin, Christoph Siomon, and Dirk Bouwmeester. Realization of Hardy's thought experiment with photons. *Physical Review Letters*, 95(3):030401(4), July 2005.

[70] E. T. Jaynes and F. W. Cummings. Comparison of quantum and semiclassical radiation theories with application to the beam maser. *Proceedings of IEEE*, 51:89–109, 1963.

[71] Jonathan A. Jones, Vlatko Vedral, Artur Ekert, and Giuseppe Castagnoli. Geometric quantum computation using nuclear magnetic resonance. *Nature*, 403:869–871, February 2000.

[72] Brian Julsgaard, Alexander Kozhekin, and Eugene S. Polzik. Experimental long-lived entanglement of two macroscopic objects. *Nature*, 413:400–403, September 2001.

[73] K. Hagiwara *et al.* (140 authors). Review of particle physics. *Physical Review D*, 66:010001(974), 2002.

[74] A. Kalev and I. Hen. The no-broadcasting theorem and its classical counterpart. November 2007. arXiv:quant-ph/0704.1754v2.

[75] B. E. Kane. A silicon-based nuclear spin quantum computer. *Nature*, 393:133, 1998.

[76] B. E. Kane, N. S. McAlpine, A. S. Dzurak, R. G. Clark, G. J. Milburn, He Be Sun, and H. Wiseman. Single spin measurement using single electron transistors to probe two electron systems. *Physical Review B*, 61:2961, 2000.

[77] Navin Khaneja, Roger Brockett, and Steffen J. Glaser. Time optimal control in spin systems. *Physical Review A*, 63(3):032308(13), 2001.

[78] Navin Khaneja and Steffen J. Glaser. Cartan decomposition of SU(2^n) and control of spin systems. *Chemical Physics*, 267:11–23, 2001.

[79] A. K. Khitrin and B. M. Fung. NMR simulation of an eight-state quantum system. 2001. arXiv:quant-ph/0101029v1.

[80] A. Yu. Kitaev. Fault-tolerant quantum computation by anyons. July 1997. arXiv:quant-ph/9707021v1.

[81] George J. Klir and Bo Yuan. *Fuzzy Sets and Fuzzy Logic: Theory and Applications.* Prentice-Hall, Englewood Cliffs, New Jersey, 1995.

[82] A. Jamiołkowski. Linear transformations which preserve trace and positive semidefiniteness of operators. *Reports on Mathematical Physics*, 3(4):275–278, 1972.

[83] B. Kraus and J. I. Cirac. Optimal creation of entanglement using a two-qubit gate. *Physical Review A*, 63(6):062309(8), 2001.

[84] R. Landauer. Irreversibility and heat generation in the computing process. *IBM Journal of Research and Development*, 5:183–191, 1961.

[85] R. B. Laughlin. Anomalous quantum Hall effect: An incompressible quantum fluid with fractionally charged excitations. *Physical Review Letters*, 50(18):1395–1398, May 1983.

[86] P. J. Leek, J. M. Fink, A. Blais, R. Bianchetti, M. Göppl, J. M. Gambetta, D. I. Schuster, L. Frunzio, R. J. Schoelkopf, and A. Wallraff. Observation of Berry's phase in a solid-state qubit. *Science*, 318:1889–1892, December 2007.

[87] Debbie W. Leung, Isaac L. Chuang, Fumiko Yamaguchi, and Yoshihisa Yamamoto. Efficient implementation of coupled logic gates for quantum computation. *Physical Review A*, 61(4):042310(7), March 2000.

[88] Daniel A. Lidar. On the quantum computational complexity of the Ising spin glass partition function and of knot invariants. *New Journal of Physics*, 6(167):1–12, 2004.

[89] Seth Lloyd. Universal quantum simulators. *Science*, 273:1073–1078, August 1996.

[90] Seth Lloyd. Quantum search without entanglement. *Physical Review A*, 61(1):010301(4), December 1999.

[91] J. Majer, J. M. Chow, J. M. Gambetta, Jens Koch, B. R. Johson, J. A. Schreier, L. Frunzio, D. I. Schuster, A. A. Houck, A. Wallraff, A. Blais, M. H. Devoret, S. M. Girvin, and R. J. Schoelkopf. Coupling superconducting qubits via a cavity bus. *Nature*, 449:443–447, September 2007.

[92] N. David Mermin. *Quantum Computer Science*. Cambridge University Press, 2007.

[93] G. J. Milburn, R. Laflamme, B. C. Sanders, and E. Knill. Quantum dynamics of two coupled qubits. *Physical Review A*, 65(3):032316(10), 2002.

[94] Charles W. Misner, Kip S. Thorne, and John Archibald Wheeler. *Gravitation.* W. H. Freeman and Company, New York, 1973.

[95] Peter J. Mohr and Barry N. Taylor. CODATA recommended values of the fundamental physical constants: 2002. *Reviews of Modern Physics*, 77:1–107, January 2005.

[96] C. Monroe, D. J. Meekhof, B. E. King, W. M. Itano, and D. J. Wineland. Demonstration of a fundamental quantum logic gate. *Physical Review Letters*, 75(25):4714–4717, December 1995.

[97] Y. Nakamura, Yu. A. Pashkin, and J. S. Tsai. Coherent control of macroscopic quantum states in a single-Cooper-pair box. *Nature*, 398:786–788, April 1999.

[98] M. A. Nielsen, E. Knill, and R. Laflamme. Complete quantum teleportation using nuclear magnetic resonance. *Nature*, 396(6706):52–55, 1998.

[99] G. Nogues, A. Rauschenbeutel, S. Osnaghi, M. Brune, J. M. Raymond, and S. Haroche. Seeing a single photon without destroying it. *Nature*, 400:239, 1999.

[100] Quinn Norton. The father of quantum computing. *Wired*, February 15 2007.

[101] J. L. O'Brien, G. J. Pryde, A. G. White, T. C. Ralph, and D. Branning. Demonstration of an all-optical quantum controlled-NOT gate. *Nature*, 426:264–267, November 2003.

[102] Jason E. Ollerenshaw, Daniel A. Lidar, and Lewis E. Kay. Magnetic resonance realization of decoherence-free quantum computation. *Physical Review Letters*, 91(21):217904(4), November 2003.

[103] Yuri F. Orlov. Origin of quantum indeterminism and irreversibility of measurements. *Physical Review Letters*, 82(2):243–246, January 1999.

[104] Yuri F. Orlov. Classical counterexamples to Bell's inequalities. *Physical Review A*, 65(4):042106(7), March 2002.

[105] Yuri F. Orlov. Quantumlike bits and logic gates based on classical oscillators. *Physical Review A*, 66(5):052324(4), 2002.

[106] Yu. A. Pashkin, T. Yamamoto, O. Astafiev, Y. Nakamura, D. V. Averin, and J. S. Tsai. Quantum oscillations in two coupled charge qubits. *Nature*, 421:823–826, February 2003.

[107] A. K. Pati and A. K. Rajagopal. Inconsistencies of the adiabatic theorem and the Berry phase. May 2004. arXiv:quant-ph/0405129v2.

[108] Wolfgang Pauli. On the connexion between the completion of electron groups in an atom with the complex structure of spectra. *Zeitschrift für Physik*, 31:765ff, 1925.

[109] Roger Penrose. *The Emperor's New Mind: Concerning Computers, Minds, and the Laws of Physics.* Oxford University Press, new edition, October 2002.

[110] Roger Penrose. *The Road to Reality: A Complete Guide to the Laws of the Universe.* Knopf, 2005.

[111] A. Peres and W. H. Zurek. Is quantum theory universally valid? *American Journal of Physics*, 50(9):807–810, September 1982.

[112] Asher Peres. *Quantum Theory: Concepts and Methods.* Kluwer Academic Publishers, Dordrecht, 1995. Reprinted 1998.

[113] Asher Peres. Separability criterion for density matrices. *Physical Review Letters*, 77(8):1413–1415, August 1996.

[114] J. H. Plantenberg, P. C. de Groot, C. J. P. M. Harmans, and J. E. Mooij. Demonstration of controlled-NOT quantum gates on a pair of superconducting quantum bits. *Nature*, 447:836–839, June 2007.

[115] B. L. T. Plourde, T. L. Robertson, P. A. Reichardt, T. Hime, S. Linzen, C.-E. Wu, and John Clarke. Flux qubits and readout device with two independent flux lines. *Physical Review B*, 72(6):060506(4), August 2005.

[116] David Poulin. Classicality of quantum information processing. *Physical Review A*, 65(4):042319(10), April 2002.

[117] J. M. Raimond and S. Haroche. Atoms and cavities: The birth of a Schrödinger cat of the radiation field. In T. Asakura, editor, *International Trends in Optics and Photonics*, pages 40–53, Berlin, 1999. ICO IV, Springer-Verlag.

[118] M. G. Raizen, J. M. Gilligan, J. C. Bergquist, W. M. Itano, and D. J. Wineland. Linear trap for high-accuracy spectroscopy of stored ions. *Journal of Modern Optics*, 39(2):233–242, 1992.

[119] D. J. Richardson, A. I. Kilvington, K. Green, and S. K. Lamoreaux. Demonstration of Berry's phase using stored ultracold neutrons. *Physical Review Letters*, 61(18):2030–2033, October 1988.

[120] M. Riebe, M. Chwalla, J. Benhelm, H. Häffner, W. Hänsel, C. F. Roos, and R. Blatt. Quantum teleportation with atoms: quantum process tomography. *New Journal of Physics*, 9:211, July 2007.

[121] T. L. Robertson, B. L. T. Plourde, P. A. Reichardt, T. Hime, C.-E. Wu, and John Clarke. Quantum theory of three-junction flux qubit with non-negligible loop inductance: Towards scalability. *Physical Review B*, 73(17):1745260(9), May 2006.

[122] Fritz Rohrlich. *Classical Charged Particles, Third Edition.* World Scientific Publishing Company, Singapore, 2007.

[123] Anders Sandberg. Quantum gravity treatment of the angel density problem. *The Annals of Improbable Research*, 7(3):5–6, May/June 2001.

[124] Sankar Das Sarma, Michael Freedman, Chetan Nayak, Steven H. Simon, and Ady Stern. Non-Abelian anyons and topological quantum computation. July 2007. arXiv:cond-mat.str-el/0707.1889v1.

[125] Ferdinand Schmidt-Kaler, Hartmut Häffner, Mark Riebe, Stephan Gulde, Gavin P. T. Lancaster, Thomas Deuschle, Christoph Bechner, Christian F. Roos, Jürgen Eschner, and Rainer Blatt. Realization of the Cirac-Zoller controlled-NOT quantum gate. *Nature*, 422:408–411, March 2003.

[126] E. Schrödinger. Die gegenwärtige Situation in der Quantenmechanik. *Naturwissenschaften*, 23:807–812, 823–828, 844–849, 1935.

[127] Peter W. Shor. Polynomial-time algorithms for prime factorization and discrete logarithms on a quantum computer. In *Proceedings of the 35th Annual Symposium on Foundations of Computer Science*, pages 124–134. IEEE Computer Society Press, 1994. arXiv:quant-ph/9508027v2.

[128] Peter W. Shor. Why haven't more quantum algorithms been found? *Journal of the ACM*, 50(1):87–90, January 2003.

[129] Mika A. Sillanpää, Jae I. Park, and Raymond W. Simmonds. Coherent quantum state storage and transfer between two phase qubits via a resonant cavity. *Nature*, 449:438–442, September 2007.

[130] Lee Smolin. *Three Roads to Quantum Gravity.* Perseus Books Group, Cambridge, Massachusetts, 2002.

[131] S. Somaroo, C. H. Tseng, T. F. Havel, R. Laflamme, and D. G. Cory. Quantum simulations on a quantum computer. *Physical Review Letters*, 82(26):5381–5384, June 1999.

[132] R. Somma, G. Ortiz, J. E. Gubernatis, E. Knill, and R. Laflamme. Simulating physical phenomena by quantum networks. *Physical Review A*, 65(4):042323(17), April 2002.

[133] Robert J. C. Spreeuw. A classical analogy of entanglement. *Foundations of Physics*, 28(3):361–374, 1998.

[134] Matthias Steffen, M. Ansmann, Radoslaw C. Bialczak, N. Katz, Erik Lucero, R. McDermott, Matthew Neeley, E. M. Weig, A. N. Cleland, and John M. Martinis. Measurement of the entanglement of two superconducting qubits via state tomography. *Science*, 313:1423–1425, September 2006.

[135] Matthias Steffen, M. Ansmann, R. McDermott, N. Katz, Radoslaw C. Bialczak, Erik Lucero, Matthew Neeley, E. M. Weig, A. N. Cleland, and John M. Martinis. State tomography of capacitively shunted phase qubits with high fidelity. *Physical Review Letters*, 97(5):050502(3), August 2006.

[136] C. H. Tseng, S. Somaroo, Y. Sharf, E. Knill, R. Laflamme, T. F. Havel, and D. G. Cory. Quantum simulation of a three-body-interaction Hamiltonian on an NMR quantum computer. *Physical Review A*, 61:012302(6), December 1999.

[137] D. C. Tsui, H. L. Störmer, and A. C. Gossard. Two-dimensional magnetotransport in the extreme quantum limit. *Physical Review Letters*, 48(22):1559–1562, May 1982.

[138] Lieven M. K. Vandersypen, Matthias Steffen, Gregory Breyta, Constantino S. Yannoni, Mark H. Sherwood, and Isaac L. Chuang. Experimental realization of Shor's quantum factoring algorithm using nuclear magnetic resonance. *Nature*, 414:883–887, December 2001.

[139] Farrokh Vatan and Colin P. Williams. Realization of a general three-qubit quantum gate. 2004. arXiv:quant-ph/0401178v2.

[140] G. Vidal and C. M. Dawson. Universal quantum circuit for two-qubit transformations with three controlled-NOT gates. *Physical Review A*, 69(1):010301(4), January 2004.

[141] Guifré Vidal. Efficient simulation of one-dimensional quantum many-body systems. *Physical Review Letters*, 93(4):040502(4), July 2004.

[142] D. Vion, A. Aassime, A. Cottet, P. Joyez, H. Pothier, C. Urbina, D. Esteve, and M. H. Devoret. Manipulating the quantum state of an electrical circuit. *Science*, 296:886–889, May 2002.

[143] W.-M. Yao *et al.* (Particle Data Group). The review of particle physics. *Journal of Physics G*, 33(1):1–1232, 2006. http://pdg.lbl.gov.

[144] Thomas A. Walker and Samuel L. Braunstein. Classical broadcasting is possible with arbitrarily high fidelity and resolution. *Physical Review Letters*, 98(8):080501(4), February 2007.

[145] Stefan Weigert. Quantum time evolution in terms of nonredundant probabilities. *Physical Review Letters*, 84(5):802–805, January 2000.

[146] Edward Witten. Perturbative gauge theory as a string theory in twistor space. 2004. arXiv:hep-th/0312171v2.

[147] Paweł Wocjan, Martin Rötteler, Dominik Janzing, and Thomas Beth. Simulating Hamiltonians in quantum networks: Efficient schemes and complexity bounds. *Physical Review A*, 65(4):042309(10), March 2002.

[148] W. K. Wootters and W. H. Zurek. A single quantum cannot be cloned. *Nature*, 299:802–803, 1982.

[149] S. L. Woronowicz. Positive maps of low dimensional matrix algebras. *Reports on Mathematical Physics*, 10(2):165–183, 1976.

[150] T. Yamamoto, Yu. A. Pashkin, O. Astafiev, Y. Nakamura, and J. S. Tsai. Demonstration of conditional gate operation using superconducting charge qubits. *Nature*, 425:941–944, October 2003.

[151] Jeng-Bang Yau, E. P. De Poortere, and M. Shayegan. Aharonov-Bohm oscillations with spin: Evidence for Berry's phase. *Physical Review Letters*, 88(14):146801(4), April 2002.

[152] Christof Zalka. Simulating quantum systems on a quantum computer. *Philosophical Transactions of the Royal Society of London A*, pages 1–11, 1996.

[153] Wojciech Hubert Zurek. Decoherence, einselection, and the quantum origins of the classical. *Reviews of Modern Physics*, 75(3):715–775, July 2003.

Index